OXFORD GUIDES TO CHAUCER

Troilus and Criseyde

Oxford Guides to Chaucer

Troilus and Criseyde

BARRY WINDEATT

CLARENDON PRESS · OXFORD
1992

Oxford University Press, Walton Street, Oxford OX2 6DP
Oxford New York Toronto
Delhi Bombay Calcutta Madras Karachi
Petaling Jaya Singapore Hong Kong Tokyo
Nairobi Dar es Salaam Cape Town
Melbourne Auckland
and associated companies in
Berlin Ibadan

Oxford is a trade mark of Oxford University Press

Published in the United States
by Oxford University Press, New York

© *Barry Windeatt 1992*

British Library Cataloguing in Publication Data
Data available

Library of Congress Cataloging in Publication Data
Windeatt, B. A. (Barry A.),
Oxford guides to Chaucer : Troilus and Criseyde / Barry Windeatt.
Includes bibliographical references and index.
1. Chaucer, Geoffrey, d. 1400. Troilus and Criseyde.   2. Chaucer
Geoffrey, d. 1400. Troilus and Criseyde—Sources.   3. Troilus
(Legendary character) in literature.   4. Trojan War in literature.
5. Love in literature.   I. Title.
PR1896.W56   1992
821'.1—dc20   92-7416
ISBN 0-19-811195-9

Typeset by Latimer Trend & Company Ltd, Plymouth
Printed and bound in Great Britain by Biddles Ltd,
Guildford and King's Lynn

To Derek Brewer

FOREWORD

The idea for a series of guides to Chaucer originated in a sense that medieval studies in general and Chaucerian studies in particular had advanced to a point where a reappraisal of his poetry was both possible and necessary. The three volumes are devoted to the shorter poetry, *Troilus and Criseyde*, and the *Canterbury Tales*. We see these books as fulfilling a role comparable to the introduction to a good edition, but at greater length than would be possible there. The kind of line-by-line expository material that the notes to an edition would contain is included only where such matters are of wider importance for an understanding of the whole text or where recent scholarship has made significant advances. We hope to provide readers of Chaucer with essential and up-to-date information, with the emphasis falling on how the interpretation of that information advances our understanding of his work; we have therefore gone beyond summarizing what is known to suggest new critical readings.

The original plan for the series was designed to provide some degree of consistency in the outline of the volumes, but it was part of the project from the start that there should be plenty of room for each author's individuality. We hope that our sense of common interests and concerns in our interpretation of Chaucer's poetry will provide a deeper critical consistency below the diversity. Such a paradox would, after all, be true to the nature of our subject.

Helen Cooper
A. J. Minnis
Barry Windeatt

ACKNOWLEDGEMENTS

The Bibliography to this volume in the Oxford Guides to Chaucer is in itself a form of acknowledgement, for it is an indication of the sheer quality and quantity of modern scholarship and criticism published on *Troilus and Criseyde*, to which anyone now attempting to discuss Chaucer's poem will be indebted. It was in the belief that the time had come for a guide to Chaucer's *Troilus*, written in the light of what has been established and suggested by such scholarship and criticism, that the present book was undertaken, and I am very grateful to Kim Scott Walwyn, English editor of Oxford University Press, not only for the original invitation but also for her patience, understanding, and support. No one can be more conscious than myself of how much I have learnt from those who have already written, which I have tried to pass on to the reader. As Chaucer remarked in the preface to his *Treatise on the Astrolabe*: 'I n'am but a lewd compilator.'

This book would not have been completed without the help and encouragement of friends, first among whom are always my mother and father. With characteristic generosity and heroic efficiency Helen Cooper has given me the benefit of her comments on more than one version of this book, and I remain profoundly indebted to her for all her constructive suggestions and for her kind encouragement. Despite many calls on her time, my colleague Ruth Morse read and commented on drafts of much of this book, which has gained greatly from her critical perceptiveness, although all the shortcomings of its final form must remain my own responsibility. I was fortunate that C. David Benson and Jill Mann so readily made available to me proofs of books by them which were to be published after my own book had been completed, and I have also been most fortunate indeed to have the help and advice of Eliseo Neuman. In the dedication of this book I can only gratefully acknowledge, and never repay, the very greatest of debts.

CONTENTS

LIST OF FIGURES   xiii

ABBREVIATIONS   xiv

Introduction   1

  Bibliographical Note   2

Date   3

Text   12

  Text, Voice, Performance 12; The *Troilus* Text: An Overview 19; The *Troilus* Text: The Questions 24

Sources   37

  Introduction: 'Lollius' 37; Boccaccio, *Il Filostrato* 50; Dares and Dictys 72; Benoît de Sainte-Maure, *Le Roman de Troie* 77; Guido de Columnis, *Historia destructionis Troiae*, and the Tradition of Troilus and Criseyde 90; Boethius, *De consolatione philosophiae* 96; Ovid 109; *Le Roman de la rose* 114; Guillaume de Machaut 118; The Story of Thebes 121; Dante 125

Genre   138

  Introduction 138; Epic 140; Romance 144; History 151; Tragedy 154; Drama 161; Lyric 163; Fabliau 169; Allegory 172

Structure   180

  Introduction 180; Symmetry 184; Structure and Setting 191; Structure of Time 198; Structure in the Stars 204

Themes   212

  Introduction 212; A Debate about Love 215, Serving and Deserving 228, Love and Religion 231, Sickness and Death 234, *Pitee* 238, Secrecy 240, Honour 244, *Trouthe* 246; Time and Change 251; Past, Present, and Future 255; Freedom and the Stars 259; Fortune and Freedom 261; Characterization 267, Troilus 275, Criseyde 279, Pandarus 288, Diomede 294; Ending(s) 298

Style   314

  The Play of Style 314, 'In science so expert' 326, 'This paynted proces' 329, 'Verray signal of martire' 337, 'Proverbes may me naught availle' 345, 'To ryme wel this book til I have do' 354

Imitation and Allusion, *c.*1385–1700   360

BIBLIOGRAPHY   383

INDEX   407

LIST OF FIGURES

1. The frontispiece picture in an early fifteenth-century manuscript of *Troilus and Criseyde*, Corpus Christi College, Cambridge, MS 61. Reproduced by courtesy of the Master and Fellows of Corpus Christi College, Cambridge.   14

2. An opening in the Phillipps manuscript of *Troilus and Criseyde* (now Huntington Library HM 114), fos. 261ᵛ–262ʳ. The right-hand leaf contains *Troilus*, iii. 1728–43, 1772–89. Troilus' song (lines 1744–71) is lacking, but the text continues from 1743 to 1772 without break. The left-hand leaf, containing the text of the song, has been inset; the other side of this leaf is blank. The paper and handwriting are the same, but the margins of the inset leaf have not been ruled and there are only four stanzas on the page. On the right-hand leaf by iii. 1743 (where the song is missing) a marginal note 'Love &*cetera* | ad t*a*le sig*num*' refers the reader to the beginning of the text of the song on the inset left-hand leaf. Reproduced by courtesy of the Huntington Library, San Marino, California.   25

3. A page in a fifteenth-century manuscript of *Troilus and Criseyde*, St John's College, Cambridge, MS L.1, fo. 83ʳ, showing iv. 1044–78. At the foot of the page is a note 'her faileth thing yᵗ is nat yt made'. Reproduced by courtesy of the Master and Fellows of St John's College, Cambridge.   29

4. The text of *Troilus*, iii. 708–35, with marginal glosses, in a page of a fifteenth-century manuscript of *Troilus and Criseyde*, British Library MS Harley 2392, fo. 64ᵛ. (For a transcription of the glosses, see p. 43.) Reproduced by courtesy of the British Library.   43

ABBREVIATIONS

EETS ES	Early English Text Society, extra series
EETS OS	Early English Text Society, original series
JEGP	*Journal of English and Germanic Philology*
PMLA	*Publications of the Modern Languages Association*
STS	Scottish Text Society

INTRODUCTION

The object of this book is to give an up-to-date summary of what is known about Chaucer's *Troilus and Criseyde* and to offer interpretation, with the aid of that diverse accumulation of modern scholarship and criticism on *Troilus* represented by the Bibliography. The poem is discussed under the following headings:

Date. The evidence of Chaucer's composition of *Troilus* in the period 1382–5/6 is reviewed.

Text. The text's projection within itself of its own composition and reception is treated by way of preface to a discussion of the questions posed by the surviving manuscripts of *Troilus* about the text of Chaucer's poem.

Sources. Summaries of sources and lists of Chaucer's borrowings are provided for reference where appropriate, together with accounts of Chaucer's use and adaptation of his various sources in composing a poem that both draws on so many 'old bokes' and yet is profoundly original and new.

Genre. As well as a variety of sources, the elements of a remarkable diversity of genres are brought together in *Troilus*, and these different generic features of the poem are described and discussed.

Structure. Chaucer's poem is a triumph of form, and the sections of this chapter analyse how *Troilus* is shaped into its distinctive structure.

Themes. Questions lie at the heart of *Troilus and Criseyde* and, for convenience of discussion, the sections of this chapter focus in turn on aspects of love, mutability, the freedom of the human will, and the enigmas of character—almost inextricably interrelated themes in a debate renewed by every reading of Chaucer's narrative.

Style. Through the sections of this chapter an account is built up of the contributions made by the sheer stylistic variousness and resource within the poem's language.

Imitation and Allusion, c. *1385–1700*. A history of subsequent reference to the Troilus and Criseyde story and to Chaucer's poem is here outlined.

Such divisions in areas of discussion and ways of approach—which conform with those in the other volumes of this series—are matters of convenience and should be seen as a point of departure. It is in the nature of so multi-faceted a poem as *Troilus and Criseyde* that to take up any single point of view or to dwell on only one aspect is always to be aware that there are others.

Bibliographical Note

In uniformity with the other Oxford Guides to Chaucer, quotation from the text of *Troilus and Criseyde*, and of Chaucer's other works, is made from *The Riverside Chaucer*, general ed. Larry D. Benson (Boston, 1987; Oxford, 1988 (paperback)), with occasional changes to punctuation. (The edition of *Troilus* in the *Riverside Chaucer*, with excellent commentary, is by Stephen A. Barney.) I am much indebted to all who have edited *Troilus and Criseyde*, and full details of the editions I have used are given below.

The Bibliography lists publications on aspects of Chaucer's *Troilus and Criseyde* used in the preparation of this book. Each section of each chapter is followed, wherever appropriate, by its own brief annotated bibliography, which records works used in compiling the section, together with suggestions for further reading. References in these section bibliographies to works listed in the main Bibliography are in the form of author's surname and short title.

Editions

Benson, Larry D. (general ed.), *The Riverside Chaucer* (Boston, 1987; Oxford, 1988 (paperback)).

Donaldson, E. T. (ed.), *Chaucer's Poetry: An Anthology for the Modern Reader* (2nd edn., New York, 1975).

Fisher, J. H. (ed.), *The Complete Poetry and Prose of Geoffrey Chaucer* (2nd edn., New York, 1989).

Robinson, F. N. (ed.), *The Works of Geoffrey Chaucer* (2nd edn., Boston and Oxford, 1957).

Root, Robert Kilburn (ed.), *The Book of Troilus and Criseyde* (Princeton, NJ, 1926).

Shoaf, R. A. (ed.), *Geoffrey Chaucer: Troilus and Criseyde* (East Lansing, Mich., 1989).

Skeat, W. W. (ed.), *The Works of Geoffrey Chaucer* (7 vols.; Oxford, 1894–1900).

Windeatt, B. A. (ed.), *Troilus and Criseyde: A New Edition of 'The Book of Troilus'* (London, 1984; 2nd edn., 1990).

Date

Chaucer was probably composing his *Troilus and Criseyde* during the early and middle years of the 1380s and had finished it by early in 1387 at the latest, as is indicated by several kinds of evidence. It is the deaths of two of Chaucer's associates, Ralph Strode and Thomas Usk, which provide evidence for the later date. Chaucer's friend Strode—the London lawyer to whom *Troilus* is co-dedicated, and submitted with a request for correction—is known to have died in 1387, although a document records him as still active on 27 April 1387. Thomas Usk, a literary disciple of Chaucer, was executed for treason on 4 March 1388, and in his prose work *The Testament of Love* he had shown a close familiarity with the text of *Troilus*, although it remains a matter of dispute whether Usk wrote the *Testament* during his final imprisonment or during an earlier period in prison, from December 1384 to June 1385. In his *Testament*—which cannot have been written later than the winter of 1387–8—Usk makes the allegorical figure of Love refer to Chaucer as 'the noble philosophical poete in Englissh' because of the 'tretis that he made of my servant Troilus' (see below, p. 361). There is another kind of response to the appearance of *Troilus* in the earlier F version of Chaucer's *Prologue to the Legend of Good Women*, thought to date from 1386–8, where the God of Love rebukes the poet for the writing of *Troilus* and the translation of the *Roman de la rose*, and imposes upon him the penance of writing the *Legend*:

> 'And of Creseyde thou hast seyd as the lyste,
> That maketh men to wommen lasse triste,
> That ben as trewe as ever was any steel.'   (F 332–4)

By presenting the composition of the *Legend* as a consequence of the circulation of *Troilus*, the *Prologue* implies that *Troilus* was a relatively recent work, and the manner in which the *Prologue* reviews Chaucer's writings without mentioning the *Canterbury Tales* has suggested that this *Prologue* must have been written by 1388.

In both versions of the *Prologue*—perhaps because of its concern for 'Good Women'—Chaucer's poem is referred to as if its title were simply 'Criseyde':

> 'Therfore he wrot the Rose and ek Crisseyde
> Of innocence, and nyste what he seyde ...'   (G 344–5)

> 'And forthren yow as muche as he mysseyde
> Or in the Rose or elles in Creseyde.'   (F 440–1)

In the little poem addressed to his scribe Adam, Chaucer refers to the poem simply as 'Troilus', while in the *Retractions* to the *Canterbury Tales* it is entitled 'the book of Troilus' (X. 1085–6). The poem does not give itself a title within its text, and the extant manuscripts differ between 'Troilus' and 'Troilus and Criseyde' in their scribal rubrics. Six of the twelve extant manuscripts that contain the conclusion (the text of four other manuscripts being lacking by that point) have explicits that in various forms and spellings refer to the poem as being of 'Troilus and Criseyde'. Usk refers to the poem as 'the boke of Troilus' (*Testament*, iii. 4), and it is the English Chaucerian John Lydgate, writing in the early fifteenth century, who first calls the poem 'Troilus and Criseyde' in a passage in his *Fall of Princes*:

> In youthe he made a translacioun
> Off a book which callid is Trophe
> In Lumbard tunge, as men may reede & see,
> And in our vulgar, longe or that he deide,
> Gaff it the name off Troilus & Cresseide.   (Prologue, 283–7)

Lydgate's hearsay information about Chaucer—whom he did not know—has to be treated carefully, and his reference here to a book entitled 'Trophe' has never been explained. If Chaucer indeed completed *Troilus* some fifteen years or more before his death in 1400, Lydgate's claim that he wrote the poem 'longe or that he deide' is not unreasonable, although Chaucer would then have been writing in his mid-forties rather than 'in youthe'.

If the date by which *Troilus* must have been circulating can be determined, the time when Chaucer began the poem—and the length of the period of composition—can be less clearly established, but it most probably lies in the years 1382–5/6. In its close engagement with Boccaccio's *Il Filostrato* and its echoes of other works of Boccaccio, and of Dante and Petrarch, *Troilus* represents the fruition of Chaucer's encounter with Italian literature, an encounter usually assumed to have begun with his first visit to Italy in 1373. But a suggestion that *Troilus* could date from as early as the later 1370s has not found favour, not least because of the number of Chaucer's other works which there are good reasons for thinking he completed before *Troilus*.

An outline chronology of Chaucer's life and works would now generally be agreed to be as shown in Table 1 (and serves to bring out the place of *Troilus and Criseyde* at the heart of Chaucer's career). There are, moreover, a number of possible pointers in references within *Troilus* which would indicate a date in the mid-1380s, together with the evidence of what was happening in Chaucer's life at the time. It was long ago suggested that when Chaucer describes Criseyde—

Right as oure firste lettre is now an A,
In beaute first so stood she, makeles.   (i. 171–2)

—he is making a graceful allusion to the fact that 'now' (following the marriage of Richard II to Anne of Bohemia on 14 January 1382) the initial of the Queen of England was an 'A'. As the date appears about right, this pleasing suggestion has tended to be accepted as an indication of when Chaucer might have been starting or at an early stage of work on *Troilus*,

TABLE 1. *An outline chronology of Chaucer's life and works*

Date	Life	Works
*c.*1340–3	Chaucer born	
1357	Page to Countess of Ulster	
1359–60	Military service, captured and ransomed	
1366	Journey to Spain	
*c.*1366	Married	
1367	Recorded as 'valettus' and 'esquier' in King's Household, with royal annuity	Before 1372: the *A.B.C.*; the *Romaunt of the Rose* (fragment A, if by Chaucer); the *Book of the Duchess*
1369	Campaign in France	
1372–3	Journey to Italy (to Genoa and Florence)	*c.*1372–80: the Life of St Cecilia (later the *Second Nun's Tale*); tragedies later used in the *Monk's Tale*; the *House of Fame*; *Anelida and Arcite*; the *Complaint unto Pity*; the *Complaint to his Lady*
1374	Appointed controller of customs	
1376–7	Journeys 'on secret business of the King' (including to France)	
1378	Journey to Italy (to Lombardy)	
1382	Additionally appointed controller of petty customs	Early 1380s: the *Parliament of Fowls*; *Boece*; 'Palamon and Arcite' (later the *Knight's Tale*)
1385	Appoints permanent deputy at customs; Justice of the Peace in Kent	*c.*1382–5/6: *Troilus and Criseyde* *c.*1385: the *Complaint of Mars* *c.*1386: the *Legend of Good Women* and its *Prologue* *c.*1387–1400: at work on the *Canterbury Tales*
1386	Knight of the Shire for Kent, then loses controllerships	
1389	Clerk to the King's Works	
1391	Relinquishes clerkship; appointed Sub-Forester of North Petherton	1391–2, with subsequent additions: *Treatise on the Astrolabe*
1394	Annuity from Richard II	1390s: Revised *Prologue to the Legend;* the later short poems ('To Scogan', 'To Bukton', 'To his Purse')
1399	Grants confirmed by Henry IV	
1400	Death of Chaucer	

although the allusion could have been made at any time after the king's marriage.

Further evidence for a possible date for *Troilus* is provided by the way the state of the heavens at a crucial point in Chaucer's poem coincides with an unusual state of the heavens at a point in the 1380s. On the night when Troilus and Criseyde consummate their love there is said to be a conjunction of Jupiter, Saturn, and the crescent moon in the sign of Cancer (iii. 624–5), and this planetary conjunction was believed to occur in May 1385, for the first time since AD 769, so that the night of the fictional consummation has been 'dated' to 13 May 1385. It was held that such conjunctions of the superior planets, Saturn, Jupiter, and Mars, were portents of great events. The conjunction of Saturn and Jupiter in the sign of Cancer was traditionally believed to have heralded the Flood, while the conjunction of Jupiter and Saturn in Aquarius in 1345 was considered to have been a forewarning of the Black Death. In this light it is possible that Chaucer may have included the conjunction of Jupiter and Saturn in Cancer within his Trojan poem as a warning of the impending fall of Troy. The occurrence of the conjunction in May 1385 was much remarked upon at the time, seen as a portent of 'maxima regnorum commotio' by the chronicler Walsingham in his *Historia Anglicana*, and, as in Chaucer's poem, was associated with heavy rain and floods: 'Eodem tempore Conjunctio duarum maximarum planetarum facta est, videlicet Jovis et Saturni, mense Maio; quam secuta est maxima regnorum commotio . . .' (At the same time in the month of May a conjunction of the two largest planets comes about, namely Jupiter and Saturn. A great upheaval of realms followed this . . .) However, it has also been argued that the phases of the moon fit better with a date of 9 June 1385, and that on that date Mars had just come into Cancer and was near to conjoining with Saturn, in a triple conjunction of Saturn, Jupiter, and Mars that was associated by astrologers with the destruction of sects and kingdoms.

Caution is, however, still needed about presuming that the poem would only refer to such a celebrated conjunction at the time or after it happened. Someone as keenly interested in astrology as Chaucer would not need to have witnessed so rare and portentous a conjunction in order to be mindful of it in the 1380s. Indeed, Chaucer's version does differ slightly from the historical conjunction in making Venus a morning star (iii. 1417). Since Venus was close to Jupiter and Saturn in May 1385, there is the possibility that Chaucer's text does not simply represent a record of the historical event, although by 9 June Venus was more than thirty degrees away. It is almost certainly too literal an application of the possible historical connections to suppose that Chaucer began Book I in 1382 and was working on Book III in May 1385. The duration of the composition process remains unclear, but may well have been begun in 1382 and brought to completion in 1385 or 1386, possibly for the festivities to mark the departure to Spain of John of Gaunt in 1386.

It is also possible that events in Chaucer's life at this time—both welcome and unwelcome—gave him more leisure for poetry and may be reflected in his writing of *Troilus*. On 17 February 1385 Chaucer had received permission to perform his duties as a controller of customs through a permanent deputy, and this release from the demands of his job must have increased Chaucer's free time for writing at this period. Less welcome release resulted from the political crisis of the following year, for by the end of 1386 Chaucer had lost his position at the customs. The parliament of October and November 1386 was attended by Chaucer as a knight of the shire, and it saw the ascendancy of Thomas of Woodstock, duke of Gloucester, and the downfall of a number of Chaucer's friends and associates. Although Chaucer was close to the Lancastrian faction, he apparently remained unharmed, but he did lose his position at this time. A petition presented at the parliament called for the removal of all controllers of customs with life terms, and a few weeks after parliament adjourned Chaucer lost both his controllerships, in December 1386.

If Chaucer was still working on *Troilus* during this difficult time, it is tempting to see some reflection of these contemporary troubles in his writing. It has been suggested that the representation of the parliamentary proceedings in Book IV of *Troilus* reflects Chaucer's jaundiced view of the parliament he had attended, just as the description of the 'peple' as fierce 'as blase of straw iset on fire' (iv. 184) has been taken as a passing allusion to the figure of Jack Straw and the Peasants' Revolt of 1381. What is undeniably true is that Chaucer has created a real sense of a parliament, and makes the exchange of Criseyde for Antenor depend on its deliberations, in a way that has no parallel in any of his sources. He dramatizes a parliamentary assembly headed by a president (iv. 211–17), with an unruly and foolish commons (iv. 183–96), in which there are majority decisions of 'lord and burgeys' (iv. 344–5), which are respected by a prince of the blood (iv. 558–60).

There was, moreover, a long tradition linking Troy with the City of London, where Chaucer lived most of his life and where he was presumably writing *Troilus*, and Trojan themes and connections were made much of at the time Chaucer was writing. It was held that Britain had been founded after the sack of Troy by a Trojan noble, Brutus, who gave the land his name. In his *Historia Regum Britanniae* Geoffrey of Monmouth told how Brutus chose to build his capital on the banks of the Thames:

Condidit itaque ciuitatem ibidem eamque Troiam Nouam uocauit. Ex hoc nomine multis postmodum temporibus appellata tandem per corruptionem uocabuli

Trinouantum dicta fuit.

There then he built his city and called it Troia Nova. It was known by this name for long ages after, but finally by a corruption of the word it came to be called Trinovantum.

By Chaucer's time the connection with Troy could proudly be seen as a continuity. The *Liber albus*, a fifteenth-century handbook of City of London customs, recalls that the city now called 'London', founded in imitation of Great Troy, was constructed and built by King Brut, the first monarch of Britain, being at first called 'New Troy' and afterwards 'Trinovant', and it also claims that the city possesses the liberties, rights, and customs of the ancient city of Troy and enjoys its institutions. Indeed, in the mid-1380s, when Chaucer was writing *Troilus*, one of his associates is reputed to have tried to get the name of the City of London changed to 'Little Troy'. According to the chroniclers Walsingham and Knighton, Nicholas Brembre, several times mayor of London and Chaucer's associate in the customs service, who was attacked in the 1386 parliament and executed in 1388, had planned to change the name of London to Little Troy and had intended—with the help of Richard II—to become 'Duke of Troy'.

Another of Chaucer's associates, the poet John Gower to whom *Troilus* was co-dedicated, recurrently refers to London in his works as 'the toun of newe Troye' (*Confessio amantis*, 37*). In his *Vox clamantis* he dreams of how New Troy (London) is pillaged by the peasants just as Old Troy was sacked, and concludes sadly of London 'No boldness of a Hector or a Troilus defeated anything then . . .' (i, ch. 13). Just as the fate of Chaucer's hero called 'Troylus' is identified with that of Troy, so in his poem celebrating the reconciliation of Richard II with the city of London in 1392 Richard of Maidstone calls London 'Trenovant' or 'Nova Troja', and calls the king 'Troilus' or 'Little Troy'. Because Troy and the Trojans were part of the imagined genealogy of the first audiences of Chaucer's poem, the qualities and the fate of the Trojan characters in *Troilus* held a special application. In one traditional interpretation the fall of Troy was understood to stem from neglect of Pallas, or wisdom, and a reversion to fleshly lust and idolatry, and it has been suggested that in *Troilus* Chaucer is drawing analogies—in the fate of the hero and his city—between this traditional moral interpretation of Troy and the political and moral instability of England in the 1380s.

It is not only its probable date but also its standing in relation to Chaucer's earlier and later works which help place *Troilus* at a focal and central point in Chaucer's development. *Troilus* stands unsurpassed artistically and unrepeated thematically in the middle years of Chaucer's poetic career, by far his longest completed single poem, his most fully achieved work of art, and in that sense his masterpiece. From the evidence of his surviving poems Chaucer was throughout his life drawn to rather brief narrative forms and shows a marked disinclination to finish what he begins, so that in terms of his own work the sheer scale and scope of *Troilus* is remarkable as a completed and fulfilled ambition.

A narrative poem so capacious in conception and form allows Chaucer to achieve things he had not attempted before and which he would not feel driven to do again. Intriguing connections link *Troilus* with Chaucer's earlier dream poems as framed narratives exploring the relation between books and experience, and the theme of love, but he also moves beyond the dream poems in narrative technique and the thematic potential this is used to explore. The potential Chaucer could see in the love-story of Troilus and Criseyde at this point of his life as a poet was apparently prompted by his happy encounter with the works of very various poets, who influenced Chaucer both in themselves and in the use he then went on to make of other works he had known long before. Chaucer's meeting with the works of the Italian trecento writers—with Dante, Petrarch, and Boccaccio—has a transformative effect on the English poet, and it is in *Troilus* that that transformation finds its supreme expression. From them Chaucer could derive an exalted sense of what it meant to be a poet, a confidence in the potential for writing in the vernacular, and a keener historical interest in the pagan past of classical antiquity and the challenge of representing it. That *Troilus* is co-dedicated to Ralph Strode—both a lawyer and an important philosophical thinker of the day—suggests the serious attention Chaucer hoped his poem would receive from readers with interests like those of himself and his friends. Chaucer came to the composition of *Troilus* after working on his translation into English prose of the *De consolatione philosophiae* of the late Roman writer Boethius, and Chaucer's imagination was evidently still steeped in Boethian ideas, images, and phrases, which suffuse his writing in *Troilus*. It would be hard to overestimate the influence of this work on the poem, and the opportunity for a sustained exploration of themes as of character was an aspect of the unparalleled length and fullness of the *Troilus* narrative. Never again would Chaucer lavish such full attention on a single story and a single set of characters.

It is the pattern of such achievements which determines the focus of the ensuing chapters of this Oxford Guide in their attention to Chaucer's presentation of the text of his narrative, the conjoining of a remarkable diversity of sources, his innovations in genre, his unique achievements in structure, and his thematic development of an old tale about the end of a love-affair into a searching exploration of human character and human purposes, in which style and language itself become a subject of the poem.

On the date of *Troilus and Criseyde*, see especially Kittredge, *Date*, and Root (ed.), *Book*. For an earlier date in the 1370s, see Tatlock, *Development* and 'Date'. On the possible allusion to Anne of Bohemia, see Lowes, 'Date'. On the date of the conjunction of Jupiter and Saturn as 13 May 1385, see Root and Russell, 'A Planetary Date'; and as 9 June, see North, *Chaucer's Universe*, 374–5. On the associations of the conjunction, see O'Connor, 'The Astronomical Dating'. On the poem's possible connections with the 1386 parliament and with other contemporary events, see McCall and Rudisill, 'Parliament', and Robertson, 'The Probable Date'.

The Ralph Strode recorded as a Fellow of Merton College before 1360 is usually thought to be the Ralph Strode subsequently recorded as common sergeant or pleader of the City of London, and standing counsel from 1386. The *Catalogus vetus* of Fellows of Merton (1422), in a note against his name, records that he was a poet and the author of an elegiac work in verse entitled 'Phantasma Radulphi'. Only fragments of Strode's logical system survive in his treatises *Consequentiae* and *Obligationes*. Strode entered into controversy with Wyclif, albeit on friendly terms. His position is now known only through Wyclif's rejoinders, but he evidently contested Wyclif's doctrine of predestination as denying free will. From 1375 Strode held rooms over Aldersgate, while Chaucer lived over Aldgate from 1374, and in 1382 they were both sureties for the peaceful behaviour of John Hende, a well-to-do London draper. See *Chaucer Life-Records*, ed. Martin M. Crow and Clair C. Olson (Oxford, 1966), 281–4. For accounts of Strode's career, see A. B. Emden, *A Biographical Register of the University of Oxford to A. D. 1500*, iii (Oxford, 1959), 1807–8, and also the article on Strode by Sir Israel Gollancz in the *Dictionary of National Biography*.

On the date of Usk's *Testament*, see Ramona Bressie, 'The Date of Thomas Usk's *Testament of Love*', *Modern Philology*, 26 (1928), 17–29, and Paul Strohm, 'Politics and Poetics: Usk and Chaucer in the 1380s', in Lee Patterson (ed.), *Literary Practice and Social Change in Britain, 1380–1530* (Berkeley, Calif., Los Angeles, and Oxford, 1990), 83–112. For Chaucer's connections and their background, see Paul Strohm, *Social Chaucer* (Cambridge, Mass., 1989), and S. Sanderlin, 'Chaucer and Ricardian Politics', *Chaucer Review*, 22 (1988), 171–84.

For the explicits to the poem in extant manuscripts, see Windeatt (ed.), *Troilus and Criseyde*, 563.

For Lydgate's reference to *Troilus and Criseyde*, see *Lydgate's Fall of Princes*, ed. H. Bergen (EETS ES 121–4; London, 1924–7).

On the rare conjunction in 1385, see Thomas Walsingham, *Historia Anglicana*, ed. H. T. Riley (Rolls Series; London, 1863–4), ii. 126.

For the documents relating to the appointment of a permanent deputy to Chaucer in the office of controller in the Port of London, see *Chaucer Life-Records*, 168–9. For documents relating to Chaucer's retirement from the controllerships in December 1386, see *Chaucer Life-Records*, 268–9.

The Historia Regum Britannie of Geoffrey of Monmouth, I, Bern, Burgerbibliothek, MS. 568, ed. Neil Wright (Cambridge, 1985). On the founding of London, see pp. 14–15.

Liber Albus: The White Book of the City of London, ed. H. T. Riley (London, 1861). On London and Troy, see p. 54. See also D. W. Robertson, jun., *Chaucer's London* (New York, 1968), and J. Clark, 'Trinovantum—The Evolution of a Legend', *Journal of Medieval History*, 7 (1981), 135–51.

On Nicholas Brembre, see Walsingham, *Historia Anglicana*, ii. 173–4, and *Chronicon Henrici Knighton*, ed. J. R. Lumby (Rolls Series; London, 1889–95), ii. 293.

On Gower's works, see below, p. 96.

Richard of Maidstone, *Concordia inter Regem Ric. II et civitatem London*, ed. T. Wright (London, 1838). For the Troy and 'Troylus' references, see pp. 31, 32, 35.

Kean comments 'Indeed, the love story, not doctrine, is the heart of the poem, and Chaucer has done with it what he was never to do again—exposed his reader directly . . . to the full impact of the action and the actors' (*Chaucer*, 177–8). It is the achievement of *Troilus* that 'like the *Canterbury Tales* it contains a world. The later work excels it only in particularity of reference, not in control of form or depth of conception' (Muscatine, *Chaucer*, 124).

Text

Text, Voice, Performance

No autograph copies of the text of Chaucer's *Troilus and Criseyde* survive, nor are there any surviving copies written near to the time of the poem's composition or before Chaucer's death in 1400. There are sixteen extant manuscripts of the poem (two of them fragmentary), all dating from the fifteenth century:

A	British Library, Additional MS 12044
Cl	The 'Campsall' MS, now Pierpont Morgan Library M817
Cp	Corpus Christi College, Cambridge, MS 61
D	University Library, Durham, Cosin MS V.II.13
Dg	Bodleian Library, Digby 181 (ends at iii. 532)
Gg	Cambridge University Library, Gg.4.27
H1	British Library, Harley 2280
H2	British Library, Harley 3943
H3	British Library, Harley 1239
H4	British Library, Harley 2392
H5	British Library, Harley 4912 (ends at iv. 686)
J	St John's College, Cambridge, MS L.1
Ph	Formerly Phillipps 8252, now Huntington Library HM 114
R	Bodleian Library, Rawlinson Poet. 163
S1	Bodleian Library, Arch. Selden B.24
S2	Bodleian Library, Arch. Selden Supra 56

Some of the earliest of these manuscripts can be dated to the first two decades of the fifteenth century (e.g. Cp and Cl), so that some twenty-five or thirty years—and an indeterminable amount of copying and recopying—may intervene between the time of the poem's composition and the earliest surviving copies.

What would medieval readers see when—with care, for books were expensive, special objects—they opened a manuscript copy of *Troilus*? Of all the sixteen manuscripts only four (Dg, Gg, H3, S1) include *Troilus* along with miscellaneous collections of other verse, while the other twelve present *Troilus* alone (apart from brief items in the endpapers). It would thus normally have been encountered by early readers as a book on its own.

In most manuscripts the text of the poem has been set out on the page with a recognition of that architectural sense in *Troilus* of structured mass and line, built from highly crafted parts, a layout suggesting that Chaucer always had in his mind's eye the contribution to his poem made by the appearance of its text on the page in manuscript copies that he would never see. The majority of manuscripts contain the Latin rubrication of incipits and explicits signalling the books and their proems. Four, five, or six seven-line stanzas may be set out on each manuscript page, although some scribes have copied the stanzas as a continuous text, with no line-break between one stanza and the next, and breaking stanzas between pages (see Figs. 2, 3, and 4). Three manuscripts contain substantial amounts of glossing, one (H4) largely in Latin, one (R) largely in English, one (S1) mixed; the other extant manuscripts are more occasionally glossed or present a plain text to their reader. Of the earliest manuscripts, several are stylishly presented. The Campsall manuscript is emblazoned on its opening page with the arms of Henry V when Prince of Wales. The Corpus manuscript, most splendid of all the *Troilus* manuscripts, contains many blank spaces for illustrations which were never executed and is prefaced by a frontispiece picture apparently depicting a poet reading to a courtly audience (see Fig. 1).

What may be deduced about the reception of Chaucer's text by its early audiences? Pandarus finds Criseyde sitting in her parlour with two other ladies who 'Herden a mayden reden hem the geste | Of the siege of Thebes, while hem leste' (ii. 83–4), and by this Chaucer contains within his poem a scene that is very likely to represent the way *Troilus* was often first encountered in its own time, as a sociable pastime. Any English group having *Troilus* read to them would thus listen to a scene in which a group of Trojans are listening to a text being read to them, and this sense of audiences within audiences is typical of a reflexive quality in *Troilus* about its nature as a text. But Criseyde is capable of reading for herself, because she later reads her lover's letter (ii. 1176–8), and she refers Pandarus to the rubrics of the manuscript being read ('And here we stynten at thise lettres rede' (ii. 103)), just as early readers of some *Troilus* manuscripts had their reading guided by rubrics and manuscript glosses which register responses to the sources, genres, and language of the poem.

Ideas about the audience of *Troilus* have been variously influenced by the frontispiece picture to the Corpus manuscript. Its design is in two component parts. The lower half presents a scene of public recital in the open air, with a figure at a pulpit surrounded by an elegantly dressed audience. The upper half presents a scene of encounter between two comparably elegant retinues, in procession through a landscape of steep rocky hills crowned with castles against a sky of gold. The figure in the

FIG. 1. The frontispiece picture to the text of *Troilus* in Corpus Christi College, Cambridge MS 61.

pulpit bears some resemblance to early likenesses of Chaucer, and this picture has been taken as the record of a particular historical occasion when Chaucer read his *Troilus* to a courtly audience, perhaps the court of Richard II itself. Yet the nature of the evidence given by such a painting for the audience of the poem needs to be carefully considered. Like the text of the poem itself, this picture exists within conventions and traditions, and needs to be interpreted in relation to these.

To set the picture in the context of contemporary manuscript illumination is to discover that, while in its subject-matter it may be conventionally determined, and its likelihood to record an actual performance of *Troilus* therefore diminished, in its time and place it is exceptional. Both scenes have been traced to various models and exemplars of illumination. The upper scene has been compared with early fifteenth-century 'Itinerary' miniatures devised by the Limbourg brothers to accompany prayers for safe journeying in manuscripts of Jean, duc de Berry; the model for the open-air recital scene appears to be depictions of a preacher and his audience. The stylistic mode of the frontispiece associates it with French and Italian work of the late fourteenth and early fifteenth centuries; and what is remarkable in an early fifteenth-century context (the costumes of the figures point to a date of not much later than *c*.1415) is that a copy of *Troilus* should be introduced by an exceptional piece in the International Gothic style of painting. This manuscript—thought to have belonged to Anne Neville (*c*.1410–80), duchess of Buckingham, and granddaughter of Chaucer's patron, John of Gaunt—is presented in the spacious format already established by continental traditions of secular book illumination but without precedent in the publication of any major English work before this time. Seen in this context, the creation of the Corpus manuscript may stand to represent the coincidence of *Troilus and Criseyde* with such refined and sophisticated English works of art of its time as the Wilton Diptych or the Bohun Psalters, products of a cultivated courtly society with strong European affiliations.

Rather than recording one particular performance of *Troilus*, the Corpus frontispiece may well represent a product of the poem's power to create a sense of a listening group, and depicts as a reality the myth of delivery that Chaucer develops so carefully in the poem. The design of the Corpus manuscript prefaces the text of the poem by a picture that implies how Chaucer presented his poem to the members of his audience, envisaging them as they are fictionalized by the poem rather than as they may actually have been. There is reason to believe that we should look beyond the entourage of king and nobility for Chaucer's audience, to the household knights and officials, career diplomats and civil servants, who constitute the 'court' in its wider sense, and Chaucer's own circle. On the evidence of the household records of medieval courts it has been questioned whether any likely courtly audience of Chaucer's would actually have contained a significant number of women, despite the poem's address 'Bysechyng every lady bright of hewe, | And every gentil womman, what she be' (v. 1772–3), and despite the presence of women in the Corpus picture.

If 'the writer's audience is always a fiction', the implication of a courtly and partly female group as the audience of *Troilus* may therefore be most significant as a pointer to how the poem asks to be read, through the presentation of the poet-figure or narrator, and the fictionalization, the inscribing, of the poem's audience within the text. Distinctions are to be drawn between Chaucer's audience(s): fictional, implied, intended, actual; and distinctions may also be made to avoid confusion of 'voice' with 'presence', so that the speaking persona is regarded not literally as an individual 'character' behind the text but as an impersonation or a kind of personification. It is no accident that it is *Troilus* of all Chaucer's poems that should have been prefaced by such a picture as the Corpus frontispiece, for it is in this of all his poems that Chaucer makes his most sustained use of the sense of a poet in the presence of an audience. The Corpus artist places the figure in the pulpit near the centre of his design, presenting an audience and an act of address as the foreground, almost the frame, to a further scene which apparently represents a moment from the poem. In this the design adapts a medieval pictorial convention in which an author is shown composing, accompanied by the events and personages within his work. In the *Troilus* picture we look through and past the framing scene of narration into a scene which probably represents the core narrative image of the traditional Criseyde story—her departure from Troy—and this can aptly stand to represent the distancing of the narrative by a foregrounding of the relationship of poet and audience in Chaucer's poem.

Many of those first encountering *Troilus and Criseyde* in Chaucer's time would have had little sense of the poem as a text in its material sense, as a physical object, because they would have heard the work rather than read the book. Yet *Troilus* presents a remarkable combination of identities and deliveries, referring to itself at various moments as being written, recited, or even 'sung' (ii. 56; iii. 1814; iv. 799; v. 1797). In the course of the poem—sometimes simultaneously—*Troilus* presents its text both as the written pages of a book, which the poet is still in the midst of composing at his desk, and as the script of a performance which is in the act of being presented to an audience; the dramatic momentum of a process under way in each case allows these two different layers of the text's existence to be superimposed. This double sense of the text as the product of both a scholar and an entertainer lends the poem its combination of painstaking, deeply pondered, bookish form with its conversational immediacy and ease, casual informality and open-endedness. Thus the first proem announces its intention to tell of the double sorrow of Troilus 'er that I parte fro ye' (i. 5), but immediately refers to composing 'thise woful vers, that wepen as I write' (i. 7). The second proem similarly refers to the processes of composition and recital as if they were concurrent, praying for help in rhyming (ii. 10) and insisting that the text is being written as a translation (ii. 14), but twice in the same passage acknowledging the presence of 'any lovere in this place | That herkneth' (ii. 30–1, 43–4). The first Canticus Troili is introduced with a line noting that, whoever wishes to 'here' the song, may find it 'next this vers', in the adjacent lines of the text (i. 398–9), and when Troilus sits down to write his letter we are invited to hear what he said (v. 1316).

As the text progresses and the book grows larger, the sense of a performance gives way somewhat to the processes of composition: in Book III the narrator refers to 'this book' and to himself as composing it (iii. 501–4) and after the consummation comments 'This joie may nought writen be with inke' (iii. 1693). In the fourth proem the poet describes the pen shaking in his hand (iv. 13–14). In its last book *Troilus* for the first time addresses itself to an imagined reader ('Thow, redere, maist thiself ful wel devyne' (v. 270)). Even in the final stanzas of the poem—with all their writerly concern for the posterity of the text—there is an overlapping sense of the poem as being simultaneously composed and recited (v. 1765–1800).

Any dramatization of the composition process would imply an audience, but Chaucer goes beyond this, so that at any reading in solitude the text re-creates for its reader a sense of being party to an occasion on which the poem is being recited. It is a distinctive stylization into permanent form of the ephemeralness of a living entertainment and the mobility of actual delivery. The superimposition of the dramatized composition process and the drama of performance is achieved because the text fictionalizes an audience with special expertise in the poem's subject of love, to whom the poet can refer his compositional concerns. It is an audience of 'lovers' that the poem ostensibly assumes throughout ('But ye loveres . . .' (i. 22; see also ii. 43–4, 1751)), addressing them as if present in the same place, sharing the same space, where the poet is reciting. Chaucer goes on to pick out imagined individuals in this audience of lovers, addressing each in the second person singular ('For to *thi* purpos this may liken *the*, | And *the* right nought' (ii. 45–6)), or forestalling individual objections (iii. 491 ff.). Here lovers are being reminded of the different style and expression of love in different ages, and it is noticeable how the fictionalized audience of lovers is involved at junctures which involve the conduct and rhetoric of love, for it is here that the dramas of composition and recital can coincide.

In a poem dedicated by name to 'moral Gower' and 'philosophical Strode', such recurrent fictionalizing of an audience of 'loveres' allows any actual audience to locate themselves in relation to that projected audience and hence to the experience the poem analyses, perhaps enjoyably watching themselves collaborating in the creation of illusion. This process is matched by a sociable and convivial manner of narration which is well mannered, tactful, anxious not to bore. It is also a deferential manner, for Chaucer developed from the traditional modesty conventions a way of poetic life. The text addresses itself rather flatteringly to the knowledge-ability of the audience it imagines to be present ('This, trowe I, knoweth al this compaignye' (i. 450; see also ii. 917)), especially flattering in the context of a narrator himself excluded from that experience assumed in the audience. In the story of a secret love, to hear is to overhear: the narrative addresses its audience as if imparting confidences, sharing secrets in a way that at times assumes the reader or hearer is an accomplice in the action ('Us lakketh nought but that we witen wolde | A certeyn houre, in which she comen sholde' (iii. 531–2)). In their inclusiveness of manner—sometimes using the first person plural ('Us thinketh hem . . .' (ii. 25))—such modes of address invite identification with an implied audience, and with qualities which place the reader within the range of the poem's judgements. Yet all the tactful invitations to identify with the fictional audience of 'ye loveres' are a device allowing the reader of *Troilus* to be aware of lovers' perspectives but also to be detached. Possible detachment had been allowed for—however ironically—in the poem's early warning to 'Ye wise, proude, and worthi folkes alle, | To scornen love . . .' (i. 233–4). Different modes of address anticipate and allow for different manners of reading, and the poem's late addresses to the women in its audience (on the subject of Criseyde), and to 'yonge, fresshe folkes', both 'he or she', suggest a development in the poem's fictionalizing of its audience of lovers. In the consummation scene the author's use of language is submitted to the correction 'of yow that felyng han in loves art' (iii. 1333), while at its conclusion the poem as a book is submitted and dedicated to the poet Gower and the philosopher Strode with a request that they should suggest corrections.

For full palaeographical descriptions of some *Troilus* manuscripts, see *Troilus and Criseyde: A Facsimile of Corpus Christi College Cambridge MS 61*, Introduction by M. B. Parkes and E. Salter (Cambridge, 1977); *Geoffrey Chaucer: The Poetical Works. A Facsimile of Cambridge University MS Gg.4.27*, Introduction by M. B. Parkes and R. Beadle (3 vols.; Cambridge, 1979–81); *St John's College Cambridge MS L.1: A Facsimile*, Introduction by R. Beadle and J. Griffiths (Cambridge, 1983).
On the Corpus *Troilus* frontispiece, see Pearsall, 'The *Troilus* Frontispiece'; Salter, 'The *Troilus* Frontispiece'; and Salter and Pearsall, 'Pictorial Illustration'. On the planned illustrations in the Corpus manuscript, see Fisher, 'The Intended Illustrations'. On the

context of the manuscript and its English and continental affiliations, see Elizabeth
Salter, *Fourteenth-Century English Poetry* (Oxford, 1983), p. 48.
On the likely historical audience for Chaucer's work in general, see Strohm, *Social Chaucer*,
and the papers in 'Chaucer's Audience: A Symposium', in the *Chaucer Review*, 18 (1983),
137–60: Richard Firth Green, 'Women in Chaucer's Audience'; R. T. Lenaghan,
'Chaucer's Circle of Gentlemen and Clerks'; and Paul Strohm, 'Chaucer's Audience(s):
Fictional, Implied, Intended, Actual'. See also Richard Firth Green, *Poets and
Princepleasers: Literature and the English Court in the Late Middle Ages* (Toronto, 1980);
Beryl Rowland, '*Pronuntiatio* and its Effect on Chaucer's Audience', *Studies in the Age of
Chaucer*, 4 (1982), 33–51; V. J. Scattergood, 'Literary Culture at the Court of Richard
II', in V. J. Scattergood and J. W. Sherborne (eds.), *English Court Culture in the Later
Middle Ages* (London, 1983), 29–43.
On aspects of the narrative voice in *Troilus*, see especially Covella, 'Audience'; Dahlberg,
'The Narrator's Frame'; Dinshaw, *Chaucer's Sexual Poetics*; Donaldson, 'Ending' and
'Criseide'; Evans, '"Making strange"'; Huppé, 'The Unlikely Narrator'; Jordan,
'Narrator'; Josipovici, *World*; Knopp, 'Narrator'; Lambert, 'Telling the Story'; Lawlor,
Chaucer; Lawton, 'Irony and Sympathy' and *Chaucer's Narrators*; Mehl, 'The Audience
of Chaucer', 'Chaucer's Audience', *Geoffrey Chaucer*, and 'Chaucer's Narrator'; Pearsall,
'The *Troilus* Frontispiece'; Rudat, 'Chaucer's *Troilus and Criseyde*'; Salter, 'Poet and
Narrator'. See also Walter J. Ong, 'The Writer's Audience is Always a Fiction', *PMLA*
90 (1975), 9–21.

The Troilus *Text: An Overview*

For the medieval poet the process of composition would usually be
succeeded and completed by a process of scribal transcription, creating
copies of the work for presentation and circulation (although no presentation copies of Chaucer's works survive). Just as *Troilus and Criseyde* has
created a sense of its text being composed in the course of a reading of the
work, so there is a fitting sense of completion that the poem at its close
anticipates within itself the process of its future (mis)transcription:

> So prey I God that non myswrite the,
> Ne the mysmetre for defaute of tonge.   (V. 1795–6)

The sheer fretful laboriousness of reproducing a text of the size of *Troilus*
in manuscript, overseeing the work of the scribes, and then tediously
checking and correcting the manuscript of their errors, is conveyed by the
poem known as 'Chaucers wordes unto Adam, his owne scriveyn':

> Adam scriveyn, if ever it thee bifalle
> Boece or Troylus for to wryten newe,
> Under thy long lokkes thou most have the scalle [scab],
> But after my makyng thow wryte more trewe;
> So ofte adaye I mot thy werk renewe,
> It to correcte and eke to rubbe and scrape,
> And al is thorugh thy negligence and rape.

The mention of copying *Boece* and *Troilus* suggests that this testy little
poem is born of Chaucer's frustration—during that busy period of his

creative life—at the irksomeness of accurately producing manuscripts of two such extensive works.

Before handing his copy over to be transcribed by a scribe such as 'Adam' on to the sheets which became the pages of a manuscript book, Chaucer may well have drafted and composed his poem by first writing on the cheap writing material of wax tablets, although some aspects of the *Troilus* manuscript variants suggest that Chaucer was drafting on sheets not tablets. In whatever shape his draft developed exactly, the variety of his sources suggests that Chaucer's composition of the poem was likely in practice to have been a series of layers—perhaps physical layers—of writing, embodying a diverse creative process of translating, transforming, and versifying into English the *Filostrato* stanzas, and of introducing other passages of borrowed and invented material which needed to be versified into the same stanza pattern.

In the absence of any autograph for *Troilus* or any manuscript possibly overseen by the poet during his lifetime, the text of what Chaucer wrote has to be retrieved by looking at the evidence of all the extant manuscripts. The anxiety in Chaucer's prayer at the end of *Troilus* that 'non myswrite the, | Ne the mysmetre . . .' (v. 1795–6) anticipates the possibility that the poem will be mistranscribed as it is reproduced. That anxiety has been justified by what has happened to the manuscripts, in which Chaucer's text gradually deteriorates through a process of successive recopying that at once miswrites and mismetres with clichéd rephrasing, omission, and substitution. The text of the poem has not been seriously corrupted, and does not need to be retrieved by conjectural emendation. Yet scribal confusions may well have transmitted in edited form some of the difficulties of Chaucer's working copy or copies of his poem. The state of the text in parts of some manuscripts may be explicable as stemming from some stage of the poet's composition process.

It is generally agreed that in their patterns of agreement with each other in variation the manuscripts of *Troilus* fall into three groups. A number of manuscripts hold very consistently throughout the poem with one group (called by editors 'γ', or Cp+, because the Corpus manuscript is one of those in this group). The other two manuscript groups are usually termed 'α' and 'β', or Ph+ and R+ respectively, after the Phillipps and Rawlinson manuscripts which in each case are the only manuscripts to hold with these groups throughout the poem. These Ph+ and R+ groupings are otherwise characterized by frequent cross-overs in allegiance, in which the same manuscript of *Troilus* is aligned with different groups for different sections of its text. During a process of multiple copying such cross-overs of allegiance between groups become very understandable, especially when some existing manuscripts of *Troilus* (A, H2, H3, R) have been copied by three or four different hands in different parts of the poem, 'corrected' in spaces left blank for the purpose (e.g. H2, Ph), or in one case (S1) quite systematically edited between the Cp + and R + traditions.

The main question posed by the nature of the *Troilus* manuscripts is whether certain differences between the manuscript groups stem from the scribes, or from the poet, or from both. Do the variations between manuscripts reflect changes of mind by Chaucer himself as to what his text should include, or the confusions of scribes, or a mixture of the two? Evaluation of the evidence for possible authorial revision of a poem written before the age of printing needs first to be set in the context of the very different conditions of 'publication' in a manuscript culture, where the processes of composition and circulation are not so clearly separated as they are by the committing of a text to printed form. In a manuscript culture it can be open to authors at any time to put into circulation an altered form of one of their works, and so to keep the text as it were 'in progress' during their lifetime. For medieval writers, this may be a way of incorporating second or even further thoughts; of updating in response to events; of responding to criticism or to plagiarism—especially as writers might well need to go on having copies of their works made long after first composition, for presentation or in response to requests.

The principal differences between the manuscripts of *Troilus* may be broadly characterized as falling into three categories:

1. omission by some manuscripts of three 'philosophical' passages: Troilus' song 'Love, that of erthe and se . . .' (iii. 1744–71), his 'predestination' soliloquy (iv. 953–1085), and his ascent to the spheres (v. 1807–27);
2. variants in some manuscripts closer to the parallel Italian lines in Chaucer's main narrative source, *Il Filostrato*, while the other manuscripts also have apparently authentic versions of the same lines;
3. other variations between manuscript groupings over several possibly authentic versions of the same line.

Some manuscripts in the same group (Ph +) both *contain* the readings closer to the Italian source and also *omit* the 'philosophical' passages (although not all the Ph + manuscripts behave consistently and there are some unexplained cross-overs). These features have often been interpreted as evidence that the text of the poem was subject to a process of rewriting, in which an earlier text closer to Chaucer's main Italian source and not containing the 'philosophical' passages was revised and supplemented by the addition of those passages. (In an extension of this interpretation, variation between the Cp + and R + groupings has been taken to represent further authorial revision.) The particular appeal of this interpretation of the manuscripts has perhaps been that it endows with the status of some kind of afterthought such parts of the poem as the predestination soliloquy and the ending, which have been a challenge to some modern taste and criticism. Unfortunately, the evidence of the manuscripts is rather more entangled and qualified, much harder to order into a pattern, than this interpretation of the manuscript variations would suggest. The questions posed by the detailed evidence are set out in the following section of this chapter, but by way of preface it may here be said that it is now difficult to determine whether the same manuscript features are the signs of authorial change made between distinct 'published' versions or editions of the poem, or whether they derive scribally and confusedly from the author's working materials, from the processes of composition and transcription.

What the three 'philosophical' passages have in common is that they are extracts closely translated from sources other than Chaucer's main source, *Il Filostrato*, and are all incorporated at points in *Troilus* where Chaucer is following *Filostrato* closely. This suggests that their absence from some manuscripts may go back to the processes of the poem's composition. The very range of his proven borrowings and innovations over any given passage of the poem indicates that Chaucer's composition of his *Troilus* was a process which involved distinctly different kinds of poetic activity, and that process may well have affected the way the poet's earliest draft developed as a material entity. The Book III song of Troilus and the ascent to the spheres in Book V are both passages interpolated between two successive and closely translated *Filostrato* stanzas, while the enormous predestination soliloquy is actually interpolated into the middle of a single *Filostrato* stanza (*Fil.* iv. 109). In all three instances it would be possible, even likely, that the poet's working draft did not read on completely smoothly from his main *Filostrato*-based narrative to any such incorporated passages and back again. The incorporated passages needed to be found from different books, and presented different exercises in composition, the Boethian passages being versified largely from Chaucer's own prose *Boece*. Such passages might have been put together separately from the work on *Filostrato* and then married to the main draft, always with the possibility of scribal misinterpretation, so that copies could be produced without some of the passages marked for inclusion. The three 'philosophical' passages were manifestly 'added' to *Troilus* (in common with numerous other passages) in the sense that they are not in *Filostrato*, but whether their absence from some manuscripts means that they were 'added' significantly later in the poem's existence remains an open question and is not indisputably indicated by the nature of the manuscripts.

The whole question of whether manuscript differences derive from authorial revision is made problematic by the intrinsic quality of many variants themselves, set against a high degree of evidently authentic finish in the poem as it exists in all manuscripts. The three groupings into which the *Troilus* manuscripts fall seem more like manuscript traditions of the text than separate authorial versions of the poem. It is important to distinguish the consistency that may mark scribal agreements between a group from the agreements that may distinguish that group as also representing an authentic state of the text. The Ph + family usually has no identity which is anything other than scribal, except in certain parts of Books I, III, and IV. Here it sporadically produces some fascinating readings. But these should be seen with the implications of their isolation and scribal context rather than allowed to endow their manuscript family as a whole with an interest of possible authenticity throughout the poem which, in themselves, the run of variants cannot support. The fluctuating, inconsistent way that some manuscripts of the Ph + and R + groups align themselves with different manuscript groups during different parts of their text probably results from the practice of sectionalized and distributed copying of parts of the same text by different scribes in the commercial manuscript production of the day.

TABLE 2. Troilus *manuscripts: Some patterns of allegiance*

Gg	←R + — [ii. 63] —Ph + — [ii. 1210] ——R +—— [iii. 398] —Ph + —→							end
H2	←———————— Ph + ———————— [iv. 196] ————————R +————→							end
H3	← R + —[ii. 1033] —Cp + —[iii. 1095] —R + —[iv. 299] —Ph + —→							end
H4	←———Ph + —— [ii. 65] ——————————————R +————————→							end
J	←————————————R +———————— [iv. 430–8] ——Ph +————→							end

Such cross-overs between manuscript groups suggest that considerable scribal editing and conflation lie behind the extant *Troilus* manuscripts. Instances of fluctuating authenticity between manuscript families present some baffling cases of independence by individual manuscripts or groupings, and such features make difficult the interpretation of the *Troilus* manuscripts as they now stand as evidence for successive, distinct, authorial versions of the poem, although some manuscripts may retain lines back to the process of composition and the poet's working copy.

On the textual questions posed by the manuscripts of *Troilus*, see Brewer, 'Root's Account' and 'Observations'; Cook, 'Revision'; Cureton, 'Chaucer's Revision'; Hanna, 'Robert K.

Root'; Owen, 'The Significance of Chaucer's Revisions', 'Minor Changes', and '*Troilus and Criseyde*'; Root, *The Textual Tradition* and (ed.), *Book*; Windeatt, 'The Text' and (ed.), *Troilus and Criseyde*.

The main cross-overs in the allegiance of certain *Troilus* manuscripts to the manuscript groups may be set out as in Table 2, although this can only represent predominant patterns of allegiance in the mixed texts of manuscripts like Gg and H3.

The Troilus *Text: The Questions*

This section sets out in more detail some of the questions raised by the main textual problems in the manuscripts of *Troilus and Criseyde*. The questions are addressed in the following order: the absent 'philosophical' passages; variants in the Ph + group of manuscripts; and variation between other manuscript groups.

The 'Philosophical' Passages

Questions raised by the three 'philosophical' passages differ from each other because of differing manuscript support and contextual implications. Only one manuscript (H2) actually omits entirely Troilus' Boethian hymn in Book III (1744–71), line 1743 being the last line at the foot of a page. In the Phillipps manuscript the text of the song has been inset on a separate leaf during correction (see Fig. 2), and the song is contained normally in the other two Ph + manuscripts here (Gg, H5). Chaucer certainly intended Troilus to sing a song at this point, all extant manuscripts reading 'And thanne he wolde synge in this manere' at iii. 1743, so the absence of the song presumably stems from scribal mistranscription of Chaucer's intentions.

On the absence from some manuscripts of Troilus' ascent to the spheres (v. 1807–27) the evidence is comparably mixed. Only one of the Ph + manuscripts at this point of the text (the Phillipps manuscript itself) omits the passage—although it has been included on an inset leaf. In this omission, however, it is joined by two other manuscripts (H2, H4) normally here agreeing with the R + manuscript group, one of a number of baffling manuscript agreements which indicate that—as a result of conflation—the differences between extant copies of *Troilus* resist interpretation as simply recording stages in a linear progress of authorial revision.

The absence from some manuscripts of Troilus' predestination soliloquy is the single most substantial difference between the manuscript groups in *Troilus*. But, once again, there is a lack of cohesive, consistent manuscript support for the absence of the passage. The evidence of the manuscripts is particularly intricate:

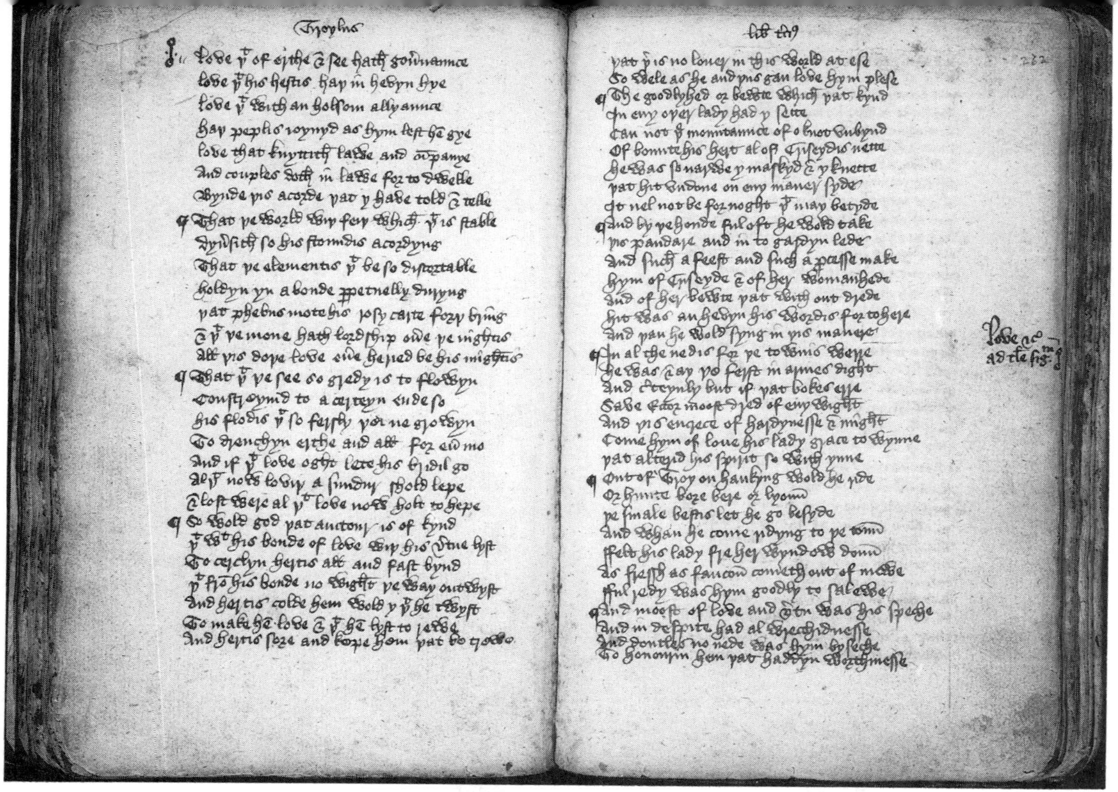

FIG. 2. An opening in the Phillipps manuscript (now Huntington Library HM 114), fos. 261ʳ–262ʳ. The right-hand leaf contains *Troilus*, iii. 1728–43, 1772–89, and by line 1743 (where Troilus' song is missing) a marginal note 'Love &cetera | ad tale signum' refers the reader to the beginning of the song (iii. 1744–71) on the inset left-hand leaf.

1. H3 and Ph do not contain lines 953–1085 of Book IV (but the passage is added separately in Ph);
2. H3 and Ph have a variant form and order for lines 950–2, not found in other manuscripts;
3. Gg omits all of the soliloquy, except the last stanza (1079–85), which is not derived from Boethius with the rest of the passage;
4. J—which holds with the Ph + manuscripts from early in Book IV— *contains* the passage, *but* between the main soliloquy and that final stanza, which Gg also contains, there is a blank page, a cancelled leaf, and another blank space; at the foot of the last written page, *after* the main body of the soliloquy, is a scribal note: 'her faileth thing y^t is nat yt made' (see Fig. 3);
5. H4—which does not at this point normally agree with the Ph + manuscripts—omits lines 953–1085 without any break in the text.

The details of fluctuating manuscript support for the three passages are set out schematically in Table 3, with the omissions signalled by dashes:

TABLE 3. Troilus *manuscripts: Absent passages*

MS	Song	Soliloquy	Ascent
Gg		—	? (text lacking)
H2	—		—
H3		—	
H4		—	—
Ph	—	—	—

These confusing patterns may stem from a confusing authorial copy, or copies, but as they stand they show every sign of subsequent scribal editing. Indeed, there is only one copy of *Troilus*—the Phillipps manuscript—which consistently omits all three of the 'philosophical' passages, in the sense that they have not been copied in their expected places; they have all then been supplied by the same hand on inset leaves—with the fact signalled in the margin at the point where they are to follow on—much as the poet himself may have worked on his text.

In brief, the undisputed fact that all three passages are added to *Filostrato*, and may therefore have been placed by Chaucer in his copy much as the Phillipps manuscript is supplemented with inset leaves, does not necessarily prove that these passages were 'later' additions to a poem previously existing and 'published' without them. The various divergences between the manuscripts over the beginning and ending of the soliloquy

point to confusions in copies behind the extant manuscripts over interpreting directions in how the passage is fitted into its context in the poem. In
the *Filostrato* stanza which Chaucer takes as his point of departure Pandaro
simply finds Troiolo thoughtful and downcast in appearance in a place
unspecified (*Fil.* iv. 109). But in Chaucer's version, in all manuscripts,
Pandarus finds Troilus alone in the temple, no longer caring for his life and
making his moan to 'the pitouse goddes' (iv. 949). To this extent all
surviving manuscripts suggest that Chaucer always conceived of his hero
concerned here with questions of man's life and death and his relation to
the gods. For the last three lines of this stanza Ph and H3 differ somewhat
in phrasing from the other manuscripts. Thus in Ph and H3, to the 'pitouse
goddes':

> He fast made his compleynt & his mone
> Bysekyng hem to sende hym oþir grace
> Or from þis world to done hym sone to pace.

In the other manuscripts these lines read:

> Ful tendrely he preyde and made his mone
> To doon hym sone out of this worlde to pace
> For wel he thoughte ther was non other grace.   (iv. 950–2)

The differences are not very substantial, but both readings may be
authentic. An interesting difference lies in the first line, where in Ph and
H3 'He fast made his compleynt': they thus insist more explicitly than the
other manuscripts that this is the action in which Troilus is engaged at this
point, even though they do not give the 'compleynt' itself.

Chaucer's handling of the remainder of the interrupted Italian stanza,
where the end of the soliloquy is connected back into the narrative, may
also be interpreted as implying the presence of the Boethian passage. In
Filostrato, when Pandaro finds Troiolo 'pensoso', he simply asks him
whether he is as miserable as he appears. The greater urgency of Chaucer's
Pandarus seems provoked in response to something he has just seen and
heard ('O myghty God . . . in trone, | I! Who say evere a wis man faren so? |
Whi, Troilus, what thinkestow to doone? | Hastow swich lust to ben thyn
owen fo?' (iv. 1086–9)). This strong response—present in all manuscripts—implies that Troilus is in a much more desperate state than in
Filostrato. Yet without the presence of the soliloquy there is nothing to
draw such a response from Pandarus.

The peculiar evidence in Gg—also towards the end of the soliloquy—
may similarly indicate that Chaucer always had the soliloquy in mind. For,
although Gg does not contain those stanzas of the soliloquy actually
translated from Boethius but only Chaucer's own last transitional stanza

(iv. 1079–85), that stanza explicitly describes the English Pandarus interrupting Troilus 'whil he was in al this hevynesse | Disputyng with hymself in this matere' (iv. 1083–4), and only the soliloquy shows him disputing with himself. Gg thus contains Chaucer's changes to prepare his context, while lacking the very material which all these changes are designed to accommodate.

But what of the mysterious comment ('her faileth thing y^t is nat yt made') in the St John's manuscript (see Fig. 3)? Even if it chanced to preserve a memo made by a scribe some three decades earlier at the time of composition in the 1380s, it still need not in itself indicate that *Troilus* was ever regarded as complete without the soliloquy or was being revised; and it may just as probably refer to the activities of other scribes in the workshop or elsewhere, engaged on comparing and conflating texts of the poem at the time of this fifteenth-century manuscript's transcription. Moreover, the note ('her faileth thing . . .') comes *after* the soliloquy, not *before* it, as would be necessary if it really stemmed from scribes who were close to the poet and who understood what he was doing. The note is suspiciously like a scribal guess at rationalizing a strange gap in an exemplar which there was no way of filling. It is unlikely to be authentic in suggesting—as it does in its location—that Chaucer was intending to translate *more* of the Boethius than he actually does. The position of the note contradicts its ostensible claim to be close to Chaucer's own intentions in composition. It is unlikely, then, that Chaucer intended his poem to have any completed, 'published' existence without the Boethian soliloquy. Indeed, without this soliloquy, which so affects the overall philosophical tone of the whole poem, there would have been rather less aptness in Chaucer's dedicating the poem to 'philosophical' Strode, a dedication present in all extant manuscripts.

Variant Readings in the Ph+ Manuscripts

If the omissions in the Phillipps manuscript—and those manuscripts fluctuatingly associated with its readings—reflect in edited form some scribal interpretations of the outcome of Chaucer's composition process, this may also be the most likely origin of the survival in those manuscripts of occasional, sporadic variants which are closer translations of the parallel lines in Chaucer's source *Filostrato* than the equivalent (but evidently authentic) lines in the other *Troilus* manuscripts. In proportion to the size of the whole poem the number of such lines is modest, although this does not diminish their interest. Resemblance is for the most part confined to single lines, or, indeed, to single words, but there are some clustered instances where certain copies have readings demonstrably closer in phrase

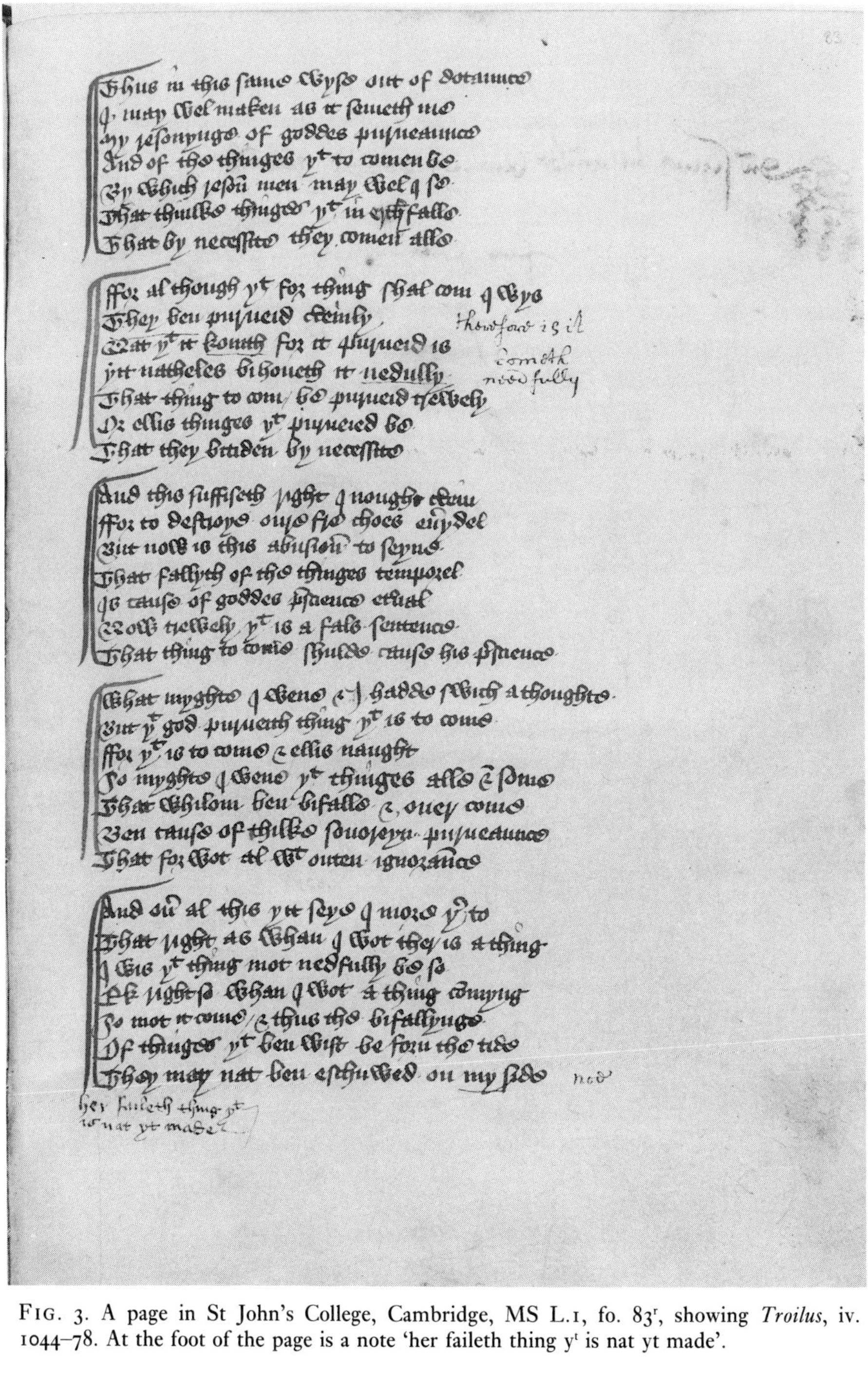

FIG. 3. A page in St John's College, Cambridge, MS L.1, fo. 83ʳ, showing *Troilus*, iv. 1044–78. At the foot of the page is a note 'her faileth thing yᵗ is nat yt made'.

or word to the Italian original. Several lines from early in Books I and IV
are the most striking cases:

> *Fil.*   da lui sperando sommo e buon consiglio.   (i. 9)
> From him hoping for excellent and good counsel.

> Ph +:   Hopyng in hym kunnyng hem to rede.   (i. 83)

> Rest: In trust that he hath konnynge hem to rede.   (i. 83)

> *Fil.*   Fu'l romor grande quando fu sentito.   (i. 10)
> Great outcry there was when it became known.

> Ph +:   Grete rumour gan whan hit was ferst aspyed.   (i. 85)

> Rest: The noise up ros whan it was first aspied.   (i. 85)

> *Fil.*   Li miseri occhi per pietà del core
> forte piangean, e parean due fontane
> ch'acqua gittassero abbondevol fore . . .   (iv. 28)

> His sad eyes in sympathy with his heart wept bitterly and
> looked like two fountains spouting water in torrents . . .

> Ph +:   His eyen two . . .
> So wepyn þat þei semyn wellis twey

> Rest: His eyen two . . .
> Out stremeden as swifte welles tweye . . .   (iv. 246–7)

At such points the Ph + manuscripts—by whatever means of transmission—suggest their links with the poet's process of composition. A
scattering of such instances—of very varying quality—marks Book I and
part of Books III and IV, precisely where Chaucer is often working most
closely from the Italian (instances are listed in a note at the close of this
chapter). Few of these instances concern signal points of characterization
or event, rather just points of expression. In some cases such readings in
Ph + are harder, because more Latinate, than that of the other manuscripts:

> Not I how longe or short it was bitwene
> This purpos and that day they *fighten* mente   (iv. 36–7)
>
> . . . *issen* . . .   (Ph, H2, H5)
>
> *incontro a' Greci uscì ne' campi piani* . . .   (*Fil.* iv. 1)
> . . . sallied forth against the Greeks on the plains
>
> And he bigan *to glade* hire as he myghte . . .   (iv. 1218)

> ... *conforte* ... (Ph, H2)
>
> *la confortò* ... (*Fil.* iv. 124)
>
> he comforted her ...

At least in some such cases the reading further from the Italian is possibly scribal, and the same holds for those instances where the *other* manuscripts are closer to the Italian than Ph +; in further cases closer approximation to Italian in some manuscripts may be accidental.

The relatively infrequent instances where there are two convincing readings for a line can offer an intriguing glimpse into the process of the poem's composition, as with one rare case of variation extending over more than a single line, in which Ph + is closer to *Filostrato*. In Book IV of *Troilus* one stanza (iv. 750–6) is copied in Ph + in a sequence which corresponds to the order of the parallel stanzas in *Filostrato*, while the other manuscripts have the stanza in a different, but equally authentic, ordering. First the situation as it exists in Ph + (here Gg, H3, J, and Ph), citing here the first three lines of the disputed stanza and the first lines of the other three stanzas involved:

> A. The salt teris from her eyen tweyne
>   Out ran as shour in Aprill ful swithe;
>   Her white brest she bet & for þe peyne ...
> B. Her ownded here þat sunnisshe was of hewe   (Ph ornyd)
> C. 'Alas,' quod she, 'out of this regioun ...'
> D. 'What shal y done, what shal he done also ...'

Now the situation as found in all other manuscripts:

> B. Hire ownded heer that sonnyssh was of hewe ...   (736 ff).
> C. 'Allas,' quod she, 'out of this regioun ...'
> A. Therwith the teris from hire eyen two
>   Down fille as shour in Ap(e)ril (ful) swithe;
>   Hire white brest she bet and for the wo ...
> D. She seyde, 'How shal he don and ich also?'

The order of the descriptive details in Ph + does follow more closely the order of the source passage, where a description of Criseida's grief (*Fil.* iv. 87) is followed by two stanzas of lamentation in direct speech:

> Ella diceva, 'Lassa sventurata,
> misera me dolente, ove vo io? ...'   (*Fil.* iv. 88)
> 'Che farò io ...'   (*Fil.* iv. 89)
>
> She said, 'Alas for me, unfortunate and wretched woman,

    whither am I bound . . .'

    'What shall I do . . .'

But in the other *Troilus* manuscripts Criseyde's tearing of her hair has been moved to become her first response, with the effect that the description of unhappiness is interspersed with Criseyde's laments, and her tears follow after her first outburst of lamentation. Description and direct speech have been recast into a closer, more expressive relation. The differences between the manuscripts represent an improvement of evidently authentic quality, but one which Chaucer may well have signalled in a working copy by some marginal indication that stanzas initially composed in the original *Filostrato* order were to be transcribed in a different sequence, always with the possibility of confusion and mistranscription.

    Quite distinct from the implications of the variant readings closer to the Italian source, it may be possible in a few contexts—especially near the beginning of the poem—to detect in some manuscripts surviving traces of Chaucer's drafting of the expression and movement of his sense:

  1. Remembreþ ȝow of old [passid] heuynesse
    ffor goddis loue & on [the] aduersite
    þat ouþer suffren þenk how sumtyme þat ȝe
    ffoundyn how loue durst ȝow displese
    Or ellis ȝe wonne hym with to grete ese   (Ph +)

    Remembreth yow on passed hevynesse
    That ye han felt, and on the adversite
    Of othere folk, and thynketh how that ye
    Han felt that Love dorste yow displese,
    Or ye han wonne hym with to grete an ese.   (i. 24–8)

  2. O verrey foles may ȝe no thing se
    Kan none of ȝow yware by oþer be

         . . . .

    But trowe ȝe not þat loue þo loked rowe
    ffor þat despite & shope how to ben wrokyn
    ȝes certein loues bowe was not ybroken
    ffor by myn heed he hit hym atte fulle   (Ph +)

    O veray fooles, nyce and blynde be ye!
    Ther nys nat oon kan war by other be

         . . . .

    At which the God of Love gan loken rowe
    Right for despit, and shop for to ben wroken.
    He kidde anon his bowe nas naught broken,
    For sodeynly he hitte hym atte fulle . . . (i. 202-3, 206-9)

However, there will always be a fine line to be drawn in discriminating between what may be simply textual deterioration due to inaccurate scribal copying, and what may be the chance survival of authorial drafting, imperfectly preserved in a manuscript grouping or tradition that more generally shows a poor level of accuracy in transcription and a neglect of metre.

There are similar problems in accepting both readings as authentic where there are some divergences over classical allusions between the Ph + group and the other manuscripts (Ph + : *wight* for 'Furie' (i. 9); omission of 'Palladion' (i. 164); *Sisiphus* for 'Ticius' (i. 786); þe *god* for 'Apollo' (iii. 543); *god* for 'Mars' (iv. 25); omission of 'Edippe' (iv. 300); *eny aungill* for 'Jove' (iv. 644); *Ther Pluto regneth* for 'That highte Elisos' (iv. 790), etc.). There is almost always something much less satisfactory about the Ph + variations over classical references, perhaps because they represent early drafts, or scribal corruption, or a mixture of both in an area of transcription where scribes often fall into bungling error. However, resemblances to Italian, and variations over classical references, are among those aspects of variation found in the Ph + manuscripts to which a decided character can be given. Most variation in Ph + is characterizable more at the level of some recurrent differences in related types of diction and in simpler syntactical patterns. It is through such typical copyists' variation that the Ph + manuscript group often differentiates itself from the other manuscripts of *Troilus*.

Variant Readings in the Cp + and R + Manuscripts

The variant readings that distinguish Cp + and R + , the other two groups of *Troilus* manuscripts, from each other are of a rather different type and extent. In practice, editors have tended to regard the line-by-line variations between these two groups as deriving from the usual processes and patterns of error in scribal transcription, which may also be seen to explain at least one of the variations in those rare cases where all three manuscript groups have variant readings of the same line. Although Cp + and R + generally agree in the course of the poem as against Ph + , in a limited run of variants in Book III Cp + and Ph + agree as against R + . In a small number of these R + variants—notably iii. 1392–3, where 'couetise' is vice and 'loue' is virtue, or iii. 1438–9, where night 'downward hasteth of malice'—there will be room for argument as to which version of the lines conveys more, or is in context the 'harder' reading and hence more likely to stem from the poet than from scribal error.

Another interesting variation would be the different positions of the two stanzas beginning 'But sooth is, though I kan nat tellen al . . .' (iii. 1324–37 in the *Riverside Chaucer*), which are found in R + (and in H3 and S1, both often associated with R +) between lines 1414 and 1415. A case can be

Text

made on contextual grounds for the inferiority of the R + reading by comparing the stanzas which precede and follow these two disputed stanzas in their two different positions in the manuscripts. In Cp + and Ph + these stanzas—concerned with the suitability of language used by an inexperienced narrator—are preceded by two stanzas explicitly concerned with the inadequate powers of the poet ('I kan namore' (iii. 1314); 'al ne kan I telle' (iii. 1323)). By contrast, in the R + manuscripts they are preceded by stanzas less directly concerned with expression. The stanzas which *follow* them in each of their positions also count against R +, for the two disputed stanzas acknowledge themselves in their (undisputed) last line to be a digression ('But now to purpos of my rather [i.e. former] speche' (iii. 1337)). In their earlier position in the text they do interrupt what seems a continuity of attention to the lovers at their union, and the stanza that follows them opens by acknowledging some intervention ('Thise ilke two, that ben in armes laft' (iii. 1338)). By contrast, the stanza which follows in R + has no relation to the disputed stanzas and contradicts their claim to be returning 'to purpos of my rather speche' by promptly introducing a new subject (the dawn and the lovers' sadness (iii. 1415 ff.)). The scribe of H4, or the scribe of a copy that stands behind his text, has hedged his bets by copying out the two stanzas in *both* positions, in another of the many signs of scribal editing and conflating of variations in the *Troilus* texts.

In general, however, the groups of *Troilus* manuscripts do not distinguish themselves from each other line by line in anything more than their own internal errors as manuscript traditions. Interesting distinctions between them exist very occasionally, within a continuum of minor divergence in vocabulary and phrasing which is more likely to reflect their transmission through different scribal traditions than to represent an authorial rewriting so indifferent and motiveless as to be distinguished with difficulty from the more generally observable characteristics of scribal copying.

Some idiosyncratic aspects of various manuscripts epitomize the perennially puzzling interrelation of authorial intent and scribal medium. What is to be made of the fact that manuscripts of the Cp + grouping—which elsewhere show no links with the possibly compositional features in other manuscripts—have a version of the text closest to *Filostrato* at the transition from Book III to Book IV, while all other manuscripts point to Chaucer's authentic changing of his source at this point? Chaucer shifts the final stanza (*Fil.* iii. 94) of the third part of *Filostrato* (a stanza on fortune) so that, rather than forming the conclusion to Book III, it forms instead the proem to Book IV of *Troilus*. But the Cp + manuscripts—by not signalling the third book to end until *after* the fortune stanzas (i.e. at iv. 28 in modern editions)—make the same division between the books as in Boccaccio's text. The resemblance to *Filostrato* in the Cp+ manuscripts depends entirely on how the text is rubricated with incipits and explicits, and it was a change from his source that Chaucer could well have made in his draft by an instruction in the margin for the scribe, with the possibility of misunderstanding and mistranscription. If so, the version in Cp+ is an intriguing case of a manuscript group apparently preserving in isolation a line right back to a moment of decision in the author's composition process.

There is a comparably baffling peculiarity in MS R, which has a special interest in being the manuscript that contains in addition to *Troilus* the only extant text of Chaucer's poem 'To Rosemounde'. It contains a stanza—of apparently authentic versification and diction—found in no other manuscript, between ii. 1750 and 1751 (although it is preceded by a repetition of lines 1576–7). Here, if authentic, it would have formed the penultimate stanza of Book II, an additional last stanza in Pandarus' arguments to Criseyde as he leads her in to visit the supposedly unwell Troilus:

> ffor ye must outher chaungen your face
> That is so ful of mercy & bountee
> Or elles must ye do this man sum grace
> ffor this thyng folweth of necessytee
> As sothe as god ys in his magestee
> That crueltee with so benigne a chier
> Ne may not last in o persone yfere.

The lines may be compared with the stanza of Pandarus' arguments to Troilus beginning 'And forthi loke of good comfort thow be' (i. 890–6) which is contained only by Ph+. Both stanzas are possibly isolated 'spot' survivals of single stanzas cancelled in Chaucer's composition process. There are unsatisfactory features to the Rawlinson stanza—its suitability for the context, and the deficient metre of the first line—which suggest that it is not part of an authentic finalized text. Yet the association in MS R with the survival of 'To Rosemounde' might indicate that this stanza is an authentic part of the poet's composition. Even more peculiar is that MS R does not contain the proems to Books II, III, and IV. Their omission compares with the omitted passages in Ph+, which are also authorial elaborations by Chaucer of the narrative line provided by *Filostrato*. The absence of the proems and the presence of the unique stanza seemingly contradict the otherwise finished character of the manuscript and could reach back close to a compositional phase. Such momentary textual flashbacks, both here and in Cp+, suggest that among the *Troilus* manuscripts the various groups have some integrity as traditions of scribal transmission, but it remains questionable whether any of them has a consistent integrity throughout in being equivalent to a distinct state of the author's text. At such points of flashback, the state of the manuscripts compares with moments in the text of the poem itself which still seem close to the poet's compositional process, as when the fourth proem apparently preserves an unrevised expectation from one stage or layer of composition that *Troilus* would be concluded in four books rather than five:

> This ilke ferthe book me helpeth fyne,
> So that the losse of lyf and love yfeere
> Of Troilus be fully shewed heere.   (iv. 26–8)

For discussion of the 'her faileth thing . . .' note, which is taken to be in the hand of the scribe, see *St John's College Cambridge MS L.1: A Facsimile*, Introduction by R. Beadle and J. Griffiths (Cambridge, 1983). On the Phillipps manuscript, see Ralph Hanna III, 'The Scribe of Huntington HM 114', *Studies in Bibliography*, 42 (1989), 120–33.

Readings closer to *Filostrato* in Ph+ include: '*con voce e con vista assai pietosa*' (*Fil.* i. 12), 'With pitous vois and tendrely wepynge' (i. 111), [Ph+: '*Wiþ chier & voys ful pytous and wepyng*']; '*Tra li qua*' fu di Calcàs la figliuola' (*Fil.* i. 19), 'Among thise othere folk was Criseyda' (i. 169), [Ph+: '*Among þe which* was . . .']; '*che'l capo e'l petto appena* gli bastava' (*Fil.* iv. 29), 'That wonder is the body may suffise' (iv. 258), [Ph+: '*þat wele vnnethe* . . .']; 'Tu non hai a *rapir* donna . . .' (*Fil.* iv. 73), 'It is no shame unto yow ne no vice' (iv. 596), [Ph+: 'Hit is no *rape* in my dome ne no vyse']; '*per vergogna* nascose la sua faccia' (*Fil.* iv. 96), 'she gan for sorwe anon | Hire tery face . . . hide' (iv. 820–1), [Ph+: 'gan *for shame* . . .']; '*che cerca disperato di morire*' (*Fil.* iv. 102, variant reading), 'For verray wo his wit is al aweye' (iv. 882), [Ph+: 'As he þat shapith hym shortly to dey']; '*di veder Troiolo afflitto*' (*Fil.* iv. 105), 'To sen that sorwe which that he is inne' (iv. 906), [Ph+: '*To se hym in þat woo* þat he is ynne']. The following readings may also be considered: 'sì mi *stringe* il disio del ritornarci' (*Fil.* iii. 46), 'Syn that desir right now so biteth me' (iii. 1482), [Ph+: '*streynith*']; '*Come* farà la mia vita dolente?' (*Fil.* iv. 33), 'What shal my sorwful lif don in this cas?' (iv. 290), [Ph+: '*How* shal . . .']; 'chi darà più conforto alle *mie* pene?' (*Fil.* iv. 36), 'Who shal now yeven comfort to *the* peyne?' (iv. 318), [Ph+: '*my* . . .']; '*parole* assai dicean da consolarla' (*Fil.* iv. 85), 'And with hire *tales* wenden hire disporten' (iv. 724), [Ph+: '*wordis*']; '*ma* quando | tempo gli parve . . .' (*Fil.* iv. 113), 'And whan that it was tyme . . .' (iv. 1124), [Ph+: '*But* . . .']; '*queste cagion* . . .' (*Fil.* vi. 34), 'the cause . . .' (v. 1028), [Ph+: 'The *causes* . . .'].

Instances where the Ph+ reading is *further* from *Filostrato* would include: 'e 'n *diversi* atti . . .' (*Fil.* i. 18), 'In *sondry* wises . . .' (i. 159), [Ph+: '*meny*']; 'Pandaro, presa la lettera' (*Fil.* ii. 108), 'This Pandare tok the lettre' (ii. 1093), [Ph+: 'This Pandare vp þerwith . . .']; '*arrabbiata fera* . . .' (*Fil.* iv. 26), 'In his *woodnesse*' (iv. 238), [Ph+: '*distresse*']; 'esci del core' (*Fil.* iv. 34), 'Fle forth *out of* myn herte and lat it breste' (iv. 306), [Ph+: 'ffle forth *anon* and do myn hert to brest']; 'di *render* Criseida' (*Fil.* iv. 43), 'to *yelden* so Criseyde' (iv. 347), [Ph+: '*chaungyn*']; Troiolo's direct address to death (*Fil.* iv. 61), preserved in iv. 506–7, is lost in the Ph+ reading; 'e' timidi *rifiuta*' (*Fil.* iv. 73), 'And *weyveth* wrecches' (iv. 602), [Ph+: '*fleeth from*'].

Sources

Introduction: 'Lollius'

What is meant by a 'source' for a poem? *Troilus and Criseyde*—which could be seen to have many sources—is presented so as to ask that question and to make the search for an answer part of the poem's meaning. In *Troilus* Chaucer makes use of a rich variety of earlier texts, but the nature of that use is almost as various as those texts themselves. Substantially indebted in outline and much detail to Boccaccio's poem *Il Filostrato*, Chaucer profoundly transforms what he takes from Boccaccio, both by free invention and by fusing that one adapted source with yet other sources. Woven into *Troilus* are extracts and excerpts from a diversity of authors, as well as close verbal echoes from small sections and single lines of other texts. There are also signs of more indirect borrowing of material, which reaches *Troilus* through the intervening medium of a third author. Beyond this again, Chaucer's poem stands in relation to more distant texts in shared conventions, shared approaches to subject-matter and form. The outcome is that Chaucer's *Troilus* reflects a plurality of diverse sources conjoined and fused—in a new unity which remakes everything it contains.

Yet, although Chaucer draws, to varying degrees, on such a diversity of sources, he acknowledges none directly, but instead claims for his source an author who almost certainly never existed, the mysterious—perhaps deliberately mystifying—Lollius. To understand Chaucer's use of sources in *Troilus* is to understand not only those sources that he can be discovered to use, but also that relationship to sources within which the poem presents—dramatizes, fictionalizes—its uses of those sources, through its references to the 'storie', 'olde bokes', 'myn auctour', and Lollius. There are two references to Lollius, one early and one late, in the course of *Troilus*. The Canticus Troili in Book I is introduced by a stanza stressing the exceptional fidelity with which its source in Lollius is being translated:

> And of his song naught only the sentence,
> As writ myn auctour called Lollius.   (i. 393–4)

The song is actually borrowed from Petrarch, but the attribution to Lollius apparently aims to identify him not only as the author of the song but as the main narrative source of *Troilus*, which was in fact Boccaccio's *Filostrato*. And near the close of the poem, when Troilus sees the brooch he gave Criseyde on the 'cote-armure' of Diomede, Chaucer works into his version of this moment an attribution to Lollius, although he is actually rendering fairly closely from *Filostrato*:

> The whiche cote, as telleth Lollius,
> Deiphebe it hadde rent fro Diomede.   (v. 1653–4)

Moreover, in the proem to the second book the audience of *Troilus* is told:

> That of no sentement I this endite,
> But out of Latyn in my tonge it write.   (ii. 13–14)

By contrast, Chaucer makes no mention of his true main source, the Italian of Boccaccio.

Who did Chaucer think Lollius was? Chaucer's only other reference to this mysterious figure comes in his description of the historians and poets who help bear up the fame of the various empires and races in the *House of Fame*, a poem usually thought to predate *Troilus*. Among those who bear up the fame of Troy is Lollius, along with Homer, Dares and Dictys, Guido de Columnis, and Geoffrey of Monmouth:

> Ful wonder hy on a piler
> Of yren, he, the gret Omer;
> And with him Dares and Tytus [Dictys]
> Before, and eke he Lollius,
> And Guydo eke de Columpnis,
> And Englyssh Gaufride eke, ywis;
> And ech of these, as have I joye,
> Was besy for to bere up Troye.   (*HF* 1465–72)

Since the other five authorities on Troy mentioned here are indeed the authors of influential texts, and since such figures as Josephus, Statius, Virgil, Ovid, Lucan, and Claudian are mentioned nearby for their support of other traditions of fame, it might well be concluded that Chaucer believed that there was an ancient authority on Trojan history called Lollius.

Such a misunderstanding might quite genuinely have arisen, through a mixture of textual corruption and misconstruing, in the interpretation of lines in one of the epistles of Horace, in which he addresses his friend Maximus Lollius:

> Troiani belli scriptorem, Maxime Lolli,
> dum tu declamas Romae, Praeneste relegi.   (*Epistles*, I. ii. 1–2)

Whilst you, Maximus Lollius, are engaged in oratory at Rome, I have been re-reading at Praeneste the writer of the Trojan War [Homer].

If *Maxime* were misread as 'greatest' instead of as part of the proper name, and if *scriptorem* were read as *scriptorum*, Lollius could be construed as 'the greatest writer about Troy'. This particular epistle of Horace is much concerned with Trojan matters (as also is the ninth ode of his fourth book, which is dedicated to a Lollius), and, although Chaucer's poems as a whole do not suggest a close acquaintance with Horace's works, the lines from the *Epistles* could be found quoted in John of Salisbury's *Policraticus* (vii. 9), with which Chaucer is known to have been familiar. In a late-twelfth-century manuscript of the *Policraticus* (Bodleian Library, MS Lat misc. c.16) Horace's line can indeed be found with just the requisite textual corruption ('Troiani belli scriptorum maxime lolli . . .'); moreover, in his French translation of the *Policraticus*, made in 1372, Jean Foullechat declares of Horace: 'il dit que Lolli fu principal escrivain de la bataille de Troye . . .'.

Chaucer cannot, however, have believed that the *Italian* poem on his desk was by the ancient author Lollius, although he might conceivably have believed that *Filostrato* was based on his work. In the early fifteenth century Chaucer's devoted disciple John Lydgate solemnly refers to Lollius as one of the historians of Troy in the prologue to his *Troy Book* (309), while in his *Fall of Princes* he declares that Chaucer derived *Troilus* from a book in the Lombard tongue called *Trophe* (see above, p. 4). While this mistaken title has never been satisfactorily explained, the fact remains that Lydgate—writing so relatively soon after Chaucer—knows that *Troilus* was translated not from a Latin source but from an Italian one.

It may be that Chaucer's 'attribution' of his poem to 'Lollius' was always a piece of deliberately transparent artifice. On the one hand, Chaucer and his first readers could well have believed in the existence of an authority called Lollius. But no one had ever seen a copy of his work, let alone read it, and it would hardly be surprising if it occurred to Chaucer that the book did not actually exist. To present *Troilus* as a translation of Lollius ostensibly amounts to a claim to be making available a major literary discovery of a lost work. This would be a triumph of bluff. Yet how seriously and for how long could Chaucer have intended such a claim to be accepted? Is it perhaps more likely that the attribution to Lollius was always an artifice designed to foreground the question of sources, and with that the role of interpretation—the relation between this Troilus story and the existing tradition of Troilus narratives? It may have started as a private joke. Gower and Strode and his other literary friends would surely have known that Chaucer had not really discovered an ancient text about Troy. What Chaucer had indeed discovered in *Filostrato* was an original modern offshoot from the Troilus story, probably as yet unknown in England to more than a few readers. The unfamiliarity of *Filostrato*'s account of Troilus' and Criseyde's love could be the measure either of its authenticity as a rediscovered ancient text or of its lack of the authority of tradition. It is Boccaccio who is Chaucer's principal source for three of his major poems which are set in pagan times—*Troilus*, the *Knight's Tale*, and the *Franklin's Tale*—yet Chaucer never mentions Boccaccio's name nor acknowledges his indebtedness to this near-contemporary author. What Gower and Strode would have understood was the need to present a new version of a modern vernacular work as if it were a translation of a classical source and to offset any contrary impression, because the claim to an 'authority' was a claim to authority for *Troilus*. Chaucer's attribution of *Troilus* to Lollius seems designed both to claim and yet to play with the tradition of attribution to an ancient authority.

It was ancient authority which medieval poets craved for their own writings, and Boccaccio's *Teseida* and *Filostrato*, with their Theban and Trojan backgrounds, also set Chaucer a comparable example of emphasis on ancient authority and silence or obfuscation on the poet's true source. In the proem to *Filostrato* Boccaccio pretends to have drawn the poem from his study of the ancient legends rather than from Benoît and Guido. In the prefatory matter to the *Teseida* Chaucer could observe how Boccaccio does not mention the name of Statius and instead says mysteriously, 'I discovered a most ancient story which was unknown to most people', and the *Knight's Tale* in turn gives no sign that it is based on a modern Italian poem but starts with an epigraph from Statius. In translating part of the *Teseida* for his poem *Anelida and Arcite* Chaucer takes over Boccaccio's stress on an ancient source while adding that it was written in Latin (*Anelida*, 10), in a parallel for the claim to a Latin source for *Troilus*. As with the *Knight's Tale*, Chaucer wishes to associate the authority of Statius with a poem on a Theban subject ('First folowe I Stace, and after him Corynne . . .' (*Anelida*, 21)). Here Chaucer invokes the authority of the names of ancient sources, not only Statius but also possibly Corinna, a Theban poetess, the friend and rival of Pindar, of whom it is unlikely that Chaucer had read a line or knew more than the name (although the *Amores* of Ovid were also known in the Middle Ages as *Corinna*). As with Lollius, the name of an appropriate classical writer or text is important for Chaucer in lending his poems the authentic associations of an identifiable ancient authority.

Although there are but two references to Lollius by name, *Troilus* maintains the fiction of its source through repeated reminders, bookishly presenting the poem as a version of a source:

> For as myn auctour seyde, so sey I . . . (ii. 18)
> Myn auctour shal I folwen, if I konne . . . (ii. 49)
> And treweliche, as writen wel I fynde . . . (iv. 1415)

The narrator mentions texts on twenty-nine occasions in the poem, and on twenty-two of these he asserts or implies his subjection to his sources.

Expressions of deference to 'myn auctour' occur when Chaucer is actually writing independently of *Filostrato* (e.g. ii. 699–700; iii. 575, 1196, 1325, 1817), and it is also noticeable that, when in Book V Chaucer's text begins to overlap with the narrative of the traditional accounts of the story of Troilus and Criseyde, references to sources become especially frequent. There is a pattern in these acknowledgements of sources which mark the borrowings from Joseph of Exeter's *Ilias* (v. 799, 834), Chaucer's reversion to Benoît de Sainte-Maure's *Roman de Troie* for Criseyde's change of heart (v. 1037, 1044, 1051), and the return of Troilus to his role in the historical events of the Trojan War as he mourns for Hector (v. 1562) and pursues his strife with Diomede (v. 1753, 1758). There is artfulness when Chaucer's uniquely compassionate revision from his sources of the story of Criseyde is declared to have no wish to be critical 'Forther than the storye wol devyse' (v. 1094). But there are also claims that sources are deficient or absent ('Ther is non auctour telleth it, I wene' (v. 1088)). It perhaps gave Chaucer a private relish—and shows his concern to raise the question of sources—when he claims his source lacks information which *Filostrato* actually contains:

> But wheither that she children hadde or noon,
> I rede it naught, therfore I late it goon . . . (i. 132–3)

or when he invents some irretrievable omission by his source which *Filostrato* nowhere claims to have made:

> For ther was som epistel hem bitwene,
> That wolde, as seyth myn autour, wel contene
> Neigh half this book, of which hym liste nought write.
> How sholde I thanne a lyne of it endite? (iii. 501–4)

At this point in Book III *Troilus* has not been following *Filostrato* at all for some while, and the acknowledgement that earlier writers have edited and abridged expresses a different estimate of the present text's relation to sources than when a transparent translation of Lollius was earlier promised:

> And of his song naught only the sentence,
> As writ myn auctour called Lollius,
> But pleinly, save oure tonges difference,
> I dar wel seyn, in al, that Troilus
> Seyde in his song, loo, every word right thus
> As I shal seyn . . . (i. 393–8)

Through such devices *Troilus* is constantly being set in relation to other writing ('I have naught herd it don er this | In story non . . .' (iii. 498–9)), with acknowledgement that it is one interpretation of a story which exists in a number of sources ('Ye may hire gilt in other bokes se' (v. 1776)). By so doing *Troilus* insistently fictionalizes its source and fictionalizes its dependence on that source, so as to bring into the foreground the question of the poem's authority and its relationship to its sources and to tradition. Some early scribes reveal by their annotations to *Troilus* manuscripts that they responded to Chaucer's use of a variety of sources for particular passages of his poem, and in that sense a recognition of 'intertextuality' was part of some very early readings of Chaucer's text.

Through these 'marginal' commentaries of the glossed manuscripts some early readers will always have been alerted to see *Troilus* in relation to sources and models, to a range of classical and medieval authorities which Chaucer has drawn upon, including Virgil, Ovid, Lucan, Statius, Boethius, Dares, 'Cato', and Alain of Lille. Not all such scribal attributions of sources are correct, although a large number are. Virgil is quoted as the source of a passage on fame (iv. 659); Statius is identified ('Stacius Thebaydos') as the source of Cassandra's exposition of Theban history (v. 1485), and of the romance of Thebes heard by Criseyde (ii. 100); Lucan is correctly identified as the source of Criseyde's maxim on power and moral virtue (ii. 167), but incorrectly as the source of the invocation of the Furies (iv. 22-4). A stream of marginal comments identify the origins of Chaucer's allusions to classical myth in Ovid's *Metamorphoses*, sometimes reproducing a line or two of Ovid's text (e.g. at iii. 721-30; iv. 791, 1138-9). There are also more speculative comments suggesting a source in Ovid's *Fasti* for Chaucer's allusion to Janus (ii. 77), or the likely Ovidian sources of such allusions as those to Procne or Niobe ('Require in Methamorphosios', 'Require in Ouidio'). When Pandarus echoes Lady Philosophy on the ass listening to the harp (i. 731), one scribal commentator correctly refers his reader to Boethius ('Baicius de consolacione philosophie'), and some manuscripts include source passages from the *Ilias* of Joseph of Exeter and a summary of Statius' *Thebaid* adjacent to or even incorporated within the parts of Chaucer's poem based on them (v. 799-840, 1498). In Book III Troilus' prayer in the 'stewe' alluding to the loves of the gods has prompted in British Library MS Harley 2392 (H4) a particularly full and careful commentary which offers explanatory notes and references to sources in various books of *Metamorphoses*, together with some brief quotation of the relevant Ovid passage (see Fig. 4).

In outcome *Troilus* reflects Chaucer's use of two distinct structures of interrelation between sources: one he receives and reinterprets, the other he constructs for himself, meshing the two structures together in his poem. Behind *Troilus* there lies the shape of the established corpus of interrelated narratives about Troy, in which the story of Troilus had already received epic, romance, and historical treatment by the time Chaucer was writing.

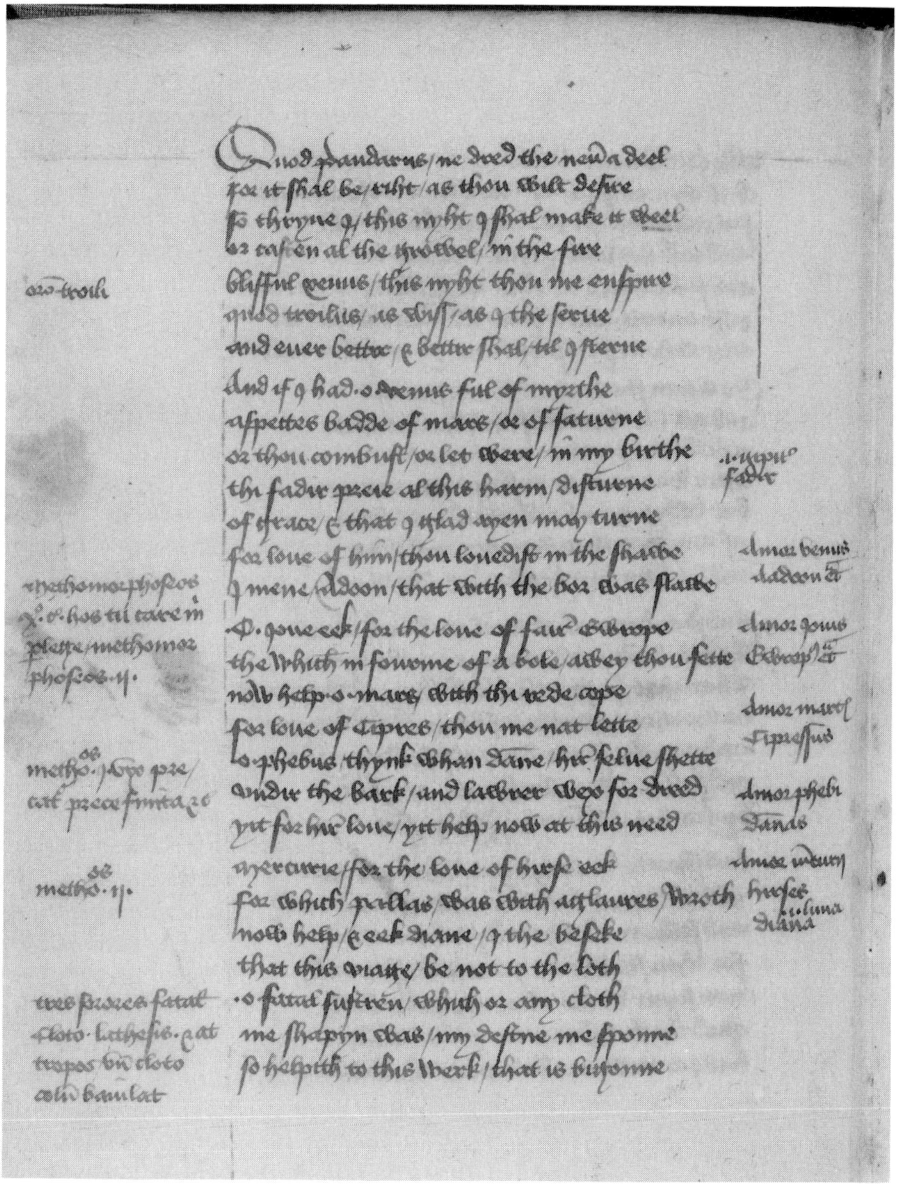

Quod pandarus / ne drede the neu*er* a deel
for it shal be / right / as thou wilt desire
so thryue I / this nyght I shal make it weel
or casten al the gruwel / in the fire
Blisful venus / this nyght thou me enspire
quod troilus / as wyss / as I the serue
and euer bettre / & bettre shal / til I sterue
And if I had . o venus ful of myrthe
aspectes badde of mars / or of saturne
or thou combust / or let were / in my birthe
thi fadir preie al thilke harm / disturne
of grace / & that I glad ayen may turne
for loue of him / thou louedist in the shawe
I mene Adoon / that with the bor was slawe
O . Ioue eek / for the loue of fair Ewrope
the which in forme of a bole / awey thou fette
now help . o . mars / with thi rede cope
for loue of cipres / thou me nat lette
o . phebus thynk whan dane / hir selue shette
vndir the bark / and lawrer wex for drede
yit for hir loue / yit help now at this need
Mercurie / for the loue of hierse eek
for which pallas / was with aglauros wroth
now help / & eek diane / I the biseke
that this viage / be not to the loth
. o fatal sustren / which or any cloth
me shapyn was / my destine me sponne
so helpeth to this werk / that is bigonne

FIG. 4. A page in British Library MS Harley 2392, fo. 64v, showing *Troilus*, iii. 708–35, with marginal glosses. The left margin reads: 'or*acio* troili' (712); 'Methomorphoseos .x°. *capitulo* hos tu care mi*hi*' (720–1); 'P*er*lege|methomorphoseos .ij.' (722–3); 'metho*morphose*os .i. Vix precat*ur* prece finita *&* *cetera*' (726–7); 'metho*morphose*os .ij.' (729); 'tres sorores fatal*es* Cloto. lathesis. & attropos|Vn*a* cloto colu*m* baiulat' (733). (The references are to *Metamorphoses*, x. 705; ii. 833–75; i. 548; ii. 722 ff.) The right margin reads: '*i.e.* iupi*ter* fadir' (718); 'Amor Venus Aadoon et*cetera*' (720–1); 'Amor Iouis Ewrop*a* et*cetera*' (722–3); 'Amor mart*is* Cipressus' (724–5); 'Amor phebi Da*n*nas' (727–8); 'Amor m*er*curij hirses' (729–30); 'Diana *i.e.* luna' (731).

But *Troilus* also reflects that unique structure of intertextuality which Chaucer fashions by conjoining with the Troilus narrative the depth and reach of his reading in a diversity of sources. It is these two kinds of interrelation which will provide the structure of the following chapter on the sources of *Troilus*.

To recall the tradition of Troy is to draw on the intertextuality of an already related group of narratives which successively reinterprets a common corpus of stories, characters, and settings. The medieval tradition of Troilus—in Dares and Dictys, in Benoît de Sainte-Maure's *Roman de Troie*, and Guido de Columnis's *Historia destructionis Troiae*—offers an assembled corpus of texts, and one which cumulatively insists on its sources through successive fabrications of authority. At the beginning of Dictys' *Ephemeridos belli Troiani libri* are two different forms of prologue— a Preface and a Letter—with differing versions of how this text was discovered. The brief 'Letter' that prefaces Dares' *De excidio Troiae historia* opens with the direction 'Cornelius Nepos sends greeting to his Sallustius Crispus', and presents Cornelius as the discoverer and translator into Latin of Dares' original. It is this prefatory letter that Benoît de Sainte-Maure embroiders upon in his turn in the opening of the *Roman de Troie*. Benoît mistakenly takes Cornelius to be the nephew (*nepos*) of the Roman historian Sallust, and 'improves' the tale of the text's discovery. In Benoît's version (*Troie*, 84ff.) the learned Cornelius is one day turning over a large number of books in a cupboard and so finds the history written by Dares, who was 'a wondrous clerk and learned in the seven arts'. One layer of authority is here being added to another, and it is the brief work of Dares that Benoît claims to be following word for word as a mere translator in his huge romance ('I shall follow the Latin closely, I shall put in nothing but as I find it written. Nor do I say or add any good word, even had I skill, but I shall follow my matter . . .'). Anyone familiar with the content and extent of Dares' account could appreciate that Benoît's romance offered an infinitely fuller version, and as such invoked Dares as a gesture towards ancient authority. Similarly, although in his *Historia* Guido is in turn translating from Benoît's French into Latin, he nowhere acknowledges his true source. In his prologue—after criticizing the shortcomings of Homer, Ovid, and Virgil as historians of Troy—Guido claims to be following the account given by Dares and Dictys. By the time Guido concludes his version of Benoît and comes to write his epilogue he seems to have discovered the extent of the discrepancy between the versions in Dares, in Joseph, and in Benoît, and he hastily invokes Dictys as his other authority. It is against the background of such attributions and invocation of authority in the story of Troy that Chaucer's introduction of the name 'Lollius' as that of his source should be seen.

But Chaucer has not read successively through the familiar corpus of Troilus texts and then—in a kind of cumulative synthesis—set down his compendium in *Troilus*. One recent, highly unusual and unconventional approach to the old subject—Boccaccio's *Filostrato*—comes into his hands, which boldly makes the familiar part of the story into the sequel of an invented narrative of how Troilus and Criseyde come together. It is through Boccaccio's poem that Chaucer works—taking it as his main narrative source—which is why *Filostrato* must come first in any account of how Chaucer uses his sources in *Troilus*, even though it is the 'last'—for Chaucer the most recent—of the Troilus tradition. *Troilus* learns from and builds on the narrative impetus that *Filostrato* gained through its focusing and excerpting from the Troy-Books, but Chaucer also seeks to reinsert his poem within the medieval tradition of the story of Troy through his varying uses of Benoît, Guido, and the 'Dares' of Joseph of Exeter. His knowledge of the traditional sources would have come before his knowledge of *Filostrato*, but he goes back to them to modify what he makes of the more recent work. It is to reflect the way that Chaucer realigns his use of the *Filostrato* narrative with the existing tradition of Troilus texts that his selective use of those much earlier texts will here be discussed *after* his use of *Filostrato*.

Unlike the Troy materials, Chaucer's other sources are not previously unified within a single literary tradition but still release new meaning into the old story. Texts as different as those of Boethius, Ovid, Dante, and the *Roman de la rose* make themselves felt in different ways and at different depths from the surface of the poem. In some cases there is close quotation of substantial extracts, while in other cases the model of form or the influence of arguments, examples, and types has been absorbed and diffused. Such use of sources in *Troilus* also needs to be set in the context of the intertextuality of manuscript culture itself. Juvenal may be rather loudly acknowledged when a few lines of his are paraphrased (iv. 197), but *Troilus* silently anthologizes many brief borrowings from other sources. The striking line on love 'In gentil hertes ay redy to repaire' (iii. 5) echoes a line of Guinizelli's, which Chaucer may have known not directly but from its quotation in Dante's *Convivio* (iv. 20). The thoughts on cultural change in the second proem (ii. 22 ff.) may recall lines from Horace (*Ars poetica*, 69–72) which Chaucer could find cited both in the *Convivio* (ii. 14, 83–9; i. 5, 55–66) and in John of Salisbury's *Metalogicon* (i. 16; iii. 3), just as Pandarus' advice on letter-writing (ii. 1041–3) may recall the opening lines of Horace's *Ars poetica* (1–5) not so much directly as through their partial quotation in John of Salisbury's *Policraticus* (ii. 18). The passage from Geoffrey of Vinsauf's *Poetria nova* 'quoted' by Pandarus (i. 1065–9) could be found quoted in other works, and it has been suggested that material like Pandarus' discussion of the sources of dreams (v. 365–85) does not derive directly from the learned authorities but comes filtered and excerpted through the pseudo-Cato, a popular school-book. The 'portraits' of the characters from Joseph of Exeter's *Ilias* are known to have been excerpted and circulated independently of the main work, and the 'Argument' to the books of the *Thebaid* drawn upon by Chaucer's Cassandra (v. 1485–1510) existed separately from the text of Statius.

It is often possible that, while Chaucer knows a particular text well and can go to it directly for some parts of *Troilus*, he may also be influenced in other contexts by the use of that same text by a third writer. Chaucer has the closest knowledge of Boethius and draws on him directly, but in some contexts his use of Boethius may have been prompted by uses of the *Consolation of Philosophy* which he could find in Boccaccio or in Machaut. Chaucer could similarly draw directly on Ovid, but his own use of Ovid may have been variously prompted and mediated by the use of Ovid in *Filostrato* or in the *Roman de la rose*, just as Chaucer's sense of the story of Thebes apparently filters not only directly from Statius, but also through Dante, Boccaccio, and French romance. For Boethius, Ovid, and Statius particularly among Roman authors, the influence of the medieval commentators has also been argued as an intervening medium which would shape and colour Chaucer's understanding and use of such sources. In responding to *Filostrato* Chaucer sometimes absorbs at one remove Boccaccio's own borrowings from Virgil, or Dante, or Petrarch. Behind Troilus' sorrowing tour of the places of Troy (v. 561–81)—adapted from *Filostrato*, v. 54–5—may lie Petrarch's Sonnet 112. When *Troilus* takes over a description of the sorrowing Troilus as a raging bull (iv. 239–41), Chaucer imitates Boccaccio (*Fil.* iv. 27) imitating Dante (*Inferno*, xii. 22–4) imitating Virgil (*Aeneid*, ii. 222–4). With the flight of Troilus' soul to the spheres, Chaucer closely translates his text from a description of the ascent of Arcita's soul in the *Teseida* (xi. 1–3) which Boccaccio wrote under the influence of such familiar authorities to Chaucer as Lucan (*Pharsalia*, ix. 1–14), Dante (*Paradiso*, xxii. 133 ff.), and the *Somnium Scipionis*. Which is the 'source'? The *Teseida* is 'textually' the source, for its narrative and diction are followed—although Chaucer may well have had Boccaccio's own sources also in mind—while Chaucer's *idea* to include the description of a flight may actually have been prompted by a passage in Boethius (iv, m. 1) describing the flight of thought.

There is also an especially strong sense of intertextuality in the mediation through successive texts of classical allusions, commonplaces and conventions of love, and various narrative motifs and type-scenes. In such contexts it becomes hard to discern one particular text as source— probably because the 'source' is the poet's use of his educated memory of accumulated reading rather than any particular book open in front of him. Chaucer's references to the Furies 'sorwynge evere in peyne' (i. 9; iv. 23) may recall Dante's picture of the Furies both as agents of torment and as themselves tormented (*Inferno*, ix. 37–51), as well as Chaucer's memory of the Furies' tearful response to the music of Orpheus in Boethius' *Consolation* (iii, m. 12), which itself recalls Ovid's *Metamorphoses*, x. 45–6.

For the problem of 'sources' for romance motifs and type-scenes, a case in point would be the debatable question of Chaucer's possible use in *Troilus* of Boccaccio's bulky prose work, *Il Filocolo*, a treatment of the romance of Floris and Blanchefleur. While the loose unity of *Filocolo* contains material which makes Boccaccio's romance as a whole read very differently from *Troilus*, there are some intriguing congruences between the two works. *Filocolo* and *Troilus* are both the romances of learned poets who approach their stories with comparable erudition, and yet also with comparable sensibility and delicacy in treating the emotions. The closest resemblances to Chaucer's poem in *Filocolo* lie in the way Florio is at last introduced secretly into the chamber of Biancofiore, and in the innocent solemnity and sense of ceremony with which the lovers move towards the consummation of their love. Whether such resemblances are more than the convergence of two romances in depicting a comparable and typical episode remains open to debate, although the question is made more significant by the way that resemblances of incident and detail exist within broader congruences between the two works in their presentation of a romance of first love in a non-Christian setting. Such congruence continues in the resemblance between the allusions to classical poets at the end of *Troilus* (v. 1791–2) and Boccaccio's closing injunction to his 'little book' *Filocolo* not to presume to be where Virgil, Lucan, Statius, Ovid, or Dante are read. This shared allusion to the classics is a fitting emblem of the possible link between *Troilus* and a work which may have been most significant as a 'source' by serving as an example of how romance could be ambitiously deployed and extended in treating the interplay between pagan and Christian in exploration of love.

Yet, although Chaucer bids his poem follow humbly after the classical poets, and although the scribes identified some of the classical sources he had drawn upon, it may well be asked to what extent perceptible allusion to other texts was part of Chaucer's design in handling his sources in *Troilus*. There is evidently a distinction to be made—not always without difficulty—between a 'borrowing' of source materials which the borrowing poet does not expect to be recognized as borrowed in their new context, and a borrowing from one text into a second text, or some form of allusion in one text to another, which is indeed to be recognized. Such recognition may open up enriching possibilities for intertextual comparisons, parallels, and contrasts between the original and the new context, but to recognize the possible sources of passages in *Troilus* through the efforts of modern scholarship is not necessarily to re-live any such recognition of allusion in the responses to Chaucer's poem by his early audiences and readers.

It remains open to question as to what Chaucer might have expected those first readers and hearers to recognize and so to compare with the earlier texts. With works such as the *Roman de la rose*, Boethius' *De consolatione philosophiae*, or such classical authors as Ovid and Statius, Chaucer might expect some of his audience to be—in varying degrees— familiar and able to make comparisons. It is to such familiar classical works as Ovid's *Heroides* and Statius' *Thebaid* that Pandarus seems to allude (i. 652–65; ii. 106–8), and it is ostensibly as a rendering of a classical text that *Troilus* is being presented. In response, it is the possible allusions to classical authors that the scribal annotators recognize and record for other readers. But in a manuscript culture where copies were few, expensive, and often inaccurate, it remains at least debatable how far an author could presume among his audience a close familiarity with the texts of secular works, as distinct perhaps from their stories. It is, of course, precisely where Chaucer is borrowing closely in *Troilus* from an Italian sonnet of Petrarch that he pretends to be following the classical 'Lollius', although in the *Clerk's Tale*—where he is working from a Latin work of Petrarch's— Chaucer respectfully acknowledges Petrarch by name as his source (IV. 31–3). In some of his other works Chaucer also acknowledges with deep respect an indebtedness to Dante, yet, although *Troilus* pervasively reflects Chaucer's reading of Dante, there is no mention within the poem of Dante's name nor any acknowledgement of his work. In that there is no extant evidence other than Chaucer's poems for the reading of Italian poetry in England until perhaps as late as the sixteenth century, it remains a matter for speculation as to whether Chaucer might expect an audience to recognize some of the sources he had so silently used in *Troilus*, and through recognition to hold the earlier context in mind alongside the new, considering similarities and contrasts which comment on the later passage. The account in this chapter of Chaucer's use of sources concerns itself with the process of how *Troilus* came to be created from the contributions of its sources, but more information would be necessary about the reading of Chaucer's early readers and audiences before it could be determined how much of this same process was perceptible to them and hence part of their response to the poem.

The following account of Chaucer's sources looks first at *Filostrato* and the Troy narratives, and then at some of the other major texts which have influenced the composition of *Troilus and Criseyde*. The sections devoted to 'Dares', Benoît de Sainte-Maure, and Guido de Columnis, illustrate how Chaucer aligns the narrative received from *Filostrato* so as to take account within his poem of the existing treatments of the Troilus story as kinds of epic, romance, and history. The text outside the Troy-Books upon which Chaucer draws most extensively as a model in both form and content is the *De consolatione philosophiae*, which is why discussion of Boethius comes first after that of the Troy-Book literature below. It is under the influence of the *Consolation* that Chaucer develops the story of the love of Troilus and Criseyde to release new 'philosophical' dimensions of implication in his narrative. But Chaucer also very much extends and amplifies the presentation of that experience of love, drawing on a tradition of related texts, and this is why discussion of Boethius will be followed by accounts of Chaucer's uses of Ovid, the *Roman de la rose*, and Guillaume de Machaut. Within such a structure of interrelated reading, particular authors—such as Ovid—will represent much more than one kind of influence and model, so constituting a 'source' in more ways than one, and this is also the case with Statius and the matter of Thebes, and with Dante, which are texts that fuelled and extended Chaucer's poetic imagination and ambition in his interpretation of the story of Troilus and Criseyde.

On Lollius, see Kittredge, 'Chaucer's Lollius'; Millett, 'Chaucer'; and Pratt, 'A Note on Chaucer's Lollius', who cites the corrupt Bodleian manuscript reading and notes Foullechat's translation.

On 'Lollius' as Chaucer's name for a figure who may stand to represent the partial and ambiguous transmission of the classics through the tradition of commentary, see Taylor, *Chaucer Reads*, 227, and Wetherbee, *Chaucer*, 25. For the suggestion of a pun by Chaucer on the plant named 'lolium', which is hard to distinguish from the true crop in the parable of the wheat and the tares in Matt. 13, see Ruth Morse, *Truth and Convention in the Middle Ages: Rhetoric, Representation, and Reality* (Cambridge, 1991), 196–7.

Chaucer's knowledge of Italian literature has traditionally been explained by his journeys to Italy on official business, although there was a community of Italian merchants and bankers in the London of Chaucer's day which may have been another means of access for him to Italian books (Schless, 'Transformations').

Beside these lines of *Troilus* the following manuscripts make the following comments: iv. 659: 'noua infra civitatem current' (H4, recalling *Aeneid*, iv. 188); ii. 100: 'Require in libro Stach Thebaidis' (R); v. 1485: 'Stacius Thebaydos' (H4); ii. 167: 'Lucanus' (S2); iv. 22: 'Herine | furie infernales unde lucanus me pronuba ducit herinis' (H4, citing *Pharsalia*, viii. 90); iv. 791: H4 quotes *Metamorphoses*, xi. 61–4; iv. 1138: H4 quotes *Metamorphoses*, x. 500–1; ii. 77: 'Require in libro Fastorum Ouidii de Jano & cetera' (R); i. 701, 713: 'Require in Methamorphosios', 'Require in Ouidio' (R); i. 731: 'Baicius | de consolacione philosophie' (R); iii. 294: 'Cato' (Dg, S2); iii. 1415: 'Nota Gallus vulgaris astrologus Alanus de planctu nature' (H4). The Latin source lines of the Book V portraits are copied in the margin of MS J, and interpolated into the English text of MS Gg. A twelve-line Latin 'argument' to the *Thebaid* stands between v. 1498 and v. 1499 in all but two manuscripts of *Troilus* (H4, R). See Benson and Windeatt, 'Manuscript Glosses'.

On Dares, see below, pp. 72–7.

The lines from Geoffrey of Vinsauf's *Poetria nova* could be found cited in Vincent of Beauvais, *Speculum doctrinale*, iv. 93, 'De maturitate', as noted by Norton-Smith,

Geoffrey Chaucer, 2; the first one-and-half lines are cited in John de Briggis, *Compilatio de arte dictandi*, as noted by J. J. Murphy, 'A New Look at Chaucer and the Rhetoricians', *Review of English Studies*, NS 15 (1964), 1–20.

On the pseudo-Cato, see R. Hazelton, 'Chaucer and Cato', *Speculum*, 35 (1960), 357–80. On the *Thebaid* 'argument', see Magoun, 'Chaucer's Summary'.

On the possible influence of Nicholas Trivet's commentary on Boethius' *Consolation*, see Gleason, 'Nicholas Trevet', and Minnis, 'Aspects', who points out the resemblance between Chaucer's stanza expanding on the Boethian idea of the ass and the harp and Trivet's commentary on the same passage of the *Consolation*:

> 'Or artow lik an asse to the harpe,
> That hereth sown whan men the strynges plye,
> But in his mynde of that no melodie
> May sinken hym to gladen, for that he
> So dul ys of his bestialite?' (i. 731–5)

Vnde asinis similes sunt homines qui audientes sermones racionabiles tantum prebent aures ad audiendum non animum ad percipiendum intellectum

Similar to asses are men who, when they hear meaningful utterances, only use their ears in order to listen, but not their mind in order to understand the meaning.

On the *Filocolo* and *Troilus*, see Wallace, *Chaucer and the Early Writings of Boccaccio*, especially chs. 3 and 4, and Young, *Origin and Development*.

Boccaccio Il Filostrato

All that is most significant, most moving, and most mysterious about Chaucer's *Troilus* distinguishes it from *Il Filostrato*, an early work of Boccaccio's, probably written about 1335 and dating from his period in Naples. To look at *Filostrato* and Chaucer's *Troilus* side by side can suggest the experience of watching over the English poet's shoulder as he works. Yet in such a comparison there is often a danger of emphasizing too much the most striking 'changes', whereas what comes through more unobtrusively from the Italian also has its own effect in the whole. In *Troilus*, its Italian source has in important ways been taken over and held within the English poem. Many large interpolations and some considerable omissions have been made, and, where it is followed more closely, *Filostrato* has been pervasively rephrased and subject to much rewriting. And since *Filostrato* has been so absorbed in *Troilus* as well as changed, comparison of the two can locate what is common to both poems, what is distinctively Chaucer's, and also what is in *Troilus* because it represents a response to the Italian.

For long stretches *Troilus* follows *Filostrato* so closely that Chaucer must have worked with a copy of the Italian in front of him as he created the draft of his poem. The distribution of narrative and dialogue into stanzas is identical in the two poems for substantial stretches: over and over again, the first line of each English stanza is very closely rendered from the parallel Italian line, stanza by stanza. One instance would be Chaucer's response in Book I of *Troilus* to the first lines of five successive stanzas in *Filostrato* (i. 35–9), and this example must stand to represent numerous others throughout *Troilus*:

> Immaginando affanno né sospiro
> poter per cotal donna esser perduto ... (*Fil.* i. 35)

Imagining that neither trouble nor sighing for such a lady could be wasted ...

> Imagenynge that travaille nor grame
> Ne myghte for so goodly oon be lorn ... (i. 372–3)

> Per che, disposto a seguir tale amore,
> pensò voler oprar discretamente ... (*Fil.* i. 36)

Thus being inclined to pursue this love, he thought of how he might set to work discreetly ...

> Thus took he purpos loves craft to suwe,
> And thoughte he wolde werken pryvely ... (i. 379–80)

> Ed oltre a questo, assai più altre cose,
> qual da scoprire ... (*Fil.* i. 37)

And besides this, he also thought of many other things: how to make his feelings known ...

> And over al this, yet muchel more he thoughte
> What for to speke ... (i. 386–7)

> E verso Amore tal fiata dicea
> con pietoso parlar: 'Signor ...' (*Fil.* i. 38)

And to Love he at such time said with piteous voice, 'Lord ...'

> And to the God of Love thus seyde he
> With pitous vois, 'O lord ...' (i. 421–2)

> 'Tu stai negli occhi suoi, signor verace
> sì come in loco degno a tua virtute ...' (*Fil.* i. 39)

'You stand within her eyes, true lord, as if in a place worthy of your might ...'

> 'Ye stonden in hir eighen myghtily,
> As in a place unto youre vertu digne ...' (i. 428–9)

The openings of some 46 stanzas, out of the 57 which make up the first part of *Filostrato*, can be found incorporated into the openings of 46 of the 78 stanzas in the first half of Book I of *Troilus* (i.e. up to i. 546). In this way, *Filostrato* provides much of the dramatic 'script' of the two romances in terms of what the characters do: when they speak and how they react is still in *Troilus* often built around the model received from *Filostrato*, in which Boccaccio's use of monologue and dialogue often forms a quasi-dramatic continuity, linked and framed by unobtrusive narrative connections. There is much syntactical closeness between passages in the two poems, and also a very real and sustained lexical influence from *Filostrato* upon the diction of *Troilus* (see below, pp. 330–1). Although there remains some question that Chaucer may have checked his understanding of the Italian against a French prose translation of *Filostrato*, the closeness of *Troilus* to the Italian not only proves that Chaucer worked from a copy of *Filostrato* but also suggests that his Italian was rather good.

Yet within most stanzas Chaucer can soon be observed to move away from translation in the modern sense of rendering like with like, and will begin to re-express, add, and replace in his re-creative and adaptive 'translacioun', as in these two stanzas:

> Né del dì trapassava nessuna ora
> che mille volte seco non dicesse:
> 'O chiara luce che 'l cor m'innamora
> o Criseida bella, Iddio volesse
> che 'l tuo valor, che 'l viso mi scolora,
> per me alquanto a pièta ti movesse;
> null'altro fuor che tu lieto può farmi,
> tu sola se' colei che puoi atarmi'.

> Ciascun altro pensier s'era fuggito
> della gran guerra e della sua salute,
> e sol nel petto suo era sentito
> quel che parlasse dell'alta virtute
> della sua donna; e, così impedito,
> sol di curar l'amorose ferute
> sollicito era, e quivi ogni intelletto
> avea posto, e l'affanno e 'l diletto.   (*Fil.* i. 43–4)

Nor did an hour of the day pass by that he did not say to himself a thousand times, 'O clear light, that moves my heart to love, O fair Criseida, would to God that your noble qualities that have made my face grow pale may move you somewhat to pity upon me. None other than you can make me happy: you alone are her who can help me.' All other thoughts of the great war and of his well-being had fled from him, and only that which spoke of the high worth of his lady was heeded within his breast, and, thus in shackles, he was only concerned to cure the wounds of love, and to this he devoted all his study and effort, and in this found all his delight.

Ek of the day ther passed nought an houre
That to hymself a thousand tyme he seyde,
'Good goodly, to whom serve I and laboure
As I best kan, now wolde God, Criseyde,
Ye wolden on me rewe, er that I deyde!
My dere herte, allas, myn hele and hewe
And lif is lost, but ye wol on me rewe!'

Alle other dredes weren from him fledde,
Both of th'assege and his savacioun;
N'yn him desir noon other fownes bredde,
But argumentes to his conclusioun:
That she of him wolde han compassioun,
And he to ben hire man while he may dure.
Lo, here his lif, and from the deth his cure!   (i. 456–69)

Even in such a short space Chaucer can be seen subtly recasting his source, re-expressing the terms and values in which Troilus conceives of his love for Criseyde, introducing those thematic elements of feudal service, illness and cure, endurance and death, which will mark the *Troilus* (see the chapter on Themes).

Chaucer himself uses for other purposes within his poem some terms that help towards defining and distinguishing the processes and layers in this creative translating. When Chaucer is working at the closest level of adopting yet adapting the source stanzas he seems engaged in a 'paynted proces' (ii. 424) of something that suggests the analogy of processes involved in painting or forms of printing and film: he overlays the existing structure with his own tones so that actions, sequences of events, are seen in a different light. But, further to this, Chaucer most pervasively works in translation through small added touches and emphases, and it is perhaps to such small insertions that the poem refers when it apologizes for words that have been 'in eched for the beste' (iii. 1329). Beyond this again, there are also much larger passages interpolated into the narrative flow of *Filostrato* but without counterpart in the Italian. One such addition (i. 218–66) concludes by self-deprecatingly referring to 'thing collateral' (i. 262), but the term is useful in suggesting how much influential material—important in the ways it alters the implications of Chaucer's account of Troilus and Criseyde—is held in relation to the narrative within the new sequence of the translated text, and thus comments upon the action.

In an attempt to convey the process and texture of how Chaucer transforms his narrative source, the nature of the changes between the two texts is set out below schematically in tabular form, by way of preface to brief analyses of each book of *Troilus*. (These analyses are themselves necessarily introductions to points about *Troilus* which will be pursued more fully in the chapters on Structure, Themes, and Style). Such a diagrammatic stylization of so finely detailed a process of 'translation' can only aim to help the reader grasp the main lineaments of creative change, as Chaucer responds in his own stanzas to his stanzaic source. Within the stanzas of *Troilus* marked as broadly corresponding to a parallel passage in *Filostrato* there may still be significant differences of idea, expression, or tone, while the simple recording of Chaucer's expansions of one Italian stanza into two or more in his version covers many various kinds of amplifying and modifying from *Filostrato* into *Troilus*.

Yet the feasibility of even beginning to set out the differences of *Troilus* from *Filostrato* in such a linear format points to how Chaucer's imagination has responded to the line and script provided by his main narrative source, so that the 'horizontal' plane of the story-line is recurrently cut across by the 'vertical' plane of new or expanded material worked into Chaucer's text, which opens up different dimensions of significance in his version of the Troilus story. It may well be that such composition through adaptive translation was a cumulative process of layers of composition which involved Chaucer in frequently going back over parts of his text. His knowledge of *Filostrato* is certainly present to Chaucer as a whole while he works on any particular context, for incidents at a later point in Boccaccio's poem may be moved earlier (as when Troiolo's faint in Part 4 is shifted forward to Book III), or a later scene in *Filostrato* may prompt Chaucer's invention at an earlier stage in *Troilus*: the family visit to the sick Troilus in Chaucer's second book has probably been suggested in part by the family visit to the sick Troiolo in Part 7 of *Filostrato*, itself rewritten by Chaucer as Cassandra's visit to interpret the dream. The following tabular format can only represent in stylized form the outcome of what may have been such a layered and cumulative composition.

In these tables roman type indicates passages in *Troilus* which broadly correspond to a parallel passage in *Filostrato*; *italic type* indicates passages in *Troilus* which represent expansion and modification of material in *Filostrato*; **bold type** indicates passages in *Troilus* which are added to *Filostrato*; contexts where Chaucer omits or abbreviates passages in *Filostrato* are signalled by square brackets. Corresponding part and stanza numbers in *Filostrato* are in parentheses at the end of entries.

BOOK I

1–56	*The Proem* (cf. *Fil.* i. 5–6)
57–63	The scene is set at Troy (i. 7)
64–77	*Calchas foresees the fall of Troy* (i. 8)

78–91	Calchas flees Troy; Trojan anger	(i. 9–10)
92–105	*Criseyde described*	(i. 11)
106–33	Criseyde seeks Hector's protection	(i. 12–15)
134–8	The course of the war: a summary	(i. 16)
138–47	**Influence of Fortune's wheel; read Homer, Dares, Dictys**	
148–54	Trojans continue to worship their gods	(i. 17)
155–68	*The festival of the Palladium*	(i. 18)
169–82	*Criseyde described at the temple*	(i. 19)
183–96	Troilus in the temple	(i. 20–1)
197–203	[Troilus mocks lovers; replacing Troiolo's antifeminist speech, (i. 22–4)]	
204–17	*Troilus smitten by Love: 'O blynde world!'*	(i. 25)
218–66	**On the power of love**	
267–73	Troilus' glance falls on Criseyde	(i. 26)
274–80	**Troilus' inward reactions**	
281–94	Criseyde described; Troilus' admiration	(i. 27–8)
295–301	**Impression on Troilus of Criseyde**	
302–8	The scorner of love is caught	(i. 29)
309–29	Troilus returns home, pretending to mock lovers	(i. 30–2)
330–50	**Troilus' speech mocking lovers**	
351–7	*Effects of love on Troilus* (from *Fil.* i. 32, ll. 7–8)	
358–92	Troilus reflects in his chamber	(i. 33–7)
393–420	**The song of Troilus, from 'Lollius'**	
421–546	Troilus' changed behaviour and outlook as a lover (i. 38–57 [omitting i. 52, 56])	
547–53	Pandarus discovers Troilus languishing	(ii. 1)
554–60	**Pandarus' teasing speech: Is Troilus being pious?**	
561–7	**Pandarus' motives explained**	
568–630	Dialogue between Pandarus and Troilus	(ii. 2–10)
631–51	**Pandarus on 'contraries'**	
652–65	**Pandarus cites Oenone's 'compleynte'**	
666–79	*Unlucky in love himself, Pandarus can still advise*	(cf. ii. 11)
680–6	Troilus need fear no reproach (ii. 12)	
687–700	**The wisdom of confiding**	
701–21	*Cease weeping; find comfort in companionship* (cf. *Fil.* ii. 13–14)	
722–35	*Troilus will not reveal his lady's identity; Pandarus alarmed* (cf. *Fil.* ii. 15)	
736–60	**Troilus reflects; tells Pandarus he will die**	
761–854	**Pandarus argues against despair**	
855–68	Pandarus urges Troilus to confess	(ii. 16–17)

869–75	[Confession; compressed from *Fil*. ii. 18–20]
876–905	*Pandarus praises Criseyde's qualities* (ii. 21–3)
906–66	**Pandarus recalls Troilus' past mockery of lovers; makes him repent; cheers and advises**
967–73	Troilus encouraged to stand fast and hope (ii. 24)
	[Pandaro acknowledges such affairs are dishonourable (ii. 25–6)]
974–87	*Pandarus foresees Criseyde's inclination to love* (ii. 27)
988–94	Pandarus declares his readiness to help (ii. 28)
995–1008	**Pandarus' 'good conceyte' of Troilus as a zealous convert**
1009–22	Troilus hopes and doubts (ii. 29–30)
1023–9	**Pandarus impatiently dismisses his fears**
1030–43	Troilus' pure intent; Pandarus' laughing reassurance (ii. 31–2)
1044–57	*Troilus embraces Pandarus, commending himself to him* (ii. 33)
1058–64	*Pandarus swears to help and departs* (ii. 34)
1065–71	**Pandarus premeditates: a blueprint for action**
1072–92	**Description of Troilus' changed nature**

In the first half of Book I Chaucer is following quite closely the narrative line provided by *Filostrato*. It is from *Filostrato* that the story-line of the book derives: how Troilus sees Criseyde during a ceremony in a temple, is smitten with love for her, and later confesses this to a friend. But first the proem has been recast—and with it the poet's relation to his subject and to his audience—developing beyond recognition *Filostrato*'s opening appeal for the prayers and compassion of lovers. Turning to his narrative, Chaucer follows Boccaccio's sequence of events: first a summary narrative setting the scene at the siege of Troy, describing Calchas' flight, and introducing Criseyde; then the temple scene, and its aftermath for Troilus until discovered by Pandarus. As the clusterings of italic and bold type in the table above suggest, Chaucer's imagination has evidently seized on particular points to develop and emphasize as he works from his source. Patterns are already emerging which will contribute to Chaucer's development of theme and character. Additions are made to emphasize: the background of the Trojan War and its known end; the description of Criseyde at her first appearances; the role of Fortune; Troilus' first dramatic encounter with love, its possible implications, his subsequent reactions and impressions in solitude. The whole extended sequence in which Troilus falls in love has been enriched in expressiveness and implications through rewriting, amplification, interpolation of a 'collateral'

TABLE 4. *The structure of* Troilus, *Book I*

Filostrato			Troilus		
Part	Stanzas	Lines	Book	Lines	Stanzas
1	57	456	I	1–546	78
2	1–33	264	I	547–1092	78

passage like that on the power of love (i. 218–66), and the inclusion of the Canticus Troili (i. 400–20). Here at work on adapting *Filostrato* is Chaucer the 'historial' poet, the pupil of the rhetoricians, the poet of 'doctryne and sentence', and the poet of a courtly mode of love.

In the second half of Book I, after the arrival of Pandarus, Chaucer's handling of his source becomes much freer—as the table reveals—developing theme and characterization through rewriting and through substantial invention of direct speech. Here is established the character of Chaucer's Pandarus as an inexhaustibly resourceful talker and strategist, as each turn in the friends' exchanges in *Filostrato* is re-expressed, amplified, sharpened, with that vivacious mixture of wit, and dash, and lore. The variousness of Chaucer's added material and its capacity to open up new perspectives on the action makes what are the same underlying transactions read very differently in the texture of their English interpretation than in *Filostrato*.

In Chaucer's reconstruction of his source material the conclusion of the first book has been moved to a point further on in the action, at the end of the first interview between Troilus and Pandarus, rather than before the first entry of Pandarus, as in *Filostrato*. Chaucer expands the friends' interview such that it forms an equal and balancing second half to Book I (see Table 4 above). The first appearance of Pandarus at the very centre of the structure of Book I is an apt reflection in the book's design of his central role as mediator and as designer of the love-affair to come; his appearance divides the book, because his presence makes the second half complementary to the first in form and theme. The division between Books I and II now coincides with a climactic moment of commitment and resolution as Pandarus undertakes to win Criseyde for Troilus, and the first book thus ends as its proem began, with a promise and the undertaking of an endeavour.

BOOK II

1–49	The Proem
50–77	Pandarus begins a new day

78–84   *Pandarus goes to Criseyde's house*   (ii. 34 ff.)

85–263   **Teasing conversation with Criseyde, hinting at good fortune**

264–80   *Pandarus gazes at Criseyde; her response* (ii. 35–6)
[Pandaro hints of a noble admirer   (ii. 37–43)]

281–94   *Pandarus on seizing the fortunate time*   (cf. *Fil.* ii. 44)

295–308   **Pandarus declares his good 'entencioun'**

309–22   *At Criseyde's entreaty, Pandarus names Troilus*   (ii. 45–6)

323–85   **Pandarus attempts to persuade**

386–9   **Criseyde's request for advice**

390–406   **Pandarus' resumed persuasions: 'carpe diem'**

407–28   *Criseyde's tears and bewilderment*   (ii. 47–8)
[Criseida thinks Troiolo's love will fade . . .   (ii. 49–51)]

429–48   *Offended, Pandarus makes as if to leave*   (ii. 52–4)
[Pandaro reminds Criseida of passing time]

449–62   **Criseyde's fears and inward reflections**

463–97   **Criseyde's conciliatory but conditional offer**
[Criseida succumbs to the thought of time passing and seeks love   (ii. 55)]

498–504   Criseyde asks how Pandarus first knew of Troilus' feelings   (ii. 55)

505–53   *Pandarus recalls overhearing Troilus complain to love*   (ii. 56–61)

554–74   *Pandarus recalls discovering Troilus languishing*   (ii. 62–3)

575–81   Pandarus urges Criseyde to save Troilus' life   (ii. 64)

582–95   **He goes too far, but smoothes over his mistake**
[Criseida feels pity, but urges discretion   (ii. 65–7)]

596–602   Pandarus leaves; Criseyde enters her 'closet' to think (ii. 68)

603–9   **Criseyde reflects, overcoming her fears**

610–65   **Troilus rides past; Criseyde's reactions**   (cf. *Fil.* ii. 82)

666–79   **Criticism of Criseyde anticipated, and indignantly rejected**

680–6   **How Venus was well disposed towards Troilus**

687–763   *Criseyde debates with herself; the attraction of loving* (ii. 69 ff.)

764–812   *She veers to the opposite view, ending in quandary* (ii. 75 ff.)

813–903   **Evening walk in her garden; Antigone's song; Criseyde's reactions**

904–31   **Criseyde retires; hears nightingale; her dream**

932–66   *Pandarus brings hopeful news to Troilus*   (ii. 79)

967–73   Troilus revives like a drooping flower   (ii. 80)

974–80   Troilus thanks Pandarus   (ii. 81)

           [Troiolo goes with Pandaro past Criseida's house; Criseida is at a window and they exchange gazes; Troiolo's desire grows ever fiercer . . .   (ii. 82–7)]

981–7   Troilus urges Pandarus to help him   (ii. 88–9)

988–1001   *Pandarus agrees to help*   (ii. 90)

1002–8   Pandarus advises Troilus to write to Criseyde   (ii. 91)

1009–22   **Pandarus plans for Troilus to pass when Criseyde is at a window of her house**

1023–43   **Pandarus advises Troilus on his letter**

1044–64   Encouraged, Troilus agrees to write   (ii. 93–5)

1065–84   *Summary of Troilus' letter*

           [The full text of Troiolo's lengthy letter   (ii. 96–106)]

1085–92   Folding and tearful sealing of letter   (ii. 107)

1093–127   *Pandarus takes the letter to Criseyde*   (ii. 108–9)

1128–41   Criseyde scruples to receive it   (ii. 110–11)

1142–62   *Pandarus forces the letter upon Criseyde*   (ii. 112–13)

           [Pandaro leaves, returning later to learn Criseida's view of the letter   (ii. 114–17)]

1163–83   *Alone in her room, Criseyde reads the letter*   (ii. 114)

1184–94   **After dinner Pandarus draws Criseyde to a window seat**

1195–218   Pandarus asks Criseyde's opinion of the letter; persuades her to reply   (ii. 118–20)

1219–25   *Summary of Criseyde's letter*

           [The full text of Criseida's letter   (ii. 121–7)]

1226–33   Criseyde gives her letter to Pandarus   (ii. 128)

           [Pandaro takes the letter straight to Troiolo   (ii. 128)]

1233–46   **Pandarus congratulates Criseyde on her concession**

1247–301   **Troilus' second ride past Criseyde's house**

1302–51   *Pandarus brings Troilus Criseyde's letter; Troilus grows more hopeful; writes often to Criseyde*   (ii. 128–31)

           [The brief final section of Part 2 of *Filostrato* now follows: Pandaro persuades Criseida to let Troiolo visit her (ii. 132–43)]

1352–757   **Pandarus promises to help Troilus; announces Poliphete's law suit against Criseyde and so contrives the dinner party at Deiphebus house, at which Criseyde is present and at which Troilus is to feign illness and retire to bed**

If Chaucer, in composing Book I, remains as it were 'within' the framework provided by the narrative sequence of events and the dramatic 'script' of speeches in *Filostrato*, in Book II the relationship may be characterized in opposite terms: a selective use of *Filostrato*, which is now held within the containing framework provided by Chaucer's construction of the book. It is from Part 2 of *Filostrato* that Chaucer derives the basis of the narrative in his Book II: how Pandarus tells Criseyde of Troilus' love, and how Criseyde moves towards accepting that love. The underlying pattern of the *Filostrato* story—in a sequence of alternating scenes as Pandarus visits each of the lovers in turn several times—still forms the narrative core of events in *Troilus*. But Chaucer's second book is restructured in a way that allows a more extended and detailed focus on the scenes of dialogue and monologue involving Criseyde, with a more developed sense of patterning in events and behaviour. As the table reveals, Chaucer draws selectively on the scenes of Pandaro's visits to Criseida, combining borrowed material with invented matter in altered sequences, so as to represent the different pace and shape of Pandarus' strategy, and the different pace and nature of Criseyde's response. The English Criseyde consequently emerges to Chaucer's reader with many qualities different from Criseida, which will prove significant as the action develops. That knowingness which Pandaro assumes in his cousin Criseida means that he can come speedily to the point with her about Troiolo. The English Pandarus must build up to the subject with his niece much more gradually, through an indirect and hinted process. There is a remarkable expansion of dialogue: the first interview between Pandarus and Criseyde grows from 32 stanzas or 256 lines (*Fil.* ii. 35–67) to 74 stanzas or 518 lines (ii. 78–595); their later dialogue is expanded from 9 stanzas (*Fil.* ii. 108–13, 118–20) to some 30 stanzas (ii. 1093–1302); and the dialogue between Troilus and Pandarus before the writing of the letter develops from 10 stanzas (*Fil.* ii. 86–95) to 18 stanzas (ii. 939–1064). Chaucer expands the account of Criseyde's quandary, and then inserts his invented sequence of the evening walk in the garden, Antigone's song, and Criseyde's dream of the eagle (ii. 813–931); and he splits the visit to Criseida's house (*Fil.* ii. 82) into the doubled event of Troilus' chance ride past and the contrived second ride past at the moment when Criseyde's letter to him has just been written and is about to be sent. Soon after this in *Filostrato* the lovers are moving towards arranging between themselves to spend a night together. This did not accord with Chaucer's interpretation of his characters and their love, and so there is no equivalent in *Filostrato* for the last phase of Book II of *Troilus*—the threatened lawsuit, the feigned illness—by which Chaucer's Pandarus brings the lovers together in Book III for their first meeting at the house of Deiphebus.

BOOK III

1–42	The Proem [excerpted and brought forward from Troiolo's song (iii. 74–9)]
43–9	**Concluding prayer to Venus and Calliope** [Leaving Criseida, Pandaro seeks Troiolo, finding him in a temple, deep in thought (iii. 3–4)]
50–210	**Criseyde's visit to Troilus' bedside; their exchanges**
211–26	**Leave-takings after the party**
227–87	*Pandarus discusses his role in the affair with Troilus* (iii. 5–10)
288–343	**Pandarus expands on the theme of secrecy, ending reassuringly**
344–57	Troilus' joyful feeling (iii. 11–12)
358–420	*Troilus reassures Pandarus* (iii. 13–19)
421–48	*Troilus' conduct and self-control at this time* (iii. 20) [Criseida tells Pandaro all is ready; Pandaro sends word to Troiolo, who on a dark night enters Criseida's house by a secret entry and waits in a hiding place for Criseida, who leads him to her room (iii. 21–30)]
449–511	**Summary account of how the affair develops over a period**
512–46	**Pandarus plans to bring the lovers together at his house**
547–94	**Pandarus visits Criseyde to invite her to supper**
595–693	**Criseyde comes for supper; a storm forces her to stay the night; she retires**
694–742	**Pandarus visits Troilus in his hiding place; Troilus prays**
743–952	**Pandarus enters Criseyde's room with a tale of Troilus' desperate jealousy; her reflections on earthly happiness; she agrees to receive Troilus**
953–1309	**Reproached by Criseyde for his jealousy, Troilus faints and Pandarus throws him into bed with Criseyde; the lovers move towards consummating their love**
1310–23	Impossible to describe that night (iii. 31–3)
1324–37	**Invitation to the audience to correct the text in the light of its experience**
1338–65	The lovers' joy; their looks, kisses, and sighs (iii. 34–7)
1366–72	**Exchange of rings; gift of a heart-shaped brooch**
1373–86	Misers cannot know such joy as lovers know (iii. 38–9)
1387–93	**May love's detractors suffer like Midas and Crassus**
1394–426	*The lovers compensate for past woes with present joys until Criseyde sorrows at the signs of dawn* (iii. 40–3)

1427–42	**Criseyde laments the departure of night**
1443–70	*Troilus curses the arrival of day* (iii. 44)
1471–91	Troilus' sorrows (iii. 45–7)
1492–519	*Criseyde's reassurances* (iii. 48–50)
1520–33	The lovers' leave-taking and parting (iii. 51–2)
1534–47	Returned to his palace, Troilus thinks only of Criseyde (iii. 53–4)
1548–54	Criseyde likewise thinks of Troilus (iii. 55)
1555–82	**In the morning Pandarus calls on Criseyde in her bedroom**
1583–96	*Pandarus comes to Troilus; their greetings* (iii. 56)
1597–617	Troilus' speech of thanks (iii. 56–9)
1618–38	*Pandarus warns Troilus not to lose his joy through indiscretion* (iii. 60)
1639–45	*Troilus reassures Pandarus* (iii. 61)
1646–66	Troilus tells of his experience and feelings (iii. 61–3)
1667–80	The lovers meet again with great joy (iii. 64–5) [Criseida expresses to Troiolo her joy and her desire (iii. 66–9)]
1681–94	**The lovers' joy is more than can be imagined or described**
1695–701	The lovers curse the coming of daylight (iii. 70)
1702–8	**Troilus complains that the sun has taken a shorter route**
1709–15	They part, arranging to meet again (iii. 71), **and thus Fortune maintained them in joy for a time**
1716–36	*How Troilus leads his life in joy and repute, his heart knit to Criseyde* (iii. 72)
1737–43	How Troilus often praises Criseyde to Pandarus (iii. 73) [Troiolo's song has already been partly used for the Proem]
1744–71	**Troilus' hymn to love, paraphrased from Boethius**
1772–85	How Troilus conducts himself in war and peace (iii. 90–1)
1786–99	*Troilus' devotion to love and lovers* (iii. 92)
1800–6	Love's beneficial moral influence on Troilus (iii. 93)
1807–20	**Venus, Cupid, and the Muses are praised as they now depart, and the book is brought to a close**

The very different characters and experience of Chaucer's Troilus and Criseyde from Boccaccio's mean that the development from the second to the third book of *Troilus* is radically recast, in Chaucer's most major departure from the plot of *Filostrato*. The third Part of *Filostrato* can move swiftly through a narrative of how Troiolo one night comes by arrangement to Criseida's house and the lovers go to bed: after a two-stanza proem

Boccaccio's narrative takes a mere 29 stanzas (232 lines) to reach the verge of the lovers' union, a point only reached in *Troilus* at iii. 1310–16, 180 stanzas (1,260 lines) after the narrative begins at iii. 50. In Chaucer's different construction of the action Pandarus contrives two social occasions, hosted by Deiphebus and then by himself, in order to bring Troilus and Criseyde together: on the first occasion for a scene of mutual declarations in which Criseyde accepts the 'service' of Troilus, and on the second for the physical consummation of their love. As the table above reveals, for the first two-thirds of Book III Chaucer accordingly makes little use of *Filostrato*. But, although Chaucer invents a very different kind of action to bring the lovers to bed together, once they have reached that point it is striking that Chaucer immediately reverts to working closely with the model of action and language provided by *Filostrato*, almost like Pandarus picking up an old romance once he has brought the lovers to bed. Yet Chaucer's reading of *Filostrato* is as adaptive as ever, turning *Troilus* into a deeply felt and pondered commentary on the story of Troilus as it was presented to him by Boccaccio. It is emblematic of the narratorial presentation of the book that the very point where Chaucer resumes a close following of *Filostrato*'s diction is at an expression of the inexpressibility of describing the lovers' joy (iii. 1310–23; cf. 1324–37). Yet Chaucer builds upon and extends the expressive dramatic script of lyrical exchanges already available in his source, developing the lovers' dawn-songs and their speeches in a sustained sequence of response to and modification of *Filostrato*. Although the remainder of Book III records that the lovers spent other nights together, Chaucer alters his source to lessen the impact of subsequent meetings and hence to throw more emphasis on to that first night, now past, at the heart of the poem's structure; his version of the second meeting shrinks to an outline report, hurried in tone and implying that description here would be merely repetitive (iii. 1676, 1684). Finally, by postponing Boccaccio's final stanza about the change of fortune that awaits the lovers (*Fil.* iii. 94)—which becomes part of the proem to Book IV—Chaucer keeps his third book intact as the span of the lovers' joy and peace, enlarging near the close on a picture of Troilus as the fulfilled lover (iii. 1772–806) which complements and develops upon the picture of Troilus as an expectant lover at the close of Book I (1072–92).

BOOK IV

1–14	*Proem: Fortune changes* (iii. 94)
15–28	**Criseyde's 'unkindness'; invocation of the Furies and Mars**

29–168	Trojan reverses; Calchas proposes exchange; discussed at Troy; Troilus' feelings (iv. 1–16)
169–75	**Troilus resolves to follow his lady's wishes**
176–96	**Hector opposes the exchange in vain**
197–210	**Antenor's betrayal of Troy anticipated**
211–17	The exchange is agreed (iv. 17)
	[Here Troiolo faints at the parley (iv. 18–21)]
218–24	Troilus speeds home to bed (iv. 22)
	[Addressing his lady, Boccaccio here identifies himself with Troiolo's woes (iv. 23–5)]
225–31	**Troilus bereft, like the leafless tree**
232–59	Troilus' misery (iv. 26–9)
260–87	*Troilus reproaches Fortune* (iv. 30–2)
288–322	Troilus' despair; **casts himself as Oedipus**; *commends his spirit to Criseyde* (iv. 33–6)
	[Troiolo could have tolerated a delayed exchange (iv. 37)]
323–9	**Troilus prays for all lovers; bids them remember him**
330–6	[Troilus curses Calchas (abbreviated from *Fil*. (iv. 38–40)]
337–43	Troilus reaches a state of collapse (iv. 41)
344–64	*Pandarus arrives sorrowing in Troilus' chamber* (iv. 42–3)
365–78	Troilus greets Pandarus (iv. 44–5)
379–92	*Pandarus laments the change of Fortune* (iv. 46)
393–406	Pandarus: Troilus has had his desire; Troy is full of ladies (iv. 47–8)
407–13	**Pandarus: Different people have different gifts**
414–27	*Expands on the theme 'New love drives out old'* (iv. 49)
428–34	**Pandarus only said all this to rouse his friend—**
435–500	*Troilus spurns Pandarus' arguments, vows himself to Criseyde* (iv. 50–60)
501–18	Troilus calls on death to kill him (iv. 60–2)
519–39	Pandarus urges Troilus to abduct Criseyde (iv. 63–5)
540–64	Troilus rejects Pandarus' arguments (iv. 66–9)
565–74	**Troilus' sense of duty to safeguard Criseyde's honour**
575–602	Pandarus urges the weeping Troilus to act (iv. 70–3)
603–16	*Criseyde will forgive her abduction* (iv. 74)
617–23	Be brave! (iv. 75)
624–30	**Pandarus will risk death to help Troilus**
631–7	Troilus will only do as his lady wishes (iv. 76)
638–44	**Pandarus says he meant this—but has Troilus asked her?**
645–58	*Pandarus will arrange a meeting* (iv. 77)
659–65	Fame spreads news of the exchange (iv. 78)

666–79	*Criseyde's response, and devotion to Troilus* (iv. 79)
680–730	Criseyde's women friends visit her (iv. 80–6)
731–60	*Criseyde retires to her chamber to lament* (iv. 86–9)
761–70	**She blames Calchas; 'To what fyn sholde I lyve?'**
771–84	She will starve herself and wear black (iv. 89–90), **in an 'order' of sorrow and complaint**
785–91	The lovers' spirits will not part (iv. 91), **united like Orpheus and Eurydice**
792–8	How will Troilus survive such grief? (iv. 92) [Criseida curses her father (iv. 93–4)]
799–805	Who could ever describe her sorrow? (iv. 95)
806–26	Pandarus finds Criseyde distraught (iv. 95–7)
827–47	**Criseyde's 'complaint'**
848–924	Exchanges between Pandarus and Criseyde (iv. 98–107)
925–38	*Pandarus urges Criseyde to help Troilus* (iv. 107)
939–45	Criseyde promises to moderate her grief (iv. 108)
946	Pandarus goes to find Troilus (iv. 109)
947–52	**Troilus is alone and despairing in a temple**
953–1085	**Troilus' Boethian soliloquy on predestination**
1086–127	Pandarus enters, scolds Troilus, summons him to see Criseyde (iv. 109–13)
1128–69	The sad meeting; Criseyde faints; Troilus' reactions (iv. 114–18)
1170–6	**Troilus' complaint and prayers**
1177–211	Troilus prepares for suicide (iv. 119–23)
1212–41	*Criseyde recovers; realizes their near escape* (iv. 124–6)
1242–53	The lovers retire to bed (iv. 126–7)
1254–372	*Criseyde argues for her going to the Greek camp* (iv. 128–36)
1373–414	**Criseyde's plans to deceive Calchas**
1415–21	**Narratorial affirmation of Criseyde's sincerity**
1422–49	Troilus' misgivings at Criseyde's plan (iv. 137–40)
1450–63	**Troilus quotes proverbs against Criseyde's plan**
1464–84	*Troilus' suspicions of Calchas* (iv. 141–2)
1485–91	**Troilus' jealousy of possible Greek suitors**
1492–526	*Troilus argues for flight* (iv. 143–5)
1527–37	Criseyde declares her true intent (iv. 146)
1538–54	**Criseyde swears fidelity in a series of oaths**
1555–61	[Criseyde warns Troilus against deserting Troy (cf. *Fil.* iv. 147–9)]
1562–8	**Criseyde anticipates loss of honour in changed circumstances**

1569–82	Criseyde anticipates criticism of Troilus and ruin to her honour (iv. 150–1) [Criseida argues that free possession would extinguish Troiolo's passion (iv. 152–3)]
1583–96	**Criseyde quotes proverbs**; urges Troilus to overcome Fortune; promises **by the moon** to return in ten days (iv. 154)
1597–603	[Troilus agrees, yet still urges flight (iv. 155–6)]
1604–52	Criseyde's further persuasions and pleas for fidelity (iv. 157–62)
1653–9	Troilus declares his faithfulness (iv. 163)
1660–6	**Criseyde's thanks; prays to requite him and behave with honour**
1667–87	*Criseyde* explains what qualities in *Troilus* inspired love (reversing *Filostrato*, where Troiolo tells Criseida (iv. 164–6))
1688–701	The lovers part at dawn (iv. 167); **the unimaginable distress of Troilus**

With the fourth part of *Filostrato* and the agreement of Criseyde's exchange, Boccaccio moves towards the events of the Troilus story that Chaucer would already know from Benoît and Guido. But in that fourth part of *Filostrato* Chaucer could find an intensely lyrical realization of the lovers' private feelings between the agreement and the occurrence of the exchange which did not correspond directly with the earlier texts. *Filostrato*'s text is here a sustainedly expressive representation of sorrow and lamentation, with monologue and dialogue building towards a dramatic continuity, linked by a connecting narrative. The length and detail of the above table reflects the close yet adaptive use Chaucer has made of *Filostrato*. Chaucer evidently appreciates the well-paced lyrical narrative of this part of *Filostrato*, in that he takes over Boccaccio's dramatic 'script' of speeches and narrative, but he constantly intervenes with modifications and additions to reinterpret and comment.

In both poems, after a summary narrative of the Trojans' unfortunate reverses in battle and the agreement to exchange Criseyde for Anthenor, the focus shifts to a lengthy scene of Troilus' grief, first alone, then visited by Pandarus. The focus then moves to Criseyde's grief, first alone, then visited by Pandarus, who leaves to find Troilus and arrange a meeting between the lovers, which takes place during the night before the day of Criseyde's exchange and forms the last scene of the book. First Troilus, then Criseyde, each in dialogue with Pandarus before meeting with each other: a sequence of scenes that makes Book IV echo the structure of the preceding three books. Chaucer has reproportioned his material to include a more balanced focus on both Criseyde and Troilus, and to build towards more of a climax at the close. As the table above reveals, the themes of Anthenor's treachery and Hector's loyalty to Criseyde are brought out by Chaucer's interpolations. The expressiveness of both lovers' grief and lamentation is heightened, and the sententiousness of Pandarus is amplified. Chaucer's most substantial departures are in making Pandarus discover Triolus in a temple complaining upon the inevitability of predestination, and then in much extending in the last scene Criseyde's arguments to Troilus for circumventing the inevitable exchange by her plan of leaving but then soon returning to Troy.

BOOK V

1–7	**The fatal destiny approaches**
8–14	**Three springs have passed since Troilus first loved Criseyde**
15–28	*Criseyde's and Troilus' distress before the exchange* (v. 1)
29–56	Troilus still thinks of abduction (v. 2–5)
57–63	Criseyde's sorrowful departure (v. 6)
	[Criseida scorns the gods and the Trojan authorities (v. 6–9)]
64–91	Troilus escorts Criseyde (v. 10–13)
92–189	**Diomede speaks of love to Criseyde**
190–6	Criseyde is greeted by Calchas (v. 14)
197–231	[Troilus returns to Troy; sorrows in his chamber (v. 15–21)]
	[Troiolo summons Pandaro, tells of his grief (v. 22–4), including . . .]
232–59	*Troilus' soliloquy; his terrible dreams* (v. 25–7)
260–6	**Troilus alternates between sorrow and optimism**
267–73	**Who could properly describe his sufferings?**
274–94	**A description of dawn;** *Troilus summons Pandarus* (cf. *Fil.* v. 22)
295–315	**Troilus gives instructions for his funeral**
316–22	**Troilus feels his dreams portend death**
323–57	*Pandarus argues for patient perseverance* (v. 29–31)
358–64	Pandarus dismisses the significance of dreams (v. 32)
365–92	**Pandarus expands on why dreams should be discounted**
393–511	The visit to Sarpedon (v. 33–49)
512–39	*They return, and go to see Criseyde's house* (v. 50–2)
540–60	*Troilus' address to Criseyde's empty house* (v. 53)
561–616	*Troilus sees memories at every place in Troy* (v. 54–9)
617–30	*He imagines he is talked about* (v. 60)

631–7	He likes to sing to himself of his woe (v. 61)
638–44	**Canticus Troili** [replaces Troiolo's song (v. 62–6)]
645–86	Troilus waits (v. 67–71)
687–8	Criseyde in the Greek camp (vi. 1)
689–707	**Criseyde's soliloquy of regret and apprehension**
708–65	Criseyde's unhappiness in the camp (vi. 1–7)
766–70	*Her change of heart anticipated* (vi. 8)
771–7	Diomede sets out to win her heart (vi. 8)
	[Diomede visits Criseida on the fourth day (vi. 9)]
778–98	Diomede's thoughts and soliloquy (vi. 9–11)
799–840	**'Portraits' of Diomede, Criseyde, and Troilus**
841–945	Diomede visits Criseyde on the *tenth* day; his conversation (vi. 12–25)
946–52	**Criseyde will allow Diomede to return next day**
953–94	Criseyde distances herself from love (vi. 26–31)
995–1008	**She permits him to return next day; declares her good intent**
	[Diomede is encouraged; his qualities described, which will weaken Criseida's resolve to keep faith to Troiolo (vi. 32–4)]
1009–50	**Summary of Diomede's progress in wooing**
1051–85	**Criseyde's soliloquy of regret and farewell**
1086–92	**Impossible to tell how soon Criseyde forsook Troilus**
1093–9	**Notoriety is punishment enough for Criseyde**
1100–232	Troilus waits (vii. 1–22)
1233–53	Troilus dreams of the boar; his distress (vii. 23–6)
	[Troiolo interprets the boar as Diomede (vii. 27–8)]
1254–74	Troilus laments Criseyde's infidelity; wishes to die (vii. 29–32)
	[Pandaro prevents Troiolo killing himself (vii. 33–9)]
1275–88	Pandarus discounts Troilus' dream (vii. 40–1)
	[Pandaro reminds Troiolo he could seek death in battle (vii. 42–7)]
1289–316	Troilus accepts Pandarus' advice to write to Criseyde (vii. 48–51)
1317–421	Troilus' letter [replacing Troiolo's (vii. 52–75)]
1422–35	**Summary of Criseyde's evasive reply**
1436–42	Troilus' woe increases; he takes to his bed (vii. 77)
1443–9	**Troilus can think of nothing but his dream**
	[His family visit Troiolo; Cassandra reproaches him for loving the low-born Criseida; they quarrel; Troiolo returns to the war; receives a disingenuous letter from Criseida (vii. 77–106)]

1450–6	**Troilus summons Cassandra to interpret his dream**
1457–519	**Cassandra links the boar with Diomede through the story of Meleager; outlines the war of the Seven against Thebes**
1520–40	*Troilus quarrels with Cassandra* (cf. *Fil.* vii. 89 ff.)
1541–7	**Fortune starts to divest Troy of its happiness**
1548–68	*The fatal death of Hector; Troilus' grief* (viii. 1)
1569–89	Troilus goes on hoping, and writing to Criseyde (viii. 2–4) [Criseida described as sending empty promises (viii. 5)]
1590–631	**Criseyde's letter**
1632–45	Troilus' doubts (viii. 6–7)
1646–66	Troilus sees his brooch on Diomede's 'cote-armure' (viii. 8–10)
1667–722	Troilus sends for Pandarus; laments Criseyde's infidelity (viii. 11–21)
1723–43	Pandarus' shame, and anger against Criseyde (viii. 22–4)
1744–64	*Troilus' sorrow; his encounters with Diomede* (viii. 25–6)
1765–71	**'Read Dares' for Troilus' martial exploits**
1772–85	**An address to ladies in the audience**
1786–92	**A farewell to the book**
1793–9	**Prayer that the text be correctly copied and understood**
1800–6	The death of Troilus at the hands of Achilles (viii. 27)
1807–27	**The ascent of Troilus' soul**
1828–41	Exclamation on Troilus' end; address to the young (cf. *Fil.* viii. 28–9) [Warning to young men about women (viii. 29–33)]
1842–8	**Commendation to love Christ**
1849–55	**Exclamations against pagans**
1856–69	**Dedication and concluding prayer** [In an envoy Boccaccio sends his poem on its way to his lady (ix)]

From the fifth Part of *Filostrato* onwards the events of Boccaccio's narrative converge with the traditional story from Benoît and Guido of the separation of Troilus and Criseyde. Here is the old tale of Criseyde's departure from Troy and Troilus to the Greek camp, the wooing of Diomede, Criseyde's acceptance of him, the sorrows of Troilus and his eventual death. But, although Chaucer here realigns his poem with tradition—supplementing *Filostrato* with episodes from the older version which Boccaccio had dropped—*Filostrato*'s focus on the sufferings of Troiolo in Troy, interrupted by one shift of scene to Criseida in the Greek camp, remains the underlying pattern in the single fifth book which Chaucer has run together out of the separate Parts 5–8 of *Filostrato* (see Table 5).

TABLE 5. *The structure of* Troilus, *Book V*

Filostrato		Troilus	
Part	Lines	Line ref.	No. of lines
5	568	15–686	671
6	272	687–1099	412
7	848	1100–1540	440
8	264	1541–764	244
		1800–6	
		1828–41	

The decisive restructuring has been in adjusting the relative lengths in *Troilus* of the portions corresponding to *Filostrato*, Part 6 (Criseida waiting in the Greek camp), and Part 7 (Troiolo waiting on the tenth day and later). In *Filostrato* this latter part devoted to Troiolo's plight attracts over three times the attention given to Criseida's circumstances in the Greek camp. But in the fifth book of *Troilus and Criseyde* Chaucer has so expanded the portion devoted to Criseyde in the camp, and so reduced that devoted to Troilus waiting, that he produces a more equally proportioned sequence of parallel scenes in the two settings. Moreover, each transition between the separate parts in his source has acted as a growth point for Chaucer's imagination, stimulating him to incorporate material in response to these junctures within the structure of his own narrative in Book V (e.g. 689–707, 1009–99, 1450 ff., 1541 ff.).

Chaucer's additions and changes alter the emphasis and meaning of his source. The inclusion from Benoît of 'sodeyn' Diomede's speech to Criseyde as they ride away from Troy, but the exclusion of Briseida's response, gives more attention to Criseyde's experience without defining or explaining her state of mind. A succession of additions emphasizes and enhances the lyric expressiveness of the waiting Troilus' helpless sorrow. The structure of Book V is made to centre climactically on the crucial tenth day: Chaucer's Criseyde is made to say and regret more in the Greek camp; Diomede's courtship and Criseyde's response are reordered and rewritten with much amplification, and with much modification of *Filostrato*'s more sympathetic account of Diomede. The purely familial, domestic occasion of Troiolo's meeting with his sister Cassandra gives Chaucer the cue for her appearance in her role as seer, interpreting Troilus' dream and contributing to that sense of a background in Trojan and Theban history and pagan belief which is so much part of Chaucer's reinterpretation of the Troilus story, as is the added alertness to the operation of fortune and fate (cf. v. 1–7, 1541–7). By this stage Criseyde is known only through the summary of an evasive letter (v. 1422–35) or the painful late letter that Troilus eventually receives (v. 1590–631). Both replace reports of letters in *Filostrato*, the Book V letters mirroring those in Book II, not least in that on both occasions Troilus' letter comes first and is some three times as long as the reply it draws from Criseyde. The whole manner of ending in *Troilus* transforms while still containing much of the ending of the eighth Part of *Filostrato*, as the table above reveals. In *Filostrato* the single stanza describing Troiolo's death (*Fil.* viii. 27) and the exclamations on that death ('cotal fine . . .' (*Fil.* viii. 28)) run on consecutively from the hero's furious encounters with Diomede (*Fil* viii. 25–6), and to compare how the English equivalents of those stanzas are now held within a structure of diverse materials and complex, shifting effects is to begin to see how far Chaucer has gone beyond his source to produce a multi-faceted ending which defies interpretation from any single viewpoint.

Boccaccio: *Il Filostrato*, ed. Vittore Branca, in *Tutte le opere di Giovanni Boccaccio*, ed. Vittore Branca, ii. (Verona, 1964). For an English translation, see Gordon (ed. and trans.), *The Story of Troilus*, and Havely (trans.), *Chaucer's Boccaccio*. For a text of *Filostrato* in parallel with *Troilus*, see Windeatt (ed.), *Troilus and Criseyde*.

On the sources of *Filostrato*, see M. Gozzi, 'Sulle fonte del *Filostrato*: Le narrazioni di argomento troiano', *Studi sul Boccaccio*, 5 (1968), 123–209. In addition to Benoît de Sainte-Maure's *Roman de Troie* and Guido de Columnis's *Historia destructionis Troiae*, Gozzi considers the *Roman de Troie en prose*, Binduccio dello Scelto's Italian version of this, the *Libro della storia di Troia*, the *Istorietta Troiana*, and the *Romanzo barberiniano*. See E. Gorra (ed.), *Testi inediti di storia Troiana* (Turin, 1887). On the place of *Filostrato* in the development of the Troilus story, see Natali, 'A Lyrical Version'.

On the date of *Filostrato*, see A. Balduino, *Boccaccio, Petrarca, e altri poeti del Trecento* (Florence, 1984), 243–7; P.G. Ricci, 'Per la dedica e la datazione del *Filostrato*', *Studi sul Boccaccio*, 1 (1963), 333–47; and L. Surdich, *La cornice di Amore: Studi sul Boccaccio* (Pisa, 1987), 107. The prose proem to *Filostrato* was formerly taken to indicate that the poem was written to be sent to Boccaccio's mistress, Maria d'Aquino, illegitimate daughter of King Robert of Naples, during her absence from Naples. Along with much other information, the Introduction to the text of *Filostrato* with parallel translation by Griffin and Myrick, *The Filostrato of Giovanni Boccaccio*, sets out this 'autobiographical' interpretation, which more recent Italian scholarship rejects as a misreading of Boccaccio's deployment of a persona in an elaborate self-fiction. See Vittore Branca, 'Giovanni Boccaccio: Profilo biografico', in *Tutte le opere*, i (1967), 2–197; 'Schemi letterari e schemi autobiografici', in *Boccaccio medievale* (3rd edn., Florence, 1970), 191–249; *Boccaccio: The Man and His Works*, trans. R. Monges (New York, 1976); *Giovanni Boccaccio: Profilo biografico* (Florence, 1977); Carlo Muscetta, *Giovanni Boccaccio* (2nd edn., Bari, 1974). On the literary context of *Filostrato* in Angevin Naples, see the introduction to Havely, *Chaucer's Boccaccio*, and Wallace, *Chaucer and the Early Writings of Boccaccio*, ch. 5.

On Chaucer's possible use of the *Roman de Troilus et de Criseida* by Beauvau, see Pratt ('Chaucer and *Le Roman de Troyle et de Criseida*') who presents a list of suggested parallels with *Troilus*, in most cases confined to single words or phrases. For a text of the *Roman*, see L. Moland and C. d'Hericault (eds.), *Nouvelles françoises en prose du XIVe siècle* (Paris, 1858). Uncertainty surrounds the identity of the French translator and when he lived and wrote. Beauvau has usually been identified with one of two seneschals of Anjou, either Pierre de Beauvau (*c.* 1380–?1436) or his son Louis (?1417/18–1462), and the text is generally attributed to the fifteenth century: see Carla Bozzolo, *Manuscrits des traductions françaises d'oeuvres de Boccacce* (Padua, 1973), 29–33; Peter Rickard (ed.), *Chrestomathie de la langue française au quinzième siècle* (Cambridge, 1976), 89. In that Beauvau's translation is generally a very faithful rendering of Boccaccio's text, it is necessarily difficult to distinguish any convincing indebtedness by Chaucer to his version rather than to Boccaccio's original. In that English is closer to French than Italian in diction and phrasing, it is inevitable that the language used by Beauvau and Chaucer in translating the Italian poem will sometimes coincide. Chaucer's reliance on French translations in some of his other works is established, and, unless the date of the *Roman de Troilus* can be finally settled, the question will remain open as to whether Chaucer used it. The more important points are that Chaucer could learn very little from the Beauvau text, and that his primary debt to *Filostrato* is evident. For a re-examination and documentation of the question, see Hanly, *Boccaccio*.

On Chaucer's use of *Filostrato* in *Troilus*, see Benson, *Chaucer's 'Troilus and Criseyde'*; Lewis, 'What Chaucer really did'; Meech, *Design*; Wallace, *Chaucer and the Early Writings of Boccaccio*; and Windeatt, 'The "Paynted Proces"', 'Chaucer and the *Filostrato*', and (ed.), *Troilus and Criseyde*. Lewis's account, in 'What Chaucer really did', of how Chaucer approached his work on *Filostrato* as a 'historial' poet, a pupil of the rhetoricians, a poet of 'doctryne and sentence', and of 'courtly love', underlies much subsequent interpretation of the relation between *Filostrato* and *Troilus*.

Dares and Dictys

But the Troian gestes, as they felle,
In Omer, or in Dares, or in Dite,
Whoso that kan may rede hem as they write.   (i. 145–7)

For medieval writers the authorities on the Trojan War were the *Ephemeridos belli Troiani libri* (A Journal of the Trojan War) by Dictys Cretensis, and the *De excidio Troiae historia* (The Fall of Troy, A History) by Dares Phrygius, which both present themselves as eye-witness accounts by participants at the siege of Troy, and which Chaucer claims to have consulted: 'Thus writen they that of hire werkes knewe' (iv. 1421). Both works are thought to have Greek originals written in the first century AD, but the extant Latin version of Dictys dates from the fourth century and that of Dares from the sixth century AD. As eye-witnesses, Dares and Dictys were felt in the Middle Ages to have the advantage over Homer, who had written later and who was anyway suspected of favouring the Greeks by a Latin West that had inherited the Romans' view of themselves as descended from the Trojans. In Dares and Dictys the roles of some protagonists differ from Homer's account. Minor characters in Homer (such as Troilus) are given more importance, while Achilles becomes lovesick and treacherous. In the *Iliad* Antenor and Aeneas play an honourable part, while in Dares they assume Sinon's role in betraying Troy, a tradition which Chaucer echoes in *Troilus*.

Indeed, although the first book of *Troilus* refers the audience to the background of its story 'In Omer, or in Dares, or in Dite', it is probable that Chaucer reveals the text he knew best when the fifth book gives a reference solely to Dares for a fuller account of the military career of Troilus ('His worthi dedes, whoso list hem heere, | Rede Dares ...' (v. 1770–1)). Not only does the supposedly Cretan Dictys support the Greek view of the Trojan War, but he only mentions Troilus once—in referring to his death, when Achilles orders his throat to be cut—and in Dictys' account Calchas is a Greek priest. By a happy chance, the papyrus fragment which preserves the only extant scrap of the original Greek text of Dictys—on the back of income tax returns for the year 206 AD— consists precisely of this reference to the death of Troilus ('No small grief came upon the Trojans when Troilus died, for he was still young and noble and handsome').

It was the *De excidio* of Dares which proved more popular in the medieval West. The author presents himself as 'Dares the Phrygian' (perhaps to be identified with the priest of Vulcan mentioned in the *Iliad*, v. 9), and the influential encyclopædist, Isidore of Seville, commented: 'Just as Moses was the first to write a Christian history from the beginning of the world, so Dares the Phrygian was in fact the first to set down a pagan history treating the Greeks and Trojans.' Such is the authority that Chaucer invokes. Dares claims the authority of one who fought at Troy, and his survival to write his history is explained in the last chapter by his association with Antenor, one of the betrayers of Troy. Although the Latin Dares is much briefer than the Latin Dictys, and so terse and spare as to suggest the style of an abridgement, some of the elements of the Troilus story as it was subsequently to be developed make their first appearance here. It is in Dares that we find descriptions of the appearance of the principal characters. In a list of descriptions, first of the Trojans and then of the Greeks, Dares includes descriptions of the appearance of Troilus and of Diomedes, and includes Briseis at the end of the list of Greeks. That Dares describes Briseis more fully than Helen of Troy—with no apparent reason, for she plays no part in his narrative—may perhaps have prompted the later invention of a story about her. It is also Dares who tells how Troilus in one of his encounters wounds Diomedes (ch. 31), and who relates how the Trojan priest Calchas meets Achilles at Delphi and—at the instruction of the oracle—throws in his lot with the Greeks (ch. 15). Although Dares does not develop Troilus as a character in love, he does show how Achilles is struck with love for Polyxena, and Troilus emerges into some prominence among Priam's sons.

The first part of Dares, being written from the Trojan point of view, is concerned to find the blame for the Trojan War in events preceding the abduction of Helen, in Jason's mission for the Golden Fleece and the Argonauts' abduction of Hesione. Priam, her brother, sends Antenor as an envoy to Greece to request her return, and when this is rejected he summons a council of his sons and allies to discuss war. At this council Troilus, who, although youngest of Priam's sons, is said to equal Hector in bravery, urged them to war and told them not to be frightened by Helenus' fearful words.

After the abduction of Helen by Paris the Greeks collect their forces and it is here that Dares provides his descriptions of the main Trojan and Greek figures, including those of Troilus, Diomedes, and Briseis. The description of Troilus comes among those of Priam's sons:

Troilum magnum, pulcherrimum, pro aetate valentem, fortem, cupidum virtutis. (ch. 12)

Troilus was tall and handsome, strong for his age, brave, and desirous of glory.

The description of Diomedes among the Greek leaders comes between those of Ulysses and Nestor:

Diomedem fortem, quadratum, corpore honesto, vultu austero, in bello acerrimum, clamosum, cerebro calido, inpatientem, audacem. (ch. 13)

Diomedes was strong, squarely built, handsome of body, austere in looks, fierce in war, clamorous, hot-tempered, impatient, and daring.

The description of Briseis is the only female figure to be included among the Greeks, although Hecuba, Andromache, Cassandra, and Polyxena are included in the Trojan list:

Briseidam formosam, non alta statura, candidam, capillo flavo et molli, superciliis iunctis, oculis venustis, corpore aequali, blandam, affabilem, verecundam, animo simplici, piam. (ch. 13)

Briseis was beautiful, not tall, blond, with soft golden hair, joined eyebrows, lovely eyes, and a well-proportioned body; she was gentle, affable, modest, simple in spirit, and pious.

After the defection of Calchas at Delphi, the Greeks arrive before Troy, and fighting begins and continues fiercely. Hector is eventually killed by Achilles. When Priam, Hecuba, Polyxena, and other Trojans visit his tomb on the first anniversary of his death, they chance to meet Achilles, who is struck by Polyxena's beauty, falls madly in love, and withdraws from fighting the Trojans. In fierce battles Troilus now emerges as the preeminent Trojan warrior, slaughtering many Greek leaders. When Menelaus argues that the Trojans cannot replace Hector, Diomedes and Ulysses answer that Troilus is the bravest of men and the equal of Hector (ch. 30).

In further battles Troilus wounds Menelaus and leads the Trojan forces with Paris; he wounds Diomedes, and Agamemnon himself (ch. 31). When the Greeks request a truce, it is Troilus who argues against allowing such a long one, but is outvoted. When the war is resumed, Troilus, fighting in the first ranks, several times puts the Myrmidons to flight, and eventually makes the Greeks flee in such confusion that Achilles re-enters the fighting, although he is quickly wounded at the hands of Troilus and has to withdraw. But the death of Troilus at the hands of Achilles is now near. Achilles draws up his Myrmidons and urges them to make an attack against Troilus. Towards the end of the day Troilus advances on horseback, exultant, and the Greeks flee with loud cries. The Myrmidons, however, launch an attack against Troilus. Troilus slays many men, but, in the midst of terrible fighting, his horse is wounded and falls, entangling and throwing him off—and swiftly Achilles is there to despatch him (ch. 33). Achilles tries to drag Troilus' body away with him, but is prevented and forced to withdraw. The Trojans have no great success after the death of Troilus, and Antenor and Aeneas eventually negotiate with the Greeks to betray the city in return for the guaranteed safety of themselves and their families, so that the fall of Troy is here very much seen in connection with that treachery of Antenor to which *Troilus* is to refer.

In including the 'portraits' of Diomede, Criseyde, and Troilus in his Book V Chaucer is thus as a historian collecting up and incorporating into his version what would be thought of as the earliest documentary evidence for the story. Yet, while the portraits in *Troilus*, Book V, stem ultimately from the idea of the series of portraits in the *De excidio*, Chaucer actually bases the text of his portraits on the rather fuller version of Dares, the *Frigii Daretis Ilias* (The Iliad of Dares the Phrygian), by the Anglo-Latin poet Joseph of Exeter (*c.* 1185). Joseph's work was often taken to be 'Dares', and when Chaucer recommends 'Rede Dares . . .' it was probably Joseph's *Yliados* that he had in mind. Chaucer's description of Diomede (v. 799–805) closely follows Joseph's portrait:

> Voce ferox, animo preceps, fervente cerebro
> Audentique ira validos quadratur in artus
> Titides plenisque meretur Tidea factis,
> Sic animo, sic ore fero, sic fulminat armis. (*Ilias*, iv. 124–7)

His voice was fierce, his temper violent. His brains boiled and his rage was daring; his limbs were massive and he stood four-square. His mighty deeds made him the worthy son of his father, Tydeus—such were the lightning bolts leaping from his spirit, his savage voice, and his arms. (Roberts, 43)

Joseph of Exeter's description of Briseis similarly underlies Chaucer's portrait of Criseyde (v. 806–26):

In medium librata statum Briseis heriles
Promit in affectum vultus. Nodatur in equos
Flavicies crinita sinus, umbreque minoris
Delicias oculus iunctos suspendit in arcus.
Diviciis forme certant insignia morum,
Sobria simplicitas, comis pudor, arida numquam
Poscenti pietas et fandi gracia lenis.   (*Ilias*, iv. 156–62)

Briseis was of medium height, and displayed a noble countenance. Her golden hair was plaited into coils of equal length. Her eye suspends in a joined arch the delights of a lesser shade. The riches of her beauty were rivalled only by the excellence of her character—sober simplicity, courteous modesty, never-failing compassion, and a kindly and gentle manner of speech. (Roberts, 43)

Indeed, one of the two *Troilus* manuscripts which includes the relevant lines from Joseph as a marginal commentary on the text of Chaucer's portraits (St John's College, Cambridge, MS L1) actually contains a variant reading of Joseph which—if it occurred in Chaucer's copy of the Latin—would seem to account for a difference between Joseph's Briseis and Chaucer's Criseyde. Where Joseph wrote that 'the marks of her character (*morum*) strive with the riches of her beauty', by scribal error this could become *amorum*, as it does in the St John's manuscript, a reading that seems to lie behind Chaucer's lines:

And with hire riche beaute evere more
Strof love in hire ay, which of hem was more.   (v. 818–19)

As Chaucer's Criseyde is later to say, 'thise bokes wol me shende' (v. 1060). Finally, Joseph's description of Troilus informs Chaucer's (v. 827–40):

Troilus in spacium surgentes explicat artus,
Mente Gigas, etate puer nullique secundus
Audendo virtutis opus, mixtoque vigore
Gratior illustres insignit gloria vultus.   (*Ilias*, iv. 61–4)

Troilus was broad and tall. In spirit he was a giant, but in age he was a boy. He was second to none in venturing upon brave deeds. Pride graced his noble features, more pleasing because it was blended with manly vigour. (Roberts, 41)

In Joseph's text Chaucer could discern the outline derived from Dares of Troilus' career as a great Trojan warrior, although Joseph's floridly full style gives more emphasis to the role of Troilus. Joseph's Troilus is spoken of as the equal of Hector: 'Troilus, the champion of the entire army, and Hector of the flashing helmet, his equal in the rush of battle, his equal in hitting the ranks like a hurricane, worked hard to outdo each other in glory . . .' (Roberts, 40); 'Troilus' heart burned within him to attack Achilles himself . . .' (p. 61). Joseph's rhetoric works to build up the military glory of Troilus to a crescendo just before the account of his death:

Alexander the Great, who came after him, and Tydeus, who came before, both had hands as deadly as his, but Troilus was mightier than either and outclassed both . . . He was greater than Hector . . . Blazing Troilus so renewed the slaughters of his brother that a great many in their bewilderment believed that Hector had been born again . . . Troilus now towered above them all, more savage than the rest, and challenging the Greeks with stronger right arm . . . Diomedes received his wound-imparting sword and roared out in distress. (Roberts, 69–71)

But eventually, when attacking the Myrmidons, Troilus' horse is wounded and throws him, whereupon Achilles is able to behead Troilus, whose death is marked by Joseph with a rhetorical lament. Although there is more than a hint of foolhardy rashness in the youthful Troilus as portrayed by Joseph of Exeter, from this 'Dares' Chaucer would have formed the impression of Troilus as a Trojan hero of very great worth and dignity. It is this epic figure that Chaucer can invoke as another aspect of his hero through his allusions to the 'Dares' of Joseph in Book V of the *Troilus*:

> His worthi dedes, whoso list hem heere,
> Rede Dares . . . (v. 1770–1)

Dares: *Daretis Phrygii de excidio Troiae historia*, ed. F. Meister (Leipzig, 1873).
Dictys: *Dictys Cretensis ephemeridos belli Troiani libri*, ed. W. Eisenhut (Leipzig, 1958).
The Trojan War: The Chronicles of Dictys of Crete and Dares the Phrygian, ed. and trans. R. M. Frazer, jun. (Bloomington, Ind., and London, 1966); for the Greek fragment of Dictys, see p. 8.
Joseph of Exeter: *Daretis Phrygii Ilias*, in *Briefe und Werke*, ed. L. Gompf (Leyden and Cologne, 1970). See also *The Iliad of Dares Phrygius*, trans. G. Roberts (Cape Town, 1970), and for the identification of Joseph as the source of the *Troilus* portraits, see Root, 'Chaucer's Dares'.
On the figure of Troilus in 'Dares' and Dictys, see Brown, 'A Separate Peace'.

Benoît de Sainte-Maure, Le Roman de Troie

It is in the *Roman de Troie* (*c.* 1155–60) of Benoît de Sainte-Maure that the story of the love of Troilus for Briseida, and his replacement by Diomedes, is first found. The taste represented by Benoît's *roman d'antiquité*—along with the *Roman de Thèbes*, the *Roman d'Eneas*, and the *Brut*—has been located in the historical interests of the Angevin court of Henry II. The underlying structure of Benoît's spacious romance of 30,316 lines is still based on Dares, although Benoît does make use of Dictys later in his work. Benoît's 'romance of Troy' begins with the adventure of the Golden Fleece, the abduction of Hesione, and the first destruction of Troy. It continues with Antenor's embassy to Greece, the abduction of Helen, the meeting of Achilles and Calchas at Delphi, and the latter's defection to the Greeks. The Greeks arrive before Troy, and the romance develops a lengthy series of battles and truces which are carefully numbered, although the pattern is still based on Dares: the death of Hector, the love of Achilles for Polyxena, the wounding of Diomedes by Troilus, the killing of Troilus by Achilles, and so on.

It is Benoît who develops three love-affairs interlinked with his story of the Trojan War: Jason and Medea, Achilles and Polyxena, and Troilus and Briseida. The romance of Achilles' love for Polyxena is immeasurably enriched from the bare account in Dares. But it is Benoît who also extends the story of Jason to include his affair with Medea, before the story of Paris and Helen is reached. The story of the love of Troilus for a daughter of the traitor Calchas is dovetailed with the history of the Trojan War so that the existence of their love is only acknowledged when Briseida is to be sent from Troy to join her father. In inventing his story Benoît effectively assigned Briseida her original Homeric role of a woman passed involuntarily from one man to another, although Benoît's Briseida has no connection—other than through the misunderstanding of a name —with Briseis, the concubine of Achilles, or with Chryseis, the daughter of Chryses, priest of Apollo, who figure in the *Iliad*. It would seem that Benoît did not recognize 'Briseida' in Dares' list of portraits as a patronymic of Hippodamia, daughter of Brises, the slave girl whom Agamemnon took from Achilles. Dares does not include the story of Hippodamia, whom he names only as 'Briseida'. Benoît apparently did not recognize that Hippodamia, whose story he took over from Dictys, was that Briseida whom he found described in Dares, and hence it would seem that he felt free to make her the object of that rivalry between Troilus and Diomedes already suggested in Dares.

In the *Roman de Troie* Benoît introduces the love of Troilus and Briseida at the point where love is to be placed under keenest pressure of external circumstance in enforced separation and distance. From its earliest extant appearance the story of Troilus' and Criseyde's love is presented within the context of its painful disappointment, its lack of enduringness, and its ending. Benoît's gift for contriving courtly but natural and graceful speech is at its best in presenting the grief of Troilus and Briseida at their parting, the wooing speeches of Diomedes, and the reflective soliloquies of Briseida in her isolated predicament. In the role of Diomede as Troilus' rival Benoît cleverly develops the potential for a larger animosity which was available in the detail in Dares that Diomede was among the Greeks wounded by Troilus.

The love of Troilus and Briseida in the *Roman de Troie* is developed against the background of the tradition inherited from Dares of Troilus as one of the most formidable warriors of Troy. But Benoît carefully establishes the moral character as well as the prowess of Troilus. When Troilus brushes aside the caution of Helenus at the Trojan council before the abduction of Helen he is described as 'no less brave than Hector was' (3992; cf. 'Save Ector most ydred of any wight' (iii. 1775)). The feats of Troilus' chivalrous prowess in successive battles are recorded by Benoît: in the third battle Troilus leads the Trojan army with Hector, and here and in the fourth battle he fights with Diomedes. In the eighth battle Troilus is unhorsed by Diomedes, who takes away his horse and sends it to Briseida, although in the same battle Polydamas gains Diomedes' horse and presents it to Troilus. In the tenth battle each of them wounds the other, and it is here too that Hector meets his end ('Mout le regrete Troïlus, | Quar rien soz ciel n'amot il plus' (*Troie*, 16399–400) (Troilus greatly mourned him, for he loved him more than anything on earth): cf. *Troilus*, v. 1562–5). After more martial successes, in the fifteenth battle Troilus wounds Diomedes seriously and addresses scornful words to his rival about Briseida as he is carried from the battlefield; he also wounds Agamemnon. In the sixteenth battle ('Trop est sa force redotee' (*Troie*, 20466) (His strength is greatly feared)) he takes one hundred knights prisoner. In the seventeenth battle Troilus does marvels; in the eighteenth he presses the Myrmidons back to their tents, and he and Achilles inflict wounds on each other. In the nineteenth battle Troilus is surrounded by the Myrmidons and slain by Achilles, who ties the body to his horse's tail.

In Benoît's account Troilus is thus among the most pre-eminent of knights. This is the understood background to the story of the love of Troilus, which Chaucer is evidently summoning up by his brief allusions to Dares and to that military career of Troilus which continues off-stage from the action of *Troilus and Criseyde* ('Ther nas but Grekes blood —and Troilus . . . He was hire deth, and sheld and lif for us . . . Whil that he held his blody swerd in honde' (ii. 198–203)). Criseyde's sight of Troilus returning triumphantly from battle, having entered by the gate of Dardanus (ii. 610 ff.) may well have been suggested to Chaucer by Benoît's account of the wounded Hector's entry into Troy amid acclamations after the second battle (*Troie*, 10201–18), as well as by Benoît's account of Dardanus among the six gates of Troy (*Troie*, 7675–86).

It is into this context of the continuing Trojan War that Benoît sets the story of the changing love of Briseida for Troilus then Diomedes, and the refinement and psychological delicacy of his treatment is foreshadowed in his handling of the list of portraits of the principals at the opening of the *Roman de Troie* which he takes over and transforms from Dares. The descriptions of both Diomedes and Briseida develop significantly in detail beyond Dares' account:

> Forz refu mout Diomedès,
> Gros e quarrez e granz adès;
> La chiere aveit mout felenesse:
> Cist fist mainte fausse pramesse.
> Mout fu hardiz, mout fu noisos,
> E mout fu d'armes engeignos;
> Mout fu estouz e sorparlez,
> E mout par fu sis cors dotez.
> A grant peine poëit trover
> Qui contre lui vousist ester:
> Rien nel poëit en pais tenir,
> Trop par esteit maus a servir;
> Mais por amor traist mainte feiz
> Maintes peines e mainz torneiz. (*Troie*, 5211–24)

Diomedes was very powerful, large, squarely built and very tall; his look was very fierce and he made many a false promise. He was very bold, very quarrelsome, and very cunning in war. He was most proud and arrogant, and was greatly feared. It would have been very hard to find anyone who would willingly oppose him; nothing could restrain him, and he was a very bad man to serve. But for love's sake he many times endured many pains and many conflicts.

> Briseïda fu avenant:
> Ne fu petite ne trop grant.
> Plus esteit bele e bloie e blanche
> Que flor de lis ne neif sor branche;
> Mais les sorcilles li joigneient,
> Que auques li mesaveneient.
> Beaus ieuz aveit de grant maniere
> E mout esteit bele parliere.
> Mout fu de bon afaitement
> E de sage contenement.
> Mout fu amee et mout amot,
> Mais sis corages li chanjot;
> E si ert el mout vergondose,
> Simple e aumosniere e pitose. (*Troie*, 5275–88)

Briseida was charming; she was neither short nor too tall. She was lovelier, fairer, and whiter than a lily or than snow upon a branch; but her eyebrows were joined, which somewhat misbecame her. She had very beautiful eyes, and was very eloquent and well-mannered and wise in conduct. Greatly was she loved, and greatly did she love, but her heart was changeable. And she was very timid, modest, generous, and compassionate.

But the exceptionally long and detailed description of Troilus distinguishes him from most other characters in Benoît's list of portraits by a lavishing of attention on his potential as a hero of romance, which complements the extensiveness of the poem's attention to him as a knight:

> Troïlus fu beaus a merveille;
> Chiere ot riant, face vermeille,
> Cler vis apert, le front plenier:
> Mout covint bien a chevalier.
> Cheveus ot blonz, mout avenanz
> E par nature reluisanz,
> Ieuz vairs e pleins de gaieté:
> Onc ne fu rien de lor beauté.
> Tant come il ert en bon talent,
> Par esguardot si doucement,
> Que deliz ert de lui veeir;
> Mais une rien vos di por veir,
> Qu'il ert envers ses enemis
> D'autre semblant e d'autre vis.   (*Troie*, 5393–406)

There follow twenty lines on the fine proportions of his face and body, then:

> Jo ne cuit or si vaillant home
> Ait jusque la ou terre asome,
> Qui tant aint joie ne deduit,
> Ne meins die qu'a autre enuit,
> Ne qui tant ait riche corage,
> Ne tant coveit pris ne barnage.
> Ne fu sorfaiz ne outrajos,
> Mais liez e gais e amoros.
> Bien fu amez e bien ama,
> E maint grant fais en endura.
> Bachelers ert e jovenceaus,
> De ceus de Troie li plus beaus
> E li plus proz, fors que sis frere
> Hector, qui fu dreiz emperere
> E dreiz sire d'armes portanz:
> Bien nos en est Daires guaranz.
> Flor fu cil de chevalerie,
> E cist l'en tint mout bien frarie;
> Bien fu sis frere de proëce,
> De corteisie e de largece.   (*Troie*, 5427–46)

Troilus was marvellously handsome; he had a cheerful expression, a rosy complexion, a clear open countenance, a broad forehead, and had all that was appropriate to a knight. He had blond hair, very becoming and naturally shining, fair eyes full of gaiety—there was never beauty like theirs. As long as he was well disposed, his look was so gentle that it was a delight to behold him—but one thing

I can tell you truly: he had a different appearance and expression for his enemies ... I do not think there can now be throughout the whole world such a valiant man, who so loves joy and delight, or who says so little to give offence to others, or who is so magnanimous or so desires fame and deeds of honour. He was neither haughty nor insolent, but cheerful, gay, and loving. He was much loved, and he himself loved much, and endured greatly for that. He was the handsomest and the most valiant knight among the youth of Troy, except for his brother Hector who was by right Emperor and by right lord of the conduct of arms, as Dares assures us. The latter was the flower of chivalry, and Troilus was most worthy to be his brother, and was well fitted to be, by virtue of his valour, courtliness, and generosity.

Benoît's story of Troilus and Briseida is a skilfully interlinked sequence of episodes of love and chivalry within his 'Romance of Troy'. Such interlacing of episodes with a wider story over a lengthy text allows the suspense of separation and of a new courtship to focus the characteristic concerns of romance, their tensions and difficulties. Even before Benoît relates the love of Troilus and Briseida he has established the enmity of Troilus and Diomedes, and the greater chivalry of Troilus: in the third battle Diomedes has gained Troilus' horse, and ignobly goes on attacking the unmounted Troilus; in the fourth battle, by contrast, Troilus chivalrously dismounts to continue fighting the unhorsed Diomedes ('Sor lui descendi Troïlus, | Qu'il haï tant qu'il ne pot plus' (*Troie*, 11285–6)) (Troilus dismounted to him, whom he hated such that he could not hate more)).

After Thoas is taken prisoner in the fourth battle, and Antenor in the fifth, an exchange of the two is arranged, and Calchas asks the Greek princes to request the transfer from Troy of his daughter Briseida. Priam fulminates against the traitor (*Troie*, 13107 ff.), commenting 'if the girl had not been noble, good, wise, and beautiful, she would have been burnt and dismembered because of him' (cf. 'And seyden he and al his kyn at-ones | Ben worthi for to brennen, fel and bones' (*Troilus*, i. 90–1)). However, Priam agrees to send Briseida, and it is here that Benoît first introduces the subject of Troilus' love for Briseida, when they have only one night left together and an indefinite prospect of separation, for there is no talk in the *Roman de Troie* of Briseida's attempting to steal back to Troy:

> Troïlus ot ire e tristece:
> Ço est por la fille Calcas,
> Quar il ne l'amot mie a guas.
> Tot son cuer aveit en li mis;
> Si par ert de s'amor espris
> Qu'il n'entendeit se a li non.
> El li raveit de sei fait don
> E de son cors e de s'amor:
> Ço saveient tuit li plusor.   (*Troie*, 13262–70)

Troilus raged and grieved for Calchas' daughter, for his love for her was no light matter. He had set his whole heart upon her and was so possessed with love for her that he had no thought for anything else. She had given him herself, her body and her love, as most people knew.

Troilus and Briseida spend together one last brief night of love and sorrow, and on the morrow Briseida is escorted out from Troy by Troilus and other sons of Priam. At parting they pledge to be true to each other, and Briseida is escorted away by Diomedes and other Greek leaders ('Qui ainz en soferra grant peine | Qu'il sol la baist ne qu'il i gise' (*Troie*, 13530–1) (He will have to suffer much before he gives her a single kiss—let alone sleeps with her ...)).

The exchanges between Diomedes and Briseida as they ride from Troy towards the Greek camp (a scene without real equivalent in Guido's *Historia* or Boccaccio's *Filostrato*) is a section of Benoît's *Roman* upon which Chaucer has drawn closely in Book V of *Troilus*, while discarding the moralization with which the scene is accompanied in the *Roman de Troie*. Benoît stresses Briseida's affliction on leaving Troy, but this is then followed by a passage of narratorial anticipation:

> Mais, se la danzele est iriee,
> Par tens resera apaiee;
> Son duel avra tost oblïé
> E son corage si müé
> Que poi li iert de ceus de Troie.
> S'ele a hui duel, el ravra joie
> De tel qui onc ne la vit jor:
> Tost i avra torné s'amor,
> Tost en sera reconfortee;
> Femme n'iert ja trop esgaree.   (*Troie*, 13429–38)

Yet although the lady is afflicted, in time she will be restored to peace again. She will soon have forgotten her sorrow, and her heart will be so altered that the Trojans will mean little to her. If today she has sorrow, she will have joy again from one who has never yet seen her. Soon she will have changed her love, and soon she will be comforted by him. A woman will never be too downcast.

Although Chaucer does build in some anticipation (cf. v. 27–8), he has entirely dropped the explicit antifeminist moral that Benoît draws from the example of Briseida:

> A femme dure dueus petit:
> A l'un ueil plore, a l'autre rit.
> Mout müent tost li lor corage.
> Assez est fole la plus sage.   (*Troie*, 13441–4)

A woman's grief lasts a short while: one eye weeps, the other smiles. Their hearts change very rapidly, and the wisest of them is foolish enough.

Even before he relates Briseida's meeting with Diomedes, Benoît inserts a digression on feminine inconstancy.

From the *Roman de Troie* Chaucer borrows the idea of Diomede's address to Criseyde, and also some of his arguments and phrases as they ride to the Greek camp (v. 99–189). In Diomede's polite enquiry as to 'whi she stood | In swich disese' and his offer of anything he can do to relieve her sorrow (v. 106–19), Chaucer is recalling but modifying Diomedes' more direct offer of his love in two passages in *Troie* (13539–43, 13604–10). In Benoît's version Diomedes also shows understanding for Briseida's sense of unfamiliarity among the Greeks (*Troie*, 13546–51; cf. *Troilus*, v. 120–3). Diomede's declaration 'I loved never womman here-biforn' (v. 155–8) recalls and fuses two similar declarations in *Troie* (13557–8, 13591–8), just as his commitment to love and defence of the possibility of love at first sight draw together different parts of the wooing speech of Benoît's Diomedes:

> 'Mais j'ai oï assez parler
> Que gent qu'onc ne s'erent veü
> Ne acointié ne coneü
> S'amoënt mout, ç'avient adès.'   (*Troie*, 13552–5)

> 'Quant Amors vueut qu'a vos m'otrei,
> Nel contredi ne nel denei:
> A son gre e a son plaisir
> Li voudrai mais dès or servir.'   (*Troie*, 13691–4)

> 'Ci sont tuit li preisié del mont
> E li plus riche qui i sont,
> E li plus bel e li meillor,
> Qui vos requerront vostre amor.
> Se de mei faites vostre ami,
> Vos n'i avreiz se honor non.'   (*Troie*, 13575–81)

> 'Leial ami e dreiturier
> Vos serai mais d'ore en avant
> A toz les jorz de mon vivant.'   (*Troie*, 13588–90)

'But I have often heard tell that people who have never seen or become acquainted or known to each other have loved each other very much—that often happens ... Since Love will have me yield to you, I shall neither gainsay nor deny him, and his will and his pleasure I wish from now on to serve. ... In this place are the most celebrated men in the world, the richest, fairest, and best, who will entreat you for your love ... If you take me as your lover you will have nothing but honour from it

... A loyal and honourable friend to you I shall be henceforth for all the days of my life.'

> 'And wondreth nought, myn owen lady bright,
> Though that I speke of love to yow thus blyve;
> For I have herd er this of many a wight,
> Hath loved thyng he nevere saigh his lyve.
> Ek I am nat of power for to stryve
> Ayeyns the god of Love, but hym obeye
> I wole alwey; and mercy I yow preye.
>
> 'Ther ben so worthi knyghtes in this place,
> And ye so fayr, that everich of hem alle
> Wol peynen hym to stonden in youre grace.
> But myghte me so faire a grace falle,
> That ye me for youre servant wolde calle,
> So lowely ne so trewely yow serve
> Nil non of hem as I shal til I sterve.'   (v. 162–75)

But in the *Roman de Troie* Briseida then replies to Diomedes with a lengthy speech (13619–80), which in its equivocation contrasts painfully with her recent declarations to Troilus, and in its poised self-possession contrasts with her recent prostration at the thought of leaving Troy. Briseida says it is not the time to promise love, in which many women are anyway deceived; she has just left home and friends, and it would be neither fitting nor prudent to have an affair in an armed camp. But she concludes not unencouragingly:

> 'Soz ciel n'a si riche pucele
> Ne si preisiee ne si bele,
> Por ço que rien vousist amer,
> Que pas vos deüst refuser:
> Ne jo nos refus autrement.
> Mais n'ai corage ne talent
> Que vos ne autre aim aparmains;
> Si poëz bien estre certains,
> S'a ço me voleie aproismier,
> Nul plus de vos n'avreie chier.
> Mais n'en ai pensé ne voleir,
> Ne ja Deus nel me doint aveir!'   (*Troie*, 13669–80)

'There is no lady under heaven so rich, or of such name, or so beautiful, that could refuse you, if she were at all inclined to love—and I am not refusing you either. But I have neither inclination nor desire to love you or anyone else at present, although you may be assured that if I decided to do so I should cherish no one more than you. But I had not thought of doing that, nor wished to—and God grant I never may!'

This painful reply of Briseida's Chaucer suppresses at this point, although he will use some of it later:

> 'If that I sholde of any Grek han routhe,
> It sholde be youreselven, by my trouthe!
>
> 'I say nat therfore that I wol yow love,
> N'y say nat nay . . .' (v. 1000–3)

Instead, Chaucer simply reports the distracted attention and distress of his heroine and the conventional courtesies of her response to Diomede's offered friendship (v. 183–9). Chaucer thus removes that pointed inconsistency in Briseida's behaviour which provides the context for Benoît's antifeminist moralizing. The French Briseida's gift of her glove to Diomedes at the close of this very first encounter was setting too fast a pace for Chaucer's Criseyde, and he delays this incident until Diomede's later wooing, where he substitutes Diomede's taking for Criseyde's giving of the glove (v. 1013).

When Briseida arrives at the Greek camp she rebukes Calchas roundly for his treachery to Troy, and suggests that his powers of divination have led him astray (*Troie*, 13721 ff.). Chaucer echoes this passage but quite alters the context, shifting the material earlier into Criseyde's long speech to Troilus on their last night together, when she describes how she will deal with her father. By this stroke the English Criseyde's remonstrations with her father remain confined to her intentions, or, at least, Chaucer prefers not to narrate them. Benoît's Briseida finds the Greek camp better than she expected, and the narrator again anticipates:

> Anceis que veie le quart seir,
> N'avra corage ne voleir
> De retorner en la cité. (*Troie*, 13859–61)

Before she sees the fourth evening there she will have neither the wish nor the desire to return to the city.

This anticipation of Briseida's change of heart—so much quicker than Chaucer will attribute to his own Criseyde—again leads on to that general antifeminist extrapolation about women which is so far from the mood of *Troilus and Criseyde*, even where Chaucer is quite closely responding to his sources:

> Mout sont corage tost müé,
> Poi veritable e poi estable;
> Mout sont li cuer vain e muable. (*Troie*, 13862–4)

Their minds are very soon changed, and have little sincerity and little constancy; their hearts are most vain and fickle.

When Troilus and Diomedes next meet on the battlefield in the eighth battle, Diomedes evidently thinks of himself as fighting Troilus for possession of Briseida (*Troie*, 14286 ff.). Having unhorsed Troilus, Diomedes sends a messenger to present the horse to Briseida, who sends back the cool message that no one who hopes for her love should mistreat Troilus. She anticipates that Troilus will soon avenge himself and concludes her message:

> 'E si li di que tort fereie,
> Puis qu'il m'aime, se jol haeie:
> Ja nel harrai, se jo n'ai dreit,
> N'ancor ne l'aim dont mieuz li seit.' (*Troie*, 14349–52)

'Tell him that I should do wrong to hate him since he loves me. I should never do right to hate him, but neither shall I love him such that he profits by it'.

When, later in the same eighth battle, Polydamas captures Diomedes' horse and presents it to Troilus, the hero rejoices to think that he will be able to do such deeds of chivalry that his beloved will hear tell of it (*Troie*, 14430–2).

Meanwhile Benoît describes the tormenting lover's malady endured by Diomedes:

> Crient sei que ja soz covertor
> Ne gise o li ne nuit ne jor:
> De ço se voudreit mout pener
> A ço tornent tuit si penser. (*Troie*, 15027–30)

He fears that he may never, night or day, lie beneath a coverlet with her—yet that is what he means to strive for, and to that end turn all his thoughts.

Diomedes urges Briseida to consider his gift to her of Troilus' captured horse, and Briseida cuttingly comments that, as he may need what he has rashly given away, she will lend him the horse. This Diomedes interprets as a hint of his eventual success, and expresses his subjection to Briseida. Much gratified by his obedience, Briseida bestows upon him her right sleeve of fine new silk, to serve as a pennant (*Troie*, 15176–8). When Diomedes furiously engages Troilus in the tenth battle, this pennant of Briseida's sleeve is to be seen attached to his ebony lance (*Troie*, 15641–4), and in the fifteenth battle Troilus taunts the wounded Diomedes with bitter words about Briseida: although Diomedes has taken his place, many others will find a welcome there before the siege comes to an end (*Troie*, 20085–98). It is in the truce following this battle, when Diomedes lies seriously wounded by Troilus, that Briseida weeps over his wounds (*Troie*, 20221) and her affections turn entirely to Diomedes. But Benoît gives his Briseida a lengthy soliloquy in which she acknowledges her wrongdoing, anticipates her notorious reputation, bids farewell to Troilus, and commits herself to Diomedes.

It is this phase of the wooing of Briseida by Diomedes in the *Roman de Troie* that Chaucer draws on both in outline and in some detail in Book V when he records how 'the storie telleth us' that Criseyde gave Diomede back the horse he had won from Troilus (v. 1037–9); that she gave him her sleeve as a token (v. 1042–3); and that she wept over his wounds (v. 1044–50). When admitting her wrongdoing in betraying Troilus ('For I have falsed oon the gentileste | That evere was, and oon the worthieste!' (v. 1056–7)), Criseyde echoes something of Briseida's thoughts if not Benoît's tone in reporting them (*Troie*, 20229–36).

For his Briseida Benoît de Sainte-Maure writes a remarkable final soliloquy. Criseyde's anticipation of her future notoriety ('Shal neyther ben ywriten nor ysonge | No good word . . .' (v. 1059–60)) closely echoes Briseida's words (20238–9), and the following texts reveal how Criseyde's soliloquy of regret and farewell has been written partly in recollection of Briseida's:

> 'Lor paroles de mei tendront
> Les dames que a Troie sont.
> Honte i ai fait as dameiseles
> Trop lait e as riches puceles.'   (*Troie*, 20257–60)

> 'E que me vaut, se m'en repent?
> En ço n'a mais recovrement.
> Serai donc a cestui leiaus,
> Qui mout est proz e bons vassaus.'   (*Troie*, 20275–8)

> 'Sovent resont ploros mi ueil:
> Ensi est or, jo n'en sai plus.
> Deus donge bien a Troïlus!
> Quant nel puis aveir, ne il mei,
> A cestui me doing e otrei.'   (*Troie*, 20316–20)

'The ladies of Troy will speak of me. I have brought the direst shame upon damsels and noble maidens . . . And what good would it do me if I repent of it? No amends can be made that way. I shall then be loyal to this man, who is a very valiant and excellent knight . . . Often my eyes are full of tears again. Thus it is now, and I can do no more about it. May God grant Troilus happiness! Since I can no longer have him, nor he me, to Diomede I give and surrender myself.'

> 'Thei wol seyn, in as muche as in me is,
> I have hem don dishonour, weylaway!
> Al be I nat the first that dide amys,
> What helpeth that to don my blame awey?
> But syn I se ther is no bettre way,
> And that to late is now for me to rewe,
> To Diomede algate I wol be trewe.
>
> 'But, Troilus, syn I no bettre may,
> And syn that thus departen ye and I,
> Yet prey I God, so yeve yow right good day,
> As for the gentileste, trewely,
> That evere I say, to serven feythfully,
> And best kan ay his lady honour kepe.'
> And with that word she brast anon to wepe.   (v. 1065–78)

Taken over its whole length, however, Briseida's soliloquy has an ambiguous mixture of defensiveness, evasiveness, and a certain rationalizing quality, which Chaucer's very selective use here in his fifth book has left behind, however much he may have been more generally influenced by this soliloquy in the characterization of his own Criseyde. The spate of references to 'the storie' in this section of Chaucer's narrative (v. 1037, 1044, 1051, 1094) indicates how Benoît's account is for Chaucer the traditional structure of events in the story of Troilus and Criseyde to which he returns, especially for the account of how Criseyde's heart comes to change. Benoît had perceived the enigmatic interest in a combination of admirable qualities with changeability in his heroine, and had created the circumstances in which her constancy would be most cruelly tested by insecurity. Chaucer has set aside Benoît's antifeminist critique, but he has absorbed the lessons available from the *Roman de Troie* in presenting the heroine at once sympathetically and critically.

Benoît de Sainte-Maure, *Le Roman de Troie*, ed. L. Constans (Societé d'Ancien Textes Français; 6 vols.; Paris, 1904–12). See also *Le Roman de Troie*, ed. A. Joly (2 vols.; Paris, 1870–1). Extracts are translated in Havely (trans.), *Chaucer's Boccaccio*, and Gordon (ed. and trans.), *The Story of Troilus*.

On Chaucer's use of the *Roman de Troie*, see Antonelli, 'Birth'; Donaldson, 'Briseis'; Lumiansky, 'Story'; Nolan, *Chaucer*; and Young, *Origin and Development*. For *Troilus*, v. 106–75 and 1000–85, in parallel with passages from the *Roman de Troie*, see Windeatt (ed.), *Troilus and Criseyde*, 452–5, 498–505.

Edward III may have owned a copy of the *Roman de Troie*, and in the list of books belonging to Thomas of Woodstock (d. 1397) there is a romance on Hector of Troy and two books in French on the Trojan War: see Viscount Dillon and W. H. St John Hope,

'Inventory of the Goods and Chattels Belonging to Thomas, duke of Gloucester', *Archaeological Review*, 54 (1897), 275–308, and E. Rickert, 'King Richard II's Books', *The Library*, 4th ser. 13 (1933), 144–7.

Guido de Columnis Historia destructionis Troiae, *and the Tradition of Troilus and Criseyde*

Benoît de Sainte-Maure's *Roman de Troie* was translated and paraphrased in Latin prose by the Sicilian judge, Guido de Columnis of Messina, whose epilogue records that he finished his version in 1287. Although so closely indebted to Benoît, Guido claims to be following Dares and Dictys. (Those matters recorded 'by Dictys the Greek and Dares the Phrygian, who were at the time of the Trojan War continually present in their armies and were the most trustworthy reporters of those things which they saw, will be read in the present little book, having been transcribed by me . . .' (Meek, 2).) Despite its manifest literary inferiority to Benoît's *Roman*, Guido's Latin *Historia* enjoyed greater influence in its pretensions to authoritative history than the vernacular romance, and was in turn much translated into European vernaculars. Such was the authority of the *Historia* that it was long assumed to be the origin of Benoît's romance, rather than the other way round. Chaucer mentions Guido not Benoît as a historian of Troy (*House of Fame*, 1464–72), and it is entirely probable that Boccaccio and Chaucer, in writing poems set at the siege of Troy, would make use of both Benoît's *Roman de Troie* and Guido's *Historia*. Chaucer's more direct debt to Benoît is proved by those instances where Chaucer uses passages in the *Roman de Troie* which Guido omits or abbreviates unsympathetically, and which Boccaccio also does not use in *Filostrato*: Diomede's speech as he escorts Criseyde from Troy (v. 99–189), Criseyde's bestowal on Diomede of the love-tokens of her glove, her sleeve, and the horse won from Troilus, and her soliloquy of regret and farewell when she abandons Troilus (v. 1009–99).

In his *Historia* Guido is less concerned than Benoît in the *Roman* to interweave the emotions of Troilus, Diomede, and Criseyde with their role in the Trojan War. His is a 'history' rather than a romance, and this explains some of his shifts of emphasis in the story, but his disregard for the links and echoes artfully established by Benoît between far-flung incidents of emotional significance has the effect of disintegrating the way the romance had related feeling to time, so that all in the *Roman de Troie* that suggested refinement and delicacy, yearning and suspense, is flattened into the two-dimensional, the predictability of another example of human sexual weakness, easy to moralize. When describing the early encounters of Troilus and Diomede in the third and fourth battles, Guido does not preserve the developing personal animosity between these eventual rivals in love which Benoît is careful to establish. As a 'historian', Guido confines himself to recording the important actions of the participants in the war, and the series of incidents between Troilus and Diomede involving their capture and loss of horses is overlooked. These differences of emphasis between the *Roman* and the *Historia* are anticipated in the way that Benoît's portraits deteriorate in Guido's versions. To the French portrait of Diomedes Guido adds the detail of his lustfulness:

Dyomedes multa fuit proceritate distensus, amplo pectore, robustus scapulis, aspectu ferox, in promissis fallax, in armis strennuus, victorie cupidus, timendus a multis, cum multum esset iniuriosus, seruiencium sibi nimis impaciens, cum molestus seruientibus nimis esset, libidinosus quidem multum, et qui multas traxit angustias ob feruorem amoris. (*Historia*, viii. 84)

Diomedes was of great height, broad-chested, square-shouldered, fierce-looking, deceitful in his promises, active in war and eager for victory. He was widely feared, since he was very aggressive, and was highly impatient with his servants for he was an extremely hard master. He was indeed very lustful and endured much torment because of the violence of his passion. (Havely, 184)

In his portrait of Briseida Guido puts more emphasis on her joined eyebrows and her inconstancy of heart:

Briseyda autem, filia Calcas, multa fuit speciositate decora, nec longa nec breuis nec nimium macillenta, lacteo perfusa candore, genis roseis, flauis crinibus, sed superciliis iunctis, quorum iunctura, dum multa pilositate tumesceret, modicam inconuenienciam presentabat. Multa fulgebat loquele facundia; multa fuit pietate tractabilis. Multos traxit propter suas illecebras amatores multosque dilexit, dum suis amatoribus animi constantiam non seruasset. (*Historia*, viii. 85)

Briseida, the daughter of Calchas, was most graceful in her beauty. She was neither tall nor short nor too thin, and had a skin of milky whiteness, rosy cheeks and golden hair. But her eyebrows were joined, and because the hairs grew very thickly at the place where they met this seemed a slight blemish. She excelled in the eloquence of her speech and was very readily moved to compassion. Her charms brought her many suitors, and she granted her favour to many, but she did not remain constant in heart towards her lovers. (Havely, 184)

Most revealingly of the emphasis in his *Historia*, Guido drastically reduces Benoît's portrait of Troilus as a hero of romance, and instead describes him almost exclusively as a warrior:

Troylus uero, licet fuerit corpore magnus, magis fuit tamen corde magnanimus, animosus multum sed multam habuit in sua animositate temperiem, dilectus plurimum a puellis, cum ipse aliqualem seruando modestiam delectaretur in illis. In viribus uero et strennuitate bellandi uel fuit alius Hector uel secundus ab ipso. In toto eciam regno Troye iuuenis nullus fuit tantis uiribus nec tanta audacia gloriosus. (*Historia*, viii. 86)

Troilus indeed, although mighty of body was made yet mightier by the greatness of his spirit. He was high-tempered and bold, but maintained a great deal of moderation in that respect. Women found him highly attractive and he, though retaining some discretion, found delight in them. Indeed, for both strength and energy in war he was equal or next to Hector. Moreover in the whole kingdom of Troy there was no young warrior of such strength or so famous for his great daring. (Havely, 184)

Guido's unsympathetic attitude to the Troilus and Briseida story inherited from Benoît emerges at once in his presentation of the lovers' response to the news of Briseida's departure. Benoît's Troilus had been prostrated, but for Guido there is a culpable excess in his emotional extremes:

quam multo amoris ardore diligebat iuueniliter, nimio calore ductus amoris in desideratiua uirtute ignee uoluptatis multo dolore deprimitur et torquetur funditus. (*Historia*, xix. 163)

He cherished her with the great fervour of youthful love and had been led by the excessive ardour of love into the intense longing of blazing passion. (Meek, 156)

Briseida's tearfulness at the thought of separation is described with a sarcastic detachment: her clothes were so drenched with her tears that if anyone had wrung them out her clothes would have poured forth a great amount of water (*Historia*, xix. 163). This is at a far remove from that sensitively described tearfulness in Benoît's *Roman* as in *Filostrato*, which comes through into *Troilus*. And when Troilus leaves Briseida at the end of their last night together, Guido inserts his own comments:

Sed, O Troile, que te tam iuuenilis coegit errare credulitas ut Briseyde lacrimis crederes et eius blanditiis deceptiuis? (*Historia*, xix. 164)

But oh, Troilus, what youthful credulity forced you to be so mistaken that you trusted Briseida's tears and her deceiving caresses? (Meek, 157)

Guido's generalizations here about feminine inconstancy are much harsher than Benoît's:

Sane omnibus mulieribus est insitum a natura ut in eis non sit aliqua firma constancia ... quarum mutabilitas et uarietas eas ad illudendos uiros semper adducit. ... Et si forte nullus solicitator earum appareat, ipsum ipse, dum incedunt. ... Nulla spes ergo est reuera tam fallax quam ea que in mulieribus residet et procedit ab eis. (*Historia*, xix. 164)

It is clearly implanted in all women by nature not to have any steady constancy ... their fickleness and changeableness always lead them to deceive men. ... If perchance no seducer appears to them, they seek him themselves. ... There is truly no hope so false as that which resides in women and proceeds from them. (Meek, 157)

It reflects Guido's view of women that, whereas in the *Roman de Troie* Briseida's affections begin to change on her *fourth* day away from Troilus, in the *Historia* Briseida's heart changes on the *same* day of her arrival (*Historia*, xix. 166), and she more or less commits herself to Diomedes when he sends her Troilus' horse as a gift (*Historia*, xx. 169).

In the next stage of the story Guido omits Benoît's account of how Briseida presents Troilus' horse back to Diomedes and makes a further present to him of her sleeve (incidents which Chaucer restores when he finds them missing from *Filostrato*). The whole section in Benoît devoted to Troilus' wounding of Diomedes in the fifteenth battle, his great bitterness towards Briseida, Briseida's concern for Diomedes' wounds, her final change of heart, and her soliloquy with its regrets, are all abbreviated by Guido. He demotes the motivation of romantic love in the *Roman de Troie* into lechery and lust: the lustfulness of Diomedes has already been mentioned several times by the point when Briseida decides 'absolute facere uelle suum, cum in eius amore tota deferueat et flagranti desiderio penitus incalescat' (xxvi. 198) (To do his will completely, since she was entirely enflamed by his love and burned hotly with blazing desire (Meek, 190)).

Guido's lack of concern in his *Historia* with the romance experience of Troilus and Briseida developed by Benoît leads to a lack of obvious connection between the *Historia* and *Troilus*. But the authentic aura of the Latin *Historia*, and that very 'historicity' which precluded some of Benoît's romance elements and excised anything of the fabulous, gave Guido's *Historia* an authority which is sometimes discernible in *Troilus*. It is possible to see Chaucer historicizing *Filostrato* as Guido had historicized the *Roman de Troie*, and perhaps with the model of the *Historia* in mind. Chaucer's alteration of *Filostrato* so that Calchas begs the Greeks 'And hem for Antenor ful ofte preyde | To bryngen hom kyng Toas and Criseyde' (iv. 137–8), may well reflect the influence of Guido. Moreover, at several other junctures the possible echo of Guido's phrasing may be detected. Criseyde's words to Diomede ('I say nat therfore that I wol yow love, | N'y say nat nay . . .' (v. 1002–3)) possibly reflect more of the brief reply to Diomedes of Guido's Briseida as they ride from Troy—'Amoris tui oblaciones ad presens nec repudio nec admitto' (*Historia*, xix. 164) (At present I neither refuse nor accept the offer of your love (Meek, 158))— than the longer speech of Benoît's heroine (*Troie*, 13669–80). At the account of the death of Hector ('For as he drough a kyng by th'aventaille, | Unwar of this . . .' (v. 1558–9)), Chaucer's phrasing may preserve something of Guido's description of this scene ('Achilles . . . accepta quadam lancea ualde forti, *non aduertente Hectore*, in ipsum irruit' (*Historia*, xxi. 175) (Achilles ... took a very strong lance, which Hector did not observe, and rushed upon him (Meek, 168))).

A more general influence may have been the model provided by Guido as the 'historian'-narrator in presenting and commenting on his material, although, if this were a model, it is one that Chaucer has much modified. Guido's digressions and condemnatory comments on pagan practice and idolatry have also suggested a connection with the outburst at the end of *Troilus* ('Lo here, of payens corsed olde rites!' (v. 1849)). Guido maintains an aloof stance as an historian-narrator, exclaiming at fateful human ignorance of the consequences of men's deeds, distancing himself from the sexual lives and religious outlooks of the protagonists in his narrative. As such, Guido provided a forceful if monotonous example of narratorial practice, in response to which Chaucer may have created his own more flexible and sophisticated handling of his historical and pagan subject-matter.

With the wide diffusion of Benoît de Sainte-Maure's *Roman de Troie*, followed by Guido's *Historia*, references to Troilus and Briseida as type figures and stock characters begin to occur. One very early such reference appears in a Provençal verse letter of about 1230, which uses the example of Briseida to persuade a woman not to finish with a man who has long loved her. By the fourteenth century the French poets Jean Froissart and Guillaume de Machaut are also found referring to the characters as type figures of lovers. In the *Paradys d'amours* (before 1370), which Chaucer imitates at the opening of his *Book of the Duchess*, Froissart places Troilus first in a list of lovers (974 ff.). In the *Dit de la fonteinne amoureuse*, an important source for *The Book of the Duchess*, Machaut includes Troilus and Briseida in a series of scenes from the Trojan War which the poet sees depicted on a marvellous fountain:

> Et Troïllus moult se traveille
> Pour la fille Calcas de Troie,
> Briseÿda ... (1338–40)

Chaucer includes the name of Troilus in a similar list of lovers seen painted in the Temple of Venus in *The Parliament of Fowls* (288–94), and, in a poem sometimes attributed to Chaucer, *Against Women Unconstant*, Criseyde is accepted as a suitable type figure to cite ('Ye might be shryned for your brotelnesse | Bet than Dalyda [Delilah], Creseyde, or Candace ...').

In the poems of Chaucer's contemporary John Gower, the co-dedicatee of *Troilus*, there are a number of references to Troilus and Criseyde which suggest how they were thought of as type figures in fourteenth-century

England. Just as Boccaccio had represented the courtly young people who tell each other the tales of the *Decameron* diverting themselves with a song about Troilus and Criseida ('Dioneo insieme con Lauretta di Troiolo e di Criseida cominciarono a cantare . . .'), so too Gower in his earliest allusion to the story represents it as the subject of a song. In his early French poem, the *Mirour de l'Omme* (*c.* 1375), the figure of Somnolence when in church, rather than praying, dreams drunkenly 'qu'il oït chanter la geste | De Troÿlus et de la belle | Creseide . . .' (5253–5). In his Latin poem, the *Vox clamantis*, most probably written before 1381, Gower includes Troilus and Criseyde in a list of classical and Biblical figures, to underline his argument that the good perish while the wicked flourish:

> Mortuus est Troilus constanter amore fidelis,
> Iamque Iasonis amor nescit habere fidem:
> Solo contenta moritur nunc fida Medea,
> Fictaque Crisaida gaudet amare duos   (vi. 1325–8)

Troilus, steadfastly faithful in love, is dead, and now Jason's kind of love does not know how to keep faith. The faithful Medea is now dead and laid out in the earth, and false Crisaida takes pleasure in loving two men.

Or again, when the speaker in one of Gower's *balades* in Anglo-Norman, collected as the *Cinkantes balades*, laments that Fortune's wheel never seems to turn for him and bring happiness in love, he uses some Trojan examples of how one man's misfortune is another man's good fortune:

> Celle infortune dont Palamedes
> Chaoit, fist tant q'Agamenon chosi
> Fuist a l'empire: auci Diomedes,
> Par ceo qe Troilus estoit guerpi,
> De ses amours la fortune ad saisi,
> Du fille au Calcas mesna sa leesce   (*Balade*, xx.17–22)

The misfortune from which Palamedes fell made it possible for Agamemnon to be chosen to command. Likewise Diomede, because Troilus was forsaken, seized possession of the fortune of his love. He had his joy of Calchas' daughter.

There are also a number of references to the story of Troilus and Criseyde in Gower's English poem *Confessio amantis*, which postdates Chaucer's *Troilus*. Diomede can provide an example for the confessor in preaching against the sin of 'supplantation' in love (ii. 2456–8); the Lover is ready to protest his wakeful attentiveness to carry out all the injunctions of his lady, among which is included the reading aloud of 'Troilus' (iv. 2794–5); and, under the heading of 'sacrilege', Gower includes the sacrilege of Troilus' falling in love with Criseyde in a temple (v. 7597–602). As the scene with the hero in the temple in Book I of *Troilus* is taken from

Filostrato and does not occur in Benoît or Guido, this instance evidently shows the influence of Chaucer's poem, although Gower has just mentioned as other instances of such 'sacrilege' Paris and Helen, and Achilles and Polyxena, whose stories derive from Benoît and Guido, so that Chaucer's contribution to the story is being added to the tradition. Finally, towards the very end of *Confessio amantis* Gower includes Troilus and Criseyde in the poet's vision of companies of lovers, where their love—as in other medieval allusions—is seen in the context of its unhappy transience and bitter conclusion:

> And Troilus stod with Criseide,
> Bot evere among, althogh he pleide,
> Be semblant he was hevy chiered,
> For Diomede, as him was liered,
> Cleymeth to ben his parconner ...   (viii. 2531–5)

Guido de Columnis, *Historia destructionis Troiae*, ed. N. E. Griffin (Cambridge, Mass., 1936). It seems most probable that Benoît's *Roman de Troie* itself, and not one of the prose versions derived from it, should be considered the immediate antecedent of Guido's *Historia* (Antonelli, 'Birth'). The *Historia* has been translated by M. E. Meek (Bloomington, Ind., 1974).

For the differences between Benoît's *Roman* and Guido's *Historia*, see Lumiansky, 'Story'.

For the influence of the *Historia* in England, and for its possible influence on Chaucer, see Benson, 'King Thoas' and *History*.

For the development of the reputation of Criseyde from the *Roman de Troie* to Chaucer and beyond, see Mieszkowski, 'Reputation'; Mieszkowski cites the Provençal verse letter from Azalais d'Altier to Clara d'Anduza, printed in V. Crescini, 'Azalais d'Altier', *Zeitschrift für Romanische Philologie*, 14 (1890), 128–32.

John Gower: *The Complete Works of John Gower*, ed. G. C. Macaulay (4 vols., Oxford, 1899–1902); *The English Works of John Gower*, ed. G. C. Macaulay (2 vols.; EETS ES 81–2; London, 1900, 1901). See also *The Major Latin Works of John Gower*, trans. E. W. Stockton (Seattle, Wash., 1962).

Boethius, De consolatione philosophiae

The *De consolatione philosophiae* (Consolation of Philosophy)—written by the late Roman public and literary figure Boethius (*c*.475–524) as he awaited an undeserved execution—is one of the most formative influences on Chaucer in *Troilus*. When Chaucer came to compose his five-book *Troilus* shortly after translating the five books of the *Consolation* into his English prose *Boece*, it is clear that the example of Boethius still lodged in his memory both in some sharp and particular detail and in an absorbed and diffused way. From his labour of love in translating the *Boece* Chaucer turns to reinterpreting the Troilus story, but with an imagination suffused with the philosophy, still inside the distinctive inner landscape of the

Consolation's preoccupations. *Troilus and Criseyde* is a re-reading of the old story of Troilus in a frame of mind suggested by the *Consolation*, and a reading of *Troilus* can seem a kind of reading of the *Consolation* at a remove: transposed, paraphrased, excerpted, it is drawn upon to focus and deepen the meaning the story of Troilus can bear, as that early reader Thomas Usk suggested, in describing *Troilus* as the work of a 'philosophical poete' (see below, p. 361).

A summary outline of the *Consolation of Philosophy* now follows, before the nature of Chaucer's use of Boethius is discussed:

The form of the *Consolation* is essentially that of a dialogue, between Boethius the prisoner and the allegorical figure of Philosophy, who visits him in his cell, and this alternation of speakers is matched by the graceful alternation in form between prose and the brief poems which both figures utter. From his self-pity and despair in his changed fortune at the opening of the work, Boethius is drawn by Philosophy to acknowledge the reason of her arguments. In a process likened to the healing of sickness by a physician, he is brought to understand the insubstantiality of Fortune, and to see that true happiness is to be sought in God rather than in the vanities of false and worldly outward things. From this he is drawn onwards to accept the real powerlessness of evil, the underlying justice of God's dispositions in this world, and finally the true relation between divine providence and the freedom of the human will.

Through Philosophy's early questions it emerges that, although Boethius does acknowledge that God watches over His creation, the prisoner has forgotten his own true nature. This loss of memory is at the root of his sickness, for, because he has forgotten the end and purpose of things, he accepts the reality of the haphazard ups and downs of Fortune. In the second book Philosophy sets out to educate Boethius into a rejection of Fortune and her goods. Philosophy dismisses the prisoner's sense of grievance at Fortune's 'change' towards him, for changeability is Fortune's essential nature, as represented by her wheel. Philosophy furthers her argument against Fortune by impersonating the arguments Fortune might use to deny Boethius' accusations: it is men's own mistake to feel they have any right of possession over the goods of Fortune, and in her very changeableness man has hope of improvement. When Philosophy bids Boethius recall his past good fortune Boethius answers that it is the worst ill-fortune to remember past happiness, but Philosophy retorts that he must blame his own misguided understanding of Fortune. Her point is 'Nothing is miserable except when you think so', and her challenge is 'Why do you mortal men seek after happiness when it lies within you?' Throughout the second and third books Philosophy proceeds to point out at length the unsatisfactoriness of such false worldly goods as riches, power, beauty or fame, before turning to link true happiness with perfect goodness and thus with God, the supreme good that governs all things.

At the opening of the fourth book Boethius exclaims against the way that evil exists and goes unpunished, but Philosophy argues that the 'power' of evil men is in reality a weakness. To Boethius' wish that evil men at least did not have the power to harm the good, Philosophy responds by arguing that the apparent success of the wicked and their escape from punishment are in truth their misery and punishment in a world governed by good.

Boethius' puzzlement that the confusions of chance seem to obscure God's ruling power in the universe prompts Philosophy's crisp exposition of the distinct operations of Providence and Destiny, and of the issue of predestination, of God's foreknowledge and human freedom of the will. Within Philosophy's interpretation all fortune is perceived as good, there is no such thing as chance; but Boethius goes on (in the passage echoed by Troilus in Book IV of Chaucer's poem) to assert the difficulty of accepting that the existence of divine foreknowledge does not contradict that of human free will, and then proceeds to the implications for reward and punishment, hope and prayer, in a predestined world. Philosophy's response develops from her point that everything is known according to the ability to know of those who do the knowing. The mind of God has the property of embracing the whole of time in one simultaneous present, such that God's essential providence does not precede or pre-empt the operation of human free will. Having led the prisoner so very far from his original despair, Philosophy's dialogue with Boethius is—on such an exalted note of acceptance and transcendence—brought to its close.

Chaucer draws on Boethius at a number of levels in his composition of *Troilus*, from phrasing and imagery to paraphrases of philosophical arguments for dramatic dialogue and situation. In terms of technique, moreover, Boethius' skill in alternating and complementing prose with verse may have offered a model for Chaucer's own incorporation of lyric set-pieces within *Troilus*. Chaucer's sense of the five-book *Consolation* informs both the structure and shape of his five-book poem and his development of its themes and characters, for many essential concerns of the *Consolation* are paraphrased and echoed by characters in *Troilus*, and many other features of Chaucer's poem suggest the suffused influence of Boethius. Almost every mythological figure mentioned in the twelfth metrum at the close of Book III of the *Consolation* is mentioned somewhere in *Troilus*: Orpheus and Eurydice (iv. 791); Calliope (iii. 45); Cerberus (i. 859); Ixion (v. 212); Tantalus (iii. 593); Tityus (i. 786); the Furies (i. 9; iv. 22–4). This metrum—which refers to the love of Orpheus 'that doublide his sorwe' and to the Furies as both tormentors and as feeling sorrow and pity—has had some effect in shaping a poem that in its first line anticipates the structure of a 'double sorwe' and which both in its first proem and then in its pivotal fourth proem invokes the Furies as sorrowing and lamenting (i. 9; iv. 23).

The patterns in Chaucer's uses of Boethius may first be brought out by means of a table, and the main Boethian contexts in *Troilus* will be discussed afterwards. However, it should be emphasized both that verbal parallels alone do not adequately suggest Chaucer's indebtedness, and also that there are inevitably resemblances between *Troilus* and the *Consolation* in various commonplaces (prefaced below by 'cf.'). References to the *Consolation* (in Chaucer's *Boece* translation) are in parentheses at the end of entries.

BOOK I

7	'Thise woful vers, that wepen ...' (cf. *Boece*, i, m. 1, 1)
637 ff.	Pandarus to Troilus: 'By his contrarie is every thyng declared' (cf. iv, pr. 2, 10 ff.)
638–9	Pandarus to Troilus: To know sweetness one must have tasted bitterness (cf. iii, m. 1, 5–6)
730	Pandarus to Troilus: 'What! Slombrestow as in a litargie?' (i, pr. 2, 18–21)
731–5	Pandarus to Troilus: 'artow lik an asse to the harpe' (i, pr. 4, 2–3)
786–8	Pandarus to Troilus: The woes of Tityos in hell (cf. iii, m. 12)
837–40	Troilus complains against Fortune (i, pr. 4; ii, m. 1)
841–54	Pandarus comforts Troilus (i. 843–4 and ii, pr. 2, 84–6; i. 845–7 and ii, pr. 3, 75–9; i. 848–50 and ii, pr. 1, 113–14; i. 851–4 and ii, pr. 2, 81–3)
857–8	Pandarus to Troilus: To be healed by a doctor one must disclose one's wound (i, pr. 4, 4–6)
891–3	Pandarus to Troilus: 'the ferste poynt is this ... A man to have pees with hymself' (cf. ii, pr. 4, 132 ff.)
960–1	Pandarus to Troilus: 'he that departed is ... Is nowher hol' (cf. iii, pr. 11, 62 ff.)

BOOK II

526–8	Troilus prays 'O god, that ... Ledest the fyn by juste purveiaunce' (iv, pr. 6, 216 ff.)
622–3	'men seyn, may nought destourbed be | That shal bityden of necessitee' (v, pr. 6, 164 ff.)

BOOK III

593	Pandarus to Troilus: Tantalus in hell (cf. iii, m. 12)
617–20	'O Fortune, executrice of wierdes' (iv, pr. 6, 60 ff.; v, m. 1, 18 ff.)
813–36	Criseyde to Pandarus: On worldly joy (ii, pr. 4)

1016–19	Criseyde to Troilus: The guiltless suffer; the guilty escape (cf. i, m. 5, 31 ff.)
1254	Troilus to Criseyde: 'O Love, O Charite!' (iii, pr. 11, 175 ff.)
1261	Troilus to Criseyde: 'Benigne love, thow holy bond of thynges' (ii, m. 8)
1625–8	Pandarus to Troilus: Misfortune in remembering past fortune (ii, pr. 4, 5–9)
1691–2	On felicity, which wise clerks commend (cf. iii, pr. 2, 8–12)
1744–71	Troilus' hymn to love (ii, m. 8)

BOOK IV

1–7	Commonplaces on Fortune (cf. ii, pr. 1, m. 1)
391–2	Pandarus to Troilus: Fortune's gifts are common (ii, pr. 2, 9–14, 84–6)
466	Troilus to Pandarus: 'Thynk nat on smert, and thow shalt fele non' (cf. ii, pr. 4, 109 ff.)
481–3	Troilus recalls Pandarus' reference to *Consolation*, ii, pr. 4, at iii. 1625–8, above.
503–4	Troilus to Pandarus: Happy is the death that comes invited and ends pain (cf. i, m. 1, 18–20)
767–70	Criseyde: A plant dies without its natural nourishment (cf. iii, pr. 11, 96 ff.)
791	Criseyde: Orpheus and Eurydice united (cf. iii, m. 12)
834–6	Criseyde: 'The ende of blisse ay sorwe it occupieth' (cf. ii, pr. 4)
958–66	Troilus: On necessity and divine prescience (cf. v, pr. 2)
974–1078	Troilus: On predestination (v, pr. 3)

BOOK V

1–4	The Fates execute Jove's decrees (cf. iv, pr. 6, 42 ff.)
212	Troilus turns like Ixion in hell (cf. iii, m. 12)
278	Phoebus with his rosy cart (ii, m. 3, 1–5)
746–9	Criseyde: 'Future tyme . . . koude I nat sen' (cf. v, pr. 6).
1541–5	To Fortune is committed the permutation of things (cf. iv, pr. 6)
1807 ff	Flight of Troilus' soul (cf. iv, pr. 1; m. 1)

Chaucer recalls in the opening lines of *Troilus* ('Thise woful vers, that wepen as I write' (i. 7)) Boethius' opening metrum: 'Allas! I wepynge, am constreyned to bygynnen vers of sorwful matere . . . and drery vers of wretchidnesse weten my face with verray teres' (*Boece*, i, m. 1), as if it is to

Boethius that Chaucer goes for the accent of one writing from within a sense of anguish. But, having echoed Boethius in the foreboding anticipation of the proem, Chaucer makes no specific use of the *Consolation* until the dialogue between Troilus and Pandarus in the second half of his first book, which draws on the *Consolation* both as a dramatic model and as a source of arguments and images. In Pandaro's visit in *Filostrato* to a languishing Troiolo—when the friend elicits the secret of Troiolo's love— Chaucer perceives a parallel with Philosophy's visit to the bedside of the prisoner Boethius, and so develops a scene of dialogue in which the English Pandarus, as 'Uncle Philosophy', applies persuasion to Troilus in a more 'philosophical' vein than in *Filostrato*, echoing many passages from Boethius and Philosophy's own brisk manner, but not necessarily her harder teachings. The embarrassed silence of Troilus (i. 722) recalls the early moments of Philosophy's visit to Boethius' bedside (*Boece*, i, pr. 2, 12–13); Pandarus' exclamation attempting to rouse Troilus (i. 730) draws on Philosophy's diagnosis of Boethius' condition, except that Pandarus applies these terms in a literal sense without the *Consolation*'s metaphysical context ('He hath a litil foryeten hymselve' (*Boece*, pr. 2, 21–3)); Pandarus challenges Troilus to pay attention (i. 731) in an echo of Philosophy's brisk manner (*Boece*, i, pr. 4, 1–3). A theme of sickness and cure, with Pandarus, like Philosophy, in the role of physician, forms a recurrent parallel with the dialogue scenes in Boethius (i. 857–8; cf. *Boece*, i, pr. 4, 4–6).

The other major borrowing in Book I occurs in the exchange between Troilus and Pandarus on the subject of Fortune. Troilus feels that 'Fortune is my fo' (i. 837) and proceeds to assert her irresistible influence on human lives: none 'may of hire cruel whiel the harm withstonde; | For as hire list she pleyeth with free and bonde' (i. 839–40). These sentiments parallel Boethius' complaint against 'the scharpnesse of Fortune, that waxeth wood ayens me' (*Boece*, i, pr. 4, 11–12) and Philosophy's characterization of Fortune and her wheel ('She, cruel Fortune, casteth adoun kynges that whilom weren ydradd ... Thus sche pleyeth, and thus sche prooeveth hir strengthes' (*Boece*, ii, m. 1)). Just as the *Consolation* shows Boethius making a false initial estimate of Fortune's power, so here Chaucer associates his Troilus with a limited understanding of human freedom. As a response, Chaucer adds Pandarus' two-stanza speech which asserts—in phrases culled from several contexts in Boethius—that the very mutability of Fortune means that joy will succeed sorrow. Pandarus' first point ('Woost thow nat wel that Fortune is comune | To everi manere wight in som degree?' (i. 843–4)) draws on the last line of the speech in which Philosophy 'impersonates' Fortune's self-defence against human reproaches (*Boece*, ii, pr. 2, 84–6). Pandarus' consolation 'That, as hire joies moten overgon, | So mote hire sorwes passen everechon' (i. 846–7) echoes Philosophy's comforting words to the prisoner that it is pointless to sorrow over past joys (*Boece*, ii, pr. 3, 75–9). Just as Pandarus points out that the essence of Fortune is the turning of her wheel (i. 848–9), so Philosophy makes her point to Boethius (*Boece*, ii, pr. 1, 113–14). For Pandarus' final point that Fortune's very mutability may bring Troilus his desire (i. 851–4), Chaucer again goes to Philosophy's impersonation of Fortune's self-defence (*Boece*, ii, pr. 2, 81–3). The distinction between Pandarus' echoing of these comments about Fortune and their implication in their original contexts lies in the persuasive purpose with which they are deployed: by Philosophy to instil in the imprisoned Boethius a detachment from all fortune; by the worldly Pandarus to persuade the hopeful lover Troilus with a pragmatic optimism that there is just as much chance of good fortune as of bad.

In Chaucer's second book the clearest Boethian echoes again associate with Troilus the language of providence and fate. The hero's complaint to Love, as reported to Criseyde by Pandarus, opens with a prayer without parallel in *Filostrato* ('O god, that at thi disposicioun | Ledest the fyn by juste purveiaunce' (ii. 526–7)) which addresses God in terms reminiscent of the *Consolation* (cf. *Boece*, iv, pr. 6, 216 ff.). In the added scene when Troilus first rides past Criseyde's house this fortunate day ('For which, men seyn, may nought destourbed be | That shal bityden of necessitee' (ii. 622–3)) is commented upon in terms recalling comparable discussions in the *Consolation* ('thilke thing that ne mai nat unbytide, it mot bytiden by necessite' (*Boece*, v, pr. 6, 164 ff.)).

In Chaucer's third book an important part of the difference from *Filostrato* in the tone and significance of the lovers' union is created by Chaucer's use of Boethius, both for narratorial passages and for speeches by the characters. The third proem, with its invocation of the harmony of the universe in love, derives at one remove from Boethius, being based on the song by Troiolo at the end of Part 3 of *Filostrato* which Boccaccio had in turn based on the concluding metrum of Book II of the *Consolation* (ii, m. 8). (By moving this song forward to the beginning of his third book Chaucer thus almost exactly 'restores' it to a position in the five-book structure of *Troilus* equivalent to the position occupied by its original in the five-book structure of the *Consolation*.) And later in Book III, when rain prevents Criseyde from leaving Pandarus' house, the apostrophe to Fortune as the 'executrice' or administering power of 'wierdes' (i.e. destinies), and the ensuing description of Fortune's role ('Soth is, that under God ye ben oure hierdes' (iii. 619)) recall the important discussion in *Consolation*, iv, pr. 6, where the operation and the relation of providence and destiny are defined and distinguished. With whatever irony the elaborate rhetoric of the apostrophe may be interpreted in its context, this narrative turning-point in *Troilus* is thus by association seen in terms of one of Boethius' explanations of the hierarchical relation of providence and destiny.

When Criseyde is later awakened by Pandarus with the news of Troilus' impending death for jealousy of Horaste, she exclaims for four stanzas (iii. 813–40) on the false felicity of worldly happiness, a speech which is a tissue of thought and phrase woven together from Philosophy's speech in *Consolation* (ii, pr. 4), where she is persuading Boethius of the unsatisfactory nature of earthly happiness by pointing to its contraries and paradoxes. The first stanza of Criseyde's complaint (iii. 813–19) draws together recollections of two separate points in the *Consolation* (*Boece*, ii, pr. 4, 118–19 and 75–8). On one level the next two stanzas (iii. 820–33) follow through closely the structure of a continuous logical argument in the same *Consolation* context (*Boece*, ii, pr. 4, 150–62), although the spare problematics of the Boethian prose have been quickened into the felt ironies of an almost desperate outburst. Criseyde's conclusion ('Ther is no verray weele in this world heere' (iii. 836)) in turn echoes the last sentence of Philosophy's argument on this topic ('How myghte thanne this present lif make men blisful?' (*Boece*, ii, pr. 4, 180–1)). It is noticeable that Criseyde feels she has developed an argument to a formal conclusion ('Wherfore I wol diffyne in this matere . . .' (iii. 834)), which suggests the special register such a Boethian speech has in the context. In the course of her later exclamations on the evils of jealousy, Criseyde reproaches Jove for allowing the innocent to suffer and the guilty to escape (iii. 1018–19), a passage that recalls the substance if not the phrasing of the early outburst of Boethius the prisoner (*Boece*, i, m. 5).

Chaucer also echoes Boethius in two exclamations by Troilus at his approaching union with Criseyde, the opening lines of two successive stanzas. The apostrophe to Cupid, Venus, and Hymen which begins 'O Love, O Charite!' (iii. 1254) recalls a phrase of Philosophy's in discussing nature and will ('And thus this charite and this love, that every thing hath to hymself, ne cometh not of the moevynge of the soule, but of the entencioun of nature' (*Boece*, iii, pr. 11, 175 ff.)). And while the following stanza derives from Dante in its latter six lines, the opening apostrophe to 'Benigne Love, thow holy bond of thynges' (iii. 1261) recalls once again the language of *Consolation*, ii, m. 8 ('al this accordaunce . . . of thynges is bounde with love . . . This love halt togidres peples joyned with an holy boond').

Echoes of the *Consolation* allow Chaucer's characters to voice both a sharper sense of the unsatisfactory nature of earthly experience and a more resonant sense of the dignity of human love, and this pattern continues in the remainder of the third book. Pandarus' warning to Troilus against that

'worste kynde of infortune', which is to remember happiness when it is
past (iii. 1625–8), is a commonplace also memorably expressed by Dante's
Francesca da Rimini in *Inferno* (v. 121–3), but Chaucer's stress upon it
here may recall Boethius' lament, 'this is a thyng that greetly smerteth me
whan it remembreth me. For in alle adversites of fortune the moost unzeely
kynde of contrarious fortune is to han ben weleful' (*Boece*, ii, pr. 4, 5–9).
Resemblance to such a Boethian passage is part of an alertness to
mutability in *Troilus* which is not present in *Filostrato*. In its original
context Boethius' complaint serves as cue for an argument by Philosophy
on rising above the depredations of Fortune. Here, as throughout, the
wisdom of the context in Boethius might be compared with the purpose the
borrowed material is made to express in the mouths of Chaucer's
characters, just as an added stanza in *Troilus* stressing the lovers' happiness
after the consummation ('Felicite ... may nought here suffise ...' (iii.
1691–2)) echoes the opening of Philosophy's definition of the sovereign
good, immediately after Boethius has been urged to 'withdrawe thy nekke
fro the yok (*of erthely affeccions*)' (*Boece*, iii, m. 1, 13–4; pr. 2, 8–12).

It is noticeable that this—in common with many of Chaucer's borrowings from Boethius—stems from a prominent, transitional point in the
Consolation. This is also the case when Chaucer uses for Troilus' song in
his moment of sexual fulfilment late in Book III (1744–71) a translation of
that metrum which brings the second book of the *Consolation* to a close
with an invocation of the cosmic harmony due to love. Although the
sequence of Boethian ideas is retained in Troilus' second and third stanzas,
Chaucer brings forward material on the unifying power of love from the
close of the metrum

'This love halt togidres peples joyned with an holy boond, and knytteth sacrement
of mariages of chaste loves; and love enditeth lawes to trewe felawes. O weleful
were mankynde, yif thilke love that governeth hevene governede yowr corages'

to form part of the opening stanza of the song:

> 'Love, that of erthe and se hath governaunce,
> Love, that his hestes hath in hevene hye,
> Love, that with an holsom alliaunce
> Halt peples joyned, as hym lest hem gye,
> Love, that knetteth lawe of compaignie,
> And couples doth in vertu for to dwelle,
> Bynd this acord, that I have told and telle.' (iii. 1744–50)

The reference to 'mariages of chaste loves' has been reformulated in a way
that Chaucer would think consistent with the outlook of an ancient pagan
such as Troilus. In that Chaucer absorbs almost every detail of the

Boethian poem, it is all the more striking that the song avoids Philosophy's climactic final lament that such an informing, cosmic sense of love does not govern the human heart, presumably because in Troilus' view it *does*.

Once Fortune's wheel has turned at the opening of Book IV—where the proem recalls Boethian iconography of Dame Fortune—some of Chaucer's principal borrowings from the *Consolation* are in all three of his characters' expression of their sense of fate and fortune. Troilus' first lamentations after Criseyde's exchange is agreed echo precisely those complaints against Fortune which Philosophy cites as unjustified if Fortune's nature be understood. Troilus' reproach of Fortune (iv. 267 ff.) recalls Philosophy's argument to Boethius ('yif thow wilt writen a lawe of wendynge and of duellynge to Fortune, whiche that thow hast chosen frely to ben thi lady, artow nat wrongful in that?' (*Boece*, ii, pr. 1, 95–8)); when Troilus vainly makes conditions with Fortune ('Nought roughte I whiderward thow woldest me steere' (iv. 282)), he echoes Philosophy's use of sailing imagery to describe subjection to Fortune ('Yif thou committest and betakest thi seyles to the wynd, thow schalt ben shoven, nat thider that thow woldest, but whider that the wynd schouveth the' (*Boece*, ii, pr. 1, 101–4)); Pandarus repeats to Troilus (iv. 391–2) that Boethian maxim on the communal nature of Fortune's gifts which he had earlier quoted to encourage the lover (i. 843–4); Troilus quotes back at Pandarus his earlier Boethian point that the worst form of misfortune is to remember lost happiness (iv. 481–3; cf. iii. 1625–8), and echoes the sorrowing Boethius of the *Consolation*'s first metrum by declaring 'For sely is that deth, soth for to seyne, | That, ofte ycleped, cometh and endeth peyne' (iv. 503–4). In her parallel scene of lamentation Criseyde's 'complaint' to Pandarus ('The ende of blisse ay sorwe it occupieth ...' (iv. 825–47)) also echoes in sentiment Philosophy's arguments to Boethius about the transient nature of worldly happiness (e.g. in *Boece*, ii, pr. 4), and at such points the Boethian example of philosophical dialogue prompts Chaucer to build a reflectiveness into his characters in their sorrow which is not part of his main narrative source in *Filostrato*.

Chaucer's most lengthy borrowing from the *Consolation*—Troilus' reflections on predestination (iv. 974–1078)—is a close translation of Boethius' text, apparently put into stanzaic verse from Chaucer's own translation in the *Boece*. Into a relatively simple scene in *Filostrato* of Troiolo's emotional distress at Criseida's prospective exchange, Chaucer has interpolated Troilus' response to what he sees as the predestinarian issue raised by his present misfortune in losing Criseyde. Both the closeness of this soliloquy to the *Consolation* and its differences are remarkable: the closeness because of the tortuously involved syntax and prosaic effect of the stanzas that result; the differences because these shift

the balance of the Boethian original towards a more resignedly predestinarian prejudice on Troilus' part, in which the *Consolation*'s arguments are
absorbed but re-expressed by Troilus with a reverent sense of God's
power, as the following selected stanzas from his soliloquy reveal:

'For yif so be that God loketh alle thinges byforn, ne God ne mai nat ben
desceyved in no manere, thanne moot it nedes ben that alle thinges betyden the
whiche that the purveaunce of God hath seyn byforn to comen. For whiche, yif
that God knoweth byforn nat oonly the werkes of men, but also hir conseilles and
hir willes, thanne ne schal ther be no liberte of arbitrie;

ne certes ther ne may be
noon othir dede, ne no wil, but thilke whiche that the devyne purveaunce, that ne
mai nat ben disseyved, hath felid byforn. For yif that thei myghten writhen awey
in othere manere than thei ben purveyed, thanne ne sholde ther be no stedefast
prescience of thing to comen,

but rather an uncerteyn
opynioun; the whiche thing to trowen of God, I deme it felonye and unleveful.'
(*Boece*, v, pr. 3, 8–25)

> 'For som men seyn, if God seth al biforn—
> Ne God may nat deceyved ben, parde—
> Than moot it fallen, theigh men hadde it sworn,
> That purveiance hath seyn before to be.
> Wherfore I sey, that from eterne if he
> Hath wist byforn oure thought ek as oure dede,
> We han no fre chois, as thise clerkes rede.
>
> 'For other thought, nor other dede also,
> Myghte nevere ben, but swich as purveyaunce,
> Which may nat ben deceyved nevere mo,
> Hath feled byforn, withouten ignoraunce.
> For yf ther myghte ben a variaunce
> To writhen out fro Goddis purveyinge,
> Ther nere no prescience of thyng comynge,
>
> 'But it were rather an opynyoun
> Uncerteyn, and no stedfast forseynge;
> And certes, that were an abusioun,
> That God sholde han no parfit cler wytynge
> More than we men that han doutous wenynge.
> But swich an errour upon God to gesse
> Were fals and foul, and wikked corsednesse.'   (iv. 974–94)

Both the laboured style and the comparison of the passage with its original
context in the *Consolation* have been taken to reflect on the inadequacy of
Troilus' grasp of the question of free will. The original speech by Boethius
disputing free will is duly followed in the *Consolation*—but not, of course,
in *Troilus*—by Philosophy's classic definition of how divine prescience and

human free will are not contradictory. The way in which Troilus' soliloquy is both more explicitly resigned to a lack of free will, and would also be judged as incompletely argued by any reference to its celebrated source, makes it one of the Boethian borrowings in *Troilus* which most turns upon a comparative sense of how the borrowed passage relates to its original context in the *Consolation*.

In Book V of *Troilus* the influence of Boethius is rather more oblique, with few close verbal echoes in relation to the length of the book. Criseyde's reflections on past, present, and future (v. 746–9) and 'felicite' (v. 763), unparalleled in *Filostrato*, may show Chaucer's recollection of Philosophy (*Boece*, v, pr. 6). Chaucer's familiarity with Boethius may even have prompted his inclusion of a heavenwards flight of the hero's soul. Although this is actually translated closely from the flight of Arcita's soul in Boccaccio's *Teseida*, Chaucer's sense of the significance for his Troilus of achieving such an altered perspective may well have been prompted by Philosophy's undertaking to show Boethius the way to his true home by a flight of thought (*Boece*, iv, pr. 1, m. 1), which will surmount the heavens and open the way to perception of the true limitations of this world (see below, pp. 210–11). Indeed, the 'pleyn felicite', which Troilus but not Arcita perceives in heaven, seems to fuse the Boccaccian narrative of the event with a Boethian understanding of its significance for the lover figure in *Troilus*.

It is recollection of the figure of Boethius—within that structure of his 'education' which determines the form of the *Consolation*—that orders and gives point to the pervasive echoes in *Troilus* of the imagery and diction of the *Consolation*. Once Philosophy has settled at the prisoner's bedside, she first laments (*Boece*, i, m. 2) how Boethius has strayed from light into darkness, driven to and fro 'with werldly wyndes', and remaining 'pressyd with hevy cheynes'—those recurrent patterns of imagery of light and darkness, of winds and seas, of binding and constraint, which colour Chaucer's imagination in his *Troilus* and prompt him to build comparable imagery into his poem. Philosophy's medicinal imagery (*Boece*, i, pr. 2, pr. 4, pr. 5) also recurs: at first in the idea of lovesickness, but also in the experience of contraries—of sweetness savoured after bitter remedies. Part of the poetic stimulus of the *Consolation* for Chaucer lay in the moral significance of its figurative language. Imagery of sickness looks forward to a spiritual cure. The *Consolation*'s resplendent imagery of cosmic order and measure—in the heavens and in the world of nature on earth—comments by eloquent contrast on the disordered state of man. Imagery of darkness and light stresses the clouding of vision and blindness in a prisoner both physically and mentally in chains ('of whom the sighte, ploungid in teeres, was dirked . . .' (*Boece*, i, pr. 1, 77–8)).

It is also worth in conclusion reversing this approach (of seeing where and how parts of the *Consolation* are held within *Troilus*) and looking instead at how the material Chaucer takes over into *Troilus* fits into the framework of the *Consolation*, for this highlights the emphases in Chaucer's use of his source.

The majority of Chaucer's borrowings from Boethius come from the earlier books of the *Consolation*. The significant exceptions to this pattern are that Chaucer has recourse to specific passages in Boethius' last two books for analysis of the relation of Providence, Destiny, and Fortune (*Boece*, iv, pr. 6) and of predestination (*Boece*, v, pr. 3), although Chaucer has also absorbed the lessons of parts of the *Consolation* which in *Troilus* he does not directly imitate and which teach a deep and detached distrust of this world and its appearances. Troilus, Criseyde, and Pandarus are made to lament and to analyse what happens in the love-story in terms of the operation of fortune, in a way which has no equivalent in Chaucer's immediate source *Filostrato*, and which recurrently echoes the argument and phrasing of the *Consolation*, in a sustained dialogue between the worlds of romance and of philosophical debate.

This pattern of echoes also suggests the limits and boundaries of their understanding. In the *Consolation* a dialogue with Philosophy leads a condemned prisoner towards enlightenment and serenity: the 'action' is the dialogue, and the progression of the work is a wholly spiritual and intellectual development. This is interpolated by Chaucer into a narrative of idealizing hope and romantic aspiration in love and its eventual disillusionment. Reflections which in the *Consolation* are part of the gathering, retrospective vision of a life with no certain earthly future are in *Troilus* quoted within the more open perspective of the story of an experience still very much being lived through by its protagonists, and are necessarily not pursued to the developments and answers that follow in the *Consolation*. As a narrative *Troilus* quotes, as it were, from but one side of the dialogue: the lamentations and questionings, but not the higher answers and explanations. There is no Lady Philosophy to engage with Troilus as she does with Boethius and explain why he is not predestined— although a medieval audience could well remember what was lacking.

In the *Consolation* Boethius progresses by remembering, as he retrieves the memory of wisdom once known but forgotten. Memory is also important both thematically and technically to the way Chaucer's use of Boethius works, for in the course of *Troilus* the reader is invited to remember more of the *Consolation* than the poem's characters quote. Their quotations are both tellingly fragmentary and inevitably shifted in context, a token of their uncompleted understanding, until the flight of Troilus' soul after death allows him to step outside time and gain detachment and perspective. Not only is the *Consolation* echoed in Troilus' first sorrow in Book I, but it is once again echoed in comparable terms when Troilus' double sorrow comes round again in Book IV, and this pattern of quotation points to the way that Troilus' life returns upon itself, by contrast with that progress Boethius makes under the guidance of Philosophy. It is this return to the accents of the opening of the *Consolation* in the fourth book of *Troilus* which suggests how the pattern of Troilus' life may be understood as the obverse of the advancing understanding of the prisoner Boethius and as moving in thematic contrast with it.

Boethius: The Theological Tractates and the Consolation of Philosophy, ed. H. F. Stewart and E. K. Rand (Loeb Classical Library; London and Cambridge, Mass., 1918, reprints to 1968), with the 1609 translation by 'I. T.'; the 1973 edition substitutes a modern translation by S. J. Tester. For Chaucer's translation, the *Boece*, see Benson (ed.), *The Riverside Chaucer*, and for Troilus' song (iii. 1744–71) and soliloquy (iv. 974–1078) in parallel with the corresponding passages in the *Boece*, see Windeatt (ed.), *Troilus and Criseyde*, 338, 406–10.

On Boethius' life and work, see Henry Chadwick, *Boethius: The Consolations of Music, Logic, Theology, and Philosophy* (Oxford, 1981). On the influence of the *Consolation*, see P. Courcelle, *La Consolation de philosophie dans la tradition littéraire: Antécédents et postérité de Boèce* (Paris, 1967), and Howard R. Patch, *The Tradition of Boethius* (New York, 1935).

For the medieval commentators on Boethius, see Gleason 'Nicholas Trevet', and Minnis, 'Aspects', and for some translations of the commentary by Nicholas Trivet, see Minnis, *Chaucer and Pagan Antiquity*, 145–50.

For a survey of Boethius' influence on Chaucer, see Jefferson, *Chaucer*, and for some particular studies of *Troilus*, see Gaylord, 'Uncle Pandarus'; Gordon, *Double Sorrow*; Huber, 'Troilus' Predestination Soliloquy'; McCall, 'Five-Book Structure'; and Stroud, 'Boethius' Influence', although Boethius is mentioned in too many other discussions of the poem to list. See also the studies cited on p. 267.

Ovid

'Ovides grans en ta poëterie': Chaucer was likened to Ovid within his lifetime in a *balade* by the contemporary French poet, Eustache Deschamps (see below, p. 362), and the influence of Ovid's works is discernible with varying degrees of definition at several different layers of the composition of *Troilus*. For medieval readers Ovid was not only a master in matters of love and a fine rhetorician, but also a thinker and philosopher, above all an *auctor*, an authority, very widely read and commented upon. From Ovid's writings stem many commonplaces on love and perhaps some models for the archetypal characters of lovers, while Ovid's poems are the likely ultimate source of many of Chaucer's allusions to classical mythology, especially a suffused Ovidian sense from the *Metamorphoses* of how the gods and the natural world are involved in processes of suffering and transformation. However, the influence of Ovid in medieval literature is so deep-rooted, accompanied by such a tradition of interpretation and commentary, that in a poem like *Troilus* some Ovidian features may well derive from intervening texts or from literary tradition. Chaucer also absorbs into *Troilus* the influence Ovid has already had on Boccaccio, for a shared tendency to draw on Ovid is one of the affinities between Chaucer and Boccaccio. Motifs from *Heroides* in the letters between Troiolo and Criseida survive into the letters of the English lovers (e.g. *Troilus*, v. 1345–7, via *Fil.* vii. 54, from *Heroides*, iii. 5–6, Briseis writing to Achilles), while some Ovidian commonplaces on the stratagems of love in *Troilus* come via *Filostrato*, ultimately from Ovid's *Ars amatoria* and *Remedia amoris*. This double example of Ovid as the poet of metamorphosis and of the 'Art of Love' is combined in the uses Chaucer makes of Ovid in *Troilus and Criseyde*.

Chaucer's easy familiarity with Ovid is indicated by his mixture of accuracy and inaccuracy in his mythological references: he slips up in describing Apollo as speaking from inside the 'holy laurer' (iii. 542); his account of Herse and Aglauros is muddled (iii. 729–30); he confuses Titan with Tithonus (iii. 1464), and Latona with Diana (v. 655). On the other hand, he also shows a sharp memory for phrases remembered from Ovid: the reference to 'Daun Phebus or Appollo Delphicus' (i. 70), for instance, probably recalls the epithet 'Delphicus' from *Metamorphoses* (ii. 543) or *Fasti* (iii. 856); Dione as the mother of Venus (iii. 1807–10) could be familiar from *Ars amatoria* (ii. 593; iii. 3, 769) or *Amores* (I. xiv. 33); and the striking description of Venus as 'Lucyfer, the dayes messager' (iii. 1417), just before Troilus laments that Tithonus has let the dawn leave him, perhaps remembers *Heroides*, xviii. 111–14, in which Leander recalls to Hero their own dawn parting. Early in Book IV is a cluster of recollections of Ovid, with the references to 'Nyghtes doughtren thre' (iv. 22), recalling *Metamorphoses*, iv. 451–2; to Quirinus (iv. 25), probably retained from a reading of *Fasti* (ii. 419, 475–6); and to 'Hercules lyoun' (iv. 32), perhaps recalling the epithet given to Hercules in a line of the *Ars amatoria* (i. 68). The allusion to the story of the walls of Troy (iv. 120–6) perhaps draws on *Metamorphoses* (xi. 194 ff.), although Paris's letter to Helen (in *Heroides*, xvi. 181 ff.) also refers to how Apollo caused the walls to rise for Laomedon, while Cassandra's account of Meleager and the boar abridges the story told in *Metamorphoses* (viii. 270–525). There is often a sound sense of relevant mythological detail in such contexts, with which Chaucer's Ovidian reading probably provided him, just as he retains an echo of the epithet for Apollo in *Ars amatoria* ('laurigero ... Phoebo' (iii. 389)) when he refers to the god as 'laurer-crowned Phebus' (v. 1107).

If the recollection of such epithets suggests Chaucer's familiarity with Ovid's text, in other cases intervening works have put Chaucer in mind of an Ovidian instance. With the story of Procne (ii. 64–6) the phrasing reveals that this allusion to *Metamorphoses* comes indirectly through Dante, as may references to the stories of Midas (iii. 1389) or of Athamas (iv. 1539). Criseyde's oath by the river 'Symois' (iv. 1548–9) probably derives indirectly through the *Roman de la rose*. Although the stories which lie behind a whole series of allusions in *Troilus* to mythological figures in hell could be found in Ovid (Cerberus, Tityos, Tantalus, Ixion, Orpheus, and Eurydice), it is likely that Chaucer had been put in mind of them by their association together in the 'Orpheus' metrum (*Boece*, iii, m. 12) of Boethius' *Consolation*. But, however directly or indirectly such references derive from Ovid, the significance for *Troilus* lies in their unifying thematic pattern as the course of the affair between Troilus and Criseyde is brought into association with such Ovidian stories of pride, obsession, and betrayal as those of Nyobe (i. 699–700), Myrrha (iv. 1139), Ascaphilus (v. 319), Phaeton (v. 664–5), Scylla (v. 1110), and the loves of the gods in Troilus' prayer in the 'stewe', including Venus and Adonis, Jove and Europa, Mars and Venus, Apollo and Daphne, and Mercury and Herse (iii. 715–35).

Yet it is in *Troilus* as an account of an exemplary love that the influence of Ovid as the poet of the 'Art of Love' is also discerned. When Pandarus urges Troilus to write a first letter to Criseyde, part of his stylistic advice (ii. 1030–6, 1041–3) stems ultimately from Horace's *Ars poetica* (355–6, 1–5), but the larger considerations of strategy (ii. 1023 ff.) recall the advice in *Ars amatoria* (i. 467 ff.). The advice to blot the letter with tears counsels cultivation of that tear-stained appearance which letters are mentioned to have in Ovid, as when Briseis, in writing to Achilles, mentions the tearstains on her letter (*Heroides*, iii. 3). Although Pandarus' maxim after the consummation ('As gret a craft is kepe wel as wynne' (iii. 1634)) may derive from the *Roman de la rose* (8231–3), its ultimate origin is in the *Ars amatoria* (ii. 11–13). Pandaro's attempt to comfort Troiolo in *Filostrato* is absorbed by Chaucer ('The newe love out chaceth ofte the olde' (iv. 415)), and so incorporates a sentiment probably recollected from *Remedia amoris*, 462 ('Successore novo vincitur omnis amor'), where it is followed and illustrated by the story of Briseis and Chryseis (465–88).

Given the sententious disposition in Chaucer's characters and narrative, it is almost inevitable that the sentiments expressed in *Troilus* about the course of love will coincide with comparable Ovidian sentiments (e.g. v. 790–1; cf. *Ars amatoria*, i. 361–2), and there is an element in Pandarus of that persona of the master and teacher in love cultivated by Ovid in his amatory poems (e.g. i. 857–8, 946–9; cf. *Remedia amoris*, 125–6, 45–6). Criseyde's comment 'men rede | That love is thyng ay ful of bisy drede' (iv. 1644–5) is just that kind of inclination to relate their own experience to general observations about love which Chaucer adds for his characters, and it is especially poignant that Criseyde's quotation of this commonplace in urging Troilus to be faithful in her absence is a recollection of faithful Penelope's observation in her letter to Ulysses in *Heroides*, i. 12: 'res est solliciti plena timoris amor' (Love is a thing ever filled with anxious fear).

Ovid's power to convey the poignancy and pathos of sorrow may have also been a model for Chaucer. So it seemed to that scribe of the Rawlinson manuscript of *Troilus* who wrote 'Require in Metamorphosios' at i. 701–7, and 'Require in Ouidio' by i. 713, where Pandarus' wry comment— 'namore harde grace | May sitte on me, for-why ther is no space'— reminded him of Ovid's *Ex Ponto*, II. vii. 41–2: 'Sic ego continuo Fortunae vulneror ictu, | vixque habet in nobis iam nova plaga locum' (I am so wounded by the continual blows of Fortune that there is scarcely room on me for a new wound). Yet Ovid is also so much the poet of wit that, even behind the irritated retort of Troilus ('What knowe I of the queene Nyobe? | Lat be thyne olde ensaumples, I the preye' (i. 759–60)), there may be an echo of an Ovidian phrase—'Quid moror exemplis, quorum me turba fatigat?' (Why do I dwell on examples, the crowd of which wearies me?)— remembered again from just before the story of Briseis and Chryseis in *Remedia amoris* (461). Ovid's epitaph as a lover in *Tristia*, III. iii. 73–6— perhaps recollected via *Teseida*, xi. 91—may also lie behind the moment when Troilus imagines how future lovers will visit his tomb (iv. 323–9).

More generally Chaucer's *Troilus* may draw on the example of lamentational stance and tone in the *Heroides*, and in the characterization of Criseyde occurs a series of echoes from the letter written to Paris by Helen of Troy in *Heroides*, xvii. Pandarus also quotes to Troilus from a certain letter of complaint to Paris from Oenone (i. 652–65), which presumably is posing as a reference to the nymph Oenone's letter to her husband Paris in *Heroides*, v, complaining of his betrayal of her in favour of Helen, and foreseeing Helen's eventual infidelity to Paris. To cite Oenone's letter to Paris as Pandarus does—only to ignore its main subject—is to bring into parallel with the *Troilus* narrative stories of the betrayed love that lies behind the Trojan War. When Criseyde swears that the River Symois may run backwards to its source on the day she is untrue to Troilus (iv. 1551–4) there may again be an ironic echo from *Heroides* (v. 29–32)—possibly recollected via the *Roman de la rose* (13195–8)—of Paris's vow that the waters of the Xanthus will flow backwards if he should not cease to breathe on spurning Oenone. Moreover, Chaucer's invention of the scenes at Deiphebus' house in Books II and III brings Helen of Troy as a character into his poem, and links between the love of Troilus and Criseyde and the fate of Troy are also links with the love of Paris and Helen, as is hinted in the fourth book, although Troilus resists Pandarus' interpretation ('Thenk ek how Paris hath, that is thi brother, | A love; and whi shaltow nat have another?' (iv. 608–9)).

In *Heroides* Helen is writing to Paris *before* she has given herself to him, so that the epistle expresses all her anxiety and uncertainty, her attempts to counter Paris's persuasions, as well as her attraction towards him. In this, a series of Helen's remarks find parallels in Chaucer's presentation of Criseyde, and especially where Chaucer departs most from his source in characterizing his heroine: in her quandary over her prospective lover in Book II, and in her sense of her reputation in Book V. In Criseyde's recurrent concern with her honour parallels may be found with Helen's own insistent anxiety about her honour (*Heroides*, xvii. 13–18, 111–14); her musing with herself over whether or not to love Troilus has parallels in thought and mood with Helen's mixture of interest and caution at Paris's advances (*Heroides*, xvii. 35–40); she acknowledges without false modesty her own beauty (ii. 746), just as Helen does (*Heroides*, xvii. 37–8); and both women are fearful of rumours and of boasters (ii. 724; v. 1611–13; cf. *Heroides*, xvii. 17–18, 147–51). The interest for Chaucer of Helen's letter was that it took the reader sympathetically into the mind of a woman— whose name subsequently became a byword for infidelity—before she took her fateful decision, poignantly showing with some pathos and irony how concerned Helen was for her good name and for acting rightly, as well as how fatally attracted to her admirer.

It is noticeable that, when Chaucer is expanding on his Criseyde's realization of her impossible position in the Greek camp, he again makes his heroine express sentiments recalling those of Ovid's Helen. The example of Helen was probably in Benoît's mind when portraying Briseida in the *Roman de Troie*. Criseyde's lament (modelled on Briseida's in the *Roman de Troie*) anticipating the eternal ruin of her reputation (v. 1054–64) and exclaiming 'O, rolled shal I ben on many a tonge!' parallels the irony of Helen's own concern for her reputation (*Heroides*, xvii. 207 ff.), while Criseyde's lament that 'Al be I nat the first that dide amys, | What helpeth that to don my blame awey?' (v. 1067–8) finds an echo in Helen's retort to Paris's argument for the usualness of love (*Heroides*, xvii. 41–2, 47). It is striking that these parallels do occur at some of the points where Chaucer has most revised the characterization of his Criseyde, so that there are telling resemblances in the common focus in Ovid and Chaucer upon a woman of notorious fame before she takes the course that will lead to notoriety.

Few authors had more influence in the Middle Ages than Ovid and it is as well to remember, by way of conclusion, how a medieval poet may have encountered Ovid not only directly but also indirectly through the work of

translators and allegorizers—although Chaucer's reliance on the mythographical tradition remains debatable (see below, pp. 174–6). Influence on
some of Chaucer's poems has been detected from such works as the *Ovide
moralisé*, a translation and allergorization of the *Metamorphoses* in French
verse, and the fourteenth-century Italian translation of the *Heroides* by
Filippo Ceffi, although in the case of *Troilus* any discernible influence is
slight. Details from Ceffi's translation of Oenone's letter to Paris in
Heroides, v, may possibly be echoed in Pandarus' account of that letter (i.
659–65), while faint resemblances to the *Ovide moralisé* have been
suggested in such classical references in *Troilus* as Criseyde's allusion to
'the feld of pite' (iv. 789) and Cassandra's to the burning of Thebes (v.
1510).

See the following convenient editions of Ovid in the Loeb Classical Library: *Ars amatoria*
    and *Remedia amoris* in *The Art of Love and Other Poems*, ed. and trans. J. H. Mozley
    (London and New York, 1929); *Fasti*, ed. and trans. Sir James George Frazer (London
    and Cambridge, Mass., 1931); *Heroides and Amores*, ed. and trans. Grant Showerman
    (London and Cambridge, Mass., 1914); *Metamorphoses*, ed. and trans. Frank Justus
    Miller (2 vols.; London and Cambridge, Mass., 1916); *Tristia and Ex Ponto*, ed. and
    trans. A. L. Wheeler (London and New York, 1924).
For general studies of Chaucer and Ovid, see Helen Cooper, 'Chaucer and Ovid: A
    Question of Authority', in C. A. Martindale (ed.), *Ovid Renewed* (Cambridge, 1988),
    71–81, and Fyler, *Chaucer and Ovid*.
For some particular aspects of Ovid and *Troilus*, see Arn, 'Three Ovidian Women'; Fyler,
    '*Auctoritee*'; Shannon, *Chaucer*; and Wetherbee, *Chaucer*, ch. 3.
The *Ovide moralisé* is edited by C. de Boer (5 vols.; Amsterdam, 1915–38). See Lowes,
    'Chaucer and the *Ovide Moralisé*', and Meech, 'Chaucer and the *Ovide Moralisé*'.
    Criseyde's reference to 'the feld of pite, out of peyne, | That highte Elisos ...' (iv.
    789–90), may be compared with the *Ovide moralisé*, xiv. 827–30 and xi. 167–8. Cooper
    ('Chaucer and Ovid ...') argues against Chaucer's knowledge of the *Ovide moralisé*.
For the Filippo Ceffi translation, see Meech, who compares the two Latin lines, now
    considered spurious, at the end of some texts of *Heroides*, v ('Ipse repertor opis vaccas
    pavisse Pheraeas | Fertur et a nostro saucius igne fuit') with Chaucer's stanza, i. 659–65
    ('Phebus, that first fond art of medicyne, ... Al for the doughter of the kyng Amete' (i.
    659, 664)), and with Ceffi's expansion of the Latin: 'E lo nominato Iddio Febo, *che da
    prima trovò la scienza della medicina*, già per amore diventoe pastore, amando *la bella
    figliuola del Re Ameto*; e sappiendo a tutte gravezze dare rimedio, da amore solamente non
    si seppe guardare' ('Chaucer and an Italian Translation', 112–3).

Le Roman de la rose

As the Prologue to his *Legend of Good Women* reveals (F 327–31), Chaucer
had himself translated into English at least a part of the *Roman de la rose*—
that seminal allegorical analysis of the young Lover's experience of
encountering love and first approaching his lady—and there are parallels
and resemblances at a number of levels between the *Roman* and Chaucer's
presentation of the Troilus story. Parallels emerge between the figure and

experience of Amant and Troilus, in stock situations and phrasing, and in the Lover figure's relations with his advisers, for the *Roman* is a poem bursting with lore, both in the opening dream in the garden of love by Guillaume de Lorris (*c*.1230), and in the encyclopædic, discursive continuation of the poem by Jean de Meun (*c*.1280). The influence of both parts of the *Roman* is felt in *Troilus*, but the distinctive character of the *Roman*'s two original parts by two different authors is overridden in *Troilus*, so that echoes of the Lorrisian vision of the experience of love and the sententious wisdom of Jean de Meun's section are thrown more closely together in Chaucer's poem.

In a sequence of changes Chaucer moves the experience of his Troilus into closer parallel with that of the generic figure of the Lover in the *Roman*. The moment when Troilus first sees Criseyde in the temple is rewritten by Chaucer in terms which recall an archetypal scene in the *Roman*, with the angered God of Love shooting Troilus with his bow (i. 206–10) as he shoots Amant in the garden of the *Roman de la rose* (cf. *Romaunt*, 1715 ff.). The ensuing excursus in *Troilus* on submission to love contains several echoes of what Amant finds in the *Roman*, such as the warning to the proud against scorning love (i. 234–5; cf. *Romaunt*, 882–4). The further warning in *Troilus* ('Men reden nat that folk han gretter wit | Than they that han be most with love ynome; | ... The worthiest and grettest of degree ...' (i. 241 ff.)), echoes the account of the properties of the Well of Narcissus in the *Roman* ('Full many worthy man hath it | Blent, for folk of grettist wit | Ben soone caught heere and awayted ...' (*Romaunt*, 1609–11)), and, when Troilus makes 'a mirour of his mynde' (i. 365–6), this recalls the God of Love's account to the Lover of the comforts of 'Swete-Thought' (*Romaunt*, 2804–8).

When Chaucer augments the conventional feelings and gestures of the lover figure from *Filostrato*, he does so in the tradition of the life of the lover in the *Roman*, whose typical isolation and whose demoralization by *dangier* (the lady's reserve) may be discerned behind the passive and tentative disposition of Troilus. By such added resemblances to the *Roman* the experience and expression of Troilus as a lover are set in more archetypal patterns. The resolution of Troilus after the arrow from Love's bow ('Criseyde for to love, and nought repente' (i. 392)) parallels the declaration of Amant after he has been hit by Love's arrow ('I wole ben hool at youre devis ... And repente for nothyng' (*Romaunt*, 1974–6)). Again, in adding Troilus' declaration after Pandarus has agreed to help his suit to Criseyde ('My lif, my deth, hol in thyn hond I leye' (i. 1053)), Chaucer uses terms comparable with those in which Amant commits himself to the God of Love ('My lyf, my deth is in youre hond; | I may not laste out of youre bond' (*Romaunt*, 1955–6)). Although strongest in Book I, these patterns of resemblance continue in the later books. Troilus' concern for secrecy parallels that of Amant; his wish to kiss the doors of Criseyde's house as an empty shrine (v. 551–3) echoes the God of Love's description to Amant of a lover's life, of watching outside the lady's house and kissing the door (*Roman*, 2521–4); and the English Troilus echoes Amant's declaration of faithfulness to Cupid ('And lyve and dye I wol in thy byleve' (v. 593; cf. *Roman*, 10337–8)).

As the friend and sententious counsellor of the lover figure, Chaucer's Pandarus shares something of the role and sentiments of Ami, the Lover's friend in the *Roman*, with some admixture from the figure of La Vieille, the old woman who knows so much about love. Pandarus' advice on secrecy and discretion (iii. 281 ff.), on prudence (iii. 1634), against divided energies (i. 960–1), and on the need to sample everything and judge by experience (i. 637 ff., 927–8), all recall the *Roman* (cf. *Roman*, 9823 ff., 8231–3, 2233–4, 21529–52). Pandarus' account to Criseyde of how he overheard Troilus complaining of his love 'In-with the paleis gardyn, by a welle' (ii. 508) provides a setting reminiscent of the *Roman* in which the role of Pandarus ('Tho gan I stalke hym softely byhynde' (ii. 519)) recalls that of the God of Love stalking his prey through the garden (*Romaunt*, 1450–4), while, in his energy and drive as the shaping power of the story in moving the lovers towards union, Pandarus may be seen in relation to the figure of Genius in the poem by Jean de Meun. It is tempting to see the distinction between the two parts of the *Roman* worked out in the distinct natures of Troilus and Pandarus: to see Troilus as the distillation of the poem by Guillaume de Lorris and Pandarus as the essence of the poem by Jean de Meun, with the interrelation of the two friends and Pandarus' scheme for Troilus in a parallel with Jean de Meun's design on the poem by Guillaume de Lorris. But in drawing on the tradition of the *Roman* Chaucer did not always keep any correspondence between Troilus and Pandarus, and the Lover and the Friend, separate from the larger influence of the *Roman* as a whole.

There is a series of passages in *Troilus* which recall moments in the *Roman* where the opposition between Love and Reason is at issue. An important speech like Amant's, reaffirming his allegiance to the God of Love and rejecting the persuasions of Reason, seems at several points to lie behind the language of commitment used by both Chaucer's lovers (v. 593; iii. 1493 ff.). When Chaucer's Troilus is described distilling into tears 'As licour out of a lambyc ful faste' (iv. 520), this may not only recall a phrase in the *Roman* but also draw ironically on the context, where Reason is encouraging the Lover to spurn Love:

'Je voi maintes foiz que tu pleures
conme alambit seur alutel . . .

. . . . . .
qu'onques hom en nule seson,
por qu'il usast d'entendement,
ne mena deul ne marrement'   (*Roman*, 6352–3, 58–60)

'Many times I see you crying as an alembic does into an aludel . . . For no man at any time, provided that he used his understanding, ever encouraged sorrow or sadness . . .' (Dahlberg, 125)

Troilus' oath as he prepares to kill himself (iv. 1208) may likewise recall a whole context, where the Lover commits himself to Love and spurns Reason (*Roman*, 10339–44). It is no accident that in lines describing Criseyde's beauty at her first appearance in the poem ('As doth an hevenyssh perfit creature, | That down were sent in scornynge of nature' (i. 104–5)) Chaucer apparently combines recollections of two descriptions in the *Roman*: of the God of Love as like an angel come down from heaven, and of the beauty of Reason, which it surpassed Nature to contrive (*Romaunt*, 916–17, 3205–11). Nor is it accidental that Criseyde in Book II should twice echo some of the aphoristic wisdom used by Reason when persuading the Lover to renounce Love (ii. 167–8, 716–18; cf. *Roman*, 5630–2, 5711–15).

Together such echoes point to the influence of the example available in the *Roman*, as a work which combines an exploration of the idealizing experience of romantic love with a serious attempt to set all love in the context of a larger understanding. As a whole, the *Roman* presents a comprehensive anatomy of love, first in the allegory of the Garden of Love, and then as that love comes to be explored and understood through the instruction of Reason and Nature and the example of Venus, as Jean de Meun draws on a range of sources, especially the allegory of Alan of Lille's *De planctu naturae* (The Complaint of Nature). Some of the moral seriousness and sense of the universal in *Troilus*, the sustained interest in the idea of nature and in Venus which Chaucer builds into his version of the story, and the way that idealization of a particular love invites broader and deeper reflection on the nature and limits of human love, may well have been in part suggested to Chaucer by the implications of the analysis of love in the *Roman de la rose*, which in turn influence later French poetry known to Chaucer.

Le Roman de la Rose, ed. Felix Lecoy (Classiques Français du Moyen Âge; 3 vols.; Paris, 1965–70). There is also a convenient edition by Daniel Poirion (Paris, 1974), and an English translation by Charles Dahlberg (Princeton, NJ, 1971). The *Romaunt of the rose*, a fragmentary Middle English translation, may be found in Benson (1987), and quotation has been made from this version wherever possible. The authorship of the *Romaunt* has

been disputed and only Fragment A is generally accepted as Chaucer's work; for discussion, see Benson (ed.), *The Riverside Chaucer*, 1103–4.
On the *Roman* and *Troilus*, see especially Lewis, *The Allegory of Love*; Muscatine, *Chaucer*; and Wimsatt, 'Realism'.

Guillaume de Machaut

Of the relation between *Troilus* and the poems of the fourteenth-century French poet Guillaume de Machaut (*c*.1300–77), it is useful to speak of the congruences between Chaucer's poem and the courtly narratives of Machaut's *dits amoureux* and his lyrics, rather than of sources and borrowings. Of Chaucer's familiarity with such poems as Machaut's *Jugement dou Roy de Behaingne* and *Remède de fortune* the verbal texture of the *Book of the Duchess* leaves no doubt. In *Troilus*, however, the example of Machaut's poems has been absorbed further below the surface of the English poem. The exception is the set-piece of Antigone's song, which draws together many phrasal borrowings from a whole set of Machaut lyrics (including two lays, two *balades*, and a virelai), and so suggests how influential a model in phrase and theme Machaut's work still was for Chaucer long after composition of the *Book of the Duchess*.

It is some of the most individual achievements of Machaut's *dits* that can be paralleled—however modified or transformed—in Chaucer's writing: their structuring of narrative and lyric; their fusion of Boethian and love themes; their characterization of a narratorial persona; and their refined and graceful representation of lovers. In the *dits amoureux* Chaucer could find accomplished examples of extended narratives within which lyric pieces play a role so important as to produce a hybrid lyric–narrative form. In the *Remède de fortune* Chaucer knew a poem where the timid lover's protracted courtship of his lady is interpolated with the counselling of the lover by the figure of Esperance, whose advice draws extensively on Boethian materials, especially from Books II and III of the *Consolation*. The *Remède de fortune* presents an allegorical inner debate constructed on the model of the *Consolation of Philosophy*, and, like Lady Esperance, Pandarus brings hope and counsel to Troilus, sometimes echoing the words of Esperance (as when i. 813–19 recalls *Remède*, 1636–48). If the dialogue between Boethius and Philosophy was one of Chaucer's models for the dialogue of Troilus and Pandarus, the *Remède* had anticipated and perhaps prompted Chaucer in making the crises of a lover's career the context for a rehearsal of Boethian themes, a lover's consolation of philosophy.

While the guise of an outsider to love is deployed with original flair in the narrating of *Troilus*, the fullest development of a comparable technique would have been known to Chaucer from the *dits* of Machaut: the poet-figure who overhears and humbly records in the *Behaingne* and the *Dit de la fonteinne amoureuse* the complaints of those who love; the naïve and innocent narrator of the *Jugement dou Roy de Navarre* who is punished for what he has written about the way women love; the characterization of the timid lover in the *Voir-Dit*. In none of these poems is Machaut's characterizing of the poet-figure straightforwardly a 'source' for the Chaucerian narrating voice, yet the persistence in Machaut's work of his interest in the possibilities of such a persona provides one of the principal congruences in the period for Chaucer's own development of the narrator as 'outsider'.

Interesting parallels also occur between the presentation of the lover's experience in the *dits amoureux* and Chaucer's characterization of Troilus and Criseyde, so that the differences of Troilus from Troiolo are frequently resemblances to the lover figures in Machaut's poems. This is especially so in the representation of the hero's first perception of his lady, and in his expectation of how and where the affair will proceed. The naïve, impressionable, idealistic sides of Troilus bear a marked resemblance to Machaut's lovers, who are wont to recollect the overpowering experience of seeing their ladies, their strong desires for modest recompense, and the speechlessness that overcame them when they tried to express their love to their ladies. Troilus' experience of first seeing and falling in love with his lady is an overwhelming experience, producing an emotional prostration very reminiscent in both attitude and phrasing of the experience of the Lover in the *Remède* and the knight in the *Behaingne*. By contrast with Troiolo, the absence of any explicit sense of sexual ambition in Troilus and his more oblique courtliness, bashfulness, and hesitancy find many parallels with the *dits amoureux*.

Although Machaut is by no means the only available source for such features, his series of poems—well known to Chaucer—are their most copious, quintessential, and accomplished expression. There are certainly verbal echoes between *Troilus* and Machaut's poems in some of the archetypal terminology and circumstances, the phases and processes of courtship. The singular impression of the lady's look or first sight, the astonishment of the lover, the drowning in tears and fainting for distress draw on a common courtly stock of that *sentement* disclaimed in the second proem of *Troilus* (ii. 13) but prayed for in the third (iii. 43–4). Patterns emerge in the experience of Troilus and the lovers of the *dits* (as in the lover's faint after his complaint in the *Remède* (1490–3), which may be compared with Troilus' swoon in Book III). There is a common emphasis in Machaut's lovers and Chaucer's Troilus on a certain distraction and abstractedness: in the lovers' first meeting in Book III most aspects of Troilus' diffident behaviour have a counterpart in the knight's recollection of his courtship of his lady in *Behaingne* (463–76), including the prepared and then forgotten speech and the intense emotional agitation.

Such distraction and sense of 'distance' from the lady allow for more emphasis on the expression of emotion by the English Troilus and Machaut's lover figures, and their disposition to 'complaint' (see below, pp. 166–9), prompts both poets to incorporation of lyric set-pieces. The model of how a courtly relationship proceeds—through confession of one's feelings to the lady, petition for her mercy, and the eventual concession of this 'grace'—is the model of procedure for Machaut's lovers, as for Troilus and other courtly lovers. In Machaut's poems the emphasis on the life of *sentement*, the lack of outward action by comparison with the focus on the feelings, mean that the *dits* point more towards the psychology, the inner debates of their lovers.

The differences between Criseida and Criseyde may reflect similar taste. The antifeminist attitude towards Criseida at the close of *Filostrato* is further from *Troilus* than the regretful attitude of the knight in *Behaingne* to the lady who has shown every virtue except loyalty to him. The ladies of Machaut's poems were perhaps for Chaucer the essence of the courtly convention of a properly distant, reserved, and cautiously deliberating lady, attended by her *daunger* or standoffishness. For Pandarus, Criseyde's very virtuousness means she will have the quality of pity (i. 897–900), a consolation similar to that offered the Lover by Esperance in the *Remède* (1671 ff.), a poem where indeed the lover receives only a look from his lady. In so far as Chaucer refines away the sensuality and impulsiveness of Criseida, and replaces it with a cautious, reflective character whose *daunger* is one spring of his radically rewritten second book, then Chaucer is moving the personality of his Criseyde much nearer to a type familiar from Machaut's poems. Likewise, her ringing identification of the *moral vertu* for which she loved Troilus (iv. 1667–80) is as comparable to the reflective moral deliberations of Machaut's ladies as it is distant from the simple sensuality of Criseida.

Chaucer's recourse to not one but a series of Machaut lyrics in his song for Antigone (ii. 827–75) suggests that he had registered the example offered in Machaut's poems of women analysing and reflecting on the virtues of their lovers, men committed to a virtuous service of their ladies. In Machaut's *Le Paradys d'amours* the lady extols the paradise in which she is living; in the *Mireoir amoureux* the lady praises her lover—as a mirror of excellence, a source of joy and worth which puts an end to all sorrow, a sapphire and a sun which makes all good things to blossom (15–20, 85–7, 122–6)—in a sequence of epithets which finds many echoes in the song of Antigone (ii. 841–5). By creating his own lyric on the model of Machaut's sustained sequence of both courtly and moral epithets, and also drawing on imagery of committed permanence and assurance voiced by the speakers in Machaut's lyrics, Chaucer here brings into *Troilus* specifically (and throughout the poem more pervasively) that superlative and idealizing but earnest rhetoric in presenting and analysing the experience of lovers, for which the poems of Machaut were such an accomplished and familiar model.

Œuvres de Guillaume de Machaut, ed. Ernest Hoepffner (Societé d'Anciens Textes Français; 3 vols.; Paris, 1908–21) includes the *Jugement dou Roy de Behaingne*, the *Remède de fortune*, the *Dit de la fonteinne amoureuse*, and the *Jugement dou Roy de Navarre*. For selected translations, see B. A. Windeatt, *Chaucer's Dream Poetry: Sources and Analogues* (Cambridge, 1982). For the *Voir-Dit*, see the edition by P. Paris (Paris, 1875).

Guillaume de Machaut: Poésies lyriques, ed. V.-F. Chichmaref (2 vols.; Paris, 1909), and *La Louange des dames by Guillaume de Machaut*, ed. Nigel Wilkins (Edinburgh, 1972). For the lay *Mireoir amoureux*, see Chichmaref, 362–70.

On the 'congruences' between Machaut's poems and *Troilus*, see Wimsatt, 'Guillaume de Machaut', which extends the work on Chaucer and Machaut begun by Kittredge, 'Antigone's Song of Love' and 'Chaucer's *Troilus*'. See also William Calin, *A Poet at the Fountain: Essays on the Narrative Verse of Guillaume de Machaut* (University of Kentucky Press, 1974), and James I. Wimsatt, *Chaucer and the French Love Poets* (Chapel Hill, NC, 1968).

The Story of Thebes

> Whan he was come unto his neces place,
> 'Wher is my lady?' to hire folk quod he;
> And they hym tolde, and he forth in gan pace,
> And fond two othere ladys sete and she,
> Withinne a paved parlour, and they thre
> Herden a mayden reden hem the geste
> Of the siege of Thebes, while hem leste.   (ii. 78–84)

So Chaucer introduces into his Trojan setting the act of reading the story of an older and fated city further in the past. The Theban war was included and dated as an historical event in the 'universal chronicles' and chronologies of the ancient world compiled by medieval historians, for whom Thebes was part of the succession of pagan empires ('as regnes shal be flitted | Fro folk in folk …' (v. 1544–5)). As Cassandra's genealogy suggests, in tracing Diomede's forebears for Troilus in Book V, it was imagined that the grandfathers and fathers of those at the siege of Troy had been involved in the war of the Seven against Thebes, which would thus be a past but not a distant event for Troilus and Criseyde. It is characteristic of Chaucer's historical sense to provide the Trojan 'present' of his narrative with its own sense of a past, and the story of Thebes may be seen as a significant source lying behind Chaucer's composition of *Troilus*. The idea and outline of the siege of Thebes is itself the most important 'source', to which Chaucer had access through several specific texts, of which he made varying uses.

The most fundamental of these was the *Thebaid*, an epic in twelve books by the Roman poet Statius (*c*.AD 45–96), which was available to medieval readers in manuscripts with an apparatus of glosses and *scholia*. There was in addition a twelfth-century French *roman d'antiquité* drawn from Statius, the *Roman de Thèbes* (which was followed by prose redactions), and Chaucer also knew well Boccaccio's *Teseida* (i.e. 'Theseid'), which he used as the main source of the *Knight's Tale*. Here Boccaccio attempts to write a twelve-book epic centred on Theseus and the loves of the two Theban knights Palamon and Arcita captured at Theseus' sack of Thebes.

Chaucer drew on all such sources in his concern to build a network of allusions to this tale of an earlier city into his own poem and, after a more selective use of the *Thebaid* in his earlier works, in *Troilus* he shows his recollection of all twelve books of the epic. On his first visit in Book II the bookish Pandarus is immediately inquisitive about Criseyde's book, which enables Criseyde to tell how they have been hearing the fateful stories of King Laius and Oedipus, and of Amphiaurus. But while the lady refers to her book as a romance (ii. 100), her bookish uncle makes a point of declaring that he knows the story in its twelve books, presumably referring to Statius' Latin poem (ii. 108). When Criseyde refers to the ancient seer Amphiaurus as 'the bisshop', her description absorbs the terminology of the *Roman de Thèbes* (where Amphiaurus is called 'evesque' (2026, 5053, 5079)), and her account of what they have so far read apparently describes the *Roman de Thèbes* rather than the *Thebaid* (which starts *in medias res* with Oedipus already blind, while the *Roman de Thèbes* and its prose redactions provide a narrative of Oedipus' earlier life and killing of Laius). Chaucer's Trojan Criseyde thus hears an account of what for her is a historical event, but she listens to it in a book which Trojan Pandarus apparently knows in a Roman version and which Criseyde has in its twelfth-century romance form. And, since manuscripts of the *Roman de Thèbes* sometimes contained copies of the *Roman de Troie* in addition, there is a sense in which Chaucer's Criseyde could be listening to a book which also contains the story of one of her selves, Benoît's Briseida.

It is Cassandra's speech (v. 1457–1519) which most strongly confirms Chaucer's interest in relating a sense of the whole story of Thebes to his own. The clotted shorthand quality of Cassandra's speech is explained in part by its origin in Chaucer's translation of a twelve-line Latin 'Argument' which serves to outline the content of all the twelve books of Statius' epic. Indeed, the Latin text of such an *argumentum* is incorporated into the text of *Troilus* as a kind of miniature *Thebaid*, inserted between two English stanzas after v. 1498 in all but two of the extant *Troilus* manuscripts, to serve as a confirmatory gloss. Other evocations of Theban history tend to be more oblique, in the form of brief allusions to Theban figures, and in some echoes of the epic machinery of Statius' poem. A number of Chaucer's allusions to classical myth may have been prompted by allusions in the *Teseida* or the *Thebaid* (the story of Apollo (i. 659–65); of Adonis (iii. 720–1); of Alcmena (iii. 1428)) rather than in other available sources. The possible influence of the *scholia* to the *Thebaid* has also been discerned in some of Chaucer's classical allusions, such as the 'blody cope' of Mars (iii. 724) or the 'feld of pite, out of peyne' (iv. 789). Chaucer also had recourse to the Theban story for the names of his minor characters, such as Criseyde's niece Antigone or her mother 'Argyve' (iv. 762), which has inevitably been linked with Cassandra's later reference to Argia, wife of Polynices ('And of Argyves wepynge and hire wo . . .' (v. 1509)). Argia, a princess of Argos, was the sister of the wife of Tideus, and thus an aunt of Diomede: it follows that Criseyde and Diomede would be cousins.

From the *Thebaid* Chaucer could also recall the epic machinery of invocation: the invocation of the fury Tisiphone in the first proem of *Troilus* (recalling Oedipus' invocation of Tisiphone at the opening of the *Thebaid*), as of Clio and Calliope in the second and third proems, and of the Furies and 'cruel' Mars in the fourth, are all paralleled in Statius, as is the allusion to the Parcae in the opening of the fifth book, or to the 'fatal sustren' at iii. 733. For from the *Thebaid* Chaucer could derive a strong sense of the terrible and violent inevitability in the downfall of Thebes and the implacable enmity of Juno, to which he adds explicit reference in *Troilus* when Criseyde recalls how Juno sent Tisiphone to afflict Athamas, king of Thebes, with madness (iv. 1538–40), and when Troilus prays that Juno be not as cruel to the 'blood of Troie' as she was to that of Thebes (v. 599–602), although with hindsight the reader knows this prayer to be in vain. Chaucer could also find in the *Thebaid* a vehement view of the vanity 'Of Jove, Appollo, of Mars, of swich rascaille!' (v. 1853), and the sceptical maxim 'Drede fond first goddes' repeated by Criseyde (iv. 1408) is found in Statius as in other contexts where Chaucer might know it (there is a possible link between Criseyde's scepticism and that of Capaneus in the *Thebaid*). This is the darkening world of the fateful end of Amphiaurus, of Capaneus, and of Oedipus as Statius recounts them, all of which are alluded to by Chaucer's characters (for Capaneus, see ii. 1145, v. 1504–5; for Oedipus, see iv. 300–1, and cf. *Thebaid*, i. 46–87). To maintain during a reading of *Troilus* a continuous comparative reference to the *Thebaid* is to highlight differences in the role of narrative, in the interpretation of fate and history, the values represented by the gods, and the effect of virtue in human nature. Yet the world of the *Thebaid* is also that of Statius' epic hero Menoeceus, who dies in the service of the commonweal and whose virtue is rewarded with a celestial flight and a place among the stars. From Statius Chaucer could draw a powerful sense of the epic of Thebes which may have coloured his view in his Trojan poem of how pagan heroes lived and met their fate.

Chaucer's more immediate source for the ascent of Troilus' soul is Boccaccio's *Teseida* (xi. 1–3), and *Troilus* is marked by a series of close verbal imitations of Boccaccio's Theban epic, particularly in Book V. Its opening line ('Aprochen gan the fatal destyne') is remembered from the opening line of the ninth book of *Teseida* ('Già s'appressava il doloroso fato'); the account of the Parcae as instruments for executing Jove's decrees (v. 3–4) could be recalled from the opening of the *Thebaid* (i. 212 ff.); and the description of time passing (v. 8–11), closely imitates the opening of the second book of the *Teseida*. Indeed, it is noticeable how the elaborate openings to the books of the *Teseida* have lodged themselves in Chaucer's memory, as when he seems to echo the opening of the first book of the *Teseida* ('O *sorelle* castalie, che nel monte | Elicona *contente dimorate*') in the address to the Muses which closes the third book of *Troilus* ('Yee sustren nyne ek, that by Elicone | In hil Pernaso listen for t'abide . . .' (iii. 1809–10)).

Chaucer's inclination to draw on the *Teseida* for heightened rhetorical descriptions of passing time continues in the fifth book when he uses the description of dawn from *Teseida* ('Il ciel tutte le stelle ancor mostrava' (*Teseida*, vii. 94)) to emphasize the early hour when Troilus sends for Pandarus after his first night of separation from Criseyde ('On hevene yet the sterres weren seene' (v. 274 ff.)). It is also to the account of Arcita's cremation in the *Teseida* (and more generally perhaps to the *Thebaid*) that Chaucer goes for details of pagan rites in the directions Troilus gives for his funeral (v. 298 ff.). Here Chaucer is drawing on his Theban sources for details of ancient custom and pagan belief, which is part of the historical texture Chaucer strives to include in his *Troilus*. When Chaucer draws to a close his own 'litel boke' by echoing a list of classical poets ('Virgile, Ovide, Omer, Lucan, and Stace' (v. 1792)) comparable to that of the *Filocolo* and of *Inferno* (iv. 85 ff.), and by recalling Statius' injunction to his *Thebaid* to follow in the footprints of the *Aeneid* (1791), the English lines convey something of that fusion through which Chaucer knows the story of Thebes both in itself and as it was mediated through the work of Boccaccio and through Dante's allusions to Statius' poem in the *Divine Comedy*.

Statius, *Thebaid*, ed. and trans. J. H. Mozley (Loeb Classical Library; London and New York, 1928). On the twelve-line Latin *argumentum* to the *Thebaid*, see Magoun, 'Chaucer's Summary', and on the *scholia*, see Clogan, 'Chaucer and the *Thebaid* Scholia'.

Le Roman de Thèbes, ed. G. Raynaud de Lage (2 vols., Paris, 1966–71). The *Roman de Thèbes* is found with the *Roman de Troie* in one manuscript (Paris, Bibliothèque Nationale, fr. 375) and with the *Roman de Troie* and the *Roman d'Eneas* in another (Bibliothèque Nationale, fr. 60), as discussed by Antonelli, 'Birth'.

Boccaccio, *Il Teseida delle nozze d'Emilia*, ed. Alberto Limentani, in *Tutte le opere di Giovanni Boccaccio*, ed. Vittore Branca, ii. (Verona, 1964). Selections are translated in Havely, *Chaucer's Boccaccio*. On Chaucer's use of the *Teseida*, see Piero Boitani, 'Style, Iconography and Narrative: The Lesson of the *Teseida*', in Boitani (ed.), *Chaucer and the Italian Trecento*, 185–99, and Pratt, 'Chaucer's Use'.

For aspects of the Theban theme in *Troilus*, see Anderson, 'Theban History'; Clogan, 'Chaucer's Use' and 'The Theban Scene', and Wetherbee, *Chaucer*, in addition to the now dated study by Wise, *Influence*.

Schibanoff ('Argus and Argyve') notes that a medieval moral allegory, *Super Thebaidem*, links the name Argia with the hundred-eyed Argus and compares Criseyde's lament at her lack of foresight (v. 744–5); Anderson suggests that the brooch Criseyde gives Troilus (iii. 1370–2) might be the 'brooch of Thebes' which belonged to Argia and is described in Chaucer's *Complaint of Mars* (245–60) as causing 'double wo and passioun' to the person who loses possession of it ('Theban History', 127–8).

Dante

What Dante meant to Chaucer as he composed his *Troilus* is suggested by the fact that Chaucer's passages of closest verbal imitation of the *Divine Comedy* (from *Paradiso*, xxxiii. 13–8, and xiv. 28–30) are placed at two of the most prominent and significant points in the poem: at the very centre of the work as Troilus nears the consummation of his love ('Whoso wol grace and list the nought honouren' (iii. 1262 ff.)), and in the lofty prayer to the Trinity which brings the poem to an end ('Uncircumscript, and al maist circumscrive' (v. 1863 ff.)). That Chaucer should here translate so closely from Dante speaks for his knowledge and appreciation of the *Comedy*. That he should make such effective borrowing but make it so briefly is also representative, however, of the relation between *Troilus* and Dante. Close verbal parallels are relatively few, but there are a range of significant parallels in both poets' sense of structure and of time, in patterns of mythological allusion, in the understanding of love, and in conceptions of the art of poetry, which suggest how Chaucer may have had the example of the *Divine Comedy* in mind as he worked on his *Troilus*.

Dante's narrative was scarcely imitable as such by Chaucer. His influence is often to be inferred through congruences and coincidings in very brief passages, in commonplaces and topoi, and this inference may be made because elsewhere in *Troilus* Chaucer shows a close grasp of the diction of passages from the *Comedy*. If Chaucer could draw on Dante as widely and selectively as even the relatively few close verbal parallels would suggest, it is likely that he had also absorbed the lesson of other aspects of the *Comedy*.

In order to clarify the distribution of possible parallels between the *Comedy* and *Troilus* it may help to set out the main resemblances in tabular form, though they are very different in type and closeness.

BOOK I

| 1 | *Purgatorio*, xxii. 56 | ('The double sorwe …') |
| 6 | *Inferno*, ix. 37–51 | ('Thesiphone …') |

BOOK II

1–3	*Purgatorio*, i. 1–3	('The boot … of my connyng')
64–70	*Purgatorio*, ix. 13–15	('The swalowe Proigne …')
925–31	*Purgatorio*, ix. 19–21	('How that an egle …')
967–71	*Inferno*, ii. 127–32	('But right as floures …')

BOOK III

39–42	*Paradiso*, xxxiii. 16	('Lady bryght, for thi benignite')
45	*Purgatorio*, i. 9	('Caliope, thi vois be now present')
1257	*Purgatorio*, i. 19	('Venus … the wel-willy planete')
1261–7	*Paradiso*, xxxiii. 13–18	('Whoso wol grace …')
1387–93	*Purgatorio*, xx. 106–8	(Midas … Crassus … *affectis wronge*)
1419–20	*Purgatorio*, xix. 4–6	('And estward … *Fortuna Major*')
1625–8	*Inferno*, v. 121–3	(Remembering past happiness)
1693	*Paradiso*, xix. 8	(Joy not to be written with ink)
1784	*Paradiso*, xix. 34	('As fressh as faukoun …')

| 1807–8 | *Paradiso*, viii. 7–8 | ('Lady bryght ... doughter to Dyone') |

BOOK IV

22–4	*Inferno*, ix. 43–51	('O ye Herynes ...')
225–7	*Inferno*, iii. 112–14	('... as in wynter leves ben bir-aft')
239–41	*Inferno*, xii. 22–4	('Right as the wylde bole ...')
473	*Inferno*, ix. 44	('But down with Proserpyne ...')
776	*Paradiso*, i. 20–1	('Til I my soule ... unshethe')
927	*Inferno*, xxxi. 4–6	('cause of flat than egge')
1188	*Inferno*, v. 4–6	('The doom of Mynos ...')
1538–40	*Inferno*, xxx. 1–12	('As wood as Athamante ...')

BOOK V

601	*Inferno*, xxx. 1–12	('As Juno ... unto the blood Thebane')
817	*Paradiso*, xviii. 21	('Paradis stood formed in hire yën')
1541–5	*Inferno*, vii. 78–81	('Fortune ... permutacioun ...')
1807–27	*Paradiso*, xxii. 133–5	('This litel spot of erthe ...')
1863–5	*Paradiso*, xiv. 28–30	('Uncircumscript ...')
1869	*Paradiso*, xxxiii. 1	('mayde and moder thyn benigne')

One pattern which emerges is in the way that Chaucer uses echoes from Dante at such prominent, structurally important points as the openings and closings of books, and at transitions within books. There are echoes of the *Comedy* in the proems to all the first four books, and Chaucer has recourse to material paralleled in Dante for part of the close of the central third book, which thus both opens and closes with Dantean echoes. It is noticeable that some of Dante's beginnings have lodged most firmly in Chaucer's mind: he uses the opening of the first canto of *Purgatorio* to start both his second and third books. A number of other passages from the *Comedy* echoed in *Troilus* are either at the very beginning of a canto or near it. The opening of *Inferno*, xxx—'Nel tempo che Iunone era crucciata ... contra'l sangue tebano' (In the time when Juno was enraged ... against the Theban blood)—is echoed twice at solemn moments of avowal and prayer (iv. 1538–40; v. 599–602). The grim opening of *Inferno*, v, where Minos stands at the entrance to the second circle of carnal sinners who subject reason to desire (*Inferno* v. 38–9), may have prompted Chaucer's added reference to Minos (who judges the suicides in *Inferno*, xiii. 94–6) when Troilus is about to kill himself (iv. 1188). The opening of *Purgatorio*, xix— 'quando i geomanti lor Maggior Fortuna | veggiono in orïente, innanzi a l'alba' (when the geomancers see their *Fortuna Major* rise in the east before dawn)—gives mysterious definition to the hour when the lovers' first night draws to its close (iii. 1419–20). The opening of *Paradiso*, viii, is echoed at the close of Chaucer's third book ('Thow lady bryght, the doughter to Dyone' (iii. 1807)) and chimes with the opening address of the book to Venus whose light 'adorneth al the thridde heven faire' (iii. 2), which derives from *Filostrato*. From the *first* line of *Paradiso*, xxxiii—'Vergine madre, figlia del tuo figlio' (Virgin Mother, daughter of thy Son)—perhaps comes the idea of having as the very *last* line of *Troilus*: 'For love of mayde and moder thyn benigne' (v. 1869).

A number of Chaucer's borrowings from the opening phases of Dante's cantos (e.g. from *Purgatorio*, i, ix, xix; *Paradiso*, viii) are representations of morning, and in a poem like *Troilus*—deeply concerned with time and concerned to mark its passing—one of the influences of Dante on Chaucer is in the expression of passing time. Other striking images, processes, or arresting scenes may lie behind passages in *Troilus*: to fly without wings (iii. 1263); unsheathing the spirit from the body (iv. 776); the sword that cures the wounds it makes (iv. 927); the covetous man compelled to drink molten gold (iii. 1391). Some of the scenes in Dante's hell and purgatory— as of the Furies, the dead souls like falling autumn leaves, Virgil addressing Statius—come back to Chaucer as he composes *Troilus* (iv. 22–4, 225–7; i.1).

For it is to Dante as the poet of hell and of heaven that Chaucer has returned. The pattern in resemblances between the *Comedy* and *Troilus* points to how Chaucer's imagination goes back to *Inferno* to etch more forcefully the hell of grief and suffering into which his two lovers fall in the fourth book. Not so clear a pattern of borrowing emerges from the other books, although there are significant resemblances to *Paradiso* at points in the third book of the lovers' bliss. In the experience of Troilus broad parallels and specific echoes emerge between Book I as hell, Book II as purgatory, and Book III as paradise, and the movement into the Christian conclusion of Book V is accompanied by references to *Paradiso*.

The influence of such cantos as *Inferno*, ix, *Purgatorio*, i, or *Inferno*, xxx, is sustained over a span of Chaucer's poem. *Inferno*, ix, suggests the Furies (iv. 22–4), Tisiphone (i. 6), and Proserpina (iv. 473–5). The opening of *Purgatorio*, i, with its reference (19–20) to the benevolent Venus as a morning star, may have reinforced her similar roles invoked in the consummation scene (iii. 1257, 1417). *Inferno*, xxx, opens with Juno's enmity towards Thebes (recalled by Troilus (v. 599–602)), proceeds to the madness of Athamas (recalled by Criseyde (iv. 1538–40)), then to Fortune's overthrow of the Trojans from high to low (*Inferno*, xxx. 13 ff.), and before long to Myrrha and her wrongful love, to which Chaucer adds an allusion (iv. 1138–9). In this canto with its series of *falsatori* Chaucer could find instances which at various levels of association struck chords with his work on the Troilus story of unfaithful love. The clustering of instances into a linked thematic treatment in Dante, and their appearance in *Troilus* rearranged and reapplied is a pointer to how the example of Dante's poem imbued the *Troilus* text with its associations.

In the opening phase of Book II there is a comparable example of how Chaucer's memory unfolds the material packed into the striking dawn opening of *Purgatorio*, ix. After describing how the Dawn, Aurora, came forth from Tithonus' arms (a reference recurring in Troilus' dawn-song (iii. 1464)) Dante relates how—at the hour when the swallow sings plaintively, perhaps in recollection of her ancient woes—he dreamed, in an hour of special insight, of an eagle preparing to swoop and catch him up in its claws:

> Nell'ora che comincia i tristi lai
>   la rondinella presso alla mattina,
>   forse a memoria de' suo' primi guai . . .   (*Purgatorio*, ix. 13–15)

> in sogno mi parea veder sospesa
>   un'aguglia nel ciel, con penne d'oro,
>   con l'ali aperte ed a calare intesa . . . (*Purgatorio*, ix. 19–21)

At the hour near morning when the swallow begins her plaintive songs, in remembrance, perhaps, of her ancient woes . . . I seemed to see in a dream an eagle poised in the sky, with feathers of gold, with open wings, and prepared to swoop.

Dante's allusion to the story of the sisters Procne and Philomela is in Chaucer's own mind when, at the beginning of Book II, he has his Pandarus half-hear the twittering of the swallow as he lies in bed in the morning:

> The swalowe Proigne, with a sorowful lay,
> Whan morwen com, gan make hire waymentynge
> Whi she forshapen was . . .   (ii. 64–6)

The resemblance between the rhymes on *tristi lai* and *sorowful lay*, and the common setting of a mortal sleeping in the morning and hearing the noise of the swallow's lamenting its transformation, are strong connections between Chaucer and Dante. Dante's scene frames and spans Chaucer's sequence of action, which begins as Pandarus hears the swallow and is drawn to a close as Criseyde lies listening to a nightingale, Philomela, outside her chamber and falls asleep to an apparently prophetic dream of an eagle which flies to her and exchanges his heart for hers (ii. 925–31).

Other parallels with Dante may similarly reflect Chaucer's concern to extend the significance of his narrative. In *Filostrato* there was a series of reminiscences of Dante in which Criseida is an unstable Beatrice, Pandaro an unreliable pseudo-Virgilian guide, and Troiolo is the uncertain pilgrim of love who climbs from hell to paradise, only to fall again into hell, his arc of progress confirmed by recourse to the language of Dante's Francesca da Rimini in both his upward and downward course (*Fil.* ii. 7; iv. 56, ll. 6–7, 59, ll. 1–4, 122, ll. 5–6). Yet Chaucer absorbs relatively little of this structure of parallels, although the extended similes of Troiolo reviving like a flower (ii. 967–70) and raging like a bull (iv. 239–41) survive into *Troilus*. In *Inferno*, v, Francesca tells how she and her lover were influenced towards their adulterous act by reading a romance about Lancelot and Guinevere, in whose story the character of Galehout acts as go-between or pander:

> 'Galeotto fu il libro e chi lo scrisse:
> quel giorno più non vi leggemmo avante'   (*Inferno*, v. 137–8)

'A Galehout was the book and he that wrote it—that day we read in it no further.'

In so far as Francesca's story comments on the influence of romance and romance traditions on behaviour—on romance itself as a go-between and pander to its reader—*Troilus* as romance has been seen as influenced by a reading of this canto of Dante's. Yet it is also possible to read *Troilus* in parallel with Dante so as to see Troilus evolving through love to become a virtual pagan counterpart to the lover-pilgrim of Dante's *Commedia*. Read in this way with reference to spiritually charged episodes in Dante, various scenes in *Troilus* implicitly comment on the partial and fragmented nature of the pagan characters' understanding. Pandarus and Criseyde in Book II show a very different level of understanding from that promised by Dante's dream of the eagle in *Purgatorio*, ix, or by the pageant of church history in *Purgatorio*, xxxii, with which the imagery of Criseyde's dream has also been compared. The leading of Troilus and the pilgrim towards their respective paradises by their guides might prompt a comparison between the meeting of Troilus and Criseyde at Pandarus' house and the ritual scene of Dante's reunion with Beatrice in the earthly paradise at the summit of Purgatory, with the sequence of accusation, swoon, and forgiveness in *Troilus*, Book III, in parallel with that in the scene with Dante and Beatrice in *Purgatorio*, xxx–xxxi.

Such parallels, and others that have been suggested, make the highest demands on the scholarly ingenuity of Chaucer's readers and their recall of Dante, for such parallels have to be perceived between poems of very dissimilar narrative surface and procedure. The experience of an uncertain narrator who defines his role in relation to old books and eventually arrives at a spiritual understanding of his story has been seen to offer a parallel in *Troilus* with the fictional role Dante creates for Statius in *Purgatorio*, as a prototype of his own assimilation of the literary past for a spiritual purpose. Chaucer draws on both the classical poets and on Dante's allusions to the classics in a way that suggests Chaucer endowed Dante too with the authority of a classic, and it is through Dante's use of Virgil, Ovid, and Statius for his spiritual purpose that some allusions to those poets enter *Troilus*—Criseyde's oath of fidelity by Athamas (iv. 1534–40) is prompted not so much by his story in Ovid's *Metamorphoses* (iv. 416–562), which has nothing to do with oath-breaking, but rather by Dante's allusion to him in the elevated opening of that thirtieth canto of *Inferno* which is devoted to the *falsatori*.

It is upon the model of poetic craft, of poetic seriousness and grandeur in Dante that Chaucer draws to express some of the seriousness of implication that he suggests in his interpretation of the story. In Book III of *Troilus* the language of prayer and invocation imports a solemnity and eloquence comparable to the Dantean passages it resembles:

> 'Donna, se' tanto grande e tanto vali,
>   che qual vuol grazia ed a te non ricorre,
>   sua disïanza vuol volar sanz'ali.
> La tua benignità non pur soccorre
>   a chi domanda, ma molte fiate
>   liberamente al dimandar precorre.'   (*Paradiso*, xxxiii. 13–18).

'Thou, Lady, art so great and so prevailing that whoso would have grace and does not turn to thee, his desire would fly without wings. Thy loving-kindness not only succours him that asks, but many times it freely anticipates the asking.'

> 'Benigne Love, thow holy bond of thynges,
> Whoso wol grace and list the nought honouren,
> Lo, his desir wol fle withouten wynges;
> For noldestow of bownte hem socouren
> That serven best and most alwey labouren,

> Yet were al lost, that dar I wel seyn, certes,
> But if thi grace passed oure desertes.'   (iii. 1261–7)

Such solemn expression, marking structurally important transitions, occurs again when Chaucer prefaces the account of Hector's death with a
recollection of Virgil's exposition of the role of Fortune (*Inferno*, vii. 73 ff.;
Troilus, v. 1541–7), or when *Troilus* is closed with the Trinitarian prayer
from early in *Paradiso*, xiv:

> Quell'uno e due e tre che sempre vive
>   e regna sempre in tre e 'n due e 'n uno,
>   non circunscritto, e tutto circunscrive.   (*Paradiso*, xiv. 28–30)

That One and Two and Three who ever lives and ever reigns in Three and in Two
and in One and uncircumscribed circumscribes all.

> Thow oon, and two, and thre, eterne on lyve,
> That regnest ay in thre, and two, and oon,
> Uncircumscript, and al maist circumscrive.   (v. 1863–5)

This prayer comes in *Paradiso* immediately after the affirmation:

> Qual si lamenta perchè qui si moia
>   per viver colà su, non vide quive
>   lo rifrigerio dell'etterna ploia ...   (*Paradiso*, xiv. 25–7)

Whoso laments that we die here to live above has not seen there the refreshment of
the eternal showers ...

and so comes with its own appropriateness after the flight of Troilus' soul
to the spheres. It is also intriguing that Chaucer ends his poem with this
passage shortly after having prayed to God (v. 1788) to send him 'myght to
make in som comedye'. The prayer for 'comedye' shortly before the flight
of Troilus' soul to the spheres (v. 1807–27) may also indicate that, while
the account of the flight is borrowed from Boccaccio's *Teseida*, the
implications of Troilus' ascent as *comedye* in transcending earlier *tragedye*
were suggested to Chaucer by a passage in the *Divine Comedy* that lies
behind Boccaccio's text: Dante's vision from the spheres in *Paradiso*, xxii.
133–8.

An awareness of Dante may have influenced Chaucer's conception of the
seriousness of the poet's vocation, his work of composition, and his
language. Among post-classical writers it is only to Dante and Petrarch
that Chaucer applies the term 'poet', and, as Chaucer refers to Petrarch as a

poet before translating one of his Latin prose works (*Clerk's Tale*, IV. 31–3), Dante is the one vernacular writer for whom Chaucer reserves the title. It is certainly noticeable how Chaucer talks of his own task of composition in terms paralleled in Dante: poetic composition as a voyage and the invocation of Calliope in *Purgatorio*, i. 1–9, are echoed in *Troilus*, ii. 1–4, and iii. 45; both poets make a single use only of the word *poesì/poesye* (*Purgatorio*, i. 7; *Troilus*, v. 1790).

For Chaucer's remarkable alertness in the second proem to the effects of change on language ('Ye knowe ek that in forme of speche is chaunge' (ii. 22 ff.)), there are also very few parallels among medieval poets beyond Dante's own discussion of such questions for the poet in his *De vulgari eloquentia* and *Convivio*. In *Convivio* (ii. 14, 83–9) Dante cites a passage on language change from the *Ars poetica* of Horace (60–72) which Chaucer may also have known directly or through its quotation in John of Salisbury's *Metalogicon* (i. 16; iii. 3); but Dante's discussions of language-change offer closer parallels. Commenting in *Convivio* on the changes in the Italian language Dante remarks:

Onde vedemo nelle città d'Italia, se bene volemo agguardare, a cinquanta anni da qua molti vocaboli essere spenti e nati e variati; onde se 'l piccolo tempo così trasmuta, molto più trasmuta lo maggiore. Sicch' io dico, che se coloro che partiro di questa vita, già sono mille anni, tornassono alle loro cittadi, crederebbono quelle essere occupate da gente strana per la lingua da loro discordante.   (I. v. 55–66)

Hence in the cities of Italy, if we will look attentively back over some fifty years, we see that many words have become extinct and have come into existence and been altered; wherefore, if a short time so changes the language, a longer time changes it much more. Thus I say that if those who have departed from this life a thousand years ago were to return to their cities, they would believe that these had been occupied by some foreign people, because the language would be at variance with their own.

Dante's conception of the changes noticed over a thousand years is paralleled by the way Chaucer introduces his discussion:

> Ye knowe ek that in forme of speche is chaunge
> Withinne a thousand yeer, and wordes tho
> That hadden pris, now wonder nyce and straunge
> Us thinketh hem, and yet thei spake hem so . . .   (ii. 22–5)

Chaucer's development of his argument by comparing changes in language over time with differences between contemporary cultures (ii. 27–8, 36–42) also finds a parallel in the *De vulgari eloquentia*:

Nec dubitandum reor modo in eo quod diximus temporum, sed potius opinamur tenendum; nam si alia nostra opera perscrutemur, multo magis discrepare videmur a vetustissimis concivibus nostris quam a coetaneis perlonginquis.   (I. ix)

I do not think there should be any doubt that language varies with time, but rather that this should be retained as certain; for if we examine our other works we see much more discrepancy between ourselves and our ancient fellow-citizens than between ourselves and our contemporaries who live far from us.

But beyond all such particular and verbal echoes, there was also the potential example of the *Divine Comedy* in its exploration and understanding of love. As Chaucer reinterprets the implications of the narrative of love inherited from *Filostrato*, his *Troilus* comes to share preoccupations discussed in the analysis of love in *Purgatorio*. There will be a difference in the mode of addressing such questions in a narrative poem and in the exchanges of Virgil and Dante, yet in few places other than the *Comedy* could there be so lucid an analysis of those problems and responsibilities in love that Chaucer's story of Troilus and Criseyde seems to raise. Behind the events of a tale of love in *Troilus* there is a pervasive concern with freedom of action, and it is freedom of the will which Beatrice ringingly affirms in *Paradiso*, v. 19–24. This freedom involves a moral responsibility, and, while human loving is free, it may be good or bad depending on whether freedom is exercised well or badly. In *Purgatorio*, xvii, which begins as Dante imagines the nightingale singing of Procne's impious deed, Chaucer could find Virgil instructing Dante in how a disordered and misdirected love is the principle of sin:

> 'Nè creator nè creatura mai,'
>   cominciò el 'figliuol, fu sanza amore,
>   o naturale o d'animo; e tu 'l sai.
> Lo naturale è sempre sanza errore,
>   ma l'altro puote errar per malo obietto
>   o per troppo o per poco di vigore.
> Mentre ch'elli è nel primo ben diretto,
>   e ne' secondi sè stesso misura,
>   esser non può cagion di mal diletto;
> ma quando al mal si torce, o con più cura
>   o con men che non dee corre nel bene,
>   contra 'l fattore adovra sua fattura.
> Quinci comprender puoi ch'esser convene
>   amor sementa in voi d'ogni virtute
>   e d'ogne operazion che merta pene.'
>
>   (*Purgatorio*, xvii. 91–105)

'Neither Creator nor creature ... was ever without love, either natural or of the mind ... the natural is always without error, but the other may err through a wrong object or through excess or defect of vigour. While it is directed on the primal good and on the secondary keeps right measure, it cannot be the cause of sinful pleasure; but when it is warped to evil, or, with more or with less concern than is due, pursues its good, against the Creator works His creature. From this thou canst understand that love must be the seed in you of every virtue and of every action deserving punishment.'

Slight parallels in the phrasing of Pandarus, when analysing Criseyde's possible inclination to love—

> 'For this have I herd seyd of wyse lered,
> Was nevere man or womman yet bigete
> That was unapt to suffren loves hete,
> Celestial, or elles love of kynde;
> Forthy som grace I hope in hire to fynde.'   (i. 976–80)

—only serve to suggest that Chaucer has absorbed such questions more deeply than any character is able to articulate within the poem, whether or not Pandarus is read as a 'false Virgil'.

The lucidities of Virgil's analysis of the nature of love are especially pertinent to the inner life of Troilus. In *Purgatorio*, xviii, Dante asks Virgil to 'expound love to me, to which thou reducest every good action and its opposite', and Virgil begins by describing how the mind responds to love:

> 'L'animo, ch'è creato ad amar presto,
>   ad ogni cosa è mobile che piace,
>   tosto che dal piacere in atto è desto.
> Vostra apprensiva da esser verace
>   tragge intenzione, e dentro a voi la spiega,
>   sì che l'animo ad essa volger face;
> e se, rivolto, inver di lei si piega,
>   quel piegare è amor, quell'è natura
>   che per piacer di novo in voi si lega.
> Poi, come 'l foco movesi in altura
>   per la sua forma ch'è nata a salire
>   là dove più in sua matera dura,
> così l'animo preso entra in disire,
>   ch'è moto spiritale, e mai non posa
>   fin che la cosa amata il fa gioire.
> Or ti puote apparer quant'è nascosa
>   la veritate alla gente ch'avvera
>   ciascun amore in sè laudabil cosa,
> però che forse appar la sua matera
>   sempre esser buona; ma non ciascun segno
>   è buono, ancor che buona sia la cera ...'   (*Purgatorio*, xviii. 19–39)

'The mind, created quick to love, is readily moved towards everything that pleases, as soon as by the pleasure it is roused to action. Your perception takes from outward reality an impression and unfolds it within you, so that it makes the mind turn to it; and if the mind, so turned, inclines to it, that inclination is love, that is nature, which by pleasure is bound on you afresh. Then, as fire moves upward by its form, being born to mount where it most abides in its matter, so the mind thus seized enters into desire, which is a spiritual movement, and never rests till the thing loved makes it rejoice. Now may it be plain to thee how hidden is the truth for those who maintain that every love is in itself praiseworthy, perhaps because its matter always seems good; but not every stamp is good, even if it be good wax.'

It enriches a reading of the early parts of *Troilus*, where Troilus' impressions of Criseyde lead on to love, to see them in the context of this discussion of the relation between perception, inclination, and spiritual movement, and to bear in mind Virgil's comments alongside that excursus on submission to the power of love which Chaucer adds to follow Troilus' first perception of Criseyde (i. 225 ff.).

Dante's response to Virgil's account of 'impression' is to wonder about the individual's freedom and responsibility. Troilus' submission of his reasoning powers to the force of his passion may be seen as a surrender of a fundamental freedom, and a reading of Dante's succinct exposition of the faculties that enable control over love (*Purgatorio*, xviii. 61–74) could have confirmed Chaucer's concern to explore the relation between love, freedom, and responsibility. The understanding of love in these *Purgatorio* cantos, and their expository clarity of statement, can be present only by implication or contrast in the dramatized narrative of a pagan world in *Troilus*, yet a reading of them may well have been—at a level imperceptible in verbal borrowing—among the most influential aspects of the *Divine Comedy* in Chaucer's reinterpretation of the Troilus story.

The Divine Comedy of Dante Alighieri, Italian text with translation and comment by John D. Sinclair (Oxford, 1971) (whose translations are used in this section). See also the edition and translation by Charles S. Singleton (3 vols.; Princeton, NJ, 1970–5).

Le opere di Dante Alighieri, ed. E. Moore and P. Toynbee (4th edn., Oxford, 1924).

Il Convivio, ed. M. Simonelli (Bologna, 1966).

De vulgari eloquentia, ed. P. V. Mengaldo (Padua, 1968).

Early studies of Dante and Chaucer by Lowes, 'Chaucer and Dante', and Praz, 'Chaucer', have been followed by the very full survey by Schless, *Chaucer and Dante*. See, in addition, Bennett, 'Chaucer', and Piero Boitani, 'What Dante Meant to Chaucer', in Boitani (ed.), *Chaucer and the Italian Trecento*, 115–39.

Among a number of particular studies of Dante and *Troilus*, see especially Fyler, '*Auctoritee*'; Garbáty, '*Troilus*'; Havely, 'Tearing or Breathing'; Kirk, '"Paradis"';

Schless, 'Chaucer and Dante' and 'Transformations'; Shoaf, 'Dante's *Commedia*'; Karla Taylor, 'A Text and its Afterlife' and *Chaucer Reads*; Wallace, 'Chaucer's "Ambages" '; Wetherbee, 'Descent', *Chaucer*, and '"Per te poeta fui"'; and Wheeler, 'Dante'.

Genre

Introduction

> Go, litel bok, go, litel myn tragedye,
> Ther God thi makere yet, er that he dye,
> So sende myght to make in som comedye! (v. 1786–8)

That Chaucer is among the first writers to use such terms in English—
although not without self-deprecatory humour—underlines the innovative
nature of his exploration in *Troilus* of the potential in using and combining
various genres within the same work. Treatments of the story of Troilus
written before Chaucer wrote his *Troilus* may be thought of as falling into
particular generic categories, whereas Chaucer's poem achieves a specially
mixed and combinative use of different genres. Chaucer's inheritance in
the Troilus story of a tradition and a succession of sources, a sequence of
alternative sources in various genres, results in a special inclusiveness of
genres in *Troilus*, and this gathering of genres in the poem—the absorp-
tion, combination, quotation, and transcendence of genres—is a distinctive
part of the nature and meaning of *Troilus and Criseyde*.

Although much formal medieval discussion or definition of vernacular
genre is lacking, it was nevertheless an age of exceptional richness in the
use of genres. The *Canterbury Tales*, as a whole and in detail, reveals
Chaucer's sense of the possibilities in combination, variety, and contrast of
genre, both in the pilgrimage tale-telling and within individual tales. And
Chaucer's realization of the fabliau as a genre in English—long after that
genre's demise in its French homeland, and without surviving precedent in
England—is only one instance of Chaucer's acuteness to the potential he
sees in genre.

There is also the evidence of those early readers of *Troilus*—the scribes
and glossators of the poem—whose notes in the margins of some of the
extant manuscripts offer a commentary on the text which reveals contemporary alertness at several levels to Chaucer's combination of classical and
medieval generic elements. The marking of the beginnings of speeches, or
the openings and closes of letters, songs, and other lyric units and divisions
within the poem is itself such a kind of formal commentary. Scribal
indications of sources (see above, pp. 42, 49–50) reveal that medieval
readers of the poem acknowledged the 'intertextuality' of Chaucer's poem,
and with that comes recognition of the various literary kinds that the poem
draws upon and still embodies in itself.

This sense of intertextuality, and of an inclusiveness of genres, tends to a mutual, interrelating critique between genres: by being included together they ultimately react and comment upon each other. An alertness to genre is signalled within the world of the poem's characters, as in the different terminology of romance and epic used by Criseyde and Pandarus to describe the story of Thebes (ii. 100, 108; see above, p. 122). In the same scene Criseyde declares that she ought to be reading saints' lives (ii. 118); in Book III Pandarus picks up an 'old romaunce' to read (iii. 980). In Book I he quotes one of Ovid's *Heroides* to Troilus as a piece of contemporary writing (i. 659–65); in Book V he apparently alludes to the world of pastoral in dismissing unrealistic hopes (v. 1174). And in the writing of letters and utterance of lyrics by the characters there is recurrent alertness to literary form and kind, as in their sense that of their experience 'Men myght a book make of it, lik a storie' (v. 585). Framing all, there is the compositional role of the narrating persona, with his concern for his poem in relation to *geste* (ii. 83; iii. 450), *storie* (ii. 31; v. 1037 ff., 1651), or *tragedye* and *comedye* (v. 1786–8).

The inclusiveness in Chaucer's *Troilus* of different literary kinds is the outcome of its place in a particular sequence of texts which represent responses to the same *matière*, and of the expectations Chaucer's audience would consequently have of a work addressing that subject-matter; generic combination and extension would be part of their perception of Chaucer's *Troilus*. The effect of such a combination of genres is that in interrelating they modify and comment upon each other in an inclusiveness which allows for multiple viewpoints and denies any dominating single perspective or interpretation. In any combination some genres are more or less dominant than others in different parts of the work, and in *Troilus* genres are modified not only by their interrelation with each other but also by elements of comedy and irony, of narrative presentation, and of the nature of the poem's ending.

It will be helpful to summarize here that distinctive conjoining of genres created by Chaucer's combinative use of diverse source elements, before proceeding to examine the contributing genres.

From a knowledge of 'Dares' (or the Anglo-Latin *Ilias* of Joseph of Exeter) Chaucer's first readers might have been aware of the martial and epic role of Troilus, and, from Benoît de Sainte-Maure's *Roman de Troie*, of the story as romance and of Troilus as lover. In Guido de Columnis' *Historia destructionis Troiae*, with its imitations in various 'Troy-Books', the story exists as part of a history, as in the 'historical' narratives of Dares and Dictys. In the *Filostrato* Boccaccio invents a beginning for the story, an ascending 'upbeat' action in the narrative of the first three books which incorporates various generic features of romance, before the story fulfils its traditional downwards course, which Chaucer can associate both implicitly and explicitly with *tragedye*. Medieval understandings of the genre of tragedy allow possible associations with contemporary reading of Seneca's plays as tragic narratives within which dialogue and monologue have an extensive role, and from this concern with dramatic speech there also arises a connection with the generic models offered by philosophical dialogue and debate, as in Boethius. Related to this 'dramatic' sense of the role of speech in the poem is the addition and extension by Chaucer of material drawing on various lyric genres. These lyrics express aspects of characterization in *Troilus* which are in turn counterpointed by Chaucer's incorporation of motifs recalling the fabliaux or related comic tales of arranged meetings and seductions, while in the course of the whole poem significant elements of allegory may also be discerned.

See H. R. Jauss, *Toward an Aesthetic of Reception*, trans. T. Bahti (Brighton, 1982), ch. 3 ('Theory of Genres and Medieval Literature'), and A. Fowler, *Kinds of Literature* (Oxford, 1982).

On medieval literary analysis of the Bible, see A. J. Minnis, *Medieval Theory of Authorship* (London, 1984), and A. J. Minnis and A. B. Scott (eds.), *Medieval Literary Theory and Criticism, c. 1100–c. 1375: The Commentary Tradition* (with the assistance of David Wallace; Oxford, 1988).

On the *Canterbury Tales*, see Helen Cooper, *The Structure of the Canterbury Tales* (London, 1983), ch. 3 ('An Encyclopaedia of Kinds').

Epic

At the beginning of the third book, which is to describe the physical union of the lovers, Chaucer invokes the assistance of Calliope, muse of epic poetry (iii. 45), and, whatever irony there may be in such an invocation in such a context, Chaucer takes steps throughout his poem to associate the narrative of *Troilus* with the writing of epic. This may seem a surprising claim, and the signals are complex. In different contexts *Troilus* both lays claim to and disclaims the status of epic. The disclaimers are familiar, but deceptive. The lines near the beginning and end of the poem (i. 145–7; v. 1771), which refer the reader to Homer, Dares, and Dictys, for accounts of Troilus' deeds in arms, are referring the reader elsewhere for the matter of epic which *Troilus* is apparently claiming to exclude.

Yet Chaucer's version of the Troilus story has a more complex relation to epic than this would suggest. Boccaccio could legitimately refer his reader elsewhere in search of an epic treatment of the story, for in *Filostrato* he filters out the possibilities for epic action and epic protagonists, passing over the long-term rivalry of Troilus and Diomede on the battlefield, and also the experience and endurance of Troilus as a seasoned warrior. Troiolo's attributes as a warrior remain confined to the stereotypical and conventional, described in language influenced by the minstrel tradition. Chaucer absorbs into his *Troilus* many intrinsically non-epic features of action and character from the Italian poem, but he also incorporates motifs of presentation, action, and characterization from earlier epic tradition with a resulting increase in complexity and richness of meaning.

Chaucer's poem has the most un-epic of subjects—a very private and intimate affair—yet this subject is framed and articulated in a series of echoes of the machinery of classical epic poems, with a special debt to the *Thebaid* of Statius, together with all the formal apparatus of books, invocatory proems, and a sustained dignity of verse-form. There are epic associations in the very way that *Troilus* opens. In the traditional announcement of subject, in a syntactical structure of complement preceding verb, of lofty inversion and delay ('The double sorwe of Troilus to tellen . . . My purpos is . . .'), there are echoes of how the epics of Virgil, Lucan, and Statius begin, although Chaucer may also have known from Ovid's *Amores* the mock-epic use of such epic announcements of subject in order to begin a poem of love. 'The double sorwe of Troilus' recalls the double sorrow of Queen Jocasta in the Theban Wars as Dante's Virgil describes it to Statius (*Purgatorio*, xxii. 56). At the close of *Troilus*, when Chaucer bids his poem follow after and kiss the footprints of such writers of epic poetry as 'Virgile . . . Omer, Lucan, and Stace' (v. 1792), he echoes the very words with which Statius bids his epic *Thebaid* to follow after in the footprints of the *Aeneid*, his revered model (*Thebaid*, xii. 816–17), and even to follow humbly in the footsteps of Virgil and the other epic poets is to claim some relation to their kind of writing.

The declaration near the close of *Troilus* ('And if I hadde ytaken for to write | The armes of this ilke worthi man . . .' (v. 1765–6)) claims to set aside the possibility of an epic treatment in *Troilus* even as it plays with the famous opening declaration of the *Aeneid* which Chaucer had once echoed in the *House of Fame* ('I wol now synge, yif I kan, | The armes and also the man' (143–4)), while the adjacent reference to 'The wrath, as I bigan yow for to seye, | Of Troilus the Grekis boughten deere' (v. 1800–1) may even echo the opening of the *Iliad*, which Chaucer perhaps knew from Latin *florilegia*. Chaucer's discernible allusions to Virgil include some figurative ideas derived via Dante and Boccaccio (e.g. iv. 225–7, 239–41, 659–60), while the consummation of love during a storm in *Troilus* Book III may recall the story of Dido and Aeneas surprised by a storm (*Aeneid*, iv. 160–72), and the reference to 'Poliphete' and Aeneas (ii. 1467, 1474), in the episode in which Deiphebus and Helen appear, may recall Virgil's reference to a Trojan priest, Polyboetes (*Aeneid*, vi. 484), seen by Aeneas in the underworld shortly before he meets Deiphebus, who tells how Helen betrayed him at the fall of Troy. Yet, despite the fleeting nature of Chaucer's possible allusions to Virgil, he may have thought of *Troilus* in a kind of complementary relation to the epic example of the *Aeneid*, and the epic of Statius was a powerful model for epic features in *Troilus* both in its form and in the presentation of its subject and action. Formal echoes of epic machinery betoken also a thematic absorption into *Troilus* of conceptions of destiny and human character in the epic-writing best known to Chaucer, who adds to his poem formal invocations (as to Tisiphone and the other Furies) characteristic of an epic as familiar to his audience as the *Thebaid*, while borrowings from the *Teseida* bring an epic manner to some moments in Book V (e.g. v. 1, 8–11, 274–9; see above, pp. 121–5).

It is in the later books that Chaucer can most associate the outward experience of his Troilus with that of epic heroes. The fatal battle at the beginning of Book IV which proves the turning-point in the lovers' lives is carefully accounted for as a military event. In Book V (298–315) Troilus' dispositions for his own funeral draw—with whatever ironic excess in context—on descriptions of the cremation of Arcita in the *Teseida* and more generally on the cremations and funeral games of the *Thebaid* (to which Cassandra alludes: 'Of Archymoris brennynge and the pleyes' (v. 1499)). Epic associations in the presentation of Troilus' experience are emphasized when his portrait is taken over directly from that in Joseph of Exeter's *Ilias*, which portrays the hero in terms of his identity as a warrior in the Trojan War. In the way Troilus' life closes there is also an epic quality distinct from *Filostrato*. Chaucer retains *Filostrato*'s extraordinarily understated account, in which a single line (v. 1806) suffices to describe the death in battle of the hero of a lengthy poem. But if the manner and report of death have nothing of epic about them, the ascent of the hero's soul after death in battle associates Chaucer's poem with—among other texts—Lucan's epic *Pharsalia*, where the soul of Pompey looks down and laughs at the sight of his own mistreated corpse (*Pharsalia*, ix. 1–14).

It is also part of Chaucer's sense of Trojan background as the 'present day' of his narrative to show the active prosecution of the Trojan War, and so to associate Troilus' valour in war with his worthiness for love, while linking the fate of the lovers with the destiny of Troy: Troilus returns triumphant from battles; he and Pandarus keep their hand in with their weapons and discuss tactics (ii. 510–13); Troilus is acclaimed as the 'holder up' of Troy and scourge of the Greeks, in which he is surpassed only by the epic feats of Hector (ii. 644). None of this is suggested in *Filostrato*, and by such changes Chaucer raises in his narrative the possibility that Troilus is simultaneously participating as a character in an epic which is concurrent with the present tale, and which could form the subject-matter of an alternative text.

In the *Aeneid* Virgil includes Troilus among the warriors whom Aeneas sees in a mural at Carthage depicting the battles at the siege of Troy, along with the exploits of 'plumed Achilles' and Diomedes, 'the blood-stained son of Tydeus'. He is 'infelix puer atque impar congressus Achilli' (i. 475)—'unfortunate youth and ill-matched in conflict with Achilles'—and the *Aeneid* includes Troilus only to mention his death as a youth, in a scene that presents him more as a figure of pathos than an epic hero. Behind this lie traditions of Troilus as a type figure of early death, and of a prophecy that the fates of Troy and Troilus were linked. In his *Bacchides* (953–5) Plautus alludes to the belief that Troilus' death would be one of the signs of Troy's downfall, and the First Vatican Mythographer comments that, if Troilus reached the age of twenty, Troy could not be overthrown.

The role of Troilus in tradition is in some sense that he is fated not to develop into a mature warrior, and hence into an epic hero. In Greek sources (such as Apollodorus, *Epitoma*, iii. 32, 151) Troilus' murder by Achilles comes about not in battle but in an ambush at a temple of Apollo, so acquiring associations of ritual and the sacrifice of a fated victim, and this is the subject of the surviving fragments of Sophocles' tragedy of *Troilus*. The strategic motive for Achilles to kill Troilus was clear enough, but to this was added a further tradition—implied in the third century BC by Lycrophon in his *Alexandra* (307–13)—which suggested that another motive for the slaying was an unrequited love felt by Achilles for Troilus (whom Lycrophon describes as 'thyself unwounded by thy victim'). It may have been this tradition which led Servius in his commentary on the *Aeneid* to set the Troilus story apart from fit subjects for heroic poetry:

Veritas quidem hoc habet: Troili amorem Achillem ductum, palumbes ei, quibus ille delectabatur, objecisse: quas cum vellet tenere, captus ab Achille, in ejus amplexibus periit. Sed hoc, quasi indignum heroico carmine, mutavit poeta.

The truth, however, is this: that Achilles, led by love of Troilus, set doves, which he delighted in, before him; when he sought to hold them, he was taken by Achilles, and perished in his embraces. But this, which is unworthy of a heroic poem, the poet has changed.

Although the Greek sources were unavailable to medieval readers, it would seem that from Virgil, Servius, and the mythographers Chaucer could derive a sense of qualification and exclusion of epic in aspects of the traditional role of Troilus, an exclusion Boccaccio has in his own way pursued in the relation between *Filostrato* and the epic Troilus of Dares. Chaucer, by contrast, explores a dimension of the Troilus story by raising questions about the relation to the epic genre of the character and experience of his hero.

On the figure of Troilus in classical literature and vase-painting, see Piero Boitani, 'Antiquity and Beyond: The Death of Troilus', in Boitani (ed.), *European Tragedy*, 1–19. For a comparison between *Troilus* and the way in which the progress of the hero in classical epic—especially the *Aeneid*—was allegorized by the mythographers and commentators, see Maresca, *Three English Epics*. On the *Iliad* and Latin *florilegia*, see Boitani, *The Tragic and the Sublime*, p. 209. On the *Filostrato* and epic, see Natali, 'A Lyrical Version'.

For brief references in classical literature to the death of Troilus, see Callimachus, fragment 363; Cicero, *Tusculan Disputations*, I. xxxix. 93; Horace, *Odes*, II. ix. 13–16; Hyginus, *Fabulae*, cxiii. 3; Seneca, *Agamemnon*, 748; Quintus Smyrnaeus, *Fall of Troy*, iv. 430–1.

For suggested allusions to Virgil, *Georgics*, i. 217–18, in *Troilus*, ii. 55, and to *Aeneid*, xi. 338, in *Troilus*, v. 804, see Clayton, 'A Virgilian Source', and Frost, 'A Chaucer–Virgil Link'.

Apollodorus: *Epitome*, in *The Library*, ed. and trans. Sir J. G. Frazer (2 vols.; Loeb Classical Library; London and New York, 1921).

Lycrophon: *Alexandra*, in *Callimachus and Lycrophon*, ed. and trans. A. W. Mair (Loeb Classical Library; London and New York, 1921).

Plautus: *Bacchides*, in *Plautus*, vol. i, ed. and trans. P. Nixon (5 vols.; Loeb Classical Library; London and New York, 1916–38).

Servius: *Servianorum in Vergilii Carmina commentariorum editionis Harvardianae*, ed. E. K. Rand *et al.* (Cambridge, Mass., 1946–65).

Sophocles: *Troilos*, in *The Fragments of Sophocles*, ed. A. C. Pearson (Cambridge, 1917), ii. 257.

Vatican Mythographers: in *Scriptores rerum mythicarum Latini*, ed. G. H. Bode (Celle, 1834), 210, p. 66: 'Cui dictum erat, quod si ad annos xx pervenisset, Troia everti non potuisset' (It was said of him that if he reached the age of twenty Troy could not be overthrown).

Romance

What of the relation of *Troilus* to romance, a genre from which Chaucer appears to distance himself in his poems? An instinctive indifference to the marvellous as theme, and an impatience with the episodic structure of repetitious exploits at arms, may underlie Chaucer's abandonment of the *Squire's Tale*; the Wife of Bath's tart assumption that the enchantments of England's Arthurian past survive no more (*Wife of Bath's Tale*, III. 857–81), and the Nun's Priest's arch assertion that his fable of cock and fox is 'also trewe, I undertake, | As is the book of Launcelot de Lake, | That wommen holde in ful greet reverence' (*Nun's Priest's Tale*, VII. 3211–13), have further seemed to reflect Chaucer's own detachment from Arthurian romance. That *Sir Thopas* is unendurable to the society of Canterbury pilgrims has suggested not only Chaucer's sophisticated distancing of himself from popular English romance, but—by extension—his amused impatience with the characteristic concerns and forms of romance throughout his works.

For Chaucer it is the unsustainable nature of human idealisms which both draws him towards some depiction of the characteristic experience of romance, and also necessarily leads him to show its qualification and exhaustion. It is not unreasonable to think of *Troilus* as being in a real and important sense a romance, for in a setting long past and exotically distant the experience of the principal young character is that of typical romance protagonists: an intense and transfiguring experience of love, as well as a kind of adventure. Dragons and enchantments are there none, yet the all-possessing nature of the experience in love of Troilus makes that experience for him a kind of inward equivalent of *aventure*, just as the new strangeness of love, and that force of idealization which he brings to his experience, give it the momentum of a quest and the quality of a marvel, the marvellous inward adventure of love.

This association with romance experience does not bring the conclusion that the romances of tradition generally have, for Chaucer's poem ends in an intensity of disappointment that invites comparison with some of the effects of tragedy. *Troilus* works through a process of disillusionment with the idealization of romance, although the disillusionment is dependent upon—and so in a sense lesser than—the idealization. It is in the wisdom of this disappointment that the poem can incorporate and transcend the typical experience of romance in testing, refining, and educating its characters. In *Troilus* style and convention reflect the aspiration of romance, which structure and context comment upon with some irony. In a poem like *Troilus*, where the outward enchantments of traditional romance have been replaced or internalized in the inward adventure of the hero's engrossing but eventually disillusioning experience in love, Chaucer's development of the genre may be seen as in every sense the disenchantment of romance.

It is Boccaccio who makes what is just one of a number of romantic episodes in the other narratives into the sole subject of his *Filostrato* and (as it were) invents 'backwards' to provide with a beginning what is a story about an ending, in the sense that its characterization of Troiolo and Criseida is designed to explain the given denouement with its intrinsic antifeminist bias. Although Boccaccio has put together the beginning of the affair through the available structures of romance archetypes, the *Filostrato* is by no means straightforwardly a romance either in treatment or in construction. It offered a shapeliness of overall structure, tightly constructed scene-sequence, and single-minded clarity of focus on one unified narrative action, very different from the way medieval romance characteristically proceeds by the multiplication and augmentation of episodes and digressions. The remarkable conveying of meaning through structure in *Troilus*—its lucid scene-structure dovetailed and co-ordinated together, and shaped into books—builds further upon the foundation of that selection and concentration. In *Filostrato* the process of wooing and winning the lady has been domesticated and urbanized both in the circumstances and the characterization of the affair, and the lovers' speech does not define their relationship in terms of the typical romance idiom of the feudal service of the lady by her knight, because Boccaccio's Troiolo is not much presented as a knight and his Criseida is not much of a lady. Nor do the pragmatic and realistic attitudes of characters like Pandaro and Criseida allow opportunity either for the marvellous and the mysteriously other-worldly, or for the idealization of this-worldly experience, which distinguish so much romance. Chaucer's *Troilus*, by contrast, is developed in a kind of referential relation to the traditions of romance, sometimes going back to the earlier romance tradition of Benoît to imitate or borrow materials that give *Troilus* more of the setting and procedures of romance, sometimes using language and representing behaviour in ways which overlap with the English romances of Chaucer's day and with romance tradition generally. *Troilus* is hence composed of material from different layers in the development of a romance tradition: its main narrative source in *Filostrato* has already reached the stage of a post-romance, the single undigressive narrative of a novella, achieved in part by excluding some generic features and subjects of romance. Chaucer accepts this singularly co-ordinated structure but not the exclusions of subject and mood which helped create it. Yet *Filostrato*'s powerful drive away from romance still moves beneath the English poem. *Troilus* thus develops in an uneasy relation to romance traditions and comes to question the presuppositions with which romances typically begin and end.

In inventing the action of his first three books Boccaccio creates a story with many archetypal features of the ways romances develop in their earlier stages, with various stock motifs of plot, episode, and characterization. Yet, although the text of a scene in *Troilus* actually derives from *Filostrato*, it might well be referred by its early audiences to a wider context in their awareness of the type-scenes of romance: Troilus' first sight of Criseyde at the temple in springtime, for instance, set against the background of those comparable moments of enamorment in pagan temples in the *Roman de Troie* (e.g. 17489 ff.). Troilus' following conversation with Pandarus is itself no less surrounded with associations as a type-scene, of the confession to a confidant of the hero's or heroine's being in love, and the confidant's plan to gain the beloved; there are a number of such sequences in the English romances (e.g. *Octavian*, 1093–116; *Floris and Blancheflour*, 385–424, 443–704; *Sir Degrevant*, 520–608, 803–939; *Ipomadon A*, 1210 ff.; and *William of Palerne*, 580–652), where there are a number of resemblances to the exchanges between Troilus and Pandarus. There are French romance parallels to Pandarus' role as go-between, and even for the confidant's presence with the lovers in the bedchamber (in *Guillaume de*

Palerne and *Florimont*). That Troilus' admission of his first love is set 'In-with the paleis gardyn, by a welle . . .' (ii. 508) recalls the Garden of Love and its well in the *Roman de la rose*; Criseyde's looking from her window at Troilus riding past as a knight returning from battle recalls the many scenes in romance, often in tournaments, which involve ladies looking down to their knights from windows and vantage-points aloft. Parallels in character and conduct have been noticed between Troilus and Guy of Warwick, one of the heroes of Anglo-Norman and English romance, who tends to despair for love and to swoon at crucial moments. Parallels have also been discerned between the predicament and behaviour of Troilus in Book III and that of Lancelot when he nearly swoons in the presence of Queen Guinevere and Gallehault.

Such echoes of the *Roman de la rose* and of romances of chivalry are pointers to Chaucer's procedures in developing his narrative in the light of romance conventions, and in strengthening the presentation of Troilus' experience as a part of romance in each phase of the poem through added reminiscences of various types of romance. In the earliest phase of Troilus' love Chaucer colours his account with recollections of the Lover in the *Roman de la rose* (see above, pp. 114–18); in the central scene the solemnities of the lovers' vows and exchanges of tokens have suggested a recollection of Boccaccio's *Filocolo*, although the point that Chaucer has made his poem here so resemble a type of Floris and Blanchefleur romance—in the innocence and earnestness of the lovers, and their gifts of rings and brooches—says more than whether an actual textual borrowing can be doubtfully established. The importance in *Troilus* of vows and promises is more like the world of romance than *Filostrato*, as is the role of objects which act as emblems of feeling: the heart-shaped brooch and exchanged rings with their inscription (iii. 1368–72), or the golden urn which is to contain the ashes of Troilus' heart 'for a remembraunce' (v. 309–15).

In the fifth book of *Troilus*, where the action runs parallel to that of Benoît's *Roman*, Chaucer has restored to his narrative some episodes from the French romance which had been excluded by Boccaccio, including brief reference to such quintessential romance incidents and motifs as the taking of the glove, the gift of a horse, and the token of the lady's sleeve. Chaucer's two extended borrowings from Benoît (v. 92–189, 995–1099) present in much greater fullness than Boccaccio the process of Diomede's wooing of Briseida, and reflect—in Chaucer's soliloquy for Criseyde—that sophisticated psychological interest in the finest French romances. Where Boccaccio had compressed Benoît's account, in order to show a perfunctory and easy victory for Diomede, it suited Chaucer's approach to his heroine to go back to the ampler treatment of this episode in the *Roman de*

Troie. This has the remarkable consequence in *Troilus* that the betrayal and disenchantment of the hero's experience of romance in the final book is presented within some of the most quintessential conventions of romance tradition. Criseyde is most sustainedly referred to by the traditional epithets of English romance as the 'lady bright' or 'free', or as the 'may', in that very fifth book where she can no longer be a heroine of romance. Both Diomede (v. 144, 162, 922) and Troilus (v. 465, 516, 669, 1362, 1390, 1405) refer to Criseyde as 'bright' or 'free', and Troilus calls her his 'lady bright' at some of his moments of keenest disenchantment (v. 1241, 1247, 1264, 1573, 1712), and only addresses her as his 'may' at moments of farewell and relinquishing (v. 1412, 1720). Indeed, the whole of *Troilus* is full of echoes from the stock diction and gestures with which the English popular romances convey the sorrows of their characters.

But, while Chaucer returns to some aspects of the *Roman de Troie* to supplement his *Troilus* with romance features, he entirely ignores that element of the enchanted and fabulous that he could find interwoven by Benoît with the story of his Troilus and Briseida, and which Boccaccio had already excluded from *Filostrato*. When Benoît's Briseida leaves Troy, she attires herself in a magnificent mantle of fabulous materials and decoration sent to Calcas by an Indian seer (*Troie*, 13341 ff.), and when she reaches the Greek camp she is escorted to Calcas' rich tent, which had once belonged to that Pharaoh who drowned in the Red Sea (*Troie*, 13818–21). When Boccaccio's Diomede is wooing Criseida he does mention the possibility that he is of divine descent (*Fil.* vi. 24), but even so slight a hint of any supernatural dimension to one of the characters is excluded by Chaucer (cf. v. 932–8). The focus of the *Troilus* narrative, at least before Chaucer adds the hero's final ascent to the spheres, is this-worldly, and the marvellous or supernatural is avoided, with the one significant exception of the moment that begins the hero's experience in love, which Chaucer does represent as a supernatural event, a confrontation with a god (i. 206). Such exclusion emphasizes by contrast that element of the marvellous that Troilus sees in his experience of love, which finds expression through the use in *Troilus* of all that superlative and hyperbolical language so typical of romance. Chaucer's emphasis on the hero's *in*experience in love makes experience itself an adventure into an unknown, something marvellously out of this world, which is reflected in that language of heaven's bliss, of love as a religion, as worship, in which Chaucer's Troilus thinks of his experience. His aspirations and conceptions are those of the idealistic hero of romance, and this restores to Chaucer's story that characteristic sense in romance that the hero's experiences include a process of learning, and that romance develops an education of the heart.

Troilus also excludes the typical outward movement and trappings of romance. Chaucer makes reference to Troilus' activities as a knight, but his knightliness is most memorably and momentarily seen through a window from indoors by a woman (ii. 610 ff.). Rewriting Boccaccio's rather unladylike heroine, Chaucer takes pains to restore his Criseyde to the refinement and delicacy of a lady, a suitable heroine of romance, and there is a gleam of luxury in her palace which suggests the same:

> She shette it, and to Pandare in gan goon,
> Ther as he sat and loked into the strete,
> And down she sette hire by hym on a stoon
> Of jaspre, upon a quysshyn gold-ybete ...   (ii. 1226–9)

Chaucer's treatment of Troilus similarly balances realistic perspectives with emphasis on his inner potential as a figure of romance. The repeated praise of Troilus as second only to Hector (ii. 158, 740; iii. 1775; v. 1804) introduces a note of qualification unusual among the superlative descriptions of heroes normal in romance. Joseph of Exeter's description of Troilus as a 'giant in spirit' is altered to include a concession in Chaucer's version that Troilus would not actually surpass a giant in strength:

> Al myghte a geant passen hym of myght,
> His herte ay with the first and with the beste
> Stood paregal, to durre don that hym leste.
> [Stood fully equal in daring to perform what he wished.] (v. 838–40)

Heroes of another kind of romance—like Arthur himself—may manage to overcome giants, but the emphasis in *Troilus* is on the hero's 'herte'.

Chaucer never refers to *Troilus*—or indeed to any of his other poems—as a romance. But he shows both Criseyde and Pandarus reading romances in the text (ii. 100; iii. 978–80), and through the presentation of Pandarus as a reader of romances the action of the poem is brought into a sustained reference to romance traditions. When Criseyde has accepted the 'service' of Troilus, Chaucer has his Pandarus exclaim 'Withouten hond me semeth that in the towne, | For this merveille ich here ech belle sowne' (iii. 188–9), which is an allusion to the tradition of the miraculous ringing of bells to signify a marvel in popular romance and ballad. Yet there are also limitations to romance, and, when Criseyde suggests sending the token of a ring to the jealous and perhaps dying Troilus, Pandarus shows his impatience with the enchantments of romance by ironically pointing out to her that it is not a ring with magical powers (iii. 891–3).

The fact that there are indeed no giants or magic rings or bells that ring without hand in the romance of *Troilus* only emphasizes how much like a romance the perception of their inward experience by the characters has become. Troilus and Criseyde both see themselves in relation to the experiences of romance tradition. In Book I Chaucer's Troilus in his despair feels that if his hopeless love were known, 'I shal byjaped ben a thousand tyme | More than that fol of whos folie men ryme' (i. 531–2), in a possible allusion to the madness of such great lovers in romance as Yvain or Tristan. A more evident self-comparison to the experience of romance characters is expressed by Criseyde when (like Lavinia gazing down at Aeneas in the *Roman d'Eneas* (8031ff.)) she has seen Troilus ride past below her window on his way back from battle, the very picture of a knight: 'To hireself she seyde, "Who yaf me drynke?"' (ii. 651). Here Criseyde associates her own first response to the sight of Troilus with the transforming effects of such magic love-potions of romance as those of the Tristan stories.

This vision of Troilus perceived by Criseyde as a knight of romance is partly stage-managed by Pandarus when he causes the vision to be repeated for effect (ii. 1247ff.), and when he contrives that her knight submits himself to Criseyde's 'service' (iii. 128ff.). At such moments Pandarus arranges the lovers' lives into the type-scenes of romance, and the enhanced role of Pandarus in *Troilus* as 'author' of the action is important to the way Chaucer plays on the relation between his narrative of the Troilus story and the conventions of romance. When Pandarus thinks the lovers are brought together, 'he drow hym to the feere, | And took a light, and fond his contenaunce, | As for to looke upon an old romaunce' (iii. 978–80), and this reading of a romance suggests a return to the sources of Pandarus' inspiration in contriving the affair.

Troilus becomes so much more of a romance than *Filostrato* that it can both contain and transcend the experience and the form of romance. Inclusion of realistic elements, as of comedy, tests and complements without subverting the world of romance in the poem. *Filostrato* had concerned an affair with an experienced and easily won widow—not the normal stuff of romance—and, although Chaucer rewrites so that his Criseyde may play the role of the archetypal romance heroine, a beautiful lady in distress, there remain contradictions between this role and her circumstances. Some of the major linking patterns in the structures of romance are the patterns of separation-followed-by-restoration, and love-followed-by-marriage. But *Filostrato* doubly defeats conventional romance expectations by presenting separation *without* restoration, and love *without* marriage, and this underlying pattern survives into *Troilus* as the end of all Chaucer's increased allusiveness to romance. Unhappy endings in death (as in the Tristan stories) are not unknown in romances, but such endings will make for a specially ambiguous and uneasy kind of romance, granted the conventional happy endings of the genre in the restoration of what has been lost and in marriage. But, again, *Troilus* doubly defeats the expectation of romance narrative, by ending with a death which is not an ending and does not actually conclude the narrative. Just as an ending with death disturbs the customarily optimistic, idealistic view of the possibilities in human life and behaviour which romance affirms, so an ending which shows death transcended is dealing with a kind of 'comedy' which transcends the way romances usually content themselves to end happily— and therefore as comedies—in this world. By following the hero beyond death the *Troilus* narrative goes beyond the normal boundaries of romance narrative, just as the narratorial virtuosity of the ending is rooted in, yet transcends, that invitation to prayerfulness which customarily frames the ending of much popular romance. It is through such narratorial engagement with an audience—sharing the problematic nature of interpretation, puzzling with a tantalizing indeterminacy of tone—that Chaucer takes perhaps further than any other medieval romance writer the enlisting of an audience as co-creators of the characteristic ambivalence of the genre.

On *Troilus* as a romance, Young, 'Chaucer's *Troilus and Criseyde*', is still a valuable introduction, as is his *Origin and Development* on relations with *Filostrato* and *Filocolo*, while Muscatine, *Chaucer and the French Tradition*, sets the poem in the context of romance convention. More recent studies include Boitani, *English Medieval Narrative*; Stevens, *Medieval Romance*; and Wallace, *Chaucer and the Early Writings of Boccaccio*, with valuable discussion of the nature of romance in *Filostrato* in the context of Italian popular tradition.

On type-scenes and motifs in romance, see Susan Wittig, *Stylistic and Narrative Structures in the Middle English Romances* (Austin, Tex., 1978). On resemblances in type-scenes and stock diction between *Troilus* and some medieval English romances, see Windeatt, '*Troilus* and the Disenchantment of Romance'. On Troilus and Guy of Warwick, see Derek Brewer, 'The Relationship of Chaucer to the English and European Traditions', in his *Chaucer: The Poet as Storyteller* (London, 1984), 12.

On *Guillaume de Palerne* and *Florimont*, see Muscatine, *Chaucer*, 140–1. On Troilus' and Lancelot's swoon, see Taylor, *Chaucer Reads*, 64. On *Filostrato*'s combination of lovers' separation without restoration, and love without marriage, see Wallace, *Chaucer and the Early Writings of Boccaccio*, 94. For Hanning, *Troilus* 'in many respects offers the definitive comment on the nature and limitations of the chivalric romance genre' ('Audience', 19).

On *Troilus* and the novel, see especially Grossvogel, 'Chaucer', and Spearing, 'Chaucer as Novelist'.

History

The emphasis in *Troilus* that the reader must seek 'In Omer, or in Dares, or in Dite' (i. 146) for Troilus' martial history need not be taken too literally. The very rehearsal of the names of these 'authorities' on the Trojan War acknowledges the keener sense in *Troilus* of how the action exists in significant relation to a sequence of 'historical' events. Chaucer's sympathetic attempt to re-create some sense of the pagan past within which his characters are acting reflects a real interest in fourteenth-century England in pagan practice and suggests Chaucer wants to associate his poem with the writing of history, just as his own distinctive alertness to time and change in the *Troilus* narrative reveals in practice a historian's interest.

This generic extension of *Troilus* may have been prompted by the way Boccaccio's innovations in *Filostrato* challenged the earlier approach of Benoît and Guido to the structure of the story and the role of time within it. In Benoît's *Roman* (followed in this by Guido's *Historia*) the main chronological structure is provided by the successive numbering of the battles during the siege of Troy, and he prepares the context of the lovers' parting in the knightly careers of Troilus and Diomede. Although Guido ignores this structure, in both *Roman* and *Historia* it remains true that the story of Troilus and Criseyde is narrated within a chronology structured by a 'historical' concern with the siege of Troy.

Boccaccio's innovation is that the structure of his poem reflects that span of time significant to the lovers: time matters in so far as it affects the lovers personally, and time is experienced as feeling. Boccaccio also invests his narrative with a keener sense of passing time by inventing Criseida's promise to Troiolo that she will return to Troy within a specified time, and with it that poignant tension in the last movement of the story as Troilus waits and waits against time—first a definite, measured period of ten days, and then an open-ended passage of time. This structure of personally experienced time is taken over into *Troilus*, although Chaucer then goes beyond Boccaccio in giving a sharper sense of the chronology, dating, and timing of the action within his poem (see below, pp. 198–204), with a feeling for the ancient past of his romance which aligns it with the *romans d'antiquité*. Frequent anticipation of the known ending of the received story in *Troilus* underlines that the narrative is a kind of history, and the overall effect is that *Troilus* is presented both as a personal history of private feeling and as part of a larger history.

Medieval historical narrative has been usefully distinguished into the three categories of History, Life, and Tale, and in these terms *Troilus* is an example of 'tale', distinguished from 'history' by its more closely defined 'scope' in subject-matter and its consequent 'scale' of closely focused and detailed narrative. The subject of *Troilus* is announced in the first stanza to be the hero's tragic love rather than his 'Life' or the 'History' of the siege at Troy, and through the course of a long and copiously detailed narrative this controlling focus is maintained on a single connected sequence of episodes developing as a thematically unified narrative. *Troilus* only selectively resembles within its own text the genre of history-writing, but it establishes itself in a collateral relation to the histories of Troy, while its allusions to the classical poets also have the power to draw on them as a corpus of historical material presenting collectively a version of the ancient world and a history of human conduct.

Chaucer may have found a model for the authorial stance adopted in *Troilus* in the literary role traditionally claimed by those medieval compilers, encyclopaedists, and commentators—like John of Wales, Vincent of Beauvais, and Ralph Higden—who regarded 'history' not only as a collection of facts but as a repository of exempla of vices and virtues, and who, like Chaucer, show a genuine interest in reporting pagan practice and belief as they conceived it. But when presenting pagan material many late-medieval compilers also emphasized historical relativity and their own Christianity, in order to distance themselves from the pagan beliefs and customs that they were describing, particularly those beliefs in fatalism and in the 'judicial art' of making predictions about a particular individual, which were commonly ascribed to pagans. It is in the approach of these compilations to 'history' and the pagan past—their fascination and their detachment—that a parallel may be discerned for the role of the *Troilus* narrator as historian. Chaucer similarly describes pagan belief with interest, but without responsibility for asserting its validity, until the end of the story reveals the limits of the pagan universe of Troilus, and the end of the poem signals the historian's detachment from pagan antiquity ('Lo here, of payens corsed olde rites!' (v. 1849)). With the ending comes a removal to the Christian present day of the poem's audience, and with Christ a sense of a different dimension, dividing Christian history from that pagan history within which dwelled pagans circumscribed by their fatalist beliefs.

It is in Chaucer's interest in an accurate chronology of passing time as experienced by Troilus and Criseyde, his care for authentic detail, his sympathetic exploration of pagan characters, and his awareness of cultural diversity and relativity, that his sense of history contributes to *Troilus*, recalling some of the concerns and emphases of history-writing. Indeed, Chaucer's naturalism may reflect concern with a kind of historical accuracy, a form of representationalism which in itself is a manifestation of the sense of the historic as present, although Chaucer's creation of a Trojan present is contained within a sense of historical change.

On *Troilus* and fourteenth-century attitudes to the past and to history-writing, see Minnis, *Chaucer and Pagan Antiquity*; A. J. Minnis, 'From Medieval to Renaissance? Chaucer's Position on Past Gentility', *Proceedings of the British Academy*, 72 (1986), 205–46, and Spearing, *Medieval to Renaissance*. On Chaucer's sense of history more generally, see Bloomfield, 'Chaucer's Sense of History'. On Chaucer and the classical poets as history, see Wetherbee, *Chaucer*. On 'History', 'Life', and 'Tale', see Burrow, *Medieval Writers*.

Tragedy

The expectation of tradition made a sad ending the characterizing part of the story of Troilus, and Boccaccio's provision of an ascending action to place before that well-known descending action, that 'fall', did not so much diminish as increase the poignant sense of loss in the lovers' emotional downfall. The structure of the whole action in rise and fall Chaucer takes over broadly from Boccaccio, who first discerned this pattern in the story, but it is Chaucer who then realizes the generic potential of the structure within which the old story is now presented. He opens with an anticipation of the arching structure of the narrative in rise and fall ('Fro wo to wele and after out of joie'), and refers to the poem at its close as a 'tragedye'. Indeed, the very precociousness of the use in English of the terms 'tragedy' and 'comedy'—set in juxtaposition with each other at the close of *Troilus* (v. 1786–8)—conveys Chaucer's interest in the new generic associations to which his own rewriting of the Troilus story might attain. The unusualness of terms, and the element of self-deprecation implicit in 'litel myn tragedye', suggest a spirit of experiment and tentativeness as well as pride in achievement.

For Chaucer, the term *tragedye* was most likely to recall that definition which Philosophy represents Fortune as giving to Boethius: 'What other thynge bywaylen the cryinges of tragedyes but oonly the dedes of Fortune, that with an unwar strook overturneth the realmes of greet nobleye? (*Glose. Tragedye is to seyn a dite of a prosperite for a tyme, that endeth in wrecchidnesse)*' (*Boece*, ii. pr. 2). To this would be added recollections of the 'Fall of Princes' tradition as in the examples in the *Roman de la rose*, in Boccaccio's *De casibus virorum illustrium* and *De claris mulieribus*, and works like Chaucer's *Monk's Tale* which are modelled upon the tradition, although it should be emphasized that Boccaccio nowhere actually describes his *De casibus* as 'tragedies'. Chaucer's Monk defines his 'tragedies' as a fall from high degree into misery:

> Tragedie is to seyn a certeyn storie
>
> . . . . .
>
> Of hym that stood in greet prosperitee,
> And is yfallen out of heigh degree
> Into myserie, and endeth wrecchedly . . .   (VII. 1973–7)

Although Chaucer in *Boece* translates without elaboration the definition of tragedy as an abrupt change of fortune, medieval commentaries on the *Consolation*, like those of the English Dominican friar Nicholas Trivet or the Pseudo-Aquinas, interpreted the misfortune as connected with a moral failing in the unfortunate person, and the focus of many 'Falls of Princes', with pride going before a fall, also pointed towards misfortune as the punishment for a sinful man.

In Chaucer's *Boece* and the *Monk's Tale* are definitions of tragedy as a fall from worldly sovereignty, wealth, and well-being through the tragedy of a blindly changeable Fortune. But, as the outlook of the worldly Monk, or as Philosophy's impersonation of Fortune's self-defence against men's complaints, both these definitions are qualified by their contexts, and suggest the irrelevance of tragedy to those who can learn to view this world with proper detachment, with a faith in the life to come and in the dispensations of a just and loving God. As pagan Trojans the characters within *Troilus* do not share in that Christian dispensation which might educate a medieval reader's sense of what is tragic. Chaucer's characters might see their experience in terms of *tragedye*, while it is possible to imagine that the definition of *Troilus* as 'litel myn tragedye' reflects ironically on the too-worldly understanding of a narratorial persona.

If tragedy was customarily defined in terms of the nature of its ending, it is not so surprising that *Troilus* should only be defined as a tragedy near its unhappy close, as if in response to it, and with the naming as part of the achievement of closure. There were, however, a range of other precedents and models familiar to Chaucer for narratives that end unhappily. If the Greek tragedians were unknown, and Aristotle's *Poetics* available only through Hermann Alemannus' translation of the version by Averroes, the tragedies of Seneca were closely studied in the detailed commentaries of Nicholas Trivet (*c.*1315). The limitations of medieval understanding of tragedy as a theatrical experience, as drama, need not discredit medieval understanding of tragedy in general, which in the Middle Ages could be thought of as a narrative form (as the Monk's attempted formal definition suggests: 'In prose eek been endited many oon, | And eek in meetre, in many a sondry wyse . . .' (1980–1)). It is as 'litel myn tragedye' that *Troilus* is directed to follow in the footsteps of 'Virgile, Ovide, Omer, Lucan, and Stace', an invocation suggesting how for Chaucer tragedy had wide associations in classical literature. To take Virgil and Ovid first: in *Inferno* (xx. 113) Dante has Virgil refer to his *Aeneid* as a tragedy, and in the *De vulgari eloquentia* (ii. 4) the most appropriate subjects for the 'tragic style' are described as the highest (*summa*), namely security (*salus*, the health of nations), love, and virtue. This link between the genre of tragedy and the subject of love is also made explicit by Ovid both in the *Tristia* (ii. 381–4, 407–8) and in the stories of unfortunate love in his *Heroides* and *Metamorphoses*, which Chaucer drew upon in his *Legend of Good Women* and other works.

Such an identification of tragedy and love in Ovid offers an important background to medieval ideas about tragedy, for, although tragedy as defined through the 'Falls of Princes' allowed small scope to love as a subject, both romance and medieval reworkings of classical stories bear witness to the popularity of a variety of tales of unfortunate love which may be considered tragic. Virgil's tale of the unfortunate love of abandoned Dido and faithless Aeneas may be seen transposed in *Troilus and Criseyde*. From Boccaccio's *Teseida* Chaucer borrowed for his *Knight's Tale* the tragic tale of Arcite's love for Emelye, in a work with its background linked to Statius' epic *Thebaid* on the doom of Thebes, and *Troilus* is also associated with those *romans d'antiquité* which deal with the unhappy loves of classical figures within a romance framework. The tale of unfortunate love is a sufficiently distinct variety for a group of such lovers—including Paris and Helen, Tristan and Iseult, and Troilus—to be depicted together in the *Parliament of Fowls* (288–94), and for the fourth day of the *Decameron* to be devoted to tales of love that end unhappily, a day under the governance of the character Filostrato, who explains the significance of his name ('the one overwhelmed by love') in terms of his own unhappy fortune in love, just as it is used to refer to the hero in the title of Boccaccio's poem about the unfortunate love of Troilus.

Chaucer incorporates at key points in *Troilus* some of the tones and motifs of a tragedy of rise and fall, as if to serve as pointers to a tragedy more deeply embodied in his development of the romance. The scene in which Troilus falls in love is elaborated as the fall and punishment of a proud man ('he that now was moost in pride above' (i. 230; cf. i. 210, 214, 225–33)). The climbing on the stair, the rising of one who does not see the ensuing descent (i. 183 ff.) apparently anticipates the structure of a tragedy of Fortune. There is recurrent emphasis on Troilus as a prince, as a king's son (cf. i. 2, 226, 261; ii. 165, 316, 708; iii. 170; v. 12), which seems to anticipate an interpretation as a tragedy of fortune (e.g. 'And thus Fortune a tyme ledde in joie | Criseyde, and ek this kynges sone of Troie' (iii. 1714–15)). At some important transitions in the poem (e.g. v. 1 ff., 1541 ff.) Chaucer adds a sense of destiny and fate, while the turning-point downwards in the lovers' lives is presented in the fourth proem as the movement of her wheel by a changeable Fortune, endowed with a spiteful laugh and a mocking grimace (iv. 7). The characters are recurrently dramatized in terms of their own sense (however mistaken) of participating in a tragedy of Fortune, as when Troilus imagines himself 'fallen' from high to low in misfortune (e.g. iv. 271–2, 323–9), or dreams of falling 'depe | From heighe o-lofte' (v. 258–9). The characters' awareness of the past fate of Thebes—interpreted by Cassandra as 'how that Fortune over-throwe | Hath lordes olde ...' (v. 1460–1)—only points for the reader to the characters' unwitting involvement in the present tragedy of Troy.

There have been various attempts to square the poem with available medieval formulations about tragedy. In his subjection to Fortune as he sees it, Troilus' worldly career has been interpreted as the tragically inverse movement to that progressive enlightenment of Boethius away from domination by Fortune in the *Consolation*. Troilus' own ultimate enlightenment beyond death has been held to make his career as a whole a Boethian comedy, Criseyde's a Boethian tragedy. If the story be interpreted as Troilus' wilful self-submission to sensual pleasure, then the *tragedye* may be seen in Christian terms as a re-enactment of the Fall, that fundamental tragedy of fallen, sinful man 'and all our woe | With loss of Eden'. Yet the differences between the *Troilus* narrative and the kinds of 'tragedy' represented by Chaucer's *Monk's Tale* or Boccaccio's *De casibus* make it difficult to explain *Troilus* adequately in these terms. It is this very difficulty that points to the specialness of what Chaucer is doing. The private emotional 'tragedy' of Troilus is not the outward political tragedy or material loss of a ruler's downfall. Although Troilus' princely rank and martial career are part of the setting of the poem at the siege of Troy, the poem's primary focus is on the hero's career as a lover and on his emotional rise and fall. In *Troilus* the nature of the hero's real loss is inward. The loss that Troilus suffers in Criseyde is scarcely comparable with that of some worldly good of prosperity, and Chaucer's treatment suggests Troilus has retained something that he can never lose, and that can never be taken from him. The end of Troilus' trust in Criseyde does not coincide with the end of his life, while his love—unlike that of Troiolo—is never described as having an end. There is every difference in scale, pace, and implication between the romance narrative of *Troilus*—with its protracted sequence of ending in stages, of multiple endings—and the abrupt but predictable turn of Fortune's wheel in the Monk's tales or in the 'Falls of Princes', grouped together as series of mutually confirmatory brief examples, repetitious in structure and theme, and sparsely limited in their detail of character, motive, and incident. If Troilus' love starts as a 'fall' from pride, this is the beginning, not the climax, of the action, and initiates a process in which Troilus learns and develops in humility. In tragedies of the 'Falls of Princes' we watch from the outside a change in the character's fortune seen in worldly material terms, and Chaucer has combined something of the objective and theoretical structure and recognizable formal conventions of such 'tragedy' with the more subjective and sympathetic narrative of unfortunate lovers, so that the rise-and-fall structure is re-created as a narrative shape of psychological, inward significance. *Troilus* can echo something of the *de casibus* tragedies, evoking them as a kind of generic reference point or foil for its own more inclusive form.

If it is the nature of the ending that defined a medieval tragedy of fortune, it is in the nature of the beginning Chaucer has provided for this inherited story about an ending, and in the new relation of beginning and ending, that *Troilus* explores its distinctive version of tragedy. The loss of the hero's love and life together (which the fourth proem apparently anticipates within the fourth book) instead takes Chaucer much longer to narrate, and this very fullness of the last books in describing the disillusion of the hero is one of those features of *Troilus* furthest from the simple consequences that connect the ends to the beginnings of conventional medieval tragedies. The accumulating sense of multiple points of view and the absence of any single dominating perspective in *Troilus* are intrinsically antithetical to the pretensions to omniscience in medieval tragedies of fortune as a genre. This fullness deepens and complicates the reader's sense of the characters as agents, with implications for the freedom the characters are understood to have. Chaucer's psychologically detailed attention to motivation and deliberate human action works on a quite different level from the standard attributions of change to the mechanisms of fate and fortune, and his focus on the characters promotes a sense of their free choice and action which we note on a different level of awareness from our prior knowledge of how their story will end.

This same fullness of detail and extension in Chaucer's presenting of his narrative has inevitably prompted comparison between the characterization in Chaucer's *tragedye* and the idea of the 'tragic flaw' in Aristotle's discussion of drama. But to search for a 'tragic flaw' in the character of Troilus or Criseyde proves inappropriate to the way character is presented in Chaucer's narrative. In order to connect and integrate Criseyde's behaviour in Book V with her nature in earlier books, her character has been explained in terms of some tragic flaw of excessive fearfulness, concern for her reputation or the opinion of society, while Troilus' tragic flaw has been located in the stubborn fixity of his absorbing passion for Criseyde. Yet Aristotle's understanding of a tragic flaw in the moral nature of an individual character proves of limited help in *Troilus and Criseyde*, and for some modern readers there will always be an uneasy question as to *whose* tragedy it is. For all the fullness of attention to the characters in *Troilus*, or perhaps because of it, the paradoxes and strains which prompt a sense of tragedy are implied to lie in the nature of earthly experience itself. Troilus and Criseyde do not have tragic flaws in the Aristotelian sense, although represented with characteristics that prove significant for the action and shown to make choices and decisions. Despite the unified focus of the action on a single tale of love, tragedy is perceived in the pattern of the poem's action as a whole, rather than as the tragedy of Troilus or of Criseyde.

It is in the distinctive nature of *Troilus* as a generic hybrid that it modifies the ways in which an audience's expectations of *tragedye* are drawn upon as the poem develops and ends. By fully realizing the nature of the 'ascending' action as a romance, Chaucer follows idealization of motive and experience with a transition to disillusion in a way that comments on the images of human experience in both romance and tragedy. That combination of pity and terror felt at the climax of ancient tragic drama is equally not part of medieval tragedy. Although the sensational climaxes of the Falls of Princes may evoke horror, the characteristic note of unhappy closure is the pathos found in something truly piteous, to which the response is complaint and lamentation. The last two books of *Troilus* are a study in the art of lamentation, and the fifth book especially becomes a sustained study in intense melancholy and the attenuation of hope. The manner in which the action slows even anticlimactically to a close allows for an intensity of pathos which may deflect attention from the errors and shortcomings of the characters but does not prevent the poem from suggesting a larger tragic interest in the problem of suffering in human life. The generic combination in *Troilus* allows for the fullness of a subjective empathy with the sufferings of particular characters in romance, as well as prompting—through the detachment of its associations with tragedy and philosophical dialogue—a more objective, moral, even metaphysical consideration of human suffering and disappointment.

There is, of course, a further complication to *tragedye* in *Troilus* in the way that Chaucer's narrative follows its slain hero beyond death in his ascent to the spheres. To show the spirit of the slain hero not only looking back on life from the other side of death, but contemning the world and even laughing at those he sees lamenting his death, can only be a 'tragedy' in a way that invokes tragedy but does not let it stand alone, in order to redefine the idea by relating it to a perspective which is that of divine 'comedy'. If Chaucer draws *Troilus* away from its source in one direction towards interpretation as a tragedy, that is complemented by his realization of the potential comedy as well as the potential romance in the story. The 'comedy' implied in Troilus' ascent succeeds and supplants tragedy in the hero's experience (and in the poet's, as perhaps he looks forward to following *Troilus* with 'som comedye'). But in this the concluding phase of *Troilus* draws on the complementary roles of comic and tragic already established in the poem: the fainted Criseyde's recovery from what had seemed a death, and the prompt cancellation of the hero's suicide over her 'corpse', produce as early as the fourth book the anticlimax of an evaded *Liebestod*, a projected ending in the lovers' double suicide together, which does not actually happen. This moment is 'anti-tragic', but it is also perhaps 'anti-romance', in that it parodies the apparent death-and-resurrection movement of romance by being in no sense restorative or regenerative. After this unrepeatable and 'anti-tragic' anticlimax at the lovers' last meeting, Boccaccio's invention of Criseida's promise to return within a specified period links the last phase of the story's structure with a dynamic of waiting and disappointment—a faltering and failing momentum, coming to rest in despair—so that the problematic ending Chaucer received from *Filostrato* could be neither happy nor tragic in a conventional sense. The very slowness of the concluding movement necessitated by the plot prompts Chaucer to pace the sequence of events in a way that allows a proper space to the tragic and comic and allows them to be related. The narrative action pursues its full course to the death of the hero, and this has the deliberateness and order of a sought death accepted in its time, rather than the sudden disorder of downfall in a Fall of Princes. The pause between the prince's earlier fall as a lover and his eventual death allows Troilus some process of discovery both before and after death. It is only after the fullest sense of a tragic 'fall'—in disillusion with the idealism and values represented in romance—that the pagan hero in the flight of his soul is moved nearer to the perspective of divine comedy which he can share with Chaucer's Christian audience.

Precisely because it is so much deepened and transcended by its new role within a hybrid genre, the notion of a *tragedye* of fortune does, therefore, have a significant role as a generic norm within *Troilus*, against which generic 'quotation' and differentiation could be made. The intensely realized sense of human value and loss in the lovers' experience bursts the bounds of the view of worldly life implicit in the definitions of *tragedye* by Chaucer's Monk or Boethius' Fortune, but these can still help define *Troilus* as a tragedy against them. It is a paradox that Chaucer shows his interest in genre by using a generic label which—as applied elsewhere, even in Chaucer's other works—can only diminish *Troilus and Criseyde*. A definition of genre is offered, yet this only suggests the greater reach and inventiveness in genre of the poem of which it is used.

For a survey of the subject, see William Farnham, *The Medieval Heritage of Elizabethan Tragedy* (Berkeley, Calif., 1936), 129–72, and for general discussions of tragedy in Chaucer, see Boitani, *The Tragic and the Sublime*, and Ruggiers, 'Notes'.

For particular aspects of tragedy in *Troilus*, see Brewer, 'Comedy and Tragedy'; Clough, 'Medieval Tragedy'; Erzgräber, 'Tragik und Komik'; Lawlor, *Chaucer*; Norton-Smith, *Geoffrey Chaucer*; and Robertson, 'Chaucerian Tragedy', who interprets the poem as a re-enactment of the tragedy of Adam. For a reading of Troilus' career as a Boethian comedy, Criseyde's as a Boethian tragedy, see McAlpine, *Genre*, chs. 5–6.

A standard medieval definition of tragedy is cited by Strohm, 'Storie', from the thirteenth-century *Catholicon* of Johannes Januensis:

Et differunt tragoedia et comoedia, quia comoedia priuatorum hominum continet facta, tragoedia regum et magnatum. Item comoedia humili stilo describitur, tragoedia alto. Item comoedia a tristibus incipit sed cum laetis desinit, tragoedia e contrario.

Tragedy differs from comedy because comedy contains the deeds of private men; tragedy of kings and magnates. Also, the comic is described in the low style, tragedy in the high style. Also, comedy begins with sad things, and ends with happy ones. Tragedy is the opposite.

Trivet's gloss on tragedy as 'carmen de magnis iniquitatibus a prosperitate incipiens et in adversitatem terminans' (a song of great iniquities beginning in prosperity and ending in adversity), and the gloss of Pseudo-Aquinas: 'tragedia est carmen reprehensivum viciorum incipiens a prosperitate desinens in adversitatem' (tragedy is a song in reproof of vices beginning in prosperity and ending in adversity), are cited by McAlpine, *Genre*, 89.

On Trivet's commentaries on Seneca, see Norton-Smith, *Geoffrey Chaucer*.

On the Averroes-Alemannus version of the *Poetics*, see H. A. Kelly, 'Aristotle-Averroes-Alemannus on Tragedy: The Influence of the "Poetics" on the late Middle Ages', *Viator*, 10 (1979), 161–209, and Judson Boyce Allen, *The Ethical Poetic of the Later Middle Ages* (Toronto, 1982), 19 ff., 121 ff., who cites Averroes' concept of tragedy as an art of praising:

> Tragedia etenim non est ars representativa ipsorummet hominum prout sunt indivi-
> dua cadentia in sensum, sed est representativa consuetudinum eorum honestarum et
> actionum laudabilium et credulitatum beatificantium. Et consuetudines comprehen-
> dunt actiones et mores' (For tragedy is not an art which represents men as perceivable
> individuals but one which represents their honest customs and praiseworthy actions
> and sanctifying beliefs. And customs include actions and manners) (p. 231).

See also A. J. Minnis and A. B. Scott (eds.), *Medieval Literary Theory and Criticism, c.1100–c.1375: The Commentary Tradition* (with the assistance of David Wallace; Oxford, 1988), ch. VII.

Drama

From the tragedies of Seneca—as these were understood and interpreted in the Middle Ages—Chaucer could also derive a powerful model of a tragic narrative, although it remains unclear how directly Chaucer was acquainted with Seneca's plays. Some of the impression now made by *Troilus* lies in its 'dramatic' quality: the way its narrative contains 'scenes' of lively dialogue and acutely observed gesture. The manner in which classical plays were studied by medieval readers possibly had some role in providing a model for Chaucer of such scenic form in narrative. The works of Roman playwrights were not known through performance but were read as narratives composed in dialogue, and the influential commentaries of Nicholas Trivet on the tragedies of Seneca carefully analyse their five-act dramatic structure, their division into scenes and chorus, and hence the basic element in dramatic composition which underlies the act-structure, namely the expressive scene. Guided by the commentaries, medieval readers would have imagined classical drama as making use of a mixture of narrative and dramatic forms. In his commentary on Seneca Trivet distinguishes three modes of dramatic presentation:

1. dramatic recitation where the poet alone speaks the text;
2. a performance where the poet narrates and speaks the parts while actors mime the action;
3. a more mixed mode where the poet narrates but introduces his characters who speak and act out the dialogues (for Trivet, Virgil's *Aeneid* is an example of this mixed mode).

The role imagined in the Middle Ages for the poet as the 'narrator' of Roman play texts, the presenter of the tragedy, is not dissimilar from the deployment of a narrator-figure by some medieval poets including Chaucer, especially when this is combined with the role of the chorus in Senecan tragedy. If Chaucer has changed *Filostrato* by developing a 'presenting' poet-figure and by fashioning expressive scenes of dramatic dialogue and encounter, then a central generic model was available in the plays-as-read of the Roman dramatists, as these were studied and commented. Medieval manuscripts of the plays of Plautus are careful to indicate scene-divisions in the dramas based on the entrances of a principal character: the shaping of scenic form was principally determined by character-grouping. As a text for reading and study, the play becomes a narrative in dialogue, with the presence of characters marking off the beginnings and endings of episodes.

It is clear from the marginal glosses in some *Troilus* manuscripts that the dialogic quality of long stretches of the poem was something many scribal commentators wished to mark, especially where the speakers are alternating in rapid exchanges of dialogue. Such scribal signalling of speech units can give some pages of some *Troilus* manuscripts the appearance of a play text. In its combination of 'presenting' narrative, monologue, and extensive, free-standing dialogues, a text of *Troilus* would resemble the medieval analysis of Senecan tragedy. Chaucer arranges his sequences of character-grouping far more tellingly than Boccaccio and with a keener sense of interchange, contrast, and climactic emphasis. He fashions a number of really lengthy scenes in his first four books, and the latter part of Book II— where he is composing free of *Filostrato*—has much of the continuity of a dramatized scene, although as the poem nears it close there is less sense of sustained dramatic form in Book V. Throughout, a balance is struck between the dilation of a scene and the momentum of the book, and Chaucer's shaping of scenes is always enfolded within a larger narrational frame. The medieval reading of classical drama as narrative, and Chaucer's own shaping of scenes of exceptionally 'dramatic' dialogue within his narrative, make for intriguing resemblances of form betweeen *Troilus* and medieval understanding of the genre of Seneca's tragedies.

But yet other models for dialogue were available to Chaucer from other genres, and it is evident that in several parts of *Troilus* the exchanges between characters reflect the influence of that form of philosophical dialogue found in Boethius' *Consolation of Philosophy*. Here the use of a source brings with it a 'quotation' of the genre of that source, for at key points Chaucer inserts and develops on the model of Boethius scenes of extended, serious, and rather formal dialogue, 'demonstration', and debate which have no place in Boccaccio's narrative. The dialogue between Troilus and Pandarus in the second half of Book I is an instance of this: it is marked by quotation of Boethian phrases and arguments, just as it develops Pandarus' role towards Troilus with echoes and quotations of Philosophy's role towards the prisoner. The scribe of the Rawlinson manuscript of *Troilus* who notes that this passage is from Boethius (see p. 42) shows how such scenes might be read in relation to the Boethian genre of philosophical dialogue constructed without narrative around the participants' exchange of direct speech in debate and song.

For a discussion of *Troilus* in relation to Senecan drama, see Norton-Smith, *Geoffrey Chaucer*; earlier attempts to discuss dramatic form—e.g. Utley, 'Scene Division'—were less convincing. On medieval understanding of the ancient theatre, see M. H. Marshall, 'Boethius' Definition of Persona and Medieval Understanding of the Roman Theater', *Speculum*, 25 (1950), 471–82, and 'Theater in the Middle Ages: Evidence from Dictionaries and Glosses', *Symposium*, 4 (1950), 1–39.

MS H5 of *Troilus* identifies speakers by writing their names at relevant points in the margin, as in these dialogues: i. 568, 'Troylus'; i. 582–3, 'Pander'; i. 596–7, 'Troylus'; i. 617, 'Pandare'; i. 820, 'Troylus'; i. 829, 'Pandarus'; i. 834, 'Troylus'; i. 841, 'Pandarus'. In H4 speech units, dialogue and monologue, are also remarked by marginal commentary, e.g. at iv. 827, 'Verba Cressaidis P.'; iv. 875, 'Verba P. C.'; iv. 897, 'Verba C. P.'; iv. 1527, 'Responsio C.'; v. 117, 'Verba Diomedis Cressaide'; v. 218–19, 'Verba T. in absentia C.' S1 also contains much signalling of speakers by marginal notes.

See also Seth Lerer, *Boethius and Dialogue: Literary Method in the 'Consolation of Philosophy'* (Princeton, NJ, 1985).

Lyric

Direct speech, dialogue, and song bring us to the lyric nature of *Troilus*— its relation to various types of lyric genres—for this is an aspect of the generic identity of Chaucer's poem which is insufficiently recognized, although the *Troilus* manuscripts bear witness that it was appreciated nearer to Chaucer's own time. Chaucer increases the formal lyric nature of his poem, both absorbing and adding to the sequence of lyric pieces already included in the Italian narrative, and he works to establish a context in which lyric set-pieces become part of the accepted self-expression of his characters: not only performed songs but also the lovers' formal laments or 'complaints', and so too by extension the courtly letters in which some of their entreaties and lamentations are expressed. All these are contained within an added emphasis on the formal lyric framework to the poem through its proems, and through references to the narrative itself as song (ii. 55–6; iii. 1814). Although Chaucer's other poems may refer to characters as singing, few include the actual text of the lyric; *Troilus* sets itself apart in the ambition and frequency with which it contains lyric pieces within the narrative, and establishing a uniquely effective relation between lyric and narrative is one of Chaucer's achievements in this poem.

Models for the mixing of narrative and lyric were available in works as familiar to Chaucer as Boethius' *Consolation*, the *dits* of Machaut, and *Filostrato* itself, but they work to varying purpose and effect. Boethius shows how lyric passages could be used to distil the spirit of a moment of enlarged perception and understanding. In the *dits amoureux* of Machaut lyric genres are used so extensively as to subordinate the narrative element to the role of a supportive framework which contains and orders the elaborate set-pieces; when given as a performance, a poem like Machaut's *Remède de fortune* would have had some of the effect of an opera. There are many echoes in lover's mood and Boethian theme between *Troilus* and the *Remède de fortune*, and Chaucer's incorporation of lyric genres can indeed give *Troilus* the impression of a sonnet sequence, a succession of lyric clusters that recall the *dits amoureux*. Yet *Troilus*' stronger narrative impetus, which can pause over but 'contain' its lyric passages, derives essentially from *Filostrato*, and improves upon its model by judicious shortening of the hero's songs (in Books III and V) and his letters (in Books II and V), so as to create a better balance between lyric and narrative.

Chaucer never ignores a cue for a lyric in *Filostrato*, even though he may substitute a different lyric text at the same spot, but he also develops cues further and provides additional lyrics. The *occasions* of Troilus' songs in Book III (1744 ff.) and Book V (638–44) are both kept, while the *text* of the songs is replaced in each case, just as in Book V the text of Troilus' letter is included at the same place but almost entirely rewritten (v. 1317 ff.). A passing mention of Troiolo as singing, of Criseida as writing, or of the lovers as sorrowing at dawn, prompts Chaucer to insert at those points a full text of Troilus' song (i. 400–20), of Criseyde's letter (v. 1590–1631), or of the lovers' *aubades* or 'dawn-songs' (iii. 1422–42, 1450–70). In the sequences of action invented by Chaucer in Books II and III a number of lyric set-pieces are incorporated: in Book II there is Antigone's song (ii. 827–75), while the development of Book III is structured through a whole series of such added or altered set-pieces as the proem (iii. 1–49), Troilus' petition to Criseyde in his sick-room (iii. 127–47), his prayer in the 'stewe' (iii. 712–35), his hymn 'O Love, O Charite!' (iii. 1254–74), the *aubades* (iii. 1422–42, 1450–70, 1702–8), and the hero's final prayer to love (iii. 1744–71). Book V develops a series of cues for lyrics provided by Boccaccio, so that the lyric nature of the third book finds a balance in the pattern of formal complaint in the last.

In the margins of some *Troilus* manuscripts various glosses in Latin or English offer definitions of the lyric genres identified by the scribes. The beginnings of the hero's two songs in Book I and V, of Antigone's song, and of the two letters in Book V, are marked by Latin headings in a large number of the manuscripts, and a few identify many more. The words *Cantus, Littera, lamentacio, compleynt, testamentum,* and *envoy* are added in the margins at appropriate points, as part of the evident interest for readers in recognizing the lyric genres. The stanza of Oenone's *compleynt* in Book I ('Cantus Oenonee', 'Littera Oenone'), the *aubades* in Book III ('nota verba .C. in aurora'), Troilus' testament in Book V ('Testamentum Troili'), and the envoy at the close of the poem ('Lenuoye Du Chaucer'), are among those pieces signalled in various manuscripts. There is also a noticeable concern in the manuscripts to note where lyric pieces end as well as begin, as if to mark them out and round them off as set-pieces within the presentation of the text.

It is partly through Chaucer's accomplishment in writing a narrative framed in stanzas that such a range of lyric set-pieces is related to narrative. The model is *Filostrato*, where the lyric passages are written uniformly in the same metrical form of *ottava rima* as the rest of the poem, so that narrative and lyric are not formally distinguished. For Boccaccio's audience the *ottava rima* was a verse-form with narrative associations, recalling the contemporary popular romances, the *cantare*, but the associations of Chaucer's rhyme-royal stanza were probably still more with lyric than with narrative forms, and, despite the uniform versification, lyric passages in *Troilus* are carefully distinguished from the surrounding narrative, whether by occasional rubrication, or by references inside the text to the beginning and ending of lyric pieces (e.g. i. 399; ii. 876; v. 645). A whole series of Chaucer's lyric interpolations are organized to fall into sets of three stanzas and thus constitute themselves as a type of *ballade*, as in the Petrarch sonnet that becomes the first Canticus Troili (i. 400–20), Troilus' petition to Criseyde (iii. 127–47), her reply (iii. 159–82), his two addresses to love (iii. 1254–74; v. 582–602), and the two dawn-songs (iii. 1422–42, 1450–70).

The epistle and the formal lamentation or 'complaint' also have established shapes within the flow of the stanzaic narrative. The customary expectations of how letters were to be structured—through an ordered sequence of topics and the formulas that linked them—contribute to giving the letters in *Troilus* a distinct identity despite their uniform versification. They are written with an eye to the rules on sequence of material in medieval manuals of letter-writing and echo the construction and many of the formulaic phrases found in medieval English correspondence, while the French phrase appended after the last stanza in some manuscripts ('Le vostre T.', 'La vostre C.') uses a polite formula to demarcate in the text the end of the inset letters. Even Chaucer's summary of Troilus' first letter conforms to a recognizable outline and rhetoric ('First he gan hire his righte lady calle' (ii. 1065 ff.)). The medieval *artes dictandi* (arts of letter-writing) divided the structure of a letter into five parts, to which Criseyde's last letter to Troilus conforms: a combined *salutatio* and *benevolentiae captatio* (i.e an attempt to gain the recipient's sympathy) (v. 1590–6); a *narratio*, relating her attitudes (v. 1597–1620); a *peticio*, asking Troilus for friendship and not to be displeased by brevity (v. 1621–6); a second *benevolentiae captatio* using *sententiae* (proverbs) (v. 1627–30); and a simple *conclusio* (v. 1631).

It is with their occasional marginal glosses identifying *compleynt* and *lamentacio* ('Nota bene de Troily how he complenit . . .' (S1 at v. 1674)) that some scribes draw attention and give formal definition to a distinctive strand in Chaucer's use of lyric genres in *Troilus*. In a poem that sets out to tell of a double sorrow, complaint is central to form and theme. In a courtly context the act of making 'complaint' will often be the impetus behind such lyric forms as songs or epistles. In the *Merchant's Tale* the love-sick Damian writes to May: 'And in a lettre wroot he al his sorwe, | In manere of a compleynt or a lay' (IV. 1880–1). In Chaucer's *Anelida and Arcita* the abandoned queen composes a complaint and and sends it as a letter to her false lover (207–8). The proem to the *Complaint of Mars* implies established expectations of 'the ordre of compleynt', and Mars sees the impetus for his own complaint as expressive and declarative rather than as a purposeful supplication which hopes to gain relief (162–3).

From *Filostrato* Chaucer inherited a succession of scenes structured in complaint which with modifications are taken over into *Troilus*:

Troilus' complaint after falling in love (i. 507–39)
Troilus' overheard complaint, as reported by Pandarus (ii. 523–39)
Complaints by Troilus (iv. 260–336) and by Criseyde (iv. 743–9, 757–98) after hearing of the exchange
Troilus' complaints after Criseyde's departure (v. 218–45, 414–27)

Troilus' address to Criseyde's empty house (v. 540–53)
Troilus' complaints on the places of Troy (v. 565–81, 606–16, 669–79)
Troilus' complaint to Cupid (v. 582–602)
Troilus' song and address to the moon (v. 638–58)
Criseyde's complaint in the Greek camp (v. 731–65)
Troilus' complaint on Criseyde's infidelity (v. 1674 ff.)

A preponderant emphasis on the hero's lamentations is a striking aspect of the inheritance of complaint from *Filostrato*, but Chaucer increases its incidence, lengthening some, and adding the following new ones:

Oenone's complaint (i. 659–65)
Criseyde's complaint on 'worldly selynesse' (iii. 813–40)
Criseyde's complaint on jealousy (iii. 988–1050)
Criseyde's 'aspre pleynte' (iv. 827–47)
Troilus' 'testament' (v. 295–322; cf. iv. 785–7)
Criseyde's complaints in the Greek camp (v. 689–707, 1054–85)

Further elements of complaint are woven into brief passages of dialogue and monologue, or into the descriptions of the characters' actions and gestures (the Book I Canticus Troili, the Book III *aubades*, the predestination soliloquy, Troilus' sorrowful letter to Criseyde in Book V).

Frequent *compleynt* thus becomes part of the lovers' characteristic disposition in both the ascending and descending action of the poem. Courtship itself is expressed through complaint (iii. 53, 104; cf. v. 160–1), and the hero's character grows in definition as a plaintiff, against love or misfortune. The poem attends seriously to the form of complaint (iv. 799 ff.), and lamenting speeches are labelled as 'complaints' (e.g. i. 541; ii. 560; iv. 742). There is the gravity and deliberateness of a ritual observance in the way that characters are pictured complaining (iv. 1170–1), for complaint is both stance and ceremonious event throughout *Troilus*. The accents of complaint recur repeatedly, heard like a *leitmotif* unifying the poem through all sorts of contexts and brief utterances in addition to the large set-pieces of lamentation (e.g. ii. 330–3, 344–7, 409–13, 780–98; iii. 617–20). As more lyric units have been identified within the text they have been seen as representing, and carrying forward within themselves, a continuous emotional life which is the most significant focus of action in the poem.

This is the measure of Chaucer's success in incorporating lyric genres, for in *Troilus* he boldly adapts and conjoins the lyric identity of his sources in interaction with other genres. The supposedly autobiographical frame within which *Filostrato* is presented means that Boccaccio's whole poem is contained within the bracket of a personal lyric. The hero is a fictionalization of the poet: Troiolo in some sense is the poet who writes about

Troiolo. The poem begins with a *demande d'amour* and at the close the poem is sent on its way to the poet's lady with a graceful envoy. The proem to *Filostrato* presents the ensuing poem as the lover's complaint—'con alcuno onesto ramarichio' (through some fitting form of complaint)—that provides the means for his grieving breast to be relieved of its misery, and the poet's sorrows during his lady's absence are described in terms that precisely anticipate the circumstances and feelings of Troiolo in the latter part of the poem. (Both haunt the places in the city associated with the lady; both gaze towards where she is and feel the wind comes to them from there.) Since the poet is a lover and identifies with the hero, the lover who represents him inside the story has some of the gifts of a poet, for both poet and hero are the characteristic speakers of lyrics.

In working from *Filostrato* Chaucer also extends the social occasions and opportunities for performance of lyric in the world of his *Troilus*: Criseyde listens to a song sung by her niece Antigone (ii. 824 ff.); at Pandarus' supper party Criseyde is entertained with songs and story-telling (iii. 614); after Criseyde has left, Troilus recalls his special delight in hearing his lady sing (v. 577–80), so that Chaucer associates lyrical skills with both hero and heroine while reserving the actively lyric role for Troilus. Chaucer retains in his hero something of Troiolo the lover as lyric artist—nothing suggests his lyrics are not his own compositions—while removing that framing context which explained the lyric pieces in *Filostrato* as a veiled expression of the poet's own 'complaint'. In *Troilus* the use of lyric genres must therefore establish its own purpose in the narrative, and as a medium through which Chaucer can suggest some of the broader implications of his narrative.

The selection and range of lyric and the comment made through its relation to the narrative become especially important in *Troilus*. Through all his incorporations of lyric pieces—from Petrarch, Boethius, Machaut, or in forms such as the *aubade* that were rare in English—Chaucer aims to use lyric both for its expressiveness and for its ability to convey something quintessential, as in the association of Troilus' emotions with that highly patterned oxymoronic paradox of the Petrarch sonnet. In some cases—as in lovers' partings at dawn in Book III, the lover's complaint overheard (ii. 505 ff.), or the lover's watching and waiting outside his lady's house (v. 529 ff.)—a type-scene or a predicament is marked by the characters' giving utterance to the lyric genre quintessentially associated with that moment. At such times, indeed, the forms are fixed because the emotions are fixed, and both are to be savoured for so being. But a critique also emerges in the tensions Chaucer creates between his use of such lyric genres and their contexts and conventions: lyrics like those of Troilus suppose a static realm where perfection depends on immobility and incompletion; the sentiments usually uttered by the male and female speakers of *aubades* are apparently transposed between Troilus and Criseyde; the associations of diligent exertion—usually with daylight toil—are transferred in the dawn-songs on to the nocturnal activities of the lovers; and part of the peculiar pain in Criseyde's last letter lies in its deployment of standard epistolary forms. The conclusion of the poem, with its hero laughing at the vanity of those lamenting his own death can also be seen as a final transcendence of the pattern of formal complaint created in the poem, and strikingly reflects how Chaucer works by setting his uses of genres into a mutual critique of the interpretations of life that they represent.

On the lyric element in *Troilus* in its French context, see Wimsatt, 'The French Lyric Element', and also John Stevens, 'The "Music" of the Lyric: Machaut, Deschamps, Chaucer', in Piero Boitani and Anna Torti (eds.), *Medieval and Pseudo-Medieval Literature* (Cambridge, 1984), 109–29. Payne, *Key*, sets lyric in *Troilus* in the context of a larger stylistic discussion, and sees the lyrics as a kind of sonnet-sequence distilling the poem's emotional progress. See also Taylor, 'Terms'.

On the narrative associations of Boccaccio's verse-form, see Wallace, *Chaucer and the Early Writings of Boccaccio*, ch. 5; and on the associations of rhyme royal with lyric forms, see Wimsatt, 'The French Lyric Element', 21.

In addition to the rubrics for songs and letters which probably derive from the poet, individual scribes also show an interest in signalling lyric units, as in the following marginal notes: i. 658: 'Littera Oenone' (S2); i. 659: 'Cantus Oenonee' (S1); ii. 875: 'Explicit Cantus' (H4); ii. 1065: 'Littera Troili Cressaid' (H4), 'Prima littera Troilus missa ad Criseid' (S1); ii. 1085: 'Her endes Troylus his first lettyr' (S1); iii. 1424: 'Nota verba .C. in aurora' (H4); v. 295: 'The testament of Troilus' (R), 'Testamentum Troili' (S2); v. 1421: 'Finis littere Troili' (H4); v. 1425: 'Risponcio Criseidis' (S1); v. 1855: 'Lenuoye Du Chaucer' (H3).

On 'complaint' in *Troilus*, cf. the following marginal comments in the manuscripts: i. 547: 'How Pandar fond Troilus compleynyng' (R); ii, 526: 'How Pandar tolde Crisseide that Troilus pleind to loue' (R); iii. 1422: 'How Cresseyde sorowed whan day gan for tapproche' (R); iv. 260: 'Here maketh Troylus his compleynt upon fortune' (S1); iv. 729: 'How Cresseyde comp[ley]nyd for she shuld departe oute of Troye' (R); iv. 743: 'lamentacio .C.' (H4, Ph, S2); v. 1674: 'Nota bene de Troily how he complenit' (S1); v. 1681: 'complent'; v. 1688: 'complent'; v. 1695: 'complent' (S1).

On the letters and use of epistolary conventions in *Troilus*, see Davis, 'The Litera Troili'; and McKinnell, 'Letters'.

On complaint, debate, and narrative in *Troilus*, see Davenport, *Chaucer*.

On the dawn-songs in Book III, see Kaske 'Aube', and for their context, see Saville, *The Medieval Erotic Alba*; Scattergood, '"Bisynesse"'; and Schelp, 'Die Tradition'.

Fabliau

The more passive character of the English lovers compared with their Italian counterparts prompts the incorporation of quite different, almost antithetical, generic elements. Chaucer needed a plot to bring his lovers together and to save appearances, and in enlarging the role of Pandarus he draws on elements of plotting and characterization which resemble the genres of fabliau and various 'arts of love', in which a go-between practises deceptions to accomplish the meeting and union of lovers.

By combining elements in the plots of just two fabliaux Chaucer might assemble the basic mechanism for bringing Troilus and Criseyde together in bed. In *D'Auberee* an old woman tricks a pretty wife into being her guest for the night, then brings her amorous male client in upon the guest. In *Du prestre et d'Alison* a mother with her wits about her makes a pretence of complying with the local priest's desire for her daughter. From the priest's point of view as he sits in the main room, it would seem that the daughter has gone to bed. However, the bedroom is provided with a false door and an adjacent room, which enables the daughter to be hidden and a prostitute called Alison to be smuggled in to take her place with the priest. In the latter fabliau, as in Pandarus' manipulation of both bedroom meetings, the trick turns on the architectural factor that a small room leading off a larger one has another means of entrance and egress than that through the main room. But such comings and goings through secret doors do occur in romance intrigues as well as in fabliaux, for ingenious stratagems and intrigues are also part of the stuff of romance. In the romance of *Eracle* by Gautier d'Arras (*c.* 1163) the Empress Athanais visits the house of the go-between, an old woman, and her guards watch her enter and see only the old woman inside; but her lover Parides is hidden in the house in a secret chamber with a secret door, and so the lovers may meet.

The strongest resemblances are to the twelfth-century Latin *Pamphilus*, the opening line of which is quoted in the *Franklin's Tale*, 1109–10, and which is an important source for Guillaume de Lorris's section of the *Roman de la rose*. Here, the Ovidian precepts of an 'Art of Love' stemming from the *Ars amatoria* and the *Remedia amoris* are applied and set in motion in a narrative of seduction, in which a key role is played by a go-between figure—an old woman, the Anus—who is an older adviser on love matters. It is in the stratagems for bringing the lovers to bed together that the closest resemblances occur between *Pamphilus* and *Troilus*, which may be summarized as follows:

> The go-betweens settle on a plan. They invite the ladies to their homes for a meal, while assuring them that they will be safe from all rumour and gossip. Meanwhile, after a speech to the lover on destiny, in which she declares that only God can foretell the future, the Anus, or go-between, causes Pamphilus to swoon and arouses his jealousy by telling him lies about a possible rival in love (cf. the action of Pandarus when he deceives Criseyde over the supposed jealousy that Troilus feels for Horaste). During preparations for the consummation of the affair, the go-between urges the lover to be a man, and while the lady is eating at

the home of the intermediary, they hear a wind and the lover abruptly makes his entrance. The go-between bursts into sentimental, sympathetic tears for the lovers and soon leaves them alone. In *Pamphilus* as in *Troilus*, the men call on the women to yield ...

In the morning, the go-between returns and with mock innocence asks whether anything has happened, and how the night has been spent? Both the Anus and Pandarus are promptly and roundly accused of deception and betrayal by the women.

Although in many respects—of detail of texture and of delicacy—Chaucer's *Troilus* is far removed from this, the corresponding scenes most recall something of the fabliaux in the way that they acquire a sharp focus on that physical setting which has such a direct bearing on the successful momentum of the action: the 'stewe' where Troilus waits, the drawn 'travers', the 'trappe-dore', the 'pryve wente', the cushion, the 'chymeneye', the curtained bed. The crucial significance of arrangements and locations for sleeping, the sheer virtuosity of manipulation and contrived appearance, the sudden actions, the darting and leaping about: all serve to associate the poem with the world of the fabliau genre which Chaucer evidently knew so well, and which he was to use in the *Canterbury Tales* to complement and comment upon romance. The tendency of romance to assert one extreme—in an idealizing vision of how nobly men may behave—is complemented by the contrary 'realism' of the comic tale's assumption that men's true motives and natural behaviour are of a grosser kind. Already in place within many romances is an ambiguous relish for successful deception which is not so far removed from the pleasure of stratagem in the fabliaux, and it is this overlap between the genres which Chaucer can exploit when relating them in his *Troilus*. Chaucer's containment of fabliau within a framing romance allows for intriguing ambiguities of reference in the action and characterization. It can make available a disturbing commentary on the role of Pandarus at many points; Criseyde can appear as gullible as a character in a fabliau, when a complicated comedy is set up as an elaborate alibi to satisfy her romantic conception of herself. But, in that Chaucer's conjoining of romance and fabliau is accomplished with the keenest comic sense and the lightest touch, it can allow the greatest ambiguity of interpretation.

On the fabliau element in *Troilus*, see Muscatine, *Chaucer*, 140, and for the background, see Charles Muscatine, *The Old French Fabliaux* (New Haven, Conn., 1986).

On the *Pamphilus* and *Troilus*, see Garbáty, 'The Pamphilus Tradition'. For the text of *Pamphilus*, see *Three Latin Comedies*, ed. Keith Bate (Toronto, 1976). See also T. J. Garbáty, '*Pamphilus, De Amore*: An Introduction and Translation', *Chaucer Review*, 2 (1967), 108–34.

On Chaucer's possible use of Jean Brasdefer's French version of *Pamphilus*, see Miesz-
kowski, 'Chaucer's Pandarus' and for the text, see *Pamphile et Galatée*, ed. Joseph de
Morawski (Paris, 1917).
On *Pamphilus*, the *Roman de la rose*, and *Troilus*, see Wimsatt, 'Realism'.

Allegory

An implication of the echoing and quoting of so many genres in *Troilus* is
that there are many different levels of potential meaning in the story. One
might not immediately think of *Troilus* as including elements of allegory
within its sense of a conjoining of diverse genres to suggest different
dimensions and layers of significance in the story. Yet there are pointers
towards allegory in the 'frame' of narratorial exposition and commentary
around the story, in the way *Troilus* may be read in relation to some of its
principal source texts and traditions, and perhaps in the mythographical
tradition of moral allegorization available for the figures and stories from
classical mythology mentioned in Chaucer's text.

The ambitious proems that Chaucer adds prepare the reader to see the
narrative of each book in terms of larger patterns than those of the literal
meaning of the story. The first proem is especially striking in this way, as
lovers in the audience are urged through reflection to relate the story of
Troilus to their own experience and sympathize with those in like
suffering, while the narrator prays for his success in representing 'Swich
peyne and wo as Loves folk endure, | In Troilus unsely aventure' (i. 34–5).
Again, in the second proem, the pointed explanation of meaning reminds
the reader that the action is in part being represented by figurative means
('This see clepe I the tempestous matere | Of disespeir that Troilus was
inne' (ii. 5–6)). Indeed, although he does not draw on it directly, the proem
to the poet's lady that opens *Filostrato* provided Chaucer with one model of
how the story of Troilus could be used both for itself and to present—
encoded—another level of application and interpretation, for Boccaccio
declares that the old tale is being used as a 'disguise' (*scudo*) for his secret
sorrow in love. In his own way Chaucer has also turned to old books and
traditions, and by associating these with his story of Troilus he enables that
story to point beyond itself. This allegorical potential stems in part from
the poem's associations with both the *Roman de la rose* and the *Consolation
of Philosophy*, although the whole tradition of the 'game' and 'service' of
courtly love is marked by an allegorical cast of expression, and the romance
genre itself tends to present character in ways that shade into the allegory
of psychological archetypes.

In the early books of the poem Troilus is in part an incarnation of the
allegorical process set forth in the *Roman de la rose*, sharing archetypal
features and experiences with Amant, the representative, generic Lover of

the *Roman*. In the presentation of the inner life of the lovers there are added sequences conveying inward processes in allegorical terms ('Than slepeth hope, and after drede awaketh' (ii. 810); 'Thus am I with desir and reson twight' (iv. 572)), while the sequence in which Troilus first sees Criseyde may serve to represent the influence of the *Roman de la rose* as an example of allegory at several levels. *Filostrato* refers briefly to Troiolo's being 'pierced' by love before he was aware, but the dead metaphor is revived by Chaucer into an episode which echoes the allegorical narrative of how the Amant is hunted and shot by Love in the Garden of the *Roman*. *Troilus* also shares with the *Roman* an ambition to 'anatomize' love in a reflection of the mediated influence of those 'cosmic allegories', particularly *De planctu naturae* (The Complaint of Nature) of Alan of Lille, which are excerpted and drawn upon by Jean de Meun. When Chaucer expatiates on necessary submission to the power of love and the law of nature ('For may no man fordon the lawe of kynde' (i. 238)), or when the action of the third book is bracketed between the opening invocation of Venus (iii. 1–49) and Troilus' final hymn to Love as a cosmic force (iii. 1744 ff.), the text suggests—with whatever irony in each context—that Troilus' love is to be seen in relation to those cosmic forces which are analysed through the allegories of Jean de Meun and his sources, including Alan of Lille and Boethius.

Complementing the allegorizing example of the *Roman* is the model provided by the *Consolation*. To read *Troilus* in mind of its Boethian model is in effect to be reading both texts concurrently, and both the direct quotations and the implicit contrasts and parallels mean that the poem, viewed in relation to the *Consolation*, has a significantly allegorical cast, or that, through the links and parallels between Troilus and the figure of Boethius, from the early image of the sick man awaiting cure onwards, Chaucer sought to exemplify the philosophical stages of the *Consolation*. In the very contrast between Boethius' growing recognition of the true good and his disengagement from the world, and Troilus' sensual happiness followed by his misery in this world, the inclusion of Boethian allusion gives the hero's story overall something of the momentum of a philosophical quest in certain respects parallel to the *Consolation*. Indeed, some of the recurrent imagery in the poem—of binding, of light and blindness, of religious observance—may be interpreted allegorically. Troilus' final achievement of perception when he ascends out of this world also gains an allegorical significance when set against Philosophy's words to 'Boethius' about 'thought' ('I schal fycchen fetheris in thi thought, by whiche it mai arisen in heighte ...' (*Boece*, iv, pr. 1, 65–7)). The literal meaning of the *Troilus* narrative may thus acquire the possibility of an allegorical dimension through its relation to the *Consolation*.

With the *Consolation* and the *Roman de la rose* there is the possibility of reading *Troilus* in the light of texts known to have engrossed Chaucer. In the case of the mythographers and allegorizing commentators on classical texts there is much less immediate evidence of Chaucer's own interest, and his poems do not suggest that he read his Ovid with much recourse to the exegetical commentaries. Chaucer does not deny mythographical interpretations, but his avoidance of them is so persistent as to be remarkable. By contrast, the allegorical figures he does use—especially Love and Fortune—are given a very strong definition. Nevertheless, the mythographers' allegorizing interpretations would be at least a part of contemporary understanding of the classical figures and stories that Chaucer has so extensively associated with the characters and events in *Troilus*. The identification of a hero called 'Troilus' with his city of Troy is part of tradition: Troy would fall if Troilus did not survive (see pp. 143–4), and there were traditional allegorical interpretations of the siege and fall of Troy. The fall of Troy could serve as a warning against following the example of Paris, who chose Venus in preference to the active life of Juno or the wise contemplation of Pallas, and it would be possible to see Paris's choice of Venus as reflected in the experience of Troilus. The city of Troy came to be interpreted—by such commentators as Bernardus Silvestris—as a type of the human body brought low by lechery, in turn represented by the Trojan Horse which the Trojans allowed within their city to their own undoing.

Against the background of such traditional moral reading of the fall of Troy Chaucer's story could allow some allegorical interpretation of the hero's fate. Troilus first sees and falls in love with Criseyde in the temple where the Trojans worship the sacred image of Pallas upon which the safety of Troy depends, and a medieval reader might see the irony that it was to be Diomede who would steal both Criseyde from Troilus and the Palladium from Troy, so destroying both the hero and the city with which he was identified. Chaucer's Troilus also claims the alibi of spending all night in vigil at Apollo's temple, although he is to spend the night at Criseyde's house (iii. 536–46); he pretends to keep watch to see 'the holy laurer quake'—that laurel interpreted by commentators as 'imperishable wisdom'—yet religious worship and pursuit of wisdom are not simply avoided in pursuit of sexual pleasure but deceptively deployed as ruse and alibi. Chaucer also includes many mythological references—often to the myths of disastrous passions and trangressions—which the mythographers had already interpreted in allegorical terms. To consider only the proems: Book I starts with an invocation to Tisiphone and Book IV to all the Furies, whom Bernardus Silvestris and most mythographers interpreted as daughters of sorrow out of ignorance of the spirit. Whether or not Troilus is to be seen, like Paris, as choosing Venus over Juno and Pallas, the third proem placed pointedly before his first experience of physical love evokes something of the doubleness of the 'two Venuses' of mythographical tradition, which represent the dual potential of human love, virtuous or wanton (see below, pp. 223–5). There is also the possibility that allusions to some of the heavenly bodies constitute a sustained astrological allegory in *Troilus* (see below, pp. 205–7).

When Chaucer works into his poem references to so many of the figures mentioned in the Boethian metrum about Orpheus' visit to the underworld (*Boece*, iii, m. 12), he was drawing on a passage in which the mythological figures already had a tradition of allegorical interpretation, and some of these may make available allegorical interpretations of the experience of Chaucer's hero as lover, would-be philosopher, or victim of fortune. The torments of Tityus (i. 786), whose liver 'foughles tiren evere moo | That hightyn volturis . . .', were usually interpreted as the pains of lust, but in his commentary on Boethius Trivet further interprets Tityus as a philosopher who, intent on searching into future things, practised the art of divination and is punished. Troilus writhes like Ixion (v. 212), who was bound to an ever-turning wheel in hell for attempting to lie with Juno, and in Trivet's commentary Ixion is the man dedicated to worldly matters, continually raised up in prosperity and cast down in adversity, for the wheel stops rotating when a man interested in wisdom contemns earthly things. The story of how Orpheus looked back to Eurydice is interpreted to show how Orpheus was drawn away from the 'cleernesse of sovereyn good' to the 'lowe thinges of the erthe' (*Boece* iii, m. 12, 60–9); when Criseyde consoles herself that Troilus will be together with her after death 'As Orpheus and Erudice, his fere' (iv. 791), it was at least open to readers to consider the readiness of Chaucer's unhappy lovers to imagine themselves reunited in a pagan underworld in terms of such moralizations.

The recurrent association of the characters of *Troilus* with mythographically interpreted figures could allow for an allegorical understanding of them as moral types, but this is only to emphasize the ways Chaucer's characterization may through such associations point beyond itself, and to neglect all the vividly particular and individual in the poem which seems essentially removed from allegory ('What knowe I of the queene Nyobe? | Lat be thyne olde ensaumples, I the preye' (i. 759–60)). Yet an elusive impression of layers of both individuality and archetype in the characters of *Troilus* will often suggest to the reader (as Troilus feels of his dream of the boar which 'is' Diomede) that the text is being 'shewed hym in figure' (v. 1449) in a poem so written to alert its reader to the need and the enigma of interpretation. This does not suffice, of course, to make *Troilus* an allegorical poem, any more than some of the other noted echoes and motifs of style and content turn *Troilus* into an epic, or a drama, or an opera, or a history book.

On allegory and *Troilus*, see the classic study by Lewis, *The Allegory of Love*, and also Robertson, who develops the identification of Troilus with Troy ('Chaucerian Tragedy' and *A Preface to Chaucer*) and with the choice of Paris ('Probable Date').
On the parallels between the *Consolation* and *Troilus*, see Stroud, 'Boethius' Influence'.
For a convenient text of Trivet's commentary on *Consolation*, iii, m. 12, see *The Poems of Robert Henryson*, ed. Denton Fox (Oxford, 1981), 384–92; on Trivet, see A. J. Minnis and A. B. Scott (eds.), *Medieval Literary Theory and Criticism, c.1100–c.1375: The Commentary Tradition* (with the assistance of David Wallace; Oxford, 1988), ch. VIII.

It will be apparent from the preceding sections of this chapter that some generic elements remain confined to particular contexts while others inform the poem as a whole. So inclusive a gathering of genres as that in *Troilus* aims to convey more than any single genre by such conjunction and interrelation. Just as the *Canterbury Tales*, the poem gains in richness and comprehensiveness of implication by containing those varying interpretations of experience represented by the genres, modified by comic perspectives and narrative presentation, and at last concluded by the striking ending or 'epilogue', itself a remarkable juxtaposition of generic features.

One of the main modifying factors on genre in *Troilus* is the comedy that is so much Chaucer's contribution and of which earlier texts of the Troilus story were almost entirely devoid. The main impetus for comedy derives from the character of Pandarus, and the main locations for comic action are the intrigues in Books II and III which Chaucer has added to *Filostrato*. The hero's anguished swoon, which is serious in Part 4 of *Filostrato*, comes to be seen as comic when Chaucer shifts it to a quite different context in the bedroom scene of Book III, where Pandarus is on hand to exploit the moment. This difference may stand to represent the mixed quality of comedy in *Troilus* more widely: it asks to be defined by such terms as 'serio-comic', 'tragi-comic', or 'melodramatic', which suggest a mixed generic identity and so recognize that shimmeringly ambiguous capacity of *Troilus* to read as at once comic and serious, now almost heroic, now almost absurd.

Such ambiguity often grows out of the way scenes have reached their present realization through a mixture and superimposing of genres. In Book III there are overlayings of romance, fabliau, lyric, Boethian 'demonstration', and even epic with elements of comedy, just as in Book IV there is a serio-comic sense of different layers in the presentation of what proves less than tragic: an almost tragic scene of accidentally averted suicide (iv. 1149 ff.). Through added complaints and effusions Chaucer here takes an already melodramatic scene in *Filostrato* to the brink of resolution into either the comic or the tragic, yet ultimately contains them both together so that each qualifies the other in a scene of pathos and anticlimax ('But hoo, for we han right ynough of this, │ And lat us rise, and streght to bedde go' (iv. 1242–3)). In a poem that—at least in terms of its characters' earthly lives—comes to a climax at its centre and ends in anticlimax, there is a significant role for anticlimax more largely in the 'double sorwe'. The narrative moves through a whole sequence of moments anticipated as climaxes which turn out anticlimactically, in a way that successively qualifies and redefines what is momentous and conclusive. Juxtaposition with elements of robust humour and farce—or with mishap and anticlimax—only highlights something special, fragile and other-worldly in the experience of love. The whole effect may be serious, or comic, or serious and comic at once, depending on point of view.

In the relationship of Pandarus with the lovers lies much of the balance between the poem's serio-comic aspects. Between his early laughter at lovers and his final laughter from the spheres Troilus is only once mentioned as smiling (at a joke by Pandarus (ii. 1639)), and, although Criseyde is often merry with her uncle, she never laughs with her lover. Indeed, an early scene shows Troilus solemnly repenting his former 'japes' against love (i. 911 ff.) and so deliberately renouncing one of his previous sources of humour before the poem's action begins, just as he associates Pandarus' attitude to love with a game—of 'raket'—only to repudiate it (iv. 460). Yet for all the high seriousness of Chaucer's earnest Troilus there is a pervasive lightness of tone and ambiguity of context in the first three books. Troilus' lyric effusions are at once serious in their intensity and sincerity, but often melodramatic in context, and offset by the comic practicality and flexibility of his friend. Yet Pandarus is so much fuller a character than a traditional go-between of domestic comedy, an arranger of seductions, and, with all his sententious knowledgeability and warmth, his character is developed in a way that leads to serio-comic effects. Any romance dramatized by Pandarus—like this one—will be a hybrid genre, for Pandarus' activities are susceptible of alternately comic or serious interpretation ('But Pandarus, that so wel koude feele │ In every thyng, to pleye anon bigan' (iii 960–1)). There is a possibility for both seriousness and comedy in the different levels of the characters' awareness in such scenes with Pandarus as the forcing of Troilus to repent formally of his offences against love (i. 932–43), the feigned illness and rumoured law suit, and the later manipulations of appearances in Book III, for in all are elements both of the solemnly ceremonious and the ludicrous.

That so much of the plot of *Troilus* is indeed a 'plot', an artifice of deception, a seduction, aligns part of the poem's structure in some sense with comedy. The generic inclusiveness of *Troilus* makes it harder than in examples of less mixed genre to define what type of action the poem represents, and so greatly extends the poem's thematic implications. Only a hybrid genre, not a single genre, could assimilate Chaucer's characters and what he wants to do with them, or could contain both the lyric and philosophical, the serio-comic doubleness of the action. This has its implications for how the poem is to be seen as tragedy or history, just as it comments on some of its lyrics and complaints. If the role of complaint in *Troilus* modifies the way the poem conforms to tragedy, both complaint and tragedy are eventually transcended when the narrative follows Troilus beyond death and thus beyond the normal bounds of tragedy or romance. The retrospective denomination of the text as 'tragedye' is immediately followed and modified by the prospect of writing 'som comedye' (v. 1786–8)—whether this looks forward to Chaucer's next poem or to the ensuing ascent of Troilus' spirit, reminiscent of Dante—and a sad *story* is bracketed within a larger *poem* that is a 'divine' comedy.

An ending which interrelates features from various genres within itself aptly contributes to the work's complexity of generic identity, and is the culminating example of that narratorial presentation which has such a role in the use of genres throughout the poem. One possible clue to that distinctive management of the *Troilus* narrative lies in its own generic identity as a commentary, with the narrative voice in the role of commentator. In the proem to his commentary on Peter Lombard's *Sentences* St Bonaventure distinguishes between the different levels of responsibility involved in making books:

Quadruplex est modus faciendi librum. Aliquis enim scribit aliena, nihil addendo vel mutando; et iste mere dicitur scriptor. Aliquis scribit aliena addendo, sed non de suo; et iste compilator dicitur. Aliquis scribit et aliena et sua, sed aliena tamquam principalia, et sua tamquam annexa ad evidentiam; et iste dicitur commentator non auctor. Aliquis scribit et sua et aliena, sed sua tamquam principalia, aliena tamquam annexa ad confirmationem et debet dici auctor.

There are four ways of making a book. Sometimes a man writes others' words, adding nothing and changing nothing; and he is simply called a scribe. Sometimes a man writes others' words, putting together passages which are not his own; and he is called a compiler. Sometimes a man writes both others' words and his own, but with the others' words in prime place and his own added only for purposes of clarification; and he is called not an author but a commentator. Sometimes a man writes both his own words and others', but with his own in prime place and others' added only for purposes of confirmation; and he should be called an author.

If the *Troilus* narrator is to be seen acting the role of a compiler this may contribute to the poem's generic links with history-writing, (see above, p. 153), while in his self-dramatization as the helpful translator of his source ('And if that ich . . . Have any word in eched for the beste' (iii. 1328–9)), the *Troilus* narrator casts himself as a commentator on his materials ('but with the others' words in prime place and his own added only for purposes of clarification'). The format of a commentary can present the narrator more as the collector, editor, and annotator of materials than as an inventing artist co-ordinating and subordinating his subject-matter from a determined and unified point of view. The story-line is presented in connection with an 'apparatus' of materials brought into relation with it as a kind of commentary. There is the provision of materials for others to judge, a tentative, scholarly sense of a study in process, with an awareness of the available alternatives within the tradition ('Take every man now to his bokes heede' (v. 1089)). Through the narrator as commentator, genres may be modified in the telling (and their usual narrative focus or centre moved), so that romance matter is narrated historically, or framed epically, to give a sense that the *Troilus* text shows a commentary in the process of being composed on pre-existing matter.

With their openness to genre, both the beginning and ending of *Troilus* represent the mode of commentary that frames the poem. At its beginning the poem apparently makes a succession of anticipations of genre in epic, romance, history, complaint, and tragedy. The effect is to open possibilities rather than close them with definition, just as the ending with its multiple endings and perspectives precludes any single unified point of view in retrospect on the preceding poem. In the different views of the beginning and ending of the poem about how to define the text—characteristic of a Gothic manuscript culture—is a further pointer to the adaptive and open nature in genre of *Troilus*. The beginning apparently promises a text focused much more on the hero than is that text which the ending is concerned to conclude, because attention to the role of Criseyde in relation to that of Troilus—matched by an interrelation of genres—has produced a poem which through its very inclusiveness of genres becomes distinctively and essentially *sui generis*.

St Bonaventure: *In Primum Librum Sententiarum*, proem, quaest. iv. In *Opera* (Quaracchi ed.), i (1882), p. 14, col. 2; cited by M. B. Parkes, 'The Influence of the Concepts of *Ordinatio* and *Compilatio* on the Development of the Book', in J. J. G. Alexander and M. T. Gibson (eds.), *Medieval Learning and Literature: Essays Presented to R. W. Hunt* (Oxford, 1976), 127–8; in turn cited by Burrow, *Medieval Writers*, 29–30.
For the play of genres in *Troilus* as a kind of Menippean satire, see Payne, *Chaucer*. The *Roman de la rose*, Boccaccio's *Teseida*, and *Filostrato* itself were texts well known to Chaucer that represented various mixtures of genre. Bayley notes how Chaucer 'subtly declassifies the poems which he is adapting' (*Characters*, 73), and Zimbardo remarks that 'the generic perspective of *Troilus* is a kind of cosmic irony, in the sense of a perspective upon perspectives' ('Creator and Created', 285).

Structure

Introduction

For everi wight that hath an hous to founde
Ne renneth naught the werk for to bygynne
With rakel hond, but he wol bide a stounde,
And sende his hertes line out fro withinne
Aldirfirst his purpos for to wynne.
Al this Pandare in his herte thoughte,
And caste his werk ful wisely or he wroughte.   (i. 1065–71)

At the end of the first book of *Troilus* Pandarus is described as premeditating his design in winning Criseyde for Troilus with all the careful forethought with which one plans and builds a house. Chaucer translates the passage from Geoffrey of Vinsauf's guide to composition or 'art of poetry', the *Poetria nova* (43–5), where a writer's need to plan his work in advance is compared with the need to establish an architectural design before beginning a building. Much later in Chaucer's poem, when the arrangements to bring Criseyde to Pandarus' house have been set in place, the preparations for the consummation are described in terms of the construction of a timber-frame house ('This tymbur is al redy up to frame' (iii. 530)). It is no accident that such architectural allusions to design and construction occur at these two significant junctures in the affair—the beginning of the wooing, and the designing of the consummation. The comparison in *Troilus* between the shaping of the love-affair and architectural form is consistent with a text in which the achievement of an original structure is an important aspect of the poem's interpretation of its subject.

There was no inevitable shapeliness in the story of Troilus and Criseyde, and the conspicuous use of structure as a vehicle of meaning and interpretation is almost entirely Chaucer's contribution. Like his Pandarus, Chaucer has acted the part of an architect, discerning and realizing structured form in his material. Much is owed to Boccaccio's *Filostrato*, but in terms of architecture the debt is in the foundations and ground-plan. Much of the superstructure, the proportions of space and emphasis, the role of symmetry and pattern, are fashioned by Chaucer, and articulated through the written poem's spatial, visual form in its layout as a text. Divided into nine 'Parts', *Filostrato* offered the model of a fluently continuous narrative, each Part in turn falling into brief segments, all prefaced and marked out by prose 'Arguments' describing the ensuing scene. As such, an inconspicuously efficient narrative structure in *Filostrato* draws no particular attention to itself. By contrast, *Troilus* reads as if a shape has been discerned in the story which governs the design and structural features of the poem at every level.

To be given an overview of the whole poem's shape in advance has a structuring effect, and the poem's announcement of a 'double' sorrow establishes the expectation of a pattern of balanced repetition, a symmetry to be fulfilled. The 'double sorrow' of Troilus—the repeating and mirroring pattern of 'wo to wele, and after out of joie' which Chaucer articulates, and anticipates as the emotional structure of his poem at its very beginning—lies behind the symmetry of a five-book structure and the patternings within and across the five books. That structure had, of course, many classical precedents and associations, not least Horace's prescribed form for five-act tragedy in the *Ars poetica* (189–91), although other aspects of the poem's narrative structure have been seen as characteristic of Gothic architecture and art, and tension between a classical and Gothic sense of form has been detected in *Troilus*. The regularly classical, architectural symmetry of its form sets *Troilus* apart, both from Chaucer's other poems and from previous romance in England, but the characteristic experience of a romance hero still brings with it an intrinsic structure: the 'observaunces' and the play and game of an elaborate and lengthy courtship. It was the invention of Troilus' story as a lover in winning Criseyde that allows this symmetrical five-part structure to contain a story of both improving and then declining fortunes. Chaucer's references to the turning of the wheel of Fortune near the beginnings of the first and fourth of his five books (i. 138–40; iv. 6–11) import into his poem the structural analogy of the arc or curve traced by the hero's rising and falling fortunes on the rim of Fortune's wheel, with the earlier books as an 'ascending' action, and the latter two as 'descending', although the hero's movement from misery through bliss to misery again may identify the poem's structure with the cyclical form of a whole revolution of Fortune's wheel, just as the key role of the centre of the poem is a focus for concentric structural patterns in *Troilus* of repetitions, parallels, and contrasts. The double sorrow is a formal and thematic design which presents the story by means of a structure of correspondences between later and earlier books, so that the remembering reader perceives patterns forming and developing in time. It is a pattern embedded into the very fabric of a stanzaic narrative that commits itself to shaping experience into a particular pattern over and again, a pattern in which the freer movement of the first half of the stanza is succeeded by a noticeable 'shape-holding and promise-keeping' in the verse-form of the second half. An interplay may be seen to develop in the course of the work between the powerful metaphor of harmony and order established in *Troilus* by the poem's symmetry, and the metaphor of disorder suggested by the ironic parallels and inversions as the narrative advances.

Chaucer may have been prompted to the borrowing from Geoffrey of Vinsauf by his recollection of a passage in Boethius about how a workman sees in his mind's eye beforehand the artefact that he is about to make:

> For ryght as a werkman that aperceyveth in his thought the forme of the thing that he wol make, and moeveth the effect of the werk, and ledith that he hadde lookid byforn in his thought symplely and presently by temporel ordenaunce; certes, ryght so God disponith in his purveaunce singulerly and stablely the thinges that ben to doone; but he amynistreth in many maneris and in diverse tymes by destyne thilke same thinges that he hath disponyd. (*Boece*, iv, pr. 6, 82–93)

The larger philosophical purpose to which the idea about planning a work is here put is a pointer to the thematic significance of narrative form and structure in *Troilus*. Boethius uses the distinction between the moment of planning the work and the extended process of bringing it forth in time to convey the distinction between God's all-seeing providence and the workings of destiny over the course of time. This is a distinction the pagan Troilus cannot perceive, but which the poem's pattern of a 'double sorwe'—foreknown from the start—makes part of the reader's experience of the narrative structure.

In *Troilus* the narrating voice of the poem must accomplish that articulation which in *Filostrato* had been partly effected by its layout in brief narrative sections. Chaucer omits Boccaccio's formal acknowledgement of internal subdivisions yet retains or increases within the books of *Troilus* the number of narrative sections in his source, with the effect that the management of the narrative structure is internalized and dramatized as the concern of the narrating voice. It is this narratorial voice—outlining, dividing, clarifying the narrative parts, drawing attention to structural joints and sections—which has been especially likened in its procedures to Gothic architectural form, and which in its conduct of the narrative shows an implicit sense of structure. *Troilus* both has digressions yet declares its wish to avoid them (i. 260–2); promises to be brief and to address the point (e.g. iii. 701), even if the 'purpos' or the 'effect' the narrative claims as its aim remains crucially undefined (ii. 1595–6; iii. 505); purports in places to present a transcript of what happened or was said (v. 952) but also moves to more summary accounts ('This is o word for al . . .' (iii. 1660)), distinguishing between single actions represented dramatically in a section of the narrative and actions described once which recurred over a period ('And al the while which that I yow devyse, | This was his lif . . .' (iii. 435–6)). In a poem where lovers spend much time apart there are many transitions to be made and much is made of them by the narrative in articulating the overlappings of time in concurrent scenes (e.g. ii. 687–9, 932–4, 1709–10), or the coinciding of transitions between narrative sections and characters' transitions between settings, their entrances and departures.

The most striking of transitions in the narrative structure of *Troilus* is that between Book II and Book III. The second book, whose action had begun with an invocation to Janus, god of thresholds and of 'entree' (ii. 77), is broken off at the point of maximum suspense as Troilus waits for Criseyde to step into his room for their first meeting ('O myghty God, what shal he seye?' (ii. 1757)). This is a highly dramatic but artificial conclusion, in the sense that the end of Book II comes right in the midst of events rather than coinciding with the close of an action as in Book IV, or the achievement of a certain stage or state as in Books I and III. The formal nature of the transition could scarcely be more marked, especially as the third book is prefaced by a lengthy proem before the narrative focus cuts straight back to Troilus at the moment where he was before, waiting anxiously for Criseyde to enter ('Lay al this mene while Troilus . . .' (iii. 50))—that 'mene while' is presumably the time the audience has taken to hear or read the intervening proem. This transition between Books II and III is managed to give just as much emphasis to the formal structure of division into books as to the powerfully continuous narrative impetus that bridges the books, drawing attention to the relation between the poem's action and its structural design.

The following analyses of aspects of the poem's structure aim to show how the formal ambition of *Troilus* is realized. The symmetrical and patterned structure of the poem as a whole is the starting-place, and the social and domestic structure reflected in the narrative is then followed by discussions of the structure of time and of astrological structure. *Troilus and Criseyde* is a triumph of artificial form, and it is part of the poem's form that it should be presented and understood as the accomplishment of a highly wrought artefact. In this accomplishment the dignity, ambition, and perfection of form reflect the seriousness and ambition of Chaucer's purpose in developing his story of the love of Troilus and Criseyde so as to explore the very shape and possibility of perfection in this life.

The fullest account of the structure of *Troilus* is by Provost, *Structure*. For a valuable brief analysis of the poem's structure, see Brenner, 'Narrative Structure'. Jordan, *Chaucer*, sets the poem in the context of Gothic ideas of form and structure, and valuable points about the shaping of the poem are made in the course of broader discussions by Norton-Smith, *Geoffrey Chaucer*, and by Lambert, 'Telling the Story', who suggests analogies between the structure of the stanzas and the poem in 'shape-holding and promise keeping'.

For the text of Geoffrey of Vinsauf's *Poetria nova*, see E. Faral (ed.), *Les Arts poétiques du xiiᵉ et du xiiiᵉ siècle* (Paris, 1924), 198:

Si quis habet fundare domum, non currit ad actum
Impetuosa manus: intrinseca linea cordis
Praemetitur opus ...   (43–5)

If anyone has a house to build, his impetuous hand does not rush into action.
The inward line—the plumb line—of his heart first measures out the work ...
(Chaucer's text may have read *praemittitur* or *praemittetur*, i.e. 'will be sent
out').

Symmetry

The achievement in *Troilus* of a remarkable symmetry of structure is
designed to reflect that fundamental image of the poem, the rise and fall of
Troilus' fortunes. The proportions and emphases are the outcome of
Chaucer's own structural design, as a glance at the proportions of the five
books in *Troilus* and the corresponding nine Parts of *Filostrato* reveals (see
Table 6).

TABLE 6. Filostrato *and* Troilus: *Divisions and proportions*

Filostrato				Troilus		
Part	Stanzas	Lines		Book	Stanzas	Lines
1	57	456		I	156	1,092
2	143	1,144		II	251	1,757
3	94	752		III	260	1,820
4	167	1,336		IV	243	1,701
5	71	568		V	267	1,869
6	34	272				
7	106	848				
8	33	264				
9	8	64				
Total		5,704				8,239

Chaucer's creation of a single fifth book out of the four separate Parts
5–8 of *Filostrato*, and his expansion of the third book, establish a balanced,
proportioned equivalence between Books II and IV, and Books III and V,
interlinked with each other in an alternating pattern after a shorter,
introductory Book I. There is now a shapelines of overall structure, within

which the English narrative has a different pattern and pace from the Italian. Chaucer's redesigning of the 'ascending' part of the action (as the hero's fortunes improve) means that both the consummation and Criseyde's departure occur much later in the course of *Troilus*, with the consummation occurring at the centre of the total number of lines in the poem. The midpoint of the 8,239 lines of *Troilus*, with its structure of rise and fall, comes at iii. 1271 (the 4,120th line), in a prayer by the hero praising the god of love for having 'me bistowed in so heigh a place'. The centre of the poem has been aligned with the hero's conception of the very highest point of his fortune, the highest point—as it will prove—of Fortune's wheel, and of the arc or curve of the poem's action. It is intriguing that in this of all stanzas Troilus thinks of himself as one 'Of hem that *noumbred* ben unto thi grace' (iii. 1269), as if to point to Chaucer's sense of number, measure, and proportion. Allusions to figures in Euclid's geometry—*dulcarnoun* and *fuga miserorum*—are also sited near the midpoint of the third book (iii. 931–3); the central third book of the lovers' 'heaven' of happiness opens with an invocation of Venus, the planet that 'Adorneth al the thridde heven faire' (iii. 2), and contains the first 'trio' of character grouping as Troilus, Criseyde, and Pandarus are all present together for the first time in the poem. Numerological principles have been rediscovered and interpreted in the poetry of the Renaissance, but that interest in the 'poetry of number' was evidently already shared by medieval poets in their appreciation of the role of the centre in poetic construction. Around this central point the second and fourth books—nearly equivalent in length—form a complementary pattern: Book II moves from unhappiness to the threshold of joy, ending just before Troilus and Criseyde talk together for the first time; Book IV moves from joy to grief and to the verge of parting, ending just before the lovers talk together for the last time. Framing these two books in turn on either side are the accounts of the lover's sorrow apart from Criseyde: Book I prefaced by the brief period when Troilus is outside the experience of love, Book V concluded by his movement outside earthly experience.

This structure is very deliberately articulated through an apparatus of prologues and book-endings, which emphasize each book as an entity while also contributing to the overall symmetry. The first four books are prefaced by ambitious proems, the first three of which are more-or-less uniform in length (56, 49, 49 lines). The fifth book has no proem but is in turn ended by an extended and complex closing sequence. The poem's symmetrical form is hence completed by the way in which the four-times repeated pattern of prologue-and-book is reversed into book-and-'epilogue' in the final book. This symmetry is also confirmed by correspondences between the first and fifth books, for, while the proems to the three central books are signalled by their own incipits and explicits (a feature with sufficiently consistent manuscript support to suggest authorial design), the proem to the first book is not set off in this way, just as there is no formal division between the fifth book and its closing 'epilogue' section. The first proem hence acts as a preface not only to the first book but to the poem as a whole, just as the fourth proem prefaces the entire declining action (and after the invocation to a single Fury in the proem to Book I comes the invocation to a quartet of all three Furies and Mars himself in the proem to Book IV). Instead of being formally divided from Book I, the first proem ends with a 'reprise' of the poem's opening lines about the 'double sorwe' of Troilus, mentioning Criseyde for the first time in the same breath as her forsaking of Troilus:

> For now wil I gon streght to my matere,
> In which ye may the double sorwes here
> Of Troilus in lovynge of Criseyde,
> And how that she forsook hym er she deyde.   (i. 53–6)

The last two lines are then echoed in turn near the close of Book V:

> And thus bigan his lovyng of Criseyde,
> As I have told, and in this wise he deyde.   (v. 1833–4)

The expanded third book, with its central experience of the lovers' union, becomes the midpoint of a concentric structural pattern of recurring and mirroring features and episodes. *Troilus* opens with the prayer of the first proem; the last book ends with a prayer. Soon after that opening prayer comes the smiling mockery of Troilus at lovers just before he is struck by Love's arrow; not long before the closing prayer comes the very different laughter of Troilus as his spirit rises up through the spheres. The two prayers and the two laughters enclose the whole poem, and the second pair are sadder and deeper in meaning because of the first. The Canticus Troili of first love in Book I is mirrored across the poem by his later song of grieving separation in Book V. The acknowledgement to Lollius early in Book I is echoed by the only other reference to Lollius late in Book V. The hero's rides past Criseyde's house in Book II are matched by his despairing ride past her empty house in Book V. Criseyde's dream of the eagle in Book II is matched by Troilus' dream of the boar in Book V, and the lovers' opening exchange of letters in Book II is complemented by their sad final exchange of letters. The central book of the lovers' union is in turn framed and patterned; it begins with an invocation of Venus but also ends with one, and the Boethian spirit of the proem is balanced by the Boethian song which Chaucer adds for his Troilus to sing very near the close. This structure of bilateral symmetry is set out schematically in Table 7.

TABLE 7. Troilus: *Symmetrical structure*

Feature	Context
Prayer	First proem
Smile	Troilus mocks lovers (i. 194)
Lollius	(i. 394)
Canticus Troili	(i. 400–20)
Ride past	Troilus rides past (ii. 619 ff., 1247 ff.)
Dream	Criseyde's dream (ii. 925–31)
Letter	Troilus' first letter (ii. 1065–85)
Letter	Criseyde's first letter (ii. 1219–25)
Venus	Third proem
Hymn to Love	Third proem, from *Cons.* (ii, m. 8, via *Filostrato*)
Consummation	(iii. 1254 ff.)
Hymn to Love	Troilus' song (iii. 1744–71, from *Cons.* ii, m. 8)
Venus	Address to Venus (iii. 1807)
Ride past	Troilus rides past (v. 519–60)
Canticus Troili	(v. 638–44)
Dream	Troilus' dream (v. 1233–43)
Letter	Troilus' letter (v. 1317–1421)
Letter	Criseyde's last letter (v. 1590–1631)
Lollius	(v. 1653)
Laughter	Troilus laughs from the spheres (v. 1821)
Prayer	Concluding prayer (v. 1863–9)

Such a concentric structure finds its model and also its resolution in the poem's very last Dantean stanza, with its evocation of perfect order in a pattern of 'oon, and two, and thre ... in thre, and two, and oon' when describing the uncircumscribed but all-circumscribing God:

> Thow oon, and two, and thre, eterne on lyve,
> That regnest ay in thre, and two, and oon,
> Uncircumscript, and al maist circumscrive ...   (v. 1863–5)

But, if it is Boccaccio who first puts together the rising and descending actions, every instance of a mirroring repetition of incidents or features in *Troilus* stems from Chaucer's reconstruction of his story as the double sorrow of Troilus. It is Chaucer who creates new patterns and parallels linking the rising and falling actions. In some cases (like the opening and closing prayers, or the Lollius references) Chaucer invents both features which go to make up the doubling. In others (as with Criseyde's dream in Book II, mirroring that of Troilus in Book V), actions and reactions occurring singly in *Filostrato* are 'doubled' in *Troilus*, whether by Chaucer's addition of a second, later occurrence or an earlier, preceding one, or even by 'splitting' an incident in *Filostrato*, as Chaucer seems to do when elements of Criseyde's added dream in Book II of *Troilus* are moved forward from Troiolo's dream in Part 7 of *Filostrato*.

Implicit in the original story from Benoît and Boccaccio is a double pattern, in which the treachery of her father Calchas in exchanging Troy for the Greeks is repeated in another form by Criseyde when she exchanges Troilus for Diomede. But Chaucer further enriches this pattern by doublings and anticipation. The Trojans' exchange of Criseyde for Antenor, who will later betray Troy, is taken over from *Filostrato*, but the earlier role of Antenor as the enemy of Criseyde is invented by Chaucer (ii. 1474), so that their exchange fulfils a pattern of connection between the two betrayers, of Troy and of Troilus. The intervention of Hector in Book IV to prevent Criseyde's exchange for Antenor is a further instance of these narrative doublings and echoes: the brief scene in Book I where Criseyde begs and receives Hector's protection is taken from *Filostrato*, but the scene in Book IV where Hector tries in vain to continue his protection of Criseyde is invented by Chaucer and points to the inevitable downward course of the lovers' fortunes. (In late Book II—in order to meet Criseyde—Troilus sends Deiphebus and Helen out of the way to read a 'tretys' concerning the fate of some hapless unnamed individual (ii. 1696–1701), while by early Book IV it is Criseyde's own fate that is being weighed by others.) The story of Criseyde's infidelity to Troilus turns on the 'doubling' involved in the successive wooing of Criseyde by Troilus and by Diomede, and Chaucer has built this more fully into the structure of his version than either Benoît, who starts with the parting of Troilus and Briseida, or Boccaccio, who gives a relatively brief treatment of Diomede's wooing. Criseyde's quotation of a proverb to herself when thinking of accepting the love of Troilus (ii. 807–8) is mirrored when Diomede quotes the very same proverb to himself when thinking of how to woo Criseyde (v. 784). These two courtships in the ascending and declining actions mirror each other across the poem, the second as a sadly distorted reflection of the first.

A pattern of symmetrical doubling is a device of thematic structuring that links together the progress of the ascending and descending action, bracketing episodes together within books, and from book to book, suggesting parallels, comparisons, and contrasts. The whole sequence of action in Books I–III—in which Troilus and Criseyde are made aware of each other and brought together as lovers—shows Chaucer working through a carefully contrived series of doubled and paralleled incidents. He complements Boccaccio's scene of the hero's first sight of the heroine with his own version of how Criseyde has her first sight of Troilus as he rides past her window (ii. 619 ff.). This is in turn 'doubled' into the two successive occasions when Troilus rides past Criseyde's house: the first by chance, the second prearranged by agreement between Troilus and Pandarus (ii. 1247 ff.).

Such doubling is also created in the invented sequence in which Criseyde first visits Troilus in his bedroom at Deiphebus' house, and Troilus later visits Criseyde's bedroom at Pandarus' house. The former visit, with the establishment of an understanding between the lovers, serves as a kind of preface to the fulfilment of the later meeting, and the two scenes are made to mirror each other as declarations and grantings of love. In the first scene Criseyde comes ostensibly as a petitioner to Troilus for his 'lordshipe' (iii. 76), although in effect the two scenes show a doubled and repeated episode of Troilus as suppliant to Criseyde. Indeed, the scene in which Troilus kneels to Criseyde (iii. 953 ff.) recalls and mirrors the earlier scene in which Criseyde kneels as a suppliant to Hector for his protection (i. 110). In both bedroom scenes in Book III the visit is associated with some invented affliction on the part of Troilus, first a feigned illness, and later a pretence of suicidal jealousy. In the first scene Troilus is comforted in bed by Helen (ii. 1667–73), and there are verbal echoes of the earlier comforting when he is later comforted in bed by Criseyde during the second scene (iii. 1128–34). At the first meeting Pandarus brings in Criseyde 'by the lappe' (iii. 59), just as at the second he will bring in Troilus 'by the lappe' (iii. 742), and in both episodes the person in bed (first Troilus, later Criseyde) is overcome with blushes, in a notable mirroring of gesture between the scenes (iii. 82, 956). Through this doubling of the bedside meetings and doubling of the hero's riding past, the progression of the love-affair is perceived through patterns, and this reflects the way that life and the narrative that represents it are more structured by a sense of ceremony and of lovers' 'observaunces' in *Troilus* than in *Filostrato*. In Book II Pandarus tells Criseyde he has overheard Troilus addressing the God of Love and repenting his rebellion against the god (ii. 525), but, when Pandarus has his opportunity, he makes Troilus repeat his confession and repentance once again in a more formal and deliberate way (i. 932–8), so that Troilus' repentance is 'doubled'.

The effect of such doubling is to emphasize, when it comes, the specialness of that union which is at the heart of the poem. Antigone's song in Criseyde's garden about the paradise of love (ii. 827 ff.) is echoed and realized in the hero's experience when Troilus in another garden sings of his own heaven of happiness in love (iii. 1744–71). But the poignancy of the pattern of doublings across the poem is that an experience Troilus imagines to be singular will in fact prove double, part of a pattern of repetitions, as the declining action of Books IV and V brings doublings and parallels of allusions as of events in earlier books. The 'kalendes' of hope which mark both the beginning of Book II and of the wooing of Criseyde (ii. 7) are eventually succeeded by the 'kalendes of chaunge' which Troilus perceives in Criseyde's last letter (v. 1634). When Pandarus arrives to win Criseyde for Troilus she has just listened to the beginning of the story of Oedipus (ii. 101–2), while when Troilus hears of the planned exchange of Criseyde he identifies with the fate of Oedipus at the close of that story (iv. 299–301). In the same 'romaunce of Thebes' Criseyde has also just heard the story of Amphiorax (ii. 104–5), which is to be cited again by Cassandra (v. 1500) in proving to Troilus that Criseyde has forsaken him for Diomede.

As with allusions, so too with scenes and incidents: patterns of doubling and concatenation are worked into the structure of *Troilus* as the double sorrow of the story unfolds and both halves are eventually seen entire. Troilus' pretended visit to the temple in Book III (533–46) is succeeded by his despairing visit in Book IV, while Criseyde's hiding of her face from Pandarus and her complaint on worldly bliss in Book IV (820–47) compare with her earlier complaint and hiding of her face in Book III (813–40, 1569). The swoon before the first love-making in Book III is balanced by Criseyde's swoon before what will prove their last night together in Book IV. But Criseyde's seeming death and Troilus' survival from near-suicide in this scene are answered eventually by Troilus' actual death and Criseyde's survival, her 'testament' scene (iv. 785–7) complemented by his (v. 298–315). The invented pretext of Troilus' jealousy of Horaste in Book III (796–8)—prompting Criseyde to reflections on mutable earthly joy—is an unfortunate anticipation of Troilus' later jealousy of Diomede, which will see such a view of earthly happiness confirmed. The brooch Criseyde gives Troilus in Book III (1370–2) is answered by the brooch Troilus gave her as a memento at parting and which he finds on Diomede's cloak (v. 1660–5). The hero's invoking the possibility of death at the hands of Achilles (iii. 374–5) is an ironic anticipation of what later proves his fate, and the 'heaven' of Book III is succeeded by the heaven towards which the soul of Troilus finally ascends. Troilus remembers in Book V the temple where he first saw his lady (v. 566–7), so arching across the poem to bracket a late sad scene of separation with the earliest moment of love.

For discussion, see McCall, 'Five-Book Structure', and for applications of numerological analysis, see Hart, 'Medieval Structuralism', and Peck, 'Numerology'. Structural implications of manuscript variations are discussed by Hanson, 'Center', and Owen, 'The Significance of Chaucer's Revisions'.

The consummation in *Filostrato* (at iii. 32, i.e. the 1,856th line out of 5,704) comes much earlier in that poem's structure than in *Troilus*, as does the lovers' parting (at *Filostrato*'s 3,688th line, rather than at the 6,370th line out of 8,239 in *Troilus*).
On numerology and Renaissance poetry, see Alastair Fowler, *Triumphal Forms* (Cambridge, 1970), with brief discussion of *Troilus* (p. 65).

Structure and Setting

Chaucer's architectural analogies (i. 1065; iii. 530) parallel the way his narrative is structured by occurring within a carefully observed domestic context, both a social and an architectural setting. The structure of the narrative in sectionalized form reflects the social structure in which the action is imagined to take place. Initiating and closing the narrative sections of *Troilus*, the formalities of greeting, inviting, and leave-taking are scrupulously noticed, so that the poem's sections are unobtrusively shaped by social structures of courteous convention and observance. Chaucer gives quietly careful attention to various ceremonies of arrival and greeting, as when Calchas flees to the Greeks (i. 81–2), when Criseyde arrives at her uncle's house (iii. 604–6), when Criseyde is reluctantly reunited with her father (v. 190–1), or, most intriguingly, when Diomede comes to visit in the Greek camp and is very ceremoniously received by Chaucer's Criseyde (v. 848–54). Greetings are exchanged when first Criseyde most courteously receives Pandarus (ii. 85–91); when Helen visits Troilus (ii. 1667 ff.); when Pandarus moves from one room to another in Deiphebus' house (ii. 1711–13); when Pandarus visits first Criseyde then Troilus after the consummation (iii. 1555–6, 1588); or when Pandarus visits Troilus on the morning after Criseyde's departure (v. 293). Such observance of polite conventions provides a norm against which Chaucer can set variations reflecting the feelings of his characters. In his excitement after his two visits to Criseyde in Book II Pandarus bursts in to see Troilus without mention of conventional greetings (ii. 940, 1308–9). In Book III the entry of Criseyde into Troilus' bedroom, and Troilus' later appearance in Criseyde's bedroom are both moments of embarrassment in which conventionally formal greetings seem faintly absurd in the context of what is being felt (iii. 61 ff., 955). In Book IV, when Pandarus arrives to see Troilus ('So confus that he nyste what to seye' (iv. 356)), or when the lovers meet together ('neyther of hem other myghte grete' (iv. 1130)), there is too much sorrow and formal greeting is choked, while, when Pandarus arrives to see Criseyde, Chaucer makes her quibble ironically about the fittingness of conventional greeting (iv. 831–3).

Through this texture of allusion to the observance of polite custom Chaucer creates a sense of how the inner lives of Troilus and Criseyde must exist within (and despite) the context of a structured fabric of social life at Troy. One series of social occasions centres on the taking of meals: Pandarus is invited to stay on at Criseyde's house while she reads Troilus' letter before they dine (ii. 1170–1, 1184); Chaucer's Pandarus plans the meeting of the lovers first by means of a lunch party at Deiphebus' house, and later a supper party at his own home; on several occasions Chaucer's Troilus has to endure impatiently the ritual of a public meal (ii. 947; v. 518). As lovers are conventionally supposed to have lost their appetite (Pandarus jests at the convention (ii. 1165–6)), Chaucer's deliberate construction of this social context of convivial eating seems designed to show the lovers' lives both within and apart from the structure of daily life that contains them.

There is a special significance in *Troilus* in the observance of leave-taking, for it reflects that separation which is the fundamental action of the poem. Crucial narrative junctures are associated with the taking of leave: having got his way, Pandarus is twice described taking his leave of Criseyde when 'al was wel' (ii. 1302; iii. 594); Criseyde ostensibly enters Troilus' sick-room in order to take her leave (ii. 1658, 1689, 1719–21) and there is further leave-taking as the scene closes (iii. 211–13); in Book III Criseyde attempts in vain to take her leave of Pandarus, but instead Pandarus 'Ful glad unto hire beddes syde hire broughte, | And took his leve, and gan ful lowe loute' (iii. 682–3). These structured social conventions are quite different from the passionate, lyrical leave-takings of the lovers in private in Book III, or the painful absoluteness of parting at the close of Book IV ('Withouten more out of the chaumbre he wente' (iv. 1701)). It is against this established structure that Chaucer can poignantly develop his account of events in Book V, so that Diomede's visit to Criseyde on the tenth day comes to its end with an exact echo of the earlier, hopeful leave-takings of Pandarus ('whan it was woxen eve | And al was wel, he roos and tok his leve' (v. 1014–15)), and Criseyde's decision to change her affections away from Troilus is dramatized as almost a private ceremony of one-sided leave-taking ('But al shal passe; and thus take I my leve' (v. 1085)). Scenes and sections of *Troilus* may in this way be opened, shaped, and closed through the structure of social exchanges and good manners, just as the action of Chaucer's poem is shaped by the nature of that domestic, architectural setting in which the action takes place.

It is a reflection of the secret and personal experience of the lovers that most of the poem's significant action takes place in a succession of small private rooms and confined spaces. When characters move about the city of Troy that contains these interiors the distances mentioned or implied do not seem extensive: Pandarus can run from Deiphebus' house to Criseyde's (ii. 1464), and Sarpedoun's home is 'nat hennes but a myle', rather than four miles away as in *Filostrato* (v. 403). *Troilus* is a poem of private interiors and gardens within the besieged city of Troy, and this domestic, architectural setting is very much part of the poem's fabric and shape.

The settings of *Troilus* can be interpreted in terms of what is known about the architectural layout of fourteenth-century London mansions. The typical form of such houses would be a rectangle, such as still survives in the design of the hall and related rooms in many Oxford and Cambridge colleges. The larger part of the design is occupied by the hall, which rises through the whole height of the building to the roof, sometimes from ground level, but sometimes from a floor above ground level and built over cellars. In the latter case an outside staircase is necessary to reach the porch, the main entrance into the house at one end of the 'screens' passage which runs between the hall on one side and the kitchens and household offices on the other. At the far end of the hall from the screens is the dais and on it the high table. Behind the wall at the back end of the dais would lie the lord's solar or private parlour, the lady's bower or closet, and other chambers. It would be normal for such rooms to open one on to another, and access to any first-floor rooms was gained sometimes by an external staircase and sometimes from the hall. In such circumstances the curtained bed offered the relative privacy of a room within a room, and an attendant might sleep on a pallet within the same chamber.

To imagine the events of Books II and III of *Troilus* as shaped by settings corresponding broadly to such a plan may clarify some of the action. When Pandarus proceeds to visit Criseyde he presumably enters through the porch, passes through the hall, and enters the paved parlour which would lie behind the dais. Here he finds the lady of the house with several other ladies. After Pandarus has left, Criseyde 'streght into hire closet wente anon' (ii. 599). A closet was a private or inner chamber, and not necessarily a small room. In all his writings Chaucer uses the word *closet* only in *Troilus* and only in connection with Criseyde. This closet must look out over the street, for it is from here that Criseyde hears the acclamations of the crowd and watches Troilus pass by. The public tumult of the street contrasts with Criseyde's thoughtful stillness, apart and unseen in her private chamber, as she sees Troilus framed in the window of her closet. The hero who has just entered through 'the yate ... Of Dardanus, there opyn is the cheyne' (ii. 617–18) enters the city of Troy and Criseyde's heart in a simultaneous movement.

At the close of her meditation Criseyde goes out into her garden for a walk in the twilight. The garden would presumably lie on the other side of the house from the street, and Chaucer notes how Criseyde descends to the garden ('Adown the steyre anonright tho she wente | Into the gardyn' (ii. 813–14)). She has perhaps returned from her closet on the street side of the house, through the paved parlour, and goes down into the garden, the outside staircase indicating that the ground floor of Criseyde's house is somewhat above ground level. Criseyde's garden itself forms part of the succession of enclosed and intimate spaces in *Troilus*: it is described as a contemporary medieval garden would be laid out, with sanded paths bordered by rails and well provided with turf benches (ii. 820–2). From her private garden Criseyde proceeds indoors to her bedchamber, which several details indicate to be on the first floor and on the garden side of the house—she goes upstairs to reach it (iv. 732), and from her bed hears 'A nyghtyngale, upon a cedre grene, | Under the chambre wal ther as she ley' (ii. 918–19).

When Troilus arrives home after his triumphal return from battle he once again retreats from public into a private room until Pandarus comes (ii. 935), just as Criseyde is concurrently sitting thinking in her closet. After supper, as soon as propriety allows, the friends retire for the night to a shared room, in order to have an opportunity for talk (ii. 946–9). When Pandarus next day returns to Criseyde's palace with a letter, the domestic setting again has a significant effect on the structure of events. Pandarus suggests they go into the garden for a confidential chat, and so they leave the 'chaumbre' in which Pandarus has found Criseyde and descend into the garden (ii. 1116–17). When they return they go into the hall and Criseyde proposes dinner, but in order to read Troilus' letter she first proceeds from the hall to her 'chambre' (ii. 1173). After dinner Pandarus draws Criseyde to a window seat on the street side of the hall in order to discuss Troilus' letter (ii. 1186, 1192). From here Criseyde retreats 'into a closet' to write her reply, returning with it to Pandarus, who is still by the window. This window is sufficiently high above the street for Troilus to have to look up to greet Pandarus and Criseyde as he rides past (ii. 1259). When Pandarus returns to Troilus, he finds him once again withdrawn to his bed (ii. 1305).

The concluding scene at the house of Deiphebus shows Pandarus shaping events by means of the kind of domestic setting already familiar from Criseyde's house. By the time the party begins, in 'the grete chaumbre' (ii. 1712), Troilus, apparently unwell, has already retired to bed in an adjacent small room. When Deiphebus and Helen go from the 'grete chaumbre' into the small room to visit Troilus, he is able to get rid of them down a staircase from his bedroom into the garden (ii. 1705). To the party in the 'grete chaumbre' it seems that, when Criseyde steps into the smaller room, she is joining Deiphebus, Helen, and Troilus. The second book fittingly moves to its climax in a meeting of the lovers that depends on a special conjunction of those architectural settings of small intimate chambers and private gardens that have already provided the background to the book. The little chamber which crowds had best not enter ('And fewe folk may lightly make it warm' (ii. 1647)) is the apt setting to contain the first secret meeting of Troilus and Criseyde.

In Book III the bringing of the lovers together is again shaped by means of a structure of settings which Chaucer has invented. Pandarus' accomplishing of the union of the lovers depends on a particular ground-plan of rooms in his house—the culmination of the interrelation of narrative and architectural structure in *Troilus*—and this becomes clearer if that night's action is imagined to take place within a structure like that of the medieval English house. The success of Pandarus' plan depends on his contrivance of the sleeping arrangements when Criseyde is prevented by the storm from going home after the supper at her uncle's house. A sense of unfamiliar surroundings and unusual arrangements heightens the hesitancy and suspense of the episode for the lovers. An important part of the plan is that Troilus has been waiting, cooped up in a 'stewe' in Pandarus' house, since the previous midnight. A 'stewe' is a small room with a fireplace, and this room is apparently on the street side of the house, because Troilus is able to watch Criseyde arriving (iii. 601–2). When Criseyde is persuaded to stay the night, Pandarus proposes the following sleeping arrangements for his niece, her women, and himself: his niece will sleep

'By God, right in my litel closet yonder.
And I wol in that outer hous allone
Be wardein of youre wommen everichone.
'And in this myddel chaumbre that ye se
Shal youre wommen slepen, wel and softe . . .' (iii. 663–7)

Pandarus may simply be indicating a sequence of connecting chambers, although it has been suggested that Pandarus' proposals may depend on the drawing of a curtain (a 'travers' iii. 674)) across the middle of the hall, to create two separate compartments for sleeping: a 'myddel chaumbre' at the end near the dais for the women, and an 'outer hous' at the far end for himself. Criseyde will sleep snugly in Pandarus' little closet behind the dais, which is connected to the hall by a door. Nightcaps are drunk, the 'travers' is drawn, perhaps to divide the hall in two, and Pandarus conducts Criseyde to her bed in the closet, assuring her that she will be able to call her women, who will be sleeping 'Here at this closet dore withoute, | Right overthwart' (iii. 684–5).

But instead of going to his 'outer hous' upon leaving Criseyde, Pandarus proceeds directly to the 'stewe' and 'gan the stuwe doore al softe unpynne' (iii. 698), for the movement towards the consummation is now accompanied by all the stealth necessary in a house full of sleeping people. Once inside Criseyde's bedroom Pandarus immediately crosses the room to close the open door into the chamber where the women are sleeping, and it is as

he is coming back from doing this that 'His nece awook and axed "Who goth there?"' (iii. 751). When Criseyde asks Pandarus how he has entered unbeknown to her women, he explains 'Here at this secre trappe-dore' (iii. 759). It has been surmised that this was some kind of concealed entrance in the panelling between the 'stewe' and the 'closet', although evidence is lacking that the term 'trap door' meant other than it does now, which would suggest that the 'stewe' and the 'closet' are on different floors. The sleeping arrangements contrived to assure Criseyde of her seeming security are breached by a small, secret, unexpected opening, which allows the lover to penetrate into her bedroom from the 'stewe' where he has been cooped. Indeed, the sense that an entry has been forced is only strengthened by Pandarus' invented explanation to Criseyde of how a jealous Troilus 'Is thorugh a goter, by a pryve wente [a secret way], | Into my chaumbre come in al this reyn' (iii. 787–8).

At the very centre of the poem's structure Troilus is at last impelled inside the curtained bed of Criseyde, which stands inside the 'litel closet' within Pandarus' house in the walled and besieged city of Troy. The most intimate experience of *Troilus* lies not only at the centre of its structure as a poem but at the centre of a succession of containing and enclosing structures in the fabric of its setting at Troy, within which the physical union of Troilus and Criseyde is a climax not only intrinsically but also as the fulfilment and completion of a pattern. It is towards this central episode that the poem moves with a 'centrifugal' energy which, once the centre is passed, becomes a centripetal force, and this is given form and shape through the setting and background of the action. At the opening of Book IV the shift of focus out to the unlucky battle, the Greek 'consistorie' and the Trojan 'parlement', is the first substantial break in the focus on those interiors in which the lovers' happiness is gained, because these outward events will indeed break the continuation of that private life. Troilus' 'derke chambre' with its deliberately closed doors and windows (iv. 232–3, 354) is an emblem of his state of mind, and for the first time Pandarus does not simply come in to his friend but an attendant is mentioned 'that for the tyme kepte | The chambre door' (iv. 351–2) and lets the visitor in. Most of the action of Book IV—like that of previous books—still takes place in bedrooms, and the closure of the book coincides exactly with Troilus' agonized departure from Criseyde's bedroom at the end of what will prove their last night together. The lover's departure from his lady's chamber corresponds in outward movement and setting to an inner movement and departure from within his heart:

> For whan he saugh that she ne myghte dwelle,
> Which that his soule out of his herte rente,
> Withouten more out of the chaumbre he wente.   (iv. 1699–1701).

With the separation of the lovers the interrelation between narrative and setting in Book V works to distinct effect from earlier books. Although Troilus still spends time lamenting in his chamber, his feelings also tend to be set against the larger background of the city of Troy, whose affairs have now impinged on his life at every level, public and private. The enclosed chambers of preceding books give way to the exposed openness of the Greek camp or the battlements of Troy. In Book II we looked out from inside as Troilus rode past Criseyde's house, so that when in Book V Troilus passes the same house there is a specially powerful sense of exclusion in looking from outside at the closed and deserted house, which stands for Troilus' separation and exclusion from Criseyde ('O paleis . . . O ryng, fro which the ruby is out falle' (v. 547–9)). That the narrative focus is no longer concentrated on interiors reflects the absence of any intimate life to be concealed. The focus is instead on the restless, anxious movement up and down of the waiting Troilus ('Fro thennesforth he rideth up and down' (v. 561); 'And up and down ther made he many a wente' (v. 605)). Chaucer's Troilus waits on the very rim of the walled and besieged city (v. 666, 1112), leaning his head far out over the edge of the wall as he cranes to see (v. 1145), 'And up and down . . . Upon the walles made he many a wente' (v. 1193–4). It is no accident that the successive disappointments and increasingly permanent separation are associated with a more sharply realized sense of setting in the walled city than at any earlier point in the poem, for the inward life of Troilus as a lover is now interrelated with the outward historic events of the siege of Troy. The shift from the curtained bed of Book III to the walls of Troy in Book V turns the setting inside out to signify the intrusion of the outward world of events on the enclosed and inner world of the lovers' lives. And, for a poem whose structure is so continually informed by a sense of space and setting, a conclusion in the ascent of Troilus' soul to the spheres is a correspondingly spatial means of conveying removal and detachment from the enclosed space of human life on earth 'that with the se | Embraced is . . .' (v. 1815–16).

For the sleeping arrangements, see Smyser, 'Domestic Background'. Lambert comments: 'The audience of *Troilus* is at least subliminally aware that the innermost and least spacious chamber here is what the Wife of Bath calls the chamber of Venus. At Pandarus' house Troilus enters the closet, enters the bed, enters Criseyde: this is the underlying rhythm of the joyously comic central episode' ('*Troilus*', 140).

Chaucer's changes to *Filostrato* develop a succession of intimate, private settings: Book II contains four scenes without parallel in Boccaccio in which characters retire to bed (62–3, 911 ff., 947, 1305) and three garden scenes, two in the garden of Criseyde's palace and the scene 'In-with the paleis gardyn' (508) which Pandarus describes to Criseyde. Similarly, in Book III, Chaucer retains the conversation about secrecy that Pandaro has with Troiolo in a temple (*Fil.* iii. 4), but shifts its setting to their shared bedroom in Deiphebus' house.

Structure of Time

In *Troilus* Chaucer attends to time and associates it with the poem's structure in a way that intensifies the narrative and develops its thematic implications. In its 'realistic' chronicling of time in relation to human feelings *Troilus* sets itself apart from the time-schemes of romance, where any number of years may disappear in the service of love (cf. *Knight's Tale*, I. 1426, 1452, 2967; *Franklin's Tale*, V. 1101–3). Although the unfolding of the whole story in *Troilus* is understood to take place over some years, the narrative focus is concentrated with sharp definition on the action occurring during a small number of particular days, while the intervals between those days or series of days are accounted for more vaguely and may be left open to differing interpretations. There is concentration in the way that, through Chaucer's plotting of time, *Troilus* becomes a poem of the night: a poem in which a series of important scenes occur at night-time, and in which there is a keener sense of the transitions represented by dawn and dusk, so that the succession of night and day—in which night is opportunity—becomes one of the underlying structural patterns in the poem. The ascending action towards the meeting of the lovers is carefully set within a series of time-references, as is the night of the consummation, while in the descending action of the later books Chaucer realizes to the full the power in the narrative as a story of waiting and hoping against time.

The passage of time from Troilus' first sight of Criseyde to the night on which they consummate their love is more precisely plotted in the structure of *Troilus* than it had been in *Filostrato*. Events in the first three books of *Troilus* are specified to take place during an April and a May. Yet, while Chaucer apparently sets this phase of the action within a few months of one year, he also builds in other time-references which suggest the gradual (and, consequently, modest) pace at which Criseyde succumbs to the persuasions of Pandarus. He starts with a specific dating:

> whan comen was the tyme
> Of Aperil, whan clothed is the mede
> With newe grene, of lusty Veer the pryme.   (i. 155–7)

The extra emphasis given by the classical *Veer* is without parallel in his poetry. Troilus also recalls the date of his first meeting with Pandarus precisely, as 'in Aperil the laste' (iii. 360–2). In this later conversation Pandarus refers to having found Troilus 'so langwisshyng to-yere [this year] | For love . . .' (iii. 241–2): presumably they are here speaking in the May following that April in which Troilus has fallen in love. The intervening weeks would be long enough for the lover's concentrated sensations and reactions 'day by day' (i. 442) and 'fro day to day' (i. 482), although it has been argued that Troilus may refer to the April of the *previous* year, and so have languished for a year between falling in love and confessing to Pandarus in Book I.

Chaucer's careful plotting of time in his Book II is crucial in the passage of time that connects the April scene in the temple, and Troilus' reference back to it in Book III. Chaucer's Pandarus is so afflicted with lovesickness on 3 May—a day traditionally associated with love—that he has to retire to bed for a restless night, and hence Chaucer's second book opens with a description of the state of the heavens and a statement of the date:

> Whan Phebus doth his bryghte bemes sprede
> Right in the white Bole, it so bitidde,
> As I shal synge, on Mayes day the thrydde.   (ii. 54–6)

It is on the next day—i.e. 4 May—that Pandarus awakes and sets off to see his niece at the beginning of another day ('Whan morwen com ...' (ii. 64–5)), so that the beginning of the courtship is associated with dawning and rising.

Chaucer's account of Pandarus' visit to Criseyde also follows the day through until her twilight walk in the garden, her listening to the nightingale, and her dream. Chaucer co-ordinates Troilus' letter with this same time-scheme: Pandarus recommends the writing of the letter on the evening of that day—4 May—that he returns to Troilus from his visit to Criseyde. On that evening he promises to deliver the letter early the following morning, and does so 'bytyme', swearing that it is already past 9.00 a.m. (ii. 992, 1093, 1095). He is invited to stay to dinner (ii. 1163), and in the afternoon (ii. 1185) persuades Criseyde to write a reply. In the evening of that day—5 May—he leaves Criseyde's house (ii. 1301–2) and delivers the letter to Troilus (ii. 1305).

After this closely timed account of events it is at an unspecified later date that Pandarus undertakes to contrive the meeting between Troilus and Criseyde (ii. 1359 ff.), but, once Pandarus starts to act, the sequence of events is again carefully plotted. He assures Troilus that 'er it be dayes two', or even within twenty-four hours (ii. 1362, 1399), he will arrange a meeting for him with Criseyde. He visits Deiphebus the same evening with his story that 'some men wolden don oppressioun' to Criseyde (ii. 1418–19), and humbly suggests that Deiphebus do him the honour 'To preyen hire *to-morwe*, lo, that she | Come unto yow ...' (ii. 1433–4). Leaving Deiphebus, Pandarus then rushes as fast as he can to Criseyde's house, and, while they are still talking together on this busy evening, Deiphebus calls on Criseyde to invite her 'To holde hym *on the morwe*

compaignie | At dyner' (ii. 1488–9). The hour of this dinner is later
indicated to be 'an houre after the prime' (ii. 1557), i.e. about 10.00 a.m.,
and it is sometime after this dinner that Criseyde is introduced into the
'sick-room' of Troilus, and the English lovers have their first interview. It
is Chaucer who has the conversation between Troilus and Pandarus about
secrecy take place at night (iii. 229; cf. *Fil.* iii. 4), just as—in a double
pattern—Pandarus makes Troilus go overnight to Deiphebus' house
before the party (ii. 1513), and then hide overnight before the supper party
for Criseyde (iii. 602).

The exact dating of the conjunction that night of Jupiter, Saturn, and
the moon in Cancer (iii. 617 ff.) has been the subject of dispute, but the
likely dates appear to be either 13 May, or 8 or 9 June. Assuming that one
or other of these dates marks the close of the process that Pandarus begins
on 4 May at the opening of Book II, then Chaucer presents several two-day
sequences within which time is precisely plotted: the events of 4 and 5
May, and the two-day sequence—presumably somewhat later in the same
May—which runs from late in Book II to early in Book III, closing on the
morning after the day of the meeting at Deiphebus' house (iii. 423). After
this the narrative summarizes the hero's experience during an unspecified
subsequent period until Pandarus invites his niece to supper with him (iii.
435–6). When Criseyde tells her uncle on 5 May that she intends to reward
Troilus 'with nothing but with sighte' (ii. 1295), Pandarus thinks to
himself 'this nyce opynyoun | Shal nought be holden fully yeres two' (ii.
1297–8), but Chaucer has not defined the duration of the unnarrated
intervals in Books II and III so as to suggest a period of years before the
lovers' union.

Although a sequence of events like the exchange of letters, the two rides
past, and the dream of the eagle apparently takes place within a mere two
days, the gaps in time between such intensively plotted sections preserve
the sense of a more gradual development in Criseyde of the idea of loving.
The precise time-references in the second book are more of an expression
of the character of Pandarus, with whose movements they are mostly
associated, than any suggestion of precipitate haste in the lovers' feelings.
In effect, there is both sharpening and blurring of time: sharpening of
focus on actual time for Pandarus, who is urgently concerned with using
time and pacing events; blurring of focus on time in relation to the lovers'
feelings. The more precise plotting of time in *Troilus* coexists with a
passage in which the narrating voice makes a show of countering any
criticism of Criseyde's unduly hasty inclination towards Troilus ('For I
sey nought that she so sodeynly | Yaf hym hire love ...' (ii. 673–4)),
and emphasizes the 'proces' by which Troilus succeeded in gaining her
love.

Boccaccio suggests that the love-affair moves from first sight to closure in the cycle of one year: from spring to following spring (*Fil.* i. 18; vii. 78). Chaucer's network of time-reference in *Troilus* locates events in a connected chronological sequence that is both structuring and thematic, in raising the issues of transience and mutability. Chaucer strengthens the poetic time-structure that suggests a parallel movement between the cycle of the affair and the seasonal cycle of one year. He also stresses that, at the time of Criseyde's departure, there had been three springs since the April of the Book I temple scene (v. 8–11), so suggesting a passage of time which begins in the latter part of Book III but must have substantially occurred between Books III and IV. The belated specification of the precise duration of Troilus' feelings up until the moment of Criseyde's departure stresses how enduring the liaison between the lovers has been: it has long outlasted the natural cycle of a single year and is endowed with more substance and permanence than the quickly flowering and fading affair in the *Filostrato*.

The progression of times and seasons is emphasized by a series of seasonal images, some adapted from the *Filostrato* but reinforced in *Troilus*. Criseyde, as yet unaware of Troilus, seems to him 'as frost in wynter moone' (i. 523–4). In Criseyde's soliloquy on whether or not to love, her sudden midway switch is likened to the alternating sunshine and cloud 'In March, that chaungeth ofte tyme his face' (ii. 765). The lovers' emotional life is linked with imagery of seasonal developments: Troilus' spirits revive like flowers in the sun after a cold night (ii. 967–72) or like the recovery in May of 'thise holtes and thise hayis, | That han in wynter dede ben and dreye' (iii. 351–2). But, once the mood has changed in Book IV, Chaucer leaves behind these springtime associations. In describing Troilus' grief after hearing of the exchange, Chaucer replaces Boccaccio's comparison of his hero to the lily drooping in the heat of the sun (iv. 18) with an image from Dante's *Inferno*:

> And as in wynter leves ben biraft,
> Ech after other, til the tree be bare,
> So that ther nys but bark and braunche ilaft ... (iv. 225–7)

A pattern of allusion to the natural cycle is then completed by Chaucer's inclusion at the very end of *Troilus* of the notion of 'This world that passeth soone as floures faire' (v. 1841).

Within this time-structure Chaucer constructs scenes and presents feelings against the pressures of felt time. Pandarus points out to Criseyde that the time for love is slipping by (ii. 393–406), while in the garden with Antigone Criseyde's quickening interest in the possibility of love is nicely poised against her sense of the ending of the day ('Ywys, it wol be nyght as faste' (ii. 898)). In Book III the lovers' first meeting can last until Deiphebus and Helen return to the room after looking at a document in the garden for the 'mountance of an houre', just as the night of the lovers' union—chosen because especially dark (iii. 550)—is framed within a structure of time by references to the state of the heavens both before and after the consummation, and is closed by lamentation at the departure of night at dawn (iv. 1427 ff.)—for night is opportunity for the lovers (iii. 1713). And at the end of Book IV the lovers' last night together is similarly set against an awareness of limited time, the sight of the burnt-down candle ('day is nat far henne' (iv. 1246)), and the grief of an ending and parting at one dawn which anticipates the greater parting to come.

It is with the fifth book that Chaucer reaches the traditional shape of the story familiar from Benoît and Guido, and revises his sources to intensify the significance of time in the structure of the declining action, with concurrent events on the same tenth day in the Greek camp and in Troy at the centre. The first half of the book is structured through the succession of ten days, and, after this climactic midpoint is passed, there follows a more loosely observed passage of time until the close of the narrative. The ten-day time-structure is carefully plotted through Chaucer's references to the state of the heavens and to the succession of particular days and nights as the characters live through them. Granted the foreknown ending of the story in Criseyde's failure to return, Chaucer's precise plotting of Troilus' alertness to the sequence of days is an exercise in pathos, underlining the pain of waiting for Criseyde. On three occasions in the ten days Chaucer notices the intervening night Troilus spends between the days: his soliloquy during the night after Criseyde's exchange and the description (borrowed from the *Teseida*) of the following dawn are added by Chaucer (v. 232 ff.), together with Troilus' hearing of the screech owl during 'al thise nyghtes two' (v. 320); Troilus and Pandarus retire for the night and then go to Criseyde's house near dawn the next morning (v. 512 ff.); Troilus endures a sleepless night before the tenth day, and the opening of that day is embellished with an account of its dawning ('The laurer-crowned Phebus with his heete' (v. 1107)).

Chaucer also constructs with care the timing of events around the crucial tenth day. Criseyde (by implication on the eve of the tenth day) expresses her resolution to return to Troy 'to-morwe at nyght' (v. 751), but Chaucer adds the comment 'er fully monthes two, | She was ful fer fro that entencioun!' (v. 766–7), so anticipating precisely that hopeless future moment when Troilus will write reproaching Criseyde for having stayed two months instead of ten days (v. 1348–51). Diomede is shown in nocturnal soliloquy over his chances before visiting Criseyde (v. 785), and his first visits to Criseyde take place over a two-day sequence (v. 995).

Diomede's first visit closes with the close of day, and with the implication of something also concluded ('And finaly, whan it was woxen eve | And al was wel, he roos ...' (v. 1014–15)). The beginning of Criseyde's change of heart is suggested during that night, with its reference to the movement of Venus (v. 1016 ff.). The next morning comes and Diomede comes with the morning, that new day which is implied to be his (v. 1030–1). Diomede comes to see Criseyde on that very tenth day when Troilus is expecting her to return to Troy (v. 842–3), rather than on the fourth day, as in *Filostrato* (vi. 9). Chaucer thus synchronizes the height of Troilus' expectations with the beginning of their end in Diomede's courtship of Criseyde. The focus returns to Troilus as he waits expectantly on the tenth day, just after Chaucer has reported Criseyde's soliloquy bidding him farewell—uttered at some unspecified time, as yet in the future for Troilus, although long past for us. It is not that Criseyde has already betrayed Troilus, for she has not and will not for an undisclosed time ahead, but the narrative has looked forward and 'remembered the future' from old books, before returning to the tenth day. A narrative of the present in Troilus' hopes (v. 1100 ff.) now succeeds a panorama of the future that will stem from what has already happened in the Greek camp and made the hopes of Troilus in vain.

This anticipation of Criseyde's future change of heart also gives a different feel to the vaguer passage of time after the tenth day has come and gone, reflecting the slackening expectation and hope in Troilus. The six days after the tenth day pass by in summary (v. 1205), and there are now references to Troilus' actions 'day by day' or on undetermined days (e.g. v. 1233, 1329, 1356, 1538, 1568, 1587, 1646); that earlier sense of 'driving' the time away towards the moment of reunion (e.g. v. 394, 405, 475, 628, 680) has disappeared. Troilus reproaches Criseyde for staying away 'two monthes' (v. 1351), which suggests how much time is understood to drift by in the latter half of the fifth book. Criseyde's letter contributes to this lengthening perspective by declaring her intention to return, yet 'what yer or what day | That this shal be, that kan I naught apoynte' (v. 1619–20).

The time represented between the opening of Book V and Troilus' eventual acknowledgement of Criseyde's betrayal corresponds to a period in the *Roman de Troie* between Briseida's arrival in the Greek camp before the eighth battle and her acceptance of Diomede after the fifteenth, a period which can hardly be less than two years because of the time Benoît records as elapsing between the intervening battles. But, if Chaucer bothered to make these calculations from the *Roman*, he has wisely not applied the result in specifying the duration of Troilus' declining hopes, preferring to suggest—without precisely defining—a long-drawn-out period of pain, and instead applying a historical sense of time to those public events of the Trojan War which form the backdrop to Troilus' experience ('the fyn of the parodie | Of Ector gan aprochen' (v. 1548–9)). The events of the poem begin at some time after the beginning of the siege of Troy and end sometime after the death of Hector, and so are concurrent with a substantial part of the Trojan War. As Troilus resumes his role in the war, the focus of the narrative is moving back towards a more historical narrative ('In many cruel bataille . . . As men may in thise olde bokes rede' (v. 1751–3)), while in the ending of *Troilus* the historical time of the narrative is in turn left behind for the present time of the poet and his audience, and the thought of eternity beyond this world.

See Provost, *Structure* 38–41, for an argument that the reference to the passing of three springs (v. 8–14) since Troilus first loved Criseyde would suggest that Pandarus finds Troilus a full year after the April in which Troilus fell in love, and that the consummation occurs in May or June of the year following the meeting of Troilus and Pandarus in Book I. On the date of the consummation in May or June, see North, *Chaucer's Universe* and Root and Russell, 'A Planetary Date'.

On the chronology of the poem, see Bessent, 'Puzzing Chronology'; Bie, 'Dramatic Chronology'; and Owen, 'The Significance of a Day'; and on the time-scheme, see Longo, 'Double Time Scheme', and Sams, 'Dual Time-Scheme'.

Root (ed.), *Book*, 549–50, calculated that between Briseida's arrival in the Greek camp and her acceptance of Diomedes in the *Roman de Troie* there is an interval which can hardly be less than twenty-one months: Briseida arrives before the eighth battle, which lasts a month (14516), followed by a truce of six months (15187) until the ninth battle, which is followed by one month's truce (15221). The eleventh battle occurs a year after Hector's death in the tenth (17489). There is a truce of two months between the thirteenth and fourteenth battles (19384), and of one week between the fourteenth and fifteenth (20060), after which Briseida succumbs.

In *Filostrato*, after Diomede visits Criseida on the fourth day, the sixth Part of the poem is swiftly rounded off by several summary stanzas anticipating in outline how Criseida switched her affections to Diomede (*Fil.* vi. 33–4). This is followed immediately by the opening of the seventh Part on the morning of the tenth day as Troiolo waits in Troy. *Filostrato* thus offered a linear narrative recounting successively the fourth day in the Greek camp and the tenth day in Troy, while between these two days there is a brief anticipatory summary of a future beyond the tenth day. In Boccaccio's summary conclusion to his Part 6, and in the narrative break between Parts 6 and 7, some of the most characteristic sections of Benoît's romance are being passed over. Chaucer seizes on Boccaccio's idea of anticipating the future between the two scenes, and interpolates his account of Diomede's courtship and Criseyde's change of heart and soliloquy, written with recollections of the *Roman de Troie*.

Structure in the Stars

And Signifer [the zodiac] his candels sheweth brighte.   (v. 1020)

Part of that structure through time which Chaucer adds to his version of the Troilus story is conveyed through particular references to the state of the heavens. Chaucer was always keenly interested in the contemporary science which combined aspects of what today would be distinguished as astronomy and astrology, and that interest informs and structures his poems. In his *Complaint of Mars*—written in the same period as *Troilus*—Chaucer composes a coded narrative about the love of the mythological Mars and Venus which is simultaneously an account of a planetary conjunction, so that the reader appreciates on different levels of awareness both the story as story and also its diagrammatic, charted structure as a reflection of the movements of the heavens. The *Complaint* offers one model of how far Chaucer's interests could take him at this period of his life in relating the structure of narrative and the structure of the heavens. There is no comparable superimposition of the structures of the story and the sky in *Troilus*, yet the way the heavens represent time in a material shape becomes a kind of informing structural correlative to the movement of time in the poem.

The pattern in distribution of reference to the skies is interestingly uneven, the fullest allusions being associated with the union of the lovers and the time of Criseyde's return to Troy. In Book I, before the action gets under way, the heavens remain unmentioned. But, once Chaucer's Pandarus sets off to see Criseyde, his movements are observed alongside those of the skies: it is May and the sun is in the sign of Taurus (ii. 54–5) when he consults an almanac and finds that the moon is in a propitious position for his enterprise (ii. 74–5). When Troilus first rides past Criseyde's house, the narrative notes that the planet Venus was propitiously positioned, as she had been in his horoscope at birth, a hint at the question of the stars' influence over men's lives to which we shall later return ('blisful Venus, wel arrayed, | Sat in hire seventhe hous of hevene tho' (ii. 680–1)). Venus is here located in one of the twelve 'houses' (or divisions of the whole sphere of the skies, through all twelve of which the planets pass once in twenty-four hours), and the seventh 'hous'—just above the western horizon—is especially associated with love and marriage. The day is drawn to a close by setting Criseyde against a backdrop of the setting sun ('gan westren faste'), and then brilliant moonlight (ii. 904–24).

The association of Criseyde with the moon—the sphere of change—is to continue and gather point throughout the poem, and it is at the time of one of the moon's changes that Pandarus arranges his supper party (iii. 549–50). Chaucer's third book includes a precise allusion to a particular state of the heavens, so that the physical union of the two lovers is made simultaneous with a rare and portentous conjunction of two planets, in which the crescent moon is also involved (iii. 624–5). Once union is achieved, the closure of this central scene is similarly marked by allusions which set the lovers' first dawn separation against the backdrop of an early morning state of the heavens, with Venus rising as a morning star ('Lucyfer, the dayes messager' (iii. 1417)), and also the mysterious—

perhaps deliberately mystifying—reference to 'Fortuna major' (iii. 1420), possibly Jupiter, perhaps a particular group of stars, such as the Pleiades. The stream of mythological and planetary references in the *aubades* of Chaucer's lovers—in which they wish the movements of the heavens could alter to suit their passion—only serves to emphasize implicitly that ordered, moving structure of the heavens and of time against which their love will in time be tested and measured.

After the climactic action in Book III the narrative focus of the story is on waiting against time, given definition and structure in relation to the state of the heavens. Criseyde links the timing of her return with the movement of the moon ('Lucina') through the heavens, its progress from its present position in the sign of Aries, through Taurus, Gemini, and Cancer, to the end of Leo, when it will be visible as a crescent:

> 'Er Phebus suster, Lucina the sheene,
> The Leoun passe out of this Ariete . . .'   (iv. 1591–2)

Chaucer's characters are made to 'tell the time' by the sky: Troilus anxiously calculates Criseyde's return by the moon, standing out every night and telling it his hopes ('Whan thow art horned newe, | I shal be glad . . .' (v. 650–1)), urging it to move faster ('ren faste aboute thy spere!' (v. 656)), or fearing that Phaeton is once more driving amiss the chariot of the sun (v. 663–5). The association of Criseyde with the moon hints at her change, even as the regularity of the moon's movement as a heavenly body through the sky gives a precise structural definition to the ten-day span of Troilus' waiting. So much depends on the tenth day, and as Criseyde goes to bed in the Greek camp on the evening of that tenth day a description of the state of the heavens confirms that the chariot of the moon ('Cynthea') has reached the end of the sign Leo in the zodiac ('And Cynthea hire char-hors overraughte | To whirle out of the Leoun, if she myghte' (v. 1018–19)). It is noticeable that after this splendid passage there is little further attempt to structure the narrative by reference to the heavens, so that the starry dimension which accompanies the affair from the moment Pandarus sets it in motion drops away once it begins to end, only to be resumed and transcended by the eventual flight of Troilus' spirit to a point beyond the sphere of the moon.

It is in these latter stages of *Troilus* that the parallel movement of the heavenly bodies and the emotional action of the poem especially suggests that Chaucer's use of the stars offers a kind of astrological correlative for the lives of the lovers. In the early scene in the temple Criseyde is seen as 'so bright a sterre' (i. 175); much of the navigational imagery in the poem by implication represents Criseyde's influence over Troilus as that of a star, although there is a poignant contrast between the association of 'bright' Criseyde with the changing moon and her lover's view of her as his unchanging pole-star (v. 232, 1392). In view of the parallel that develops between Criseyde and the moon, it is also striking how many transitions are undertaken with careful attention to the position of the moon (ii. 74–5; iii. 549). By the terms of her oath to return Criseyde completes the pattern that associates her own changeable nature with that of the moon. Because the moon is the planet with the smallest orbit she is the most rapidly changing of the planets and the one most influential in effecting change in the sublunary world. It is therefore very apt that Criseyde's change of heart should be signalled—and paralleled on another plane—by the passing of the moon out of the sign Leo (described as 'the beest roial, | The gentil Leon' (*Squire's Tale*, V. 264–5)), which may very fittingly be associated with Prince Troilus, who is described as 'hardy as lyoun' (v. 830), while 'in the feld he pleyde tho leoun' (i. 1074). But the moon cannot remain in any sign of the zodiac, and, in that her movement out of Leo after a given point is inevitable and irreversible, the link with Criseyde's movements implies that the moment of opportunity for Criseyde's return has passed just as irreversibly as a planet leaving a zodiacal sign.

The astrological structure of the poem has also been read in terms of the structure of the contemporary skies, for, if the conjunction of Jupiter, Saturn, and the moon in Cancer implies that the action of Book III occurs in 1385, then the later statement that the lovers separate after three winters (v. 8–14) might mean that the events of Books IV and V are understood to occur in 1388, and refer to the state of the heavens in that year. Chaucer's fourth proem had invoked 'cruel Mars' to preside over the completion of the poem (iv. 25–6) and in the early months of 1388 Mars is in a position of power, almost precisely at his exaltation on 2 April 1388, while Venus is moderately close to an unfriendly quartile aspect with Mars. The subsequent ten-day period of waiting in Book V has similarly been located to early May 1388, in accordance with Criseyde's promise to return before the moon 'the Leoun passe out of this Ariete' (iv. 1592), for on 3 May 1388 the moon was in Aries and on the evening of 13 May she was indeed leaving Leo or 'whirling out of the Lion'. Before her departure Criseyde had prayed to Jupiter (iv. 1683–4) and Chaucer's fifth book opens by alluding to Jove as the disposer of destinies (v. 1–2), while on 13 May 1388 Jupiter was in quartile (i.e. unfavourable) aspect with Venus. In astrology Jupiter is concerned with affairs of judgement and religion; it would follow that the position of Mars in early Book IV signifies that the influence of Venus and love is threatened, while in Book V love is being judged.

The death in 1387 of Ralph Strode, one of the poem's dedicatees, implies that the poem was completed at least sometime before, but such an astrological structure as has been suggested for the poem is a matter not only of observation but of 'calkulynge', of calculated symmetries and configurations. As such, it is something Chaucer might plan and calculate on the model both of what had happened or was foreseeably to happen. Indeed, by the later stages of Book V a reader alert to astrology might well anticipate that the poem had already incorporated into its course such a full set of allusions to those planets that Troilus invokes in his Book III prayer (712–32) that only a role for Mercury was still to be included, in order to complete the full complement of planets that Chaucer's hero had so carefully enumerated.

The symbolic role of the planets is often as important as their position in the zodiac, and, when Troilus feels his death is near after Criseyde's departure, he prays to Mercury to fetch and guide his soul (v. 321–2), for Mercury is the psychopompus or guide of souls, and reappears in this role in the passage describing the ascent of the hero's soul after his death (v. 1827). The full complement of the planets is thus suggested in Chaucer's astrological structure in his *Troilus*. The description of the soul's ascent is important in part as the culmination of the preceding structure of astrological references: once the story is completed the change wrought by death in the awareness of Troilus takes the form and shape of his own movement towards the heavens. The ascent of the soul through the elements fulfils its part in a symmetrical pattern as the final, fitting reversal of the soul's descent through the elements at the beginning of life, and perhaps there is also a symmetry between the ascent of Troilus' soul near the end of *Troilus* and the account of Criseyde near the beginning, whose 'aungelik' beauty made her seem like 'an hevenyssh perfit creature, | That down were sent in scornynge of nature' (i. 104–5). As the soul of Troilus ascends away from the earth—and hence first through the progressively lighter elements: earth, water, air, and fire—all the earlier, particular references to the individual planets, their spheres and powers, give way to a vision which can enfold all previous allusions within its structure, until the hero is in a position to see all the planets ('the erratik sterres'), to hear the music of the spheres, and to look down on the earth.

The soul ascends 'Up to the holughnesse of the eighthe spere' (v. 1809), and Chaucer's imagination follows the change in Troilus' viewpoint as he no longer sees the heavens as they are seen from the earth, as concave, but 'in convers letyng everich element': that is, leaving behind and looking back at the further, convex side of the spheres, so that each reader experiences something of that instant of transformation of perspective experienced by the hero. This conveys a transcendence of the previous distinction of earthly and heavenly planes in a movement from partial to completed vista, but it is only a temporary vantage-point, and afterwards 'forth he wente' to where Mercury assigned him to dwell (v. 1827).

The meaning of this conclusion to the astrological structure of Chaucer's *Troilus* is now surrounded with ambiguity. The uncertainties are threefold: (1) Did Chaucer write 'seventh' or 'eighth' sphere? (2) Which sphere is meant by seventh or eighth, counting outwards or inwards? (3) What significance attaches to whichever sphere is meant?

Most extant *Troilus* manuscripts read 'seventh' sphere (only two manuscripts, J and R, and Caxton's edition read 'eighth'), but Chaucer's source for the passage in Boccaccio's *Teseida* reads 'cielo ottava', so that it has been generally assumed that Chaucer followed Boccaccio, especially in view of the ease with which scribes mistake the Roman numerals they normally use in writing numbers. Yet it is not impossible that Chaucer altered his source, or that, while using the same words, Chaucer intended something different from Boccaccio. If Chaucer is counting outwards from the earth, then in the eighth sphere Troilus would arrive in the sphere of the fixed stars, while in the seventh he would arrive in the sphere of Saturn. If Chaucer is counting inwards towards the earth, then the eighth sphere is that of the moon, while the seventh is that of Mercury (but as Mercury acts as the guide of souls in other spheres than his own it is not necessary for Troilus to be in the sphere of Mercury to receive his guidance). There have been learned arguments in support of all four of these destinations for Troilus, but most discussion concerns the spheres of the moon or the fixed stars.

The argument that Chaucer is counting *inwards* and that Troilus therefore reaches the sphere of the moon has arisen because in Cicero's *Somnium Scipionis* the planets are numbered inwards, and the *Teseida* is held to be influenced by the *Somnium*. Yet, although Cicero's text does number the planets inwards, his hero takes up his stand in the Milky Way, i.e. in the sphere of the fixed stars, and it is there that the music of the spheres is heard. In support of Chaucer's counting inwards it can also be argued that in his prayer (iii. 712 ff.) Troilus refers to the planets in an order that corresponds to counting inwards. But Troilus is here addressing the gods more than the planets, and, since—not unreasonably in his circumstances—he disrupts the planetary sequence to put Venus first and omits Saturn, this passage is not very convincing evidence for Chaucer's disposition elsewhere to count the planets inwards. It can be argued that *Teseida*'s description of the hero's looking at the planets and then down (*in giù*) at the earth implies that he is at a point between the earth and the planets, i.e. the sphere of the moon. Yet the description of the hero looking at the planets and then at the earth might equally imply that he is at a point beyond both, and that the earth is simply further below the planets. That Troilus can see the spot where he was slain seems a poetic licence whether he looks from the moon or the fixed stars, although, when Pompey's spirit

looks down at his slain body from above in Lucan's *Pharsalia* (ix. 1–14), he
is at the edge which divides the transient air from the ether. It might also
be held that, since Chaucer's Troilus is described as rising up through the
four elements which lie between the earth and the moon, but not explicitly
described as rising through each of the planetary spheres beyond, this
could imply that he goes no further than the sphere of the moon, although
it might be argued that 'in convers letyng' suggests going further, and there
is evidence that 'element' could have the sense of 'planetary sphere'. Below
the moon is the sublunary realm of mutability, so that Troilus need rise no
further than the sphere of the moon to gain enlightened detachment from
the world of flux and change, although the moon does not necessarily
represent an absolute boundary, for the planets above ('the erratik sterres')
can also be associated with change.

   The argument that Chaucer is counting the spheres *outwards* in
describing Troilus' ascent depends on various kinds of evidence for his
likely practice in numbering the spheres and for the appropriateness of the
hero's consequent destination. In referring to the 'thridde heven' or third
sphere as that of Venus in the third proem (iii. 2) Chaucer is numbering the
spheres outwards from the earth, and it has been argued that his
professional contemporaries took the enumeration of the planets from the
Moon outwards as a matter of course. Troilus' destination would therefore
be the seventh sphere (of Saturn) or the eighth sphere (of the fixed stars).
Some corroboration of this outward numbering is provided by the example
of Dante in *Paradiso*, xxii—well known to both Boccaccio and Chaucer—
where the pilgrim ascends from the seventh sphere of Saturn to that of the
fixed stars and looks at the 'vil sembiante', the paltry semblance, of the
globe (xxii. 133–5). In Dante's scheme the seventh sphere is associated
with divine devotion, and it has been argued that Troilus reaches the same
sphere as St Benedict in *Paradiso* because he is part of an 'order' of lovers.
It has also been suggested that a passage in Boethius about the flight of
thought associates the spheres of Saturn and the fixed stars in a way
pertinent to the *Troilus* passage:

Whanne the swift thoght hath clothid itself in tho fetheris, it despiseth the hateful
erthes, and surmounteth the rowndenesse of the gret ayr; and it seth the clowdes
byhynde his bak, and passeth the heighte of the regioun of the fir, that eschaufeth
by the swifte moevynge of the firmament, til that he areyseth hym into the houses
that beren the sterres, and joyneth his weies with the sonne, Phebus, and
felawschipeth the weie of the olde, colde Saturnus; and he, imaked a knyght of the
clere sterre (*that is to seyn, whan the thought is makid Godis knyght by the sekynge of
trouthe to comen to the verray knowleche of God*)—and thilke soule renneth by the
cercle of the sterres in alle the places there as the schynynge nyght is ypainted. . . .
And whan . . . he hath gon there inoghe, he schal forleten the laste point of the

hevene, and he schal pressen and wenden on the bak of the swifte firmament, and
he schal be makid parfit of the worschipful lyght ... of God ...   (*Boece*, iv. m. 1)

If the 'houses that beren the sterres' and the 'cercle of the sterres' signify
the sphere of the fixed stars, this Boethian account of ascent might be seen
to support a reading of the *Troilus* passage in which Troilus ascends first to
the seventh sphere and is thence led forth into the eighth.

There are a number of further indications that by 'eighthe spere'
Chaucer would mean that of the fixed stars. All Chaucer's other references
to the eighth sphere, both in his poetry (*Franklin's Tale*, V. 1280) and in
the astrological prose works (*Treatise on the Astrolabe*, i. 17, 21; *Equatorie of
the Planetes*, C/34), clearly refer to the sphere of the fixed stars, and this
was indeed the standard contemporary meaning of the term 'eighth
sphere'. The principal meanings of *eighth* emerging from Biblical exegesis
were the completion of a cycle or a return to the beginning; purification;
and immortality, eternity, and eternal salvation. In *Paradiso*, xxii, the
eighth sphere is that of the Christian mystics, and Dante's ascent to the
sphere of the fixed stars brings true understanding of spiritual and material
realities. Against a background of some controversy in Chaucer's day over
the possibility of salvation for virtuous pagans, it is especially apposite that
in the Neoplatonic and Hermetic tradition of the *ogdoad* souls possessed of
gnosis return to the eighth sphere at death, and it may well be that by a
reference to the eighth sphere Chaucer intended to invoke the possibility
allowed by this tradition of a suitable point to which the soul of Troilus
might ascend, although he is careful not to say that this was the permanent
resting-place of the hero's soul.

For the fullest account of astrological themes in *Troilus*, see North, *Chaucer's Universe*, who
discusses the Pleiades (pp. 238–43), possible parallels between references in *Troilus* and
the state of the heavens in 1388 (pp. 367–95), enumeration of the planets inwards or
outwards (pp. 29–32), and contemporary understanding of the 'eighth sphere' (pp.
395–8). See also North, 'Kalenderes'.

On allusions to the heavens in *Troilus* as an 'astrological allegory or analogy to events on
earth', see Stokes, 'Moon'.

On the ending of *Troilus* and tradition, see Kellogg, 'Tradition', and Steadman, *Disembodied Laughter*, and for an account of the medieval cosmological scheme, see Edward
Grant, *Physical Science in the Middle Ages* (Cambridge, 1971), ch. V ('Earth, Heavens
and Beyond'). On 'element' in the sense of a planetary sphere, see the *Middle English
Dictionary*, s.v. *element* 2[a].

On the seventh sphere, see Cope, 'Chaucer', and Wood, *Chaucer and the Country of the
Stars*, 187. On *eighth* in Biblical exegesis, see Conlee, 'Meaning'. On the *ogdoad*, see
Bloomfield, 'The Eighth Sphere'. On the question of the seventh or eighth sphere in
Troilus, see further Clark, 'Dante'; Drake, 'The Moon'; Manzalaoui, 'Roger Bacon';
Morgan, 'Ending'; Scott, 'The Seventh Sphere'; and Wood, *Chaucer and the Country of
the Stars*, 180–91.

Themes

Introduction

> Thanne thoughte he thus, 'O blisful lord Cupide,
> Whan I the proces have in my memorie
> How thow me hast wereyed on every syde,
> Men myght a book make of it, lik a storie.'   (v. 582–5)

The themes of *Troilus and Criseyde* are questions: the poem asks questions about the most important things in human life. But the thematic concerns of the poem remain as questions much more than they are answered within the text, and it is this sense of questioning arising from the 'proces' of experience that gives the 'storie' of *Troilus* its continuing life as a poem.

Questions are inseparably built into that extraordinary inclusiveness of source materials and genres, and that remarkable ambition in design, which earlier chapters have noticed as distinguishing *Troilus*. Such a designedly inclusive structure, such a range of added reference, is aiming to take further and see more in its material than any previous treatment of the *Troilus* story. Yet precisely by such an inclusiveness of traditions—and such a foregrounding of the compositional process—Chaucer at the same time resists any simple interpretation, or introduces as many signals as he can to educate the reader's sense of the difficulty of interpreting a variety of materials set very much in relation with each other. To see *Troilus* in terms of how many kinds of sources have been drawn upon is to wonder whether Chaucer has fully integrated all he has borrowed and invented—or indeed whether he needed or aimed to do so. With inclusiveness comes ambivalence and irony. *Troilus* allows for and invites multiple points of view on its material, and, like a Gothic work of art, it is not unified by the perspective achieved from any single viewpoint.

In asking its questions *Troilus* develops a complex instrument in the way the poem presents its narrative, for it is the narratorial medium that contains and most insistently raises questions inherent in the story and so develops the poem's concern with the nature of love, with time and change, with human freedom and individual purpose. Chaucer's change in that relationship between the narrator and love which he found in *Filostrato* is a shift of fundamental thematic significance for *Troilus*. In the Italian poem a narrator professedly experienced in love is writing of a love-affair between a hero also experienced in love and a widow. Chaucer turns this round, so that a narrator professedly quite inexperienced in love himself writes about a hero's very first experience of love:

> Forwhi to every lovere I me excuse,
> That of no sentement I this endite,
> But out of Latyn in my tonge it write ... (ii. 12–14)

He recurrently defers to the two authorities of his 'auctour' and of his audience, inviting them to adapt what he has written in the light of their experience. The dramatized anxiety of the narrator over 'authority' and the intimate nature of what he is narrating are placed in comic juxtaposition in the treatment of the consummation scene. The slightest descriptive detail is attributed—with a deliciously absurd intrusiveness—to what have suddenly become not one but multiple sources:

> Criseyde, which that felte hire thus itake—
> As writen clerkes in hire bokes olde—
> Right as an aspes leef she gan to quake ... (iii. 1198–1200)

Through such constantly professed reference to sources a sense of the process and responsibility of creating a narrative is kept part of the audience's experience of that narrative, and submitted to their discretion ('Doth therwithal right as youreselven leste' (iii. 1330)). The customary authority of the writer normally assumed by readers is supposedly abdicated—or a game is made of abdicating—and a distinction is drawn between loving, and reading and writing about love, which will prove significant for the kind of love experienced by Chaucer's Troilus.

If a narrative is being presented by a narrator who professes ignorance of his subject, more attention will be paid to how the narrative is conducted. The 'character' of the narrator in *Troilus* has sometimes been taken as a much more continuous and complete characterization than is presented or needed by the text, and any particular narratorial intervention has to be examined in terms of its immediate rhetorical purpose and effect. His is a rather limited persona, fictionalized as an outsider to love, yet inexperience is not presented as ensuring an unrelieved detachment and objectivity. The narrating voice of *Troilus* grows fondly involved with what he narrates, shares the language, and seemingly the aims and values, of the lovers. Through his engagement and indulgent partiality the narrator acts as a surrogate for an audience's tendency to sympathetic identification with the story of love in a romance narrative. But in this Chaucer is dramatizing inside his poem one mode of responding to the poem's subject, so that this response can itself be contained within the poem's scrutiny.

Troilus' nature and experience are no less far-reaching in their implications. Chaucer's emphasis on the process of courtship serves to raise questions about the conventions and 'observances' of that ritual process, and the conceptualizations of love that it represents. For the inexperienced Troilus everything is being done for the first time, with consequences for the patience of hero, narrative, and reader. It is a unique, unparalleled, wholly possessing experience. That he has no previous experience to tell him what to do allows for a vast deal of talking and thinking about how he is to act and about how he feels (he is a sort of personified *Canzoniere*). This lends the poem that discursive and reflective dimension which makes it seem a consideration of love so much more than the story of its hero and heroine, and indeed it can be called a kind of *summa* of approaches to love. A particular style and art of romantic love—'fyn lovynge' as Chaucer calls it (*The Legend of Good Women*, F 544)—is being analysed implicitly and explicitly in relation to other kinds of love. *Troilus* presents at once a celebration of love and a critique.

The combination of an inexperienced hero and a bookish and professedly unsuccessful confidant results in a love-affair pursued with an almost Quixotic degree of reference to literary models. Troilus is only a 'typical' courtly lover in the sense that he is quite extraordinary within the tradition, a study in a literary archetype taken to the brink of absurdity, but perhaps therefore all the more admirable. A context for him is provided by Pandarus, who becomes the 'author' of the affair, a surrogate for the narrator within the action just as the narrator is a surrogate within the poem for the author. His role as author shapes the 'romance' of the lovers' successful experience in the first three books—and as such corresponds with the parts of *Troilus* where Chaucer as author has innovated most freely with his own sources. But Pandarus cannot shape the more 'historical' materials when outside forces impinge in the later books, and his authorial role is confined to the creating for Troilus of a certain kind of love-affair on literary models.

It is Chaucer's Pandarus who, in the manner of an author, invents a plot by arranging correspondence and meetings; he generates action from feeling, invents motivation for the characters (as in the fiction of Troilus' jealousy of Horaste), and sees further into them than they do themselves ('Thow thynkest now, "How sholde I don al this . . .?"' (ii. 1506)). His activity is set within a larger disposition to bookish interests: he reads and has opinions on books; his speech is marked by reference to books and writing, and in the stanza at the close of Book I about planning a structure the artfully delayed attribution—'Al this Pandare in his herte thoughte' (i. 1070)—apparently allows the thoughts of Pandarus and the narrator to overlap. His account to Criseyde of overhearing Troilus in the garden—whatever its relation to the truth—is a fine piece of narrative and character-description, and Chaucer creates the impression that the love-affair of Troilus and Criseyde is shaped and patterned with arranged sequences and effects through being conformed by Pandarus to familiar models. He stage-manages in advance Troilus' second ride past her house so as to impress his niece (the narrative here likens Pandarus to a smith (ii. 1275–7)), supervises the lovers' letters, brings Troilus as a suppliant to kneel at his lady's bedside, and, having brought the lovers together, retires to the fireside as if to look over 'an old romaunce' (iii. 980), although it is not altogether clear whether he is reading or watching.

The love-affair in Chaucer's *Troilus* is made so much like 'an old romaunce' that the conventions by which love is heightened and idealized are used ultimately to question and explore the love which they express. There is a strong sense in the poem of lovers as a sect apart (i. 34, 48, 303, 319, 328), and of love as a craft and an art to be practised by them (i. 379). The practice of love is in part the practice of composition, and as a lover Troilus as well as Pandarus is an artist. The third book finds him learning up the 'lesson' (iii. 51) which he is to repeat to Criseyde; Diomede similarly claims he has yet to learn how to 'complain' in love properly (v. 160–1), so that both Criseyde's lovers identify the earliest phase of courtship with learning and composition. Diomede has even been reading books about love,

> 'For wise folk in bookes it expresse:
> "Men shal nat wowe a wight in hevynesse"',   (v. 790–1)

just as Pandarus has:

> 'And ek, as writ Zanzis, that was ful wys,
> "The newe love out chaceth ofte the olde"'.   (iv. 414–15)

The line actually comes from Ovid by way of *Filostrato*; just as the narrator of *Troilus* apparently fabricates an authority in 'Lollius', so within his narrative the 'author' Pandarus apparently fabricates his own authority on love. The point is that the love-affair of Troilus and Criseyde should constantly be seen in the context of traditional literary representation of love, because this is how their experience is both enhanced and celebrated, and ultimately measured and judged.

On Pandarus as 'author', see especially Carton, 'Complicity and Responsibility', and Fyler, 'Fabrications'.

A Debate about Love

In the lengthy process which leads to the lovers' union, and in what happens to that love in the longer term, Chaucer includes so much discussion and reflection on the nature, conception, and value of loving that *Troilus* comes to constitute a form of debate about love. Some questions are raised implicitly through the nature of the story, its narrator and its hero, and how the poem ends, but questions are also voiced explicitly in the passages of reflection and dialogue. As such questions arise through the medium of a story, they are not raised systematically nor answered directly, with an effect of openness which invites any reader to participate in the debate.

To summarize those passages where questions arise about love is to review much of the structure of *Troilus* and its principal scenes and episodes. Questions are explicitly raised about the nature of love in the excursus after Troilus is struck by love's arrow (i. 218ff.); in the hero's first song, the Canticus Troili (i. 400–20); in the first interviews between Pandarus and Troilus and then between Pandarus and Criseyde; in Criseyde's monologue after discovering that she is loved (ii. 687 ff.); in the garden scene of Antigone's song about love and Criseyde's response (ii. 827–903); and in the exchanges between the characters in Book III. They are implied by many other contexts, such as the prayer which forms the first proem; the representation of Troilus' responses to love after returning home from the temple (i. 358 ff.); the richly ambiguous invocation to Venus in the third proem; the hero's various addresses to love in Book III; and the whole presentation of the disintegration of love and understanding in the later part of the poem.

The context for this questioning is a treatment of his story by Chaucer which leaves important matters intriguingly undefined. As he received it, the story is one of an illicit love-affair, but in its approach to sexual love his poem is strikingly different from its prototypes. In the *Roman de Troie* Diomede is shown looking forward explicitly to the time when he will sleep with Briseida, and the whole implication of Benoît's account is that Diomede supplants Troilus in the possession of Briseida. In *Filostrato* the physical nature of the satisfaction Troiolo seeks for his love is clear all along to the reader, as it is to Criseida. In *Troilus*, by contrast, the eventual culmination of courtship in physical union is never overtly acknowledged as an aim during that lengthy process seen retrospectively to have led towards consummation in Chaucer's poem. The relative innocence of both the lovers precludes knowingness, and the inexperience of Chaucer's Troilus allows instead for the fullest idealization of what love may mean. The effect is to produce a tentative approach, in which love unfolds as a discovery for the lovers and as an exploration for the reader of the questions raised by the experience of love.

The ambivalence of the invitations to pray for lovers in the first proem ('That Love hem brynge in hevene to solas . . .' (i. 31)) establishes at the outset an uncertainty about how to interpret love. How Troilus falls in love is dramatized as a rashly unequal encounter with a god, and this suggests the essential mystery and dominion in the beginning of love as it will seem to human beings. It is also a significant groundnote to the presentation of Troilus' love that it is a humbling for the foolish pride of imagining himself different from others in being invulnerable to love. In the reflections at this point in *Troilus* upon 'surquidrie' (arrogance) and 'foul presumpcioun' (i. 213), and in the likening of the hero to 'proude Bayard' (i. 218–24), the horse forced to draw with others in the traces, there is hardly a suggestion that love is to be avoided as Troilus has tried to avoid it.

There follows a narratorial excursus that propounds the wisdom of not resisting the irresistible power of love, and balances a rehearsal of that power with an insistence that love is a beneficial and morally improving force (i. 211–66). The examples offered of love's improving effects are unexceptionable:

> And ofte it hath the cruel herte apesed . . .
> And causeth moost to dreden vice and shame . . .   (i. 250, 252)

But the most pressing argument is the necessity in love's power, to which the virtuousness of love is a supplementary point:

> Now sith it may nat goodly ben withstonde,
> And is a thing so vertuous in kynde,
> Refuseth nat to Love for to ben bonde
> Syn, as hymselven liste, he may yow bynde . . .   (i. 253–6)

There remains something resistible in this exhortation not to resist, and the key lies in the hortatory tone of this professed outsider to love, through which Chaucer can include fully within his poem particular interpretations of the nature of love, while allowing through tone and context for a certain distance to be kept.

In representing the early stages of Troilus' love Chaucer constructs a double effect which both fully dramatizes the urgency of being in love yet allows for a reflective consideration of such love in general. In that this first experience of love for Troilus is sincere, uncalculated, wholly engrossing, Chaucer can present it as a model of its kind. This is a type case of first love, presented so as to show how such love characteristically channels and expresses itself through a power of idealization. *Troilus* has been painstakingly set in a pagan world, and one advantage of this is to give Chaucer a freedom to present in his Troilus an example of the universal experience of love, but outside the contemporary moral context in which such love would be considered and judged in his own time. In Christian terms the propriety of earthly love is determined by the attitude with which it is approached by any individual, for it should not usurp the love to be directed to spiritual things, nor should the will usurp the reason. *Troilus* presents the 'natural' love of pagans in a pre-Christian world, and sets it in a comparative relation to Christian discriminations between loves in a way that sharpens analysis of what earthly love can be.

The interview between Pandarus and Troilus in the second half of Book I, the interview between Pandarus and Criseyde in Book II, and Criseyde's monologue, serve to set this particular love in the context of discursive, theorizing, and reflective scenes in which the nature of love is further debated. Chaucer's inclusion as the first Canticus Troili of a sonnet of Petrarch's gives to his hero a lyric which exclaims wonderingly at the paradoxical quality of love. This Canticus is also characterized by a series of rhetorical questions about love, asked in solitude and never as such answered ('If love be good, from whennes cometh my woo?' (i. 402)). The song serves to suggest the need for definitions of the nature and experience of love, and the difficulty of achieving definitions ('If no love is . . . | And if love is . . .' (i. 400–1)).

It is Pandarus—who has already put the view that love is irresistible (i. 685–6)—who introduces into the poem's debate what will prove a signal distinction between types and aims of love, when to Troilus he expounds the view that all mankind is susceptible to love in one of two ways, spiritual or earthly:

> 'For this have I herd seyd of wyse lered,
> Was nevere man or womman yet bigete
> That was unapt to suffren loves hete,
> Celestial, or elles love of kynde;
> Forthy som grace I hope in hire to fynde.'   (i. 976–80)

With this speech Chaucer transforms the moral horizons of Pandaro's brutal comment here in *Filostrato* that every woman is amorous at heart and only restrained by fear—if Criseida denied it he would disbelieve her (ii. 27). Pandarus then proceeds to consider his niece's case in relation to the two loves:

> 'It sit hire naught to ben celestial
> As yet, though that hire liste bothe and kowthe;
> But trewely, it sate hire wel right nowthe
> A worthi knyght to loven and cherice,
> And but she do, I holde it for a vice.'   (i. 983–7)

It is characteristic of the spectrum of possibilities in love that Criseyde is judged by her uncle to be perfectly capable of 'love celestial'—and even inclined towards it—although Pandarus may feel that the present has other claims upon her. With this the way of 'celestial' love is set aside for the time being, but with all delicacy and without disrespect. The 'love of kynde' (that is, natural love) is instead to be pursued, and the consideration of love is set throughout *Troilus* in the context of an acknowledgement of the role of 'kynde' in conduct and character (e.g. ii. 1374; iv. 1096). Yet the very act of distinguishing between the two loves in a discussion—and choosing one rather than the other for the present—implicitly conserves the thought of 'love celestial' for the future, as a potential way of love whose time may come.

Just as the moral expectation of Criseyde is revised in Chaucer's poem, so are the hero's expectations in love. His resonant statement of good intentions in love—

> 'For dredeles me were levere dye
> Than she of me aught elles understode
> But that that myghte sownen into goode'   (i. 1034–6)

—closes the first book, emphasizing the transformative effect of love in changing the moral character of the hero for the better (i. 1076–85). This motif of the improving effect of love on the whole character of Troilus—reported on the testimony of more witnesses than the narrator—is to recur in an equally prominent and symmetrically complementary position at the close of Book III, and is significant evidence in the poem's evaluation of love.

Chaucer's reworking of the first visit of Pandarus to Criseyde and its aftermath draws us through a sequence of moments which review aspects of love. The proem alerts us to cultural difference and historical change in modes of love; Pandarus suffers an attack of lovesickness before he visits Criseyde; he engages in banter with her and her ladies, enquiring about her book ('Is it of love? O, som good ye me leere!' (ii. 97)). Here is love as the absorbing object of enquiry and study, a game in which one may learn something to one's advantage, and an art to which there may be a key ('But tho that ben expert in love it seye ...' (ii. 1367)), but it is also a prospect that is surprising and deeply unsettling to Criseyde ('Unhappes fallen thikke | Alday for love ...' (ii. 456–7)). In his account of overhearing Troilus Pandarus can project to Criseyde an archetypal romance scene of her admirer languishing in a garden for love of her, but when he apparently anticipates by innuendo the physical consummation of their love ('Whan ye ben his al hool as he is youre' (ii. 587)), Criseyde's objection ('"Nay, therof spak I nought, ha, ha!" quod she' (ii. 589)) suggests she wishes to avoid any translation of love into physical reality. Her proposals to concede to Troilus no more than 'sighte' (ii. 1295), or 'as his suster, hym to plese' (ii. 1224), may be stages in playing a courtly game of courtship, but they also bring into consideration within Chaucer's poem other possible versions of love.

Sisterly love is one such possibility; 'love of frendshipe' (ii. 371) is another. When Pandarus assures Criseyde that the pervasive 'love of

frendes' at Troy can be exploited as a concealing cloak for a discreet
relationship with Troilus (ii. 379–80), he points to the potentially confusing boundaries between love and friendship in a way that contributes to a
larger exploration of ambiguous interconnections between love, friendship,
and kinship, in which 'friends' and 'sisters' may be at some indeterminate
point on the way to becoming 'loveres', and Troilus can be begged to be a
'friend' to Criseyde (ii. 1550, 1677). The connections between Troilus,
Pandarus, and Criseyde are set in an extended social group of friends (ii.
379), from which elders and parents, and attached or married people, are
for the moment absent. The poem opens with Calchas' abandonment of
Criseyde, and the affair will be undone by the father's wish to reclaim his
daughter, when the lovers' private emotional world—where Criseyde is the
source of direction and authority to Troilus—gives way to the reassertion
of a world of patriarchal hierarchy that thwarts and encumbers the lovers.

The quasi-fraternal relationship of Troilus and Pandarus focuses attention on a different kind of friendship, which is emphasized by their
habitual way of addressing each other as 'frend' and 'brother'. The nature
of friendship and the obligations of friends to offer each other moral
guidance were the subject of a substantial literature in the Middle Ages.
Parallels may be discerned between the way Pandarus brings Troilus and
Criseyde to bed together and the Old Testament story of how Jonadab
helps his lovesick friend Amnon to sleep with Tamar (2 Sam. 13:1–20), a
Biblical instance quoted in medieval treatises on friendship precisely as an
example of what friendship should not include. Yet Chaucer's readers
could scarcely miss that the friendly brotherliness of Troilus and Pandarus
is most fully and formally expressed when their friendship involves the one
friend's winning of a sexual partner for the other (iii. 239, 1597–8), or when
offering each other their sisters (i. 860–1; iii. 407–13).

Pandarus' relationship with Criseyde is presented as one of friendship as
much as one of kinship. They enjoy meeting socially as friends do (ii. 152;
iii. 605), and each refers to the other as an especially valued friend (ii. 240,
263), so that Criseyde regards her uncle's recommendation of Troilus as a
violation of their friendship (ii. 411–13). By giving Pandarus the more
responsible role of an uncle (to whom Criseyde turns for advice (ii.
213–14)), but also by giving him a more active role in winning Criseyde for
his 'Lord, and frend, and brother dere' (ii. 1359), Chaucer suggests that in
the world of his poem the demands of friendship are as strong as they are
ambiguous. In response to the crisis of his friend's extreme distress, an
uncle takes advantage of his influence over his niece, and Chaucer shows
Pandarus fully aware in one uneasy moment of how his actions might be
open to interpretation as the sexual procuring of his kinswoman for his
friend (iii. 271–80). *Troilus* is exploring the ambiguities whereby one may

ask for the friendliness of a loved lady, and use 'love of frendshipe' as a disguise for a love which is not friendship.

Criseyde's monologue, with its two sections of attraction and of trepidation, together with Antigone's song, bring into *Troilus* a review of diverse aspects of love. Criseyde's initial hope for detachment (ii. 607–9) is modified by her diminishing self-control at the first sight of Troilus ('So fressh, so yong, so weldy semed he' (ii. 636)). Her assumption that 'love of kynde' will be one of the ends in life of those who are not given over to 'love celestial' (ii. 757–9) gives way to a complementary enumeration of the disadvantages of love: its nature as 'the mooste stormy lyf . . .' (ii. 778); the constraints of avoiding gossip; and the faithless, shifting attentions of men in love. 'Thus, bitwixen tweye, | She rist hire up, and wente hire for to pleye' (ii. 811–12), and to hear her niece Antigone sing a song written by another Trojan lady in praise of a love seen as essentially joyous, secure, and lasting (ii. 831–3), and a lover whose love is the expression of his moral virtue:

> 'As he that is the welle of worthynesse,
> Of trouthe grownd, mirour of goodlihed,
> Of wit Apollo, stoon of sikernesse,
> Of vertu roote, of lust fynder and hed . . .' (ii. 841–4)

Composed by a 'humble subgit' of love who rejects the idea that love is 'vice or thraldom', the song acclaims love for its morally transformative effects (ii. 851–4). For Criseyde—who has earlier remarked 'To what fyn is swich love I kan nat see' (ii. 794)—the song's vision of love becomes yet another interpretation of love to be questioned and considered (ii. 885–6), and the affirmations in the song (ii. 862–75) have been seen as answering point by point the questions about love raised by Criseyde in her previous soliloquy (ii. 780–91):

> 'For evere som mystrust or nice strif
> (A) Ther is in love, som cloude is over that sonne.
> (B) Therto we wrecched wommen nothing konne,
>   Whan us is wo, but wepe and sitte and thinke;
> (C) Oure wrecche is this, oure owen wo to drynke.
>
> (D) 'Also thise wikked tonges ben so prest
> (E) To speke us harm; ek men ben so untrewe,
>   That right anon as cessed is hire lest,
>   So cesseth love, and forth to love a newe.
>   But harm ydoon is doon, whoso it rewe:
> (F) For though thise men for love hem first torende,
>   Ful sharp bygynnyng breketh ofte at ende.' (ii. 780–91)
>
> (A) 'What is the sonne wers, of kynde right,

> Though that a man, for feeblesse of his yen,
> May nought endure on it to see for bright?
> (B) Or love the wers, though wrecches on it crien?
> (C) No wele is worth, that may no sorwe dryen.
> (D) And forthi, who that hath an hed of verre,
> Fro cast of stones war hym in the werre!
>
> (E) 'But I with al myn herte and al my myght,
> As I have seyd, wol love unto my laste
> My deere herte and al myn owen knyght,
> In which myn herte growen is so faste,
> And his in me, that it shal evere laste.
> (F) Al dredde I first to love hym to bigynne,
> Now woot I wel, ther is no peril inne.'   (ii. 862–75)

In closing with Criseyde's dream this sequence of interchanges in Book II about deciding between loving or not loving, Chaucer acknowledges the contribution of unconscious and unexpressed factors in the movement towards love, while the contrast between the violent imagery of the dream and the cosily amicable milieu inhabited by Criseyde suggests how the experience of love will be utterly different from any other connection. The woman's dream of a dominant and predatory male eagle is a suggestive image of the ambiguities in sexual initiative and surrender. The symbolic act of exchanging hearts is mutual and painless, violent-seeming but controlled. Yet it remains the vision of a violation, of a male creature acting upon a female, even if in order to effect an exchange. Criseyde is not reported to be awakened, distressed, or puzzled: the experience seems not discontinuous with the pleasurable state of mind in which Criseyde listens to a nightingale outside her bedchamber (ii. 921–2). In the background to this sequence there is the earlier allusion to Philomela, raped by her warrior brother-in-law Tereus (ii. 64–6). Into a dream symbolic of sexual encounter the associations of the eagle as a royal bird and bird of prey import the larger context of the love-story in the midst of the Trojan War, a struggle started—as Troilus recalls—'for ravysshyng of wommen so by myght' (iv. 548). Criseyde's dream of the eagle and Troilus' later dream of the boar represent both Criseyde's sexual partners as fierce male animals, emblems of two warriors who fight fiercely in the war fought over a woman between the Greeks and Trojans, just as they will fight to their uttermost over Criseyde (v. 1757–64), in a continuous overlapping between love and war. Criseyde's fears as to the cruelty of men (ii. 457–8) are eventually to be fulfilled in the masculine figure of Diomede, both an enemy warrior and her lover, whose proffers of love are accompanied by menacing assurances of the utter destruction of Troy and its citizens. By contrast, the modification of the conventionally aggressive warrior in Troilus' role as a lover is evidently an expression of good qualities, of forms of self-restraint, which Criseyde approves: she seems attracted by Troilus' physical vigour ('For bothe he hadde a body and a myght | To don that thing . . .' (ii. 633–4)), but later acknowledges the reasoned self-control in his 'delit' with her (iv. 1678), and it is a humbled lover who enters her presence as a petitioner at the start of that third book which is at the centre of the poem's debate about love.

At the core of the whole poem, its climactic centre, Chaucer presents the lovers' physical union, not only so as to acknowledge the bodily nature of love in a way that is as beautiful as it is frank, but also to celebrate the joy of that sexual fulfilment:

> Hire armes smale, hire streghte bak and softe,
> Hire sydes longe, flesshly, smothe, and white
> He gan to stroke, and good thrift bad ful ofte
> Hire snowissh throte, hire brestes rounde and lite.
> Thus in this hevene he gan hym to delite,
> And therwithal a thousand tyme hire kiste,
> That what to don, for joie unnethe he wiste.   (iii. 1247–53)

Without saying so in words, Chaucer's poem may be seen to admit through its very form the centrality to human life of such experience of love, in a way that no subsequent expression of changed moral emphasis or explicit re-evaluation in the poem can wholly revise—although in its wise inclusiveness *Troilus* does actually dramatize at its conclusion just such an attempt in retrospect to rewrite earlier engagement and identification with the fulfilment of sexual love.

'Thus in this hevene he gan hym to delite': grouped around the lovers' physical union the third book presents materials for the poem's debate about love and transcendence, and, when Criseyde accepts Troilus with the cry 'Welcome, my knyght, my pees, my suffisaunce!' (iii. 1309), she gives expression to the poem's exploration of how far human love can indeed give true peace and fulfilment in a world of change, for 'suffisaunce' is a word associated with Boethius' discussion of the nature of true happiness. An intensely human and immediate joy is perceived within the commentary furnished by a framing structure of symmetrically positioned hymns and prayers and by the characters' discussions of a love comprehended as the supreme principle and order in a universe created and governed by a god which is love. By using Troiolo's Boethian song to love from *after* the consummation in *Filostrato* as the proem *before* the third book of *Troilus*, and by replacing the song in its original place by his own versification of the same Boethian passage that Boccaccio had used, Chaucer doubles the claim to associate the love of Troilus and Criseyde with that power of love which unifies and orders all things, and so makes the poem's narrating voice already anticipate in the proem that claim for love which Troilus makes after the lovers' union.

The consideration of love in the third proem, invoking Venus as the presiding, enabling deity of the book, is richly ambivalent both in the proem itself and through its context. Part of the ambiguity is intrinsic to the potent medieval conception of the two Venuses, both the beneficent influence of the planet, and the sexual licence associated with the goddess. As Bernardus Silvestris had put it in his commentary on the *Aeneid*:

Veneres vero legimus duas esse, legitimam et petulantiae deam. Legitimam Venerem dicimus esse mundanum musicam, i.e. aequalem mundanorum propor-tionem, quam alii Astream, alii naturalem iustitiam vocant. Haec enim est in elementis, in sideribus, in temporibus, in animantibus. Impudicam autem Venerem, petulantiae deam, dicimus esse carnis concupiscentiam quia omnium fornicationum mater est . . .

We read that there are indeed two Venuses, one lawful, and the other the goddess of wantonness. The lawful Venus is the harmony of the world, that is, the even proportion of worldly things, which some call Astraea, and others call natural justice. This subsists in the elements, in the stars, in the seasons, in living beings. The shameless Venus, however, the goddess of wantonness, is carnal concupis-cence since that is the mother of all fornications . . .

This doubleness in the potential of human love for both harmony and disorder is emphasized in the third proem by the way Chaucer alters his model in *Filostrato* to include both the moral or divine dimension in love as well as sensual impulse. He associates love with 'gentil hertes' (iii. 5), part of a concern with the nature of *gentilesse* that runs through *Troilus*; hails Venus as 'cause *of heele and of gladnesse*' and praises 'thy myght *and thi goodnesse*' (iii. 6–7). The second stanza of the proem includes the ringing declaration 'God loveth, and to love wol nought werne' (iii. 12), which introduces a further dimension into the sense of love's transcendent influence and power, while in the third stanza (iii. 15–21) one informing pattern—the descent of God into the world through love—is being overlaid with the mythological associations of Jove's ravishment 'in a thousand formes' of earthly partners. The effect of placing this proem's evocation of both the compelling and the ordering power of love—

> Ye holden regne and hous in unitee . . .   (iii. 29)
>
> Ye folk a lawe han set in universe . . .   (iii. 36)

—as preface to the book that contains the lovers' union is only to double the question-raising ambiguity of associating the famously mutable love of Troilus and Criseyde with such ideas of enduring love. At the very heart of the poem's structure and of his sexual desire, Troilus' prayer to 'Benigne

Love, thow holy bond of thynges' (iii. 1261) understands love as the cohesive order of the universe, and invokes it in religious terms as grace mercifully bestowed despite human undeservingness (iii. 1262–7). The terms are reverent, yet Troilus' transcendent view of what love comprehends can bring together in one exclamation ('O Love, O Charite!') love which is both Cupid, the son of Venus, yet also 'charite', a term used to describe a state of Christian love. Much modern interpretation of the poem and its hero has taken the view that the fundamental error in his understanding of human love is his overcharging it beyond the limits of human nature. The resonant beauty of language conveys the idealizing longing with which Troilus wishes to understand his love as an experience that leads the individual into communion with the ordering principles of the universe, although the developing 'proces' of the poem will come to place Troilus' understanding of love in a different perspective.

At a different level, the lovers' progress towards union in Book III is presenting materials which reveal and comment on various understandings of the love experienced by Troilus and Criseyde. The approach to consummation in the third book—full of delays and anticipations—allows for a sequence of speeches in which love is reviewed in different lights. The first meeting at the house of Deiphebus is an opportunity for declarations of what is understood between the lovers about their love ('What that I mene, O swete herte deere?' (iii. 127)), and this meeting acts as an eloquent prologue to the action of the book, to be followed by the dialogue between Pandarus and Troilus which by contrast allows for reference to the affair's more questionable aspects (iii. 239–343). The prayer by Troilus in the 'stewe' (iii. 715–35), in which, beginning with Venus, he invokes the aid of the gods before he enters Criseyde's bedchamber, also bears on the poem's exploration of love, despite being promptly deflated by Pandarus. Troilus' allusions to the amorous escapades of the gods—Venus and Adonis, Venus and Mars, Jove and Europa, Apollo and Daphne, and so on—recall instances of 'love' as oppressive, disastrous, and one-sided obsession, and also convey his pagan sense of himself as a fearful, powerless supplicant in a universe swayed by a love variously menacing, irrational, and violent. Criseyde's speech on the deceptive nature of mutable earthly happiness (iii. 813–40) prefaces their consummation, together with her lengthy disquisition on the relation between love and jealousy when the supposedly jealous Troilus is introduced into her chamber ('as wolde the excellence | Of love' (iii. 988–9)).

It is the charge of jealousy, of tainting with jealousy a true understanding of love, that brings on Troilus' swoon, the extreme instance of that setting aside of conventional masculine initiative and assertiveness that has characterized Troilus as a lover: male fear and shame trigger sexual consequences rather than a male 'advance'. As a deflation of male stereotypes of sexual conquest and mastery, the swoon is both comically absurd, yet also an endorsement of value in Chaucer's hero. As a response to Criseyde's accusations, the swoon represents a form of absolute submission and renunciation on one side, but it generates the lovers' mutual requests for each other's forgiveness in which Criseyde requites Troilus in humility. Troilus' following request to Criseyde to yield to him, or the likening of the male lover to the sparrowhawk taking the lark (iii. 1191–2), can only be reassertions of the archetypal masculine role in a newly qualified context that remakes understanding. The passivity of Chaucer's lovers and the delegation of all arrangements to Pandarus means that Troilus and Criseyde arrive in bed together, and on the verge of physical union, having made no direct sexual initiative to each other. The outcome of this separation of means from end is that, when sexual fulfilment does come, it is paradoxically a physical experience that seems miraculous and transcendent. The pleasures of the consummation are inextricable from a mutual obedience to each other's pleasure ('For ech of hem gan otheres lust obeye' (iii. 1690)) and convey the sense that their first night of love remakes the world ('But now I feele a newe qualitee— | Yee, al another than I dide er this' (iii. 1654–5)).

Chaucer's version of the Troilus story therefore presents an elaborate process of courtship which defers and hesitates over union, but also includes an acknowledgement and poetic re-creation of the fullness and intensity of bodily experience. The poem's openness to the whole of what love can be gives point to the various conceptions and idealizations through which love is projected in *Troilus*: as a kind of service or feudal relationship; as a religion; as a sickness to be cured, or a suffering that requires mercy; as an emotion read differently in the private and public worlds; as an expression of *trouthe*. These come to be understood not simply as polite fictions but as fictions through which the most intense experiences may be given expression through analogies of form, process, and ritual.

It is by its inclusiveness of approach that *Troilus* can have the cumulative effect of a debate, for Chaucer has so broadened within his poem the spectrum of experience shown, and implications considered, in relation to love. It is a pointer to the direction in which Chaucer is moving *Troilus* as an exploration of love that, where in *Filostrato* Troiolo writes a first letter to Criseida and the full text of his wordy letter is included, in *Troilus* at the equivalent point Pandarus gives Troilus a set of *instructions* on how to write a love-letter (ii. 1023–43), but the text of the letter is replaced by a summary (ii. 1065–85). This difference nicely characterizes the theoretical dimension of Chaucer's poem, its concern with how love ought to be composed and practised, and Troilus' letter is described as a quintessentially typical letter of 'thise loveres':

> First he gan hire his righte lady calle,
> His hertes lif, his lust, his sorwes leche,
> His blisse, and ek thise other termes alle
> That in swich cas thise loveres alle seche;
> And in ful humble wise, as in his speche,
> He gan hym recomaunde unto hire grace;
> To telle al how, it axeth muchel space.
>
> And after this ful lowely he hire preyde
> To be nought wroth, thogh he, of his folie,
> So hardy was to hire to write, and seyde
> That love it made, or elles most he die,
> And pitousli gan mercy for to crye;
> And after that he seyde—and leigh ful loude—
> Hymself was litel worth, and lasse he koude;
>
> And that she sholde han his konnyng excused,
> That litel was, and ek he dredde hire soo;
> And his unworthynesse he ay acused;
> And after that than gan he telle his woo—
> But that was endeles, withouten hoo—
> And seyde he wolde in trouthe alwey hym holde;
> And radde it over, and gan the lettre folde.   (ii. 1065–85)

'First he gan hire his righte lady calle ...', and this introduces the tendency in Chaucer's poem to express love in terms of a feudal relationship between the lady and 'man'. In referring to Criseyde as 'His blisse' Troilus anticipates the variety of religious language with which love will be associated in *Troilus*. The address to Criseyde as 'his sorwes leche' (i.e. the physician to his sorrow) and the plea not to be angry 'or elles most he die' are part of that pervasive association of the lovers' feelings with sickness and nearness to death, just as the abject petition ('And pitousli gan mercy for to crye ...') reflects the way that Chaucer represents the lovers' relationship in terms of the feeling of pity and the granting of mercy, within an atmosphere of heightened pathos. Beyond this, in its discretion, its humility, and its declaration of faithfulness the letter aptly looks forward to that concern for secrecy, honour, and *trouthe* with which Chaucer develops his version of the Troilus story as an exploration of the nature of love. In their declared typicality the 'terms' in Troilus' letter may serve to introduce some of the typical aspects of love in *Troilus* to which the following sections of this chapter are devoted.

On aspects of love in *Troilus*, see especially Aers, *Chaucer, Langland, and the Creative Imagination* and *Chaucer*; Bayley, *Characters*; Benson, *Chaucer's 'Troilus and Criseyde'*;

Bishop, *Chaucer's 'Troilus and Criseyde'*; Borthwick, 'Antigone's Song'; Brewer, 'Love and Marriage'; Diamond, *'Troilus and Criseyde'*; Dronke, 'Conclusion'; Dunning, 'God and Man'; Gordon, *Double Sorrow*; Howard, 'Courtly Love'; Kean, 'Chaucer's Dealings' and *Chaucer*; Kelly, *Love and Marriage*; Lewis, 'What Chaucer really did'; Morgan, 'Natural and Rational Love' and 'Significance'; Robertson, *A Preface to Chaucer*; Rowe, *O Love O Charite!*; Shanley, 'The *Troilus* and Christian Love'; Wetherbee, 'Descent' and *Chaucer*.

On love in *Filostrato*, see apRoberts, 'Love in the *Filostrato*'; apRoberts and Seldis, *Giovanni Boccaccio: Il Filostrato*; Kelly, *Love and Marriage*; Natali, 'A Lyrical Version'; Wallace, *Chaucer and the Early Writings of Boccaccio*.

On friendship, see Cook, 'Chaucer's Pandarus'; Freiwald, 'Swych Love of Frendes'; and Gaylord, 'Friendship'. See Muscatine, 'Feigned Illness', on parallels with the story of Amnon and Tamar, which is cited by Aelred of Rievaulx, *De spirituali amicitia*, ed. Migne (Patrologia Latina, 195: 675A), by Peter of Blois, *De amicitia Christiana*, ed. M. M. Davy (Paris, 1932), 154–6, and by pseudo-Augustine, *De amicitia*, ed. Migne (Patrologia Latina, 40: 836). On the ambiguities of 'love and frendshipe', see Lambert, '*Troilus*'.

On her niece's song as answering the points in Criseyde's soliloquy about love in Book II, see Borthwick, 'Antigone's Song'.

On Criseyde's dream of the eagle, see Gallagher, 'Criseyde's Dream'. On the form of the consummation scene, see Howard, 'Literature and Sexuality', and Owen, 'Mimetic Form'. See also Mann, 'Troilus' Swoon'.

For Bernardus Silvestris on Venus, see *Commentum super sex libros Eneidos Virgilii*, ed. G. Riedel (Greifswald, 1924), 9. See also George D. Economou, 'The Two Venuses and Courtly Love', in Joan M. Ferrante and George D. Economou (eds.), *In Pursuit of Perfection: Courtly Love in Medieval Literature* (Port Washington, Wis., 1975), 17–50, and Earl G. Schreiber, 'Venus in Medieval Mythographic Tradition', *Journal of English and Germanic Philology*, 74 (1975), 519–35. On the third proem of *Troilus*, see especially Gordon, *Double Sorrow*, 33 ff., and Kean, *Chaucer*, 175 ff.

On courtship, see especially Richard F. Green, 'Troilus'; J. D. Burnley, '*Fine Amor*: Its Meaning and Context', *Review of English Studies*, NS 31 (1980); and E. Reiss, '*Fin' Amors*: Its History and Meaning in Medieval Literature', *Medieval and Renaissance Studies*, 8 (1979), 74–99; see also Gaylord, 'Gentilesse'.

Serving and Deserving

It is a part of Chaucer's thematic exploration of love in *Troilus* that he shows the love-affair pursued within the conventions of love as 'service', presenting strongly a 'feudalization' of the procedures and imagery of the love-affair and a conception or model of a distinct kind of love. At the core of this conception is the presentation of the hero as committed to the feudal 'service' of his 'lady' Criseyde:

> 'But as hire man I wol ay lyve and sterve . . .' (i. 427)

> 'For myn estat roial I here resigne
> Into hire hond, and with ful humble chere
> Bicome hir man, as to my lady dere.' (i. 432–4)

> And he to ben hire man while he may dure. (i. 468)

For Troilus, Love 'held hym as his thral lowe in destresse' (i. 439), and the language of service contributes to a poem concerned with suffering and endurance, and with constraints on human freedom. While such language of service remains a fiction and play of the lovers' private life and imagination—Troilus does not really relinquish his royal rank—it is part of that urgent striving that he pours into his emotional life, and all the commitment involved in service naturally seems in the world of Chaucer's poem to hope for reward: serving implies deserving.

The idiom of love-service is significant in defining love from the hero's point of view, for it is noticeable that, where Boccaccio's Criseida in a conversation about love talks to Diomede of how she 'served' her late husband (vi. 29), Chaucer in translating suppresses any such notion of the woman serving the man in love (v. 974–6). Service is very much the characteristic disposition of Chaucer's hero: for him Criseyde remains— even in intimate and emotional contexts—his 'lady' (e.g. iii. 1157; iv. 316, 1214). When Troilus takes one long last gaze at Criseyde at their dawn parting in Book IV (1691), he beholds her as 'his lady' rather than by name or by some endearment; and the title becomes all the more poignant in the beginnings of suspicion and disillusionment, as Troilus dreams how by the boar 'Lay kyssyng ay his lady bryght, Criseyde' (v. 1241), and exclaims 'My lady bryght, Criseyde, hath me bytrayed' (v. 1247).

Complementing Troilus' conception of Criseyde as his lady is both lovers' view of Troilus as Criseyde's 'knight', and the obligations this brings within such a conception of love. Criseyde's first sight of Troilus in the poem is as a knight returning from battle (ii. 610 ff.); in her song Antigone hails 'My deere herte and al myn owen knyght' (ii. 871), and this pairing of terms is twice echoed by Criseyde in addressing Troilus (iii. 176, 996–7). When Criseyde accepts Troilus as her servant in love, she makes a point of setting aside within the private world of their relationship his princely rank in society (iii. 170–2), and from their first interview she claims suzerainty as her lover's 'lady'. At their moment of union she hails Troilus before anything else as her knight (iii. 1309), while the possibility for action after news of the exchange is governed for Troilus by his obligation as her knight to the honour of Criseyde (iv. 569–70).

Through such rhetoric of feudal service Chaucer is able to explore both the self-sacrificing and the expectant aspects of the experience of love, but not without layers of humour, as when Troilus, introduced into Criseyde's bedroom in the middle of the night, kneels to his lady with grave ceremony (iii. 955). To Pandarus the lengthy kneeling is evidently absurd and the object of sport ('Nece, se how this lord kan knele!' (iii. 962)), but by contrast—and with an air of owlish solemnity—the narrating voice ponders the impossibility of determining whether Criseyde simply forgot to ask Troilus to rise, or whether she took such gestures as appropriately dutiful 'observaunce' by her servant (iii. 967–70). The effect of this double perspective is to suggest the possibilities for seeing both the solemn and the comic sides of such idealized 'observaunce'.

There is much stress in *Troilus* on the demands and effort of such knightly service in love. Troilus addresses Criseyde as the one 'to whom *serve I and laboure*' (i. 458), and his Dantean address to Love in the consummation scene refers to those lovers 'That serven best and most alwey labouren' (iii. 1265). It is such ever renewing, ever diligent service that Troilus promises to Criseyde at their first meeting (iii. 143–4), and he goes so far as to see his life finding its created purpose in the service of Criseyde (iii. 1290). For the English Troilus loving and serving, loving as serving, are inseparable, for service gives expression to devotion and hence to the suffering and self-sacrificing disposition of Chaucer's hero (iii. 1793–4). Yet, while such service may be accepted as an endless, open-ended commitment, there is also a sense that to serve is to deserve, and to lead on to fulfilment. Quick with points about love-theory, Pandarus asserts that the lover should feel that just to serve his lady recompenses him a thousandfold more than he can deserve (i. 817–19). But in other contexts in Chaucer's poem serving is undertaken with thoughts of deserving:

> the longe nyght
> He lay and thoughte how that he myghte serve
> His lady best, hire thonk for to deserve.   (iii. 439–41)

The language of service lends structure to the lovers' progress towards union, as also to Troilus' obedience to his lady's wishes when the exchange is agreed, and his patient endurance after her departure. Troilus declares his continued service to Criseyde up to and even beyond death (iv. 321, 447), but Diomede slips easily into the same language of love service (v. 173), and Criseyde eventually uses it in return (v. 972–3). In her soliloquy of farewell Criseyde praises Troilus precisely for his pre-eminence in faithful service (v. 1075–6), and in his letter he urges her to write to him in reward for that service (v. 1389). Against this accumulated background there is an edge to Troilus' rhetorical question to his absent lady ('Was ther non other broch you liste lete | To feffe with youre newe love ...' (v. 1688–9), and there is also a special poignancy in his final address to the absent Criseyde declaring that his service has not deserved her desertion:

> 'But trewely, Criseyde, swete may,
> Whom I have ay with al my myght yserved,
> That ye thus doon, I have it nat deserved.'   (v. 1720–2)

For a comparative study of ideas of feudal service in *Filostrato* and *Troilus*, see Meech, *Design*, 271–89.

Love and Religion

One of the questions in the debate about love raised by Chaucer's treatment of the Troilus story stems from the association of the processes and experiences of loving with religious terms. The devotion of the lover to his lady becomes the practice of a quasi-religious devotion, in which the lover's conversion is followed by repentance of former sin and by identification with an 'order' of lovers. But, as with the notion of love as feudal service, the association of love with religious concepts expresses the lover's idealization of his experience more than the lady's view of hers. The relationship between lover and lady, the suppliant and the source of all mercy, is structured by frequent analogy with the Christian concept of grace, while the eventual attainment of sexual fulfilment is set about with references to heaven. The familiar structures and aims of religious devotion in this way act as a correlative to the devotion within the love-affair, lending it the force and value of religious associations, yet also cumulatively suggesting the ways in which Troilus' love is not in itself a religion.

From the very manner of its opening—through a first proem which takes the form of an adapted 'bidding prayer'—*Troilus* invites its audience to approach and analyse the subject of love through the model by which the priest leads and structures the prayers of the congregation by asking them to pray successively for different things. This familiar rhetorical format of a bidding prayer is deployed to organize and lend structure to a cataloguing of the variety of well-being and misfortune in love:

> And biddeth ek for hem that ben despeired . . .   (i. 36)
>
> And ek for hem that falsly ben apeired . . .   (i. 38)
>
> And biddeth ek for hem that ben at ese . . .   (i. 43)

Yet from the start there is an ambiguity of tone in interpreting experience of love in terms of religious models ('And preieth for hem that ben in the cas │ Of Troilus . . . That Love hem brynge in hevene to solas' (i. 29–31)). Which *hevene* is meant here? The poem ostensibly leaves it to each reader whether the invitation is to pray for the kind of heaven Troilus achieves in Book III or Book V, and emphasis on which is the true heaven will shift as the poem develops. There is an ambiguous humility in the narrator's taking to himself of the Pope's title of 'servant of the servants of God' (i. 15), and in his casting of himself in a quasi-priestly role as a kind of lovers' clerk (i. 49). But if there is a certain conscious artfulness in some of the associations of love with religious ideas in *Troilus*, there is also a complementary and serious sense of how the unchanging applications of those religious ideas retain the power to make themselves felt as circumstances change: one opening prayer ('So graunte hem soone owt of this world to pace, | That ben despeired out of Loves grace' (i. 41–2)) comes to apply to the Troilus who passes from the world at the close of the poem.

How one begins, belongs, and proceeds in love is in *Troilus* presented very much in terms of religious models. The moment when Troilus falls in love—in a temple, during a religious ceremony—is likened to a religious conversion when it occurs (i. 308) and later, more artfully, by Pandarus as a conversion from heresy to right belief: Love has made him like

> thise wise clerkes
> That erren aldermost ayeyn a lawe,
> And ben converted from hire wikked werkes
> Thorugh grace of God that list hem to hym drawe,
> Thanne arn thise folk that han moost God in awe,
> And strengest feythed ben, I undirstonde.   (i. 1002–7)

When Pandarus first finds Troilus languishing, Chaucer has him tease his friend with the suggestion that his sorrows stem from a consciousness of his sins ('Or hastow som remors of conscience?' (i. 554)), deploying the technical language of confession (cf. also 'attricioun' (i. 557)). In the interview that ensues there is some submerged sense of Pandarus as a confessor figure putting the convert through a ritual of repentance of his past sins against the God of Love (i. 932–8) whom he used to blaspheme as 'Seynt Idiot' (i. 910). Pandarus also presents to Criseyde the new lover's regret for past mistakes in terms of the available Christian model of repentance, its rhetoric and gestures ('Now, *mea culpa*, lord, I me repente!' (ii. 525)).

Having repented and confessed, the lover like the convert looks towards salvation, and it is possible to see in the progress of Troilus as lover a regenerative pattern in which he passes through contrition, penance, absolution, and eventually receives from his lady the forgiveness of his sins. For the lover his salvation is at the disposition of his lady, determined by her rather than by his deservingness, and by extension the relation of the lover with the lady is presented in terms of the Christian theology of grace. It is a sign of the central importance of this concept that Troilus invokes the theology of grace at the very heart of the poem, just before the lovers' union. This is that stanza borrowed by Chaucer from a prayer to the Virgin in Dante's *Paradiso*, which in adapted form Troilus applies to the search for the grace of love ('Benigne Love … Whoso wol grace and list the nought honouren' (iii. 1261–2)). The point made by the ever-humble Troilus is that such grace is independent of merit in the lover ('Yet were al lost … | But if thi grace passed oure desertes' (iii. 1266–7; cf. also 1282)) just as Criseyde makes the surrender of herself in the religious language of mercy and forgiveness ('Of gilt misericorde' (iii. 1177)).

Through such grace heaven may be gained, and in *Troilus* the lovers'
achievement of union in the third book is presented as the attainment of a
kind of heaven, just as possible echoes of the ceremonial, prayers, and
liturgy of the mass have been detected in the lovers' two meetings in Book
III, with Pandarus in a quasi-priestly role. Antigone's song had already
anticipated the 'parfit blisse of love' (ii. 891); Pandarus asks Troilus, as one
already in bliss, to pray for him (iii. 342); however playfully, Pandarus
represents the hero's entering into Criseyde's bedroom as the entry into
heaven's bliss (iii. 704). When Troilus and Criseyde are at last in each
other's arms, the narrating voice comments—with whatever irony—'Thus
sondry peynes bryngen folk in hevene' (iii. 1204), and later exclaims 'And
lat hem in this hevene blisse dwelle' (iii. 1322), while Troilus' delighted
exploration of Criseyde's body is itself identified with an experience of
heaven (iii. 1251). Throughout *Troilus* Criseyde is associated with heaven:
in Book I she is likened to 'an hevenyssh perfit creature' (i. 104); in Book
IV sorrow alters 'Hire face, lik of Paradys the ymage' (iv. 864); in Book V
her portrait includes the notion, perhaps remembered by Chaucer from
Dante's *Paradiso*, 'That Paradis stood formed in hire yën' (v. 817). After
the consummation Chaucer's Troilus likens his state to that of the soul
brought to rest in paradise (iii. 1599; cf. ii. 894–5; iii. 1657–9).

Yet even while such religious associations are being imported to lend
their solemnity to the lovers' understanding of their union, Chaucer's text
is also making ambiguous use of other religious ideas: Troilus impatiently
asks Pandarus 'How devel maistow brynge me to blisse?' (i. 623); the
narrator wishes he could have bought the lovers' night of sexual happiness
at the cost of his own soul (iii. 1319); Christ's forgiveness of his murderers
is lightly invoked in the same breath as the account of how Criseyde
forgave Pandarus on the morning after the consummation (iii. 1577–8).
Pandarus had implicitly compared a visit to Criseyde by Troilus with a
visit to a temple (ii. 372–3); and of course Troilus first sees Criseyde during
worship of a 'relik' in a temple, and goes home to dwell on her image in his
mind's eye with what may be taken as idolatry. Such association of
religious ideas with the events of the love-affair may cause the reader to
hesitate, and anticipate the gathering sense in the course of the poem that
the use of religious language in *Troilus* is an expression of both the
ambition and the limits of the understanding of love presented through
Chaucer's characters. When the lovers are forced to part, they think of
themselves as fallen from paradise into hell (iv. 712–13; v. 1396). The
deserted house of Criseyde becomes for Chaucer's Troilus an empty shrine
which lacks the relics of its saint, although a shrine he still longs to
reverence (v. 551–3), and this expresses the intensity and the hollowness of
what his devotion has become. It is a fitting culmination to the association

of religious language with the experience of love throughout the poem that, when Troilus comes eventually to feel that his faith in Criseyde has been misplaced, he expresses this in terms of a misreading of a sacred text ('God wot, I wende, O lady bright, Criseyde, | That every word was gospel that ye seyde!' (v. 1264–5)). This cumulative sense of possible misinterpretation in the application of the language of heaven and of religious belief to the experience of earthly love is one of the ways in which Chaucer points towards the use of religious language at last in the affirmations at the conclusion of the poem.

For a comparison of *Filostrato* and *Troilus*, see Meech, *Design*, 262–70. See also Blamires, '"Religion of Love"'; Burnley, *Chaucer's Language*; W.G. Dodd, *Courtly Love in Chaucer and Gower* (Boston, 1913), 189–208; and Slaughter, 'Love and Grace'. On echoes of the mass in Book III, see Devereaux, 'A Note', and Dronke, 'Conclusion'.

Sickness and Death

In his Sonnet 132 Petrarch had written of the lover's paradoxical sensations in love, concluding:

> 'e tremo a mezza state, ardendo il verno.'
>
> 'and I shiver in midsummer, burn in winter.'

When Chaucer borrows the sonnet for the first Canticus Troili, this last line becomes the last two of the song:

> 'Allas, what is this wondre maladie?
> For hote of cold, for cold of hote, I dye',   (i. 419–20)

and the added identification of the lover's state with a special form of illness, and the association of his extreme feelings with death, are part of a recurrent pattern in *Troilus*.

In keeping with the tradition of love as a sickness or even a madness for the lover, a sickness to which the lady—as a kind of physician—may bring healing and cure, Troilus sees the pity of the lady as the prospective cure for his mortal illness (i. 461–2), and the partial narrating voice is soon echoing the same idea of pity as cure (i. 469). This allows Pandarus—who had identified himself with Apollo, the physician who could not cure himself (i. 666–72)—to play the part of a physician to the malady of Troilus (i. 726–8), in an echo of Philosophy's role in the *Consolation*, the characteristic difference of outlook between the friends emerging in Troilus' insistence that his illness is incurable (i. 757–8), and Pandarus' insistence that he should not discount the possibility of cure (i. 783, 789–91). The revealing of the lady's identity to a friend can be likened to the patient's uncovering of his wound for inspection by a doctor (i. 857–8);

Troilus can be teased with a reminder of how he once mocked at lovers' maladies and their 'blaunche fevere', which he now endures for himself (i. 916–17); and the first book is brought to a close with the image of Troilus as a patient whose process of healing is begun but not yet brought to a cure (i. 1090–1).

It is such familiarity with the image of the lover's sickness that is played upon when Pandarus uses the ploy of a pretended illness on Troilus' part in order to bring the lovers together at Deiphebus' house, except that Troilus protests that he will not need to pretend, 'For I am sik in ernest, douteles' (ii. 1529). A sequence of ironical moments arises: a man with one serious malady pretends to have another, and, when unwitting bystanders offer medical tips for curing the sickness of Troilus (ii. 1578–9), this only highlights Criseyde's inward admission that 'Best koud I yet ben his leche' (ii. 1582). The last words of Pandarus to Criseyde as he introduces her into the 'sick-room' of the supposedly fevered Troilus are 'Bryngeth hym to hele' (ii. 1750). The lovers' meetings in Book III continue this theme (iii. 61, 792–4, 1137). At the climax of the book the joy of the lovers is prefaced by a stanza associating the consummation of their love after earlier griefs with the curing of a 'fevre or other gret siknesse' through the drinking of some bitter medicine (iii. 1212–18), and the contented sighs of Troilus when united with his lady are specified to be quite unlike the sighs of those who are either unhappy or sick (iii. 1361–2). Even in their last letters Troilus begs Criseyde to restore to him 'hele swich that, but ye yeven me | The same hele, I shal non hele have' (v. 1415–16), to which Criseyde replies that—being herself 'heleles' (without health)—she is unable to send health to Troilus (v. 1596). Both the stiff formality and the hopelessness expressed in these conventional phrases mark the decline and division of the lovers' understanding. In the course of Book V ideas of love as a sickness fade, to be replaced by the 'grevous maladie | Aboute his herte' and desire for death which Troilus suffers (v. 617–30, 1231–2).

There is a parallel development in the pervasive references to death in the poem. Chaucer associates some of the most important emotional transitions and experiences for the poem's characters with death and dying. The moment when Troilus is first smitten by Criseyde's look is represented as a kind of death:

> That sodeynly hym thoughte he felte dyen,
> Right with hire look, the spirit in his herte: (i. 306–7)

and his lovesick state is frequently likened to death (i. 420, 469, 728, 875; cf. ii. 1755). Criseyde's tears, her reproaches, and his fear of her disfavour, are so painful to Troilus that they seem to him like the death of his heart (iii. 1070–1, 1081, 1171). It is a fitting culmination to such a pattern that the turning-point for Troilus towards union with Criseyde is represented as a condemned man's sudden deliverance from death (iii. 1240 ff.). The sorrow of their subsequent parting at dawn is to the lovers like 'dethis wownde' (iii. 1697), as is their enforced separation in the later books (iv. 1149, 1692).

Chaucer also develops the idea that Troilus will die for love as the means by which Pandarus puts pressure on Criseyde: 'But ye helpe, it wol his bane be' (ii. 320), and the choice is hers although Pandarus will kill himself if she allows Troilus to die (ii. 322–5). The same choice is relentlessly pressed during the rest of the interview and beyond (ii. 384–5, 441, 536, 566, 1075), and he secures the admission of Troilus to her bedroom by the same means (iii. 905). This sense of passions so intense that they may lead to death gives his lady a life-and-death power over the lover: for Troilus Criseyde is 'hire that to the deth me may comande' (i. 1057), and so she remains to the end. In his letter in Book V Troilus still thinks of Criseyde as 'she that lif or deth may me comande' (v. 1413), in a sad echo of their earlier understanding which may soon prove all too literally true.

The idea of dying for love is apparently accepted at face value by the characters, in what seems a designedly puzzling mixture. On the one hand it would appear—setting *Troilus* in the context of other courtly literature—that the notion of a man's dying for love would be recognized as a well-established erotic fiction, one of the serio-comic verbal extravagances of romantic 'play' within the 'game of love'. Indeed, Pandarus in places seemingly delights in toying with the motifs of the lover as sick (ii. 1313–15), or as dying (ii. 1310–11, 1638). Yet he as well as Troilus and Criseyde are shown as acting on the belief that lovers can die for love (e.g. iii. 262–3; iv. 524–5; ii. 466–72; iii. 361–4). The gravity with which the characters think of dying for love and yet the hint of absurdity always at the edge of such contexts are part of a challengingly equivocal mixture of the serious and the absurd in the poem's presentation of love.

The memory remains that Troilus will in fact die in battle at the siege of Troy. His death is a foreknown historical event, but it is also foreshadowed by the sustained association of his emotional life with death and dying. Chaucer, however, plays down the possibilities that the characters will actually die in the course of the story. The 'avoided' suicide of the lovers in Book IV is retained from *Filostrato*, but elsewhere Chaucer removes opportunities for suicide while retaining or even increasing the lovers' invocations of death. In Book I Troilus wishes he were 'aryved in the port | Of deth' (i. 526–7, 535–6), but Chaucer then omits the following stanza in *Filostrato*, where Troiolo declares that, if Criseida were to tell him to kill himself, he would do it to please her (*Fil.* i. 56). Chaucer's Troilus tells Pandarus that love so afflicts him 'That streight unto the deth myn herte sailleth' (i. 606), but this replaces Troiolo's admission that love 'afflicts me so much that a thousand times I have come close to taking my life' (*Fil.* ii. 7). Chaucer later adds a stanza in which Troilus rejects suicide as 'unmanhod and a synne' (i. 820–6), and in Book IV Pandarus tells Troilus 'Thenk ek, thi lif to saven artow holde' (iv. 417). After his dream of Criseida lying with a boar, Boccaccio's Troiolo runs to a knife and tries to kill himself until Pandaro overpowers him (*Fil.* vii. 33–9), but, although Chaucer retains Troilus' wish for death here (v. 1271–4), he completely omits the active suicide bid which follows in his source. It is as if Chaucer wishes to take his characters' feelings to an extreme of intensity, an intensity almost suicidal, but in the end always theoretically and figuratively so.

Like so much of the courtly understanding of love in the poem, this pattern of ready association and identification of feelings with dying undergoes a change in perspective as the lovers' fortunes change. When Troilus declares he would rather die than be indiscreet (iii 372 ff.) and swears may his heart be pierced by the spear of Achilles, his unwitting anticipation of his future death at Achilles' hands places in perspective his expressed readiness to die in the cause of love. Criseyde too expresses her readiness to die as a token of her love (iii. 1049, 1502; iv. 771 ff.), but this only contrasts with our foreknowledge of 'how that she forsook hym er she deyde' (i. 56). In the light of the foreknown ending of the story the characters' frequently expressed dedication until death becomes all the more poignant. Troilus may be ready to die rather than let Criseyde be exchanged (iv. 163), but obedience to his lady's wishes reduces him to a kind of living death ('But evere dye and nevere fulli sterve' (iv. 280)). In Book V he continues to imagine his death for love in a way which is romantically at variance with his destined end: he plans his funeral so as to send his lady a 'remembraunce' of himself (v. 309–15), and feels he should die when he sees his lady's empty house (v. 545). For a while the sorrows of Troilus and Criseyde seem equivalent, as he feels he is nearing death unless she returns (v. 641), and she longs for death in the Greek camp (v. 690–1), but little more is heard of Criseyde's readiness to die, whereas the sorrows of Troilus continue to anticipate (v. 1648), and eventually find dissolution in, death.

In Chaucer's *Troilus* the classical idea of Troilus as a type of early death in battle lies behind the exploration of love and death in courtly convention. Troilus defends Criseyde against Cassandra by referring to Alcestis (v. 1527–33), concentrating on the fact that she chose to die herself in order to save her husband from death, but ignoring both her subsequent rescue by Hercules and the mythographers' interpretation of her as representing spiritual courage. The implicit contrast with Criseyde could scarcely be greater, even if Troilus cannot see it, and the idea of a sacrificial death through love anticipates the mention of the sacrifice of Christ's death at the close, which gives a meaning to death for the audience which the development of the theme of death in *Troilus and Criseyde* shows is lacking for the Trojan lovers.

For comparison of *Filostrato* and *Troilus*, see Meech, *Design*, 301–19; see also Durham, 'Love and Death', and Wack, 'Lovesickness'. On the background, see Mary F. Wack, *Lovesickness in the Middle Ages: The 'Viaticum' and its Commentaries* (Philadelphia, 1990).

On Alcestis, see Wetherbee, *Chaucer*, 142, citing Fulgentius, *Mitologiae*, i. 22 (*Opera*, ed. R. Helm (Leipzig, 1898), 33–5), and the Vatican Mythographers, i. 92, ii. 154, iii. 13.3 (*Scriptores rerum mythicarum Latini*, ed. G. H. Bode (Celle, 1834), 31, 128–9, 247–8).

Pitee

> 'For in good herte it mot som routhe impresse,
> To here and see the giltlees in distresse.'   (ii. 1371–2)

The impression of *routhe* or *pitee* on the 'good herte' becomes a theme of growing importance in the course of *Troilus*. When misfortune comes, the narrative in turn invites the *pitee* of the audience towards the characters, and Chaucer presents the later books with such an emphasis on the pitifulness of events and experiences as to produce a sustained study in pathos. At key points, in the first and fourth proems, *Troilus* invokes the Furies, not only as cruel but also as sorrowing and complaining (i. 9; iv. 23), recalling the Furies in the *Consolation of Philosophy* (iii, m. 12) who 'wepyn teeris for pite' at the music Orpheus plays to win back his beloved.

When Chaucer's Troilus first sees Criseyde he exclaims 'O mercy, God' (i. 276), and he is recurrently shown praying to Love for pity and mercy (ii. 523; v. 591). Pandarus is described as motivated by pity for his friend (ii. 564; 1355–6), and it is very much as a concession of pity and mercy that Pandarus persuades Criseyde to show favour to Troilus. The point is not lost on Criseyde, for when Troilus rides past her house she reflects that this knight will die 'But I on hym have mercy and pitee' (ii. 655; cf. ii. 1269–70). If the lady is to grant her 'pitee', the lover must make petition to her for that concession of favour: Troilus does so in his first letter (ii. 1076), the petition for mercy at the house of Deiphebus, and the begging and receiving of mercy on the night of consummation. The lovers' first meeting can occur because Criseyde's earlier need to petition Hector for mercy (i. 112) is now complemented by her apparent need to petition for royal favour against a hostile lawsuit. In both the lovers' meetings in Book III there is some playing on the idea of who exactly is the petitioner, both lovers having prepared themselves to act the part of the humble supplicant to the other. At Deiphebus' house the prince regrets illness prevents him rising 'To knele and do yow honour in som wyse' (iii. 70), and, when Criseyde humbly craves his lordly protection, Troilus is so embarrassed that his own prepared supplication to his lady flies out of his head and he is reduced to murmuring abjectly 'Mercy, mercy, swete herte!' (iii. 98). In the lovers' later meeting Troilus is so distressed by her reproaches of his jealousy that he begs, and receives, Criseyde's mercy (iii. 1173), but, in an unexpected reversal, she also asks Troilus for his forgiveness if she has pained him (iii. 1177–83).

The emphasis on the idea of the lady's *pitee* as what is sought and granted is—beyond mere euphemism—part of Chaucer's thematic enrichment of his story. It starts early, when Pandarus encourages Troilus with the argument that Criseyde's general virtuousness must include the virtue of *pitee* (i. 899–900). To Criseyde Pandarus condemns a beauty that is without the quality of pity ('Wo worth that beaute that is routheles!' (ii. 346)). In the consummation scene itself the *pitee* of the lady towards her lover is strikingly associated with the deeper resonances of traditional religious language of mercy (e.g. iii. 1177, 1267, 1282–3), and Troilus, gazing at Criseyde's face, envisages it as a kind of book in which the text expressing the message of mercy is inscribed but difficult to discover ('Though ther be mercy writen in youre cheere, | God woot, the text ful hard is, soth, to fynde!' (iii. 1356–7)).

This association of compassion in love with religious ideas of mercy is contained within a world of belief where the characters are much disposed to prayer for divine mercy in their difficulties, especially in the later books. As the story progresses from happiness towards sorrow the poem comes to invite its audience increasingly to feel compassion and pity for the characters, just as the petition for *pitee* has earlier been one of the motifs of the lovers' lives. That Troilus' suffering was so extraordinarily pitiful that it would provoke pity from anybody recurs like a refrain (iii. 113–14; iv. 1140–1; v. 260–6, 560). Most famously, the narrator feels such pity for Criseyde in her impossible position in the Greek camp that he 'wolde excuse hire yet for routhe' (v. 1099), and implicitly invites his reader to do so too.

In his reinterpretation of the story in his fourth and fifth books, Chaucer studiedly draws attention to the *pitous* nature of what takes place: Pandarus' pity for the piteous state of each of the lovers (iv. 368, 823–4); the lovers' view of the gods (iv. 789, 846–7, 949, 1174–6); their piteous feeling at their final interview and parting (iv. 1249, 1438; v. 79); Troilus' lamentations after his return to Troy (v. 216, 243, 313, 451); Troilus' reactions to Criseyde's deserted house, a scene of extraordinary pathos as Chaucer presents it (v. 522, 555, 559–60); Criseyde's gazing back from the Greek camp at the skyline of Troy (v. 729). Troilus himself is aware of how his state invites compassion (v. 624–6), though Achilles will show none in killing him (v. 1806). The pity felt at the heart of the love of Troilus and Criseyde within the story is built by Chaucer into the response to that story ('If any drope of pyte in yow be . . . ' (i. 23)), because in that intensity of pathos is the *tragedye*, against which may be set the insistent references in the closing stanzas of *Troilus* to the mercy available in Christ (v. 1861, 1867–8).

For a general study, see Douglas Gray, 'Chaucer and "Pite"', in Mary Salu and Robert T. Farrell (eds.), *J. R. R. Tolkien, Scholar and Storyteller: Essays in Memoriam* (Ithaca, NY, 1979), 173–203.

Secrecy

It is part of the inheritance of romance conventions from *Filostrato* to *Troilus* that in both poems it is felt essential that the love-affair be kept secret from society at large. But comparison of the poems reveals that in *Troilus* Chaucer draws a society that allows much less privacy to his lovers, while at the same time he gives to his characters a much increased sense that the love-affair must remain a secret. In Chaucer's poem the private life of the individual still has to be won from a surrounding society. From this arises the particular sense of tension and difficulty in observing that secrecy traditional to romance that emphasizes the intensity and exclusiveness of inward experience. In *Troilus* the concern to preserve secrecy has an effect in defining the selves of both Troilus and Criseyde; yet secrecy in a society involves pretence, and the affair necessarily becomes implicated with some of the dissimulation that allows it existence. There will also be ironies for the narrator and for the reader or hearer of a narrative which divulges an intensely secret and personal experience.

Chaucer portrays in *Troilus* a society in which the characters lead an accompanied life, waited upon by the attendants who are part of the royal and noble households in which the action takes place. The first thought of Pandarus on waking is to call for a servant to attend him (ii. 71), and, when Troilus retires to his bedroom to grieve, Chaucer describes him as 'allone, | But if it were a man of his or two | The which he bad out faste for to go' (iv. 220–2). When Criseyde strolls in her own garden, she is attended by a company of her women, and, when she goes to dinner with her uncle, she is accompanied by a considerable retinue. If Pandarus invites Criseyde into Troilus' sick-room, he accepts as natural that she should take with her another woman in attendance (ii. 1716–17), just as Criseyde immediately wants to call one of her women when she finds her uncle in her bedroom (iii. 760).

With so little expectation of privacy, a private or even a secret life must be won from the surrounding society, often by dissimulation or deception.

Chaucer's Pandarus encounters two surrounding circles of attendants before he can talk alone with Criseyde (ii. 79–80, 215–17), and, when he returns with Troilus' first letter, he can only contrive the privacy to discuss it on the pretence of having a state secret to confide (ii. 1111–20). He even has to thrust the letter into his niece's bosom (ii. 1155), daring her to throw it away and so attract her ladies' attention, but she calmly retorts that she can wait until they have gone before she discards it (ii. 1158). Chaucer shows his characters susceptible both to the presence of others and to the influence such presence may have in determining their behaviour for appearances' sake.

In such an attended life it is no accident that so many important conversations in *Troilus* take place in bedrooms, for, although the act of retiring itself may be accompanied and marked by a certain ceremony (ii. 911–14; iii. 680–3), it is in bedrooms that relative privacy may be achieved: Pandarus' report to Troilus on his first visit to Criseyde (ii. 946–8); the first meeting of the lovers at Deiphebus' house; several conversations in Book III, including, of course, the consummation scene (e.g. iii. 239 ff., 1555–82, 1583 ff.); Pandarus' visit to Troilus after Criseyde's departure (v. 292–4). The first scene in the poem where the lovers meet entirely alone (in Book IV) is also their last night together before parting.

The difficulty of obtaining privacy is matched by the lovers' concern for secrecy. Pandarus subjects Troilus to a lengthy discourse on the theme (iii. 281 ff.), only just holding himself back from a 'thousand olde stories' of ladies betrayed by indiscretion. The characters' preoccupation with honour and shame is at the root of this, for the unmarried love of Troilus and Criseyde would bring shame upon Criseyde (and upon her uncle, because of his role in the affair), despite both the rank of Troilus and the way that Chaucer has increased the social status of his Criseyde. It remains so, despite the aura of quasi-matrimonial solemnity that pervades the consummation scene, with its exchange of rings and undertakings, and which has invited comparisons with the 'clandestine marriages' of medieval society, in which a private agreement between a couple was at some later stage recognized as marriage. Concern for secrecy is both the condition for, and the characteristic expression of, a private and personal experience more intense than anything in the public, outward world, and is part of the distinctive nature of Troilus and Criseyde as lovers.

The narrative reflects this concern for secrecy by focusing on a sequence of scenes which involve problems of disclosure. In Book I Troilus is turned from one who tries mockingly to detect love in others into a new lover who tries, out of shame, to conceal his change of feelings from his companions, and whose agony at disclosing his lady's name to Pandarus is compared to entering hell (i. 871–2). In Book II Pandarus defers the revelation of the nature of his errand to Criseyde and the identity of her admirer. Both Troilus and Criseyde are introduced in ways that establish their characters in terms of the secret and inward side of their natures, the character they do not show outwardly.

Troilus consistently struggles to conceal his emotions from the surrounding society: he 'gan dissimilen and hide' his woe (i. 322); burnishes up his looks and talk so as to cover his inward distress (i. 327); mocks at other lovers in order to conceal himself (i. 329); and considers both what he will speak and what he will hold back (i. 387). His inner life defines him and sets him apart from others, even from Pandarus, to whom he behaves 'With sobre chere, although his herte pleyde' (i. 1013). The intensity of his private emotions produces several trance-like states, as when he has a kind of waking dream of the Criseyde he saw earlier in the temple (i. 362–7); the force of his idealizing imagination can isolate his inner life for a while from outward circumstances (cf. i. 722–8). The secrecy of love is part of its force, as of the covered fire that burns more fiercely (ii. 538–9). The self-discipline of Troilus in dissembling his true feelings is presented as a kind of virtue, for it safeguards that private world which is most real to the lovers ('From every wight as fer as is the cloude | He was, so wel dissimilen he koude . . .' (iii. 433–4)), and it becomes one of the most important parts of his 'observaunce' and 'service' as a lover. Even his emotions outside Criseyde's deserted house (v. 537–9, 551–2, 556), or his lamentations in her absence (v. 635 ff.), are carefully concealed, in a reflection of the special inwardness of his feelings and his obedient respect for his lady.

For Criseyde, the importance of secrecy is part of her concern with her reputation and her recurrent anxiety over what people may say or think. Pandarus has to assure her of the secrecy of the affair ('And wre yow in that mantel evere moo . . .' (ii. 380)), and she is pleased and reassured when she notices the discretion of Troilus (iii. 477–80). So important is secrecy to her honour that Criseyde would rather leave Troy than endure the disclosure of the affair. In opposing disclosure, Boccaccio's Criseida made the openly sensual argument that furtive love gave more pleasure and excitement (iv. 153); Criseyde's concern for secrecy remains an expression of her fear of harm to her name. In *Troilus* secrecy does allow sensual fulfilment, but it is argued for only in terms of its social necessity in winning personal happiness by deception from a disapproving society (ii. 1743; 'Now al is wel, for al the world is blynd | In this matere . . .' (iii. 528–9)).

The secret love of Troilus and Criseyde is given all the elaborate furtiveness of the great adulterous affairs of romance, yet the moral transgression represented by this love-affair between two unmarried pagans must be of a different order, and the second proem perhaps acknowledges that the poem presents a distinctive version of secrecy (ii. 38–40). The trappings and anxieties of a secret affair are some of the conventions of *fin' amors* that Chaucer explores and tests. Its secrecy is both the strength and weakness of the affair—its intensity and its vulnerability—for what is not acknowledged to exist cannot be taken account of when Criseyde's exchange becomes an issue. Such paradoxes in the secrecy of the love of Troilus and Criseyde are also built into the ambiguities of the narrative process. A poem in which the characters are so preoccupied with a secret experience prompts an awareness in readers or hearers that they are party to a disclosure of something intimate and personal. We watch Troilus confess what we already know to Pandarus; we watch Pandarus toy with Criseyde's curiosity about a secret love we already know of; we share Criseyde's pleasure in her secret knowledge about what will cure Troilus (ii. 1581–2, 1590–1), but we also know the secret —as she apparently does not—of how she is to be deceived into meeting Troilus alone. The narrative enables us to overhear Criseyde in her private thoughts, when she acknowledges to herself that she is the most beautiful woman in Troy but adds 'Al wolde I that noon wiste of this thought' (ii. 745). In a tête-à-tête in their shared bedroom Pandarus urges Troilus to keep the affair secret, explaining

> 'I sey nought this for no mistrust of yow,
> Ne for no wis-man, but for foles nyce ...'   (iii. 323–4)

Yet any 'foles nyce' who may overhear this private conversation can only be among the audience of the poem. The arrangements for the consummation are a secret from everyone at Troy, yet shared with every reader or hearer of the poem, who is drawn in, much as an accomplice, awaiting the last secret details to be finalized:

> Now al is wel, for al the world is blynd
> In this matere, bothe fremde and tame.
> This tymbur is al redy up to frame;
> Us lakketh nought but that we witen wolde
> A certeyn houre, in which she comen sholde.   (iii. 528–32)

There is something ironic in the public performance of a poem about an experience so essentially private that it depends on secrecy for its continuance, especially in the foreknowledge that what once was secret is now a well-known story. Indeed, the one moment of intimacy Troilus and Criseyde are forced to show in public suffices to give them away to Diomede (v. 80–90) and justifies Criseyde's fears: once she is known to have a lover, she is cast as a loose woman. It becomes a privilege and a responsibility to participate as reader or hearer in the secret which the narrative discloses, and this is one of the many devices by which attention is drawn in *Troilus* to question the usual assumptions and procedures of narrative and its reception and appraisal.

On secrecy in *Troilus*, see Windeatt, ‘“Love That Oughte Ben Secree”’. See also Kelly, ‘Marriage’, and *Love and Marriage*; Maguire, ‘Clandestine Marriage’; Wentersdorf, ‘Observations’.

Honour

The story of Criseyde as Chaucer knew it from Benoît, Guido, and Boccaccio is the tale of a woman who loses her good name. Chaucer accepts this established ending, yet in his version of the story he creates his character of Criseyde as a woman concerned above all to keep her good name and her honour, whose name comes close to her identity. Reputation bears on the characters’ understanding of themselves and of their freedom of action, and also prompts questions about the relation between the subject and the form of the poem.

Criseyde has to maintain her honour in Troy as best she can, despite ‘hire fadres shame’ and treason (i. 107, 128–31), and Troilus first discerns in her ‘Honour, estat, and wommanly noblesse’ (i. 287). A concern with honour is much to the point when Troilus reveals the name of his lady and Pandarus plans the winning of his kinswoman for his friend, while commending her reputation (i. 880–1) and concern for honour:

> ‘In honour, to as fer as she may strecche,
> A kynges herte semeth by hyrs a wrecche.’   (i. 888–9)

Her disposition towards honour is a recommendation to her would-be lover and not, as in *Filostrato*, something inconvenient to Troiolo’s sexual aspirations (*Fil.* ii. 23). Pandarus emphasizes his concern for Criseyde’s good name both to Troilus and to herself (i. 902–3; ii. 295), yet his protestations are inevitably at variance with the fundamentally dishonourable nature of his intentions as a go-between, and he accurately analyses his own role in attempting to distance himself from it:

> ‘I am thyn em; the shame were to me,
> As wel as the, if that I sholde assente
> Thorugh myn abet that he thyn honour shente.’   (ii. 355–7)

After Criseyde has seen Troilus ride past in triumph, the honour to herself of an honourable connection with the prince lingers in her thoughts:

> ‘It were honour with pley and with gladnesse
> In honestee with swich a lord to deele,
> For myn estat, and also for his heele.’   (ii. 705–7)

For Criseyde the worst thing that can be imagined is that people would guess at the affair, and in her soliloquy she persuades herself of a satisfactory accommodation of love with honour: if she loves Troilus 'and kepe alwey myn honour and my name, | By alle right, it may do me no shame' (ii. 762–3). This is psychologically apt, for it catches the way excitement stimulates a boldness not quite consistent with the character more largely: Criseyde rather overlooks how sensitive she will prove to the thought of what people will say. At Deiphebus' house Pandarus urges Criseyde with an idealized description of Troilus' motives ('this man wol nothing yerne | But youre honour . . .' (iii. 152–3)), and Criseyde, with the proviso 'Myn honour sauf' (iii. 159), accepts Troilus' service. The lofty and honourable language of the lovers' meeting is pointedly followed by the acknowledgement of the disreputable role of the pander Pandarus and the reliance of Criseyde's continued honour upon the discretion of Troilus. A lover's indiscreet boast will be his lady's dishonour, and Pandarus lays extraordinary emphasis on the present reputation of Criseyde in Troy, which rests on the fact that simply no one knows of any misdeed on her part (iii. 267–70). It is in part her conviction of the utter secrecy of the occasion which persuades Criseyde to receive Troilus in her bedroom, although she is still concerned to remind Pandarus to balance the safeguarding of her reputation against the satisfaction of Troilus ('So werketh now in so discret a wise | That I honour may have, and he plesaunce' (iii, 943–4)).

In the discretion of Troilus' loyal 'service' there is no danger to Criseyde's honour, while the influence of love on the hero's way of life leads to an increase in his honour (iii. 1724). But once Criseyde's exchange is agreed, Troilus seems to equate his fall from fortune as a lover with a kind of fall from honour (iv. 270–2), while the whole question of whether the lovers should accept the exchange or take action to evade it turns on the issue of Criseyde's reputation. Troilus 'than thoughte he thynges tweye: | First, how to save hire honour' and then how to withstand the exchange (iv. 158–60). Abduction 'mooste be disclaundre to hire name. | And me were levere ded than hire diffame' (iv. 564–5); and it becomes impossible both to evade the exchange and to maintain the lady's honour, Troilus stressing that Criseyde's honour is more important than his own life (iv. 566–71). The turning point of the story is made to hinge on the value Criseyde places upon what society will think of her, her 'name' and hence her social identity. The irony is that by keeping her name Criseyde is to lose it: she preserves her name for a while by not eloping with Troilus, only to lose her reputation for ever in the 'shame' of what is to happen in the Greek camp. In the lovers' last private conversation Criseyde tries to persuade Troilus to accept her going by arguments about honour: ten days is a short time to wait in order to preserve her good name (iv. 1328–9); and she points out the harm to his reputation as well as her own in the discovery of their love and any possible elopement (iv. 1560–75). Yet there is little indication that Criseyde has internalized her conception of honour: she no longer possesses honour in the sense of chastity, but nor is she to imitate the honour which a married woman keeps through fidelity to her husband. Instead, her idea of honour resides in preserving the outward semblance of chastity, and into this there enters a concern for appearances. Moreover, she assures Troilus at the close of Book IV that—in recompense for his faithfulness to her—she will act in such a way as will redound to her own honour (iv. 1665–6).

That Chaucer goes back to Benoît for Criseyde's soliloquy in the Greek camp regretting her loss of reputation (v. 1054 ff.; omitted in *Filostrato*) suggests how concerned he was to focus on that moment of awareness in which the heroine sees that, by trying to preserve her good name, she has not only lost that name but will become a byword for shame. It is also Chaucer who adds the pathetic irony of the faithful Troilus' continuing to wish in his letter for the increasing of Criseyde's honour (v. 1359–61), and his late exclamation of regret at the loss of her name (v. 1686–7). The wisdom of Chaucer's interpretation of the character of Criseyde and her concern for honour lies in his suggestion both of the constraints in depending on the opinion others have of us, and also of the inevitability that identity is to some degree related to a sense of what others believe. This is ironically conveyed by the very existence of this poem and its predecessors as the story of Criseyde. These have perpetuated the memory of her shame, until Chaucer—by attempting to understand sympathetically Criseyde's concern for her name—explores the limits of the concept of honour that Troilus and Criseyde have.

On honour in Chaucer, including *Troilus*, see Brewer, 'Honour'.

Trouthe

> 'Thow shalt be saved by thi feyth, in trouthe.'   (ii. 1503)

To the traditional story of Criseyde as a faithless woman who betrays the love of Troilus, the theme of loyalty, of trust and *trouthe*, is central. But this story of personal and emotional betrayal is set in the context of political and historical instances of betrayal in the Trojan War, a conflict itself caused by infidelity, while the pagan background allows for many references to mythological stories which turn upon some betrayal. In the background of the story from Benoît to Boccaccio is the treachery of Criseyde's father, who betrays his own people, and it is this betrayal of trust which is stressed at the opening of Chaucer's poem (i. 87, 89, 107, 117). Yet Calchas betrays his *trouthe* because he prophesies truly what will happen to Troy. In the background as well, becoming more prominent later in Chaucer's version, is Antenor's betrayal of Troy. In taking over this traditional story Chaucer adapts and adds in a way that brings sharply into focus the way the received story turns on issues of *trouthe* between the characters.

Chaucer structures his version of the story so that faith, trust, and loyalty are strongly identified with the early and successful love between Troilus and Criseyde, and it is against this established sense of *trouthe* that the older story of Criseyde's failure in *trouthe* may more poignantly be set. The shape of the story, its moments of transition and climax, are structured by oaths and promises. The whole texture of conversation, throughout the poem, is intensified by oaths such as 'by my trouthe' (e.g. i. 676, 770, 831, 906, 995), pervasively inviting and challenging a trial of fidelity and constancy in actions and intentions. From Pandarus, such ambiguous claims serve as a reminder of how any understanding that *trouthe* is at the heart of friendship as well as of love (i. 584–7) is continuously tested by the untruthfulness and shifty stratagems through which he puts his interpretation of loyalty into effect in winning his niece for his friend. Each step in *Troilus* is taken with an alertness to the keeping of faith: Book I concludes with Pandarus' pledging of faith as he undertakes his friend's suit (i. 1054, 1061); Pandarus is anxious that Criseyde will keep her promises (ii. 493); and Troilus' first letter to Criseyde assures her of his constant fidelity (ii. 1084).

Against this background of promised faith the two encounters between the lovers presented in Book III are to turn on questions of *trouthe*. The characters evidently look back on their earlier meeting at Deiphebus' house as representing a plighting of troth between them (iii. 782). The commitments which the lovers make to each other at this first meeting make sustained reference to what they are promising in terms of *trouthe*, acting as *trewe* people *trewely* (iii. 133, 141–2, 162–4). This is what Criseyde has been led to expect, for Pandarus has been concerned to present Troilus to her by identifying him with the quality of *trouthe* (ii. 159–61, 331, 339). 'Moral vertu, grounded upon trouthe' is hence an important part of the whole moral character of Troilus with which Criseyde later claims to have fallen in love (iv. 1672–3), echoing the resonant language in which the female composer of Antigone's song addresses her lover ('trewe in myn entente . . . Of trouthe grownd . . .' (ii. 828, 842)).

Yet the demands of secrecy and honour involve dissimulation towards others, and the consummation of their love can only be achieved for Troilus and Criseyde through deception and manipulation by Pandarus, who is repeatedly untruthful in order loyally to promote his friend's interest as he sees it. He arrives on his first visit to his niece to find her listening to a Theban story of betrayal in the fate of Amphiorax, a soothsayer like Criseyde's father (ii. 64–70, 103–5). He is conscious that he may have betrayed his niece's trust ('And I hire em, and traitour eke yfeere!' (iii. 273)), yet he curses himself to the fate of Tantalus (iii. 593)— who served a treacherous meal to the gods—just when he is inviting Criseyde to a supper party where entertainments include the tale of Wade (iii. 614), which may allude to the winning of a woman by deception.

It is by a deliberately untrue accusation of unfaithfulness that Pandarus applies pressure to Criseyde to give proof of her *trouthe* to the jealous Troilus. Aghast at the suggestion that she would ever 'falsen Troilus' (iii. 806), her initial idea is to send him a blue ring, traditional token of constancy; she goes on to give Troilus a resounding reassurance of her fidelity, avowing that it was because of his quality of *trouthe* that she first took pity on him (iii. 992), and assuring him of her true heart (iii. 1001). Troilus receives forgiveness for a crime he did not commit but must go along with the pretence of having committed. When Troilus faints, Criseyde's reaction is to swear fidelity in order to revive him (iii. 1111); her final acceptance of him as a lover is identified with her acceptance of his *trouthe* (iii. 1227–9), and he in return pledges his own committed faith (iii. 1297–8). It is thus by means of untruthfulness and deception that Criseyde is persuaded to accept her lover's integrity, but it is part of the poem's irony, or wisdom, to grant us a different perspective from Criseyde and then to leave to the reader the question of how to weigh the means against the end.

The known ending of the story in Criseyde's future unfaithfulness lends a particularly poignant irony to the concern with faithfulness of both the lovers. Both in word and deed Chaucer's Troilus is a truer lover than his prototypes, and he is certainly made to express his sense of faithfulness more often and more resonantly in the course of a love-affair only famous for the way that fidelity is eventually not returned. There is a special poignancy in the anxiety of Criseyde that Troilus be true to her, as she bids farewell to him after their first night together with the urgent entreaty 'Beth to me trewe, or ellis were it routhe, | For I am thyn, by God and by my trouthe!' (iii. 1511–12).

It is against this much fuller presentation of the characters' identification with *trouthe* in Chaucer's version of the ascending action that the traditional story-shape of Criseyde's enforced departure from Troy and her failing allegiance to Troilus can be set and contrasted. The stream of references to Antenor which Chaucer adds serve to bring into association with the decision over Criseyde's exchange an awareness of Antenor's future betrayal of Troy (iv. 149, 177, 189, 196, 202–5, 212, 347, 378, 665, 792, 878, 1315). Chaucer never asserts that exchange implies the inter-changeability of Criseyde and Antenor, but he insistently pairs her with the future traitor for whom she is to be exchanged, and recalls the treason of her father Calchas (iv. 663–5; 761), cursed by Troilus in the same scene where he prays for the lasting constancy of lovers (iv. 325, 330).

The lovers' concern with *trouthe* is most markedly developed in their final dialogue in Book IV, but it is ironically set against the reader's foreknowledge of the *untrouthe* which is to come. Criseyde's plan is to keep faith with Troilus through deceiving her father the seer, but this stratagem for deception proves a self-deception: she will betray Troilus after Diomede convinces her of the truth of what her traitorous father foresaw. In Book IV the narrating voice is moved to emphasize the sincerity of Criseyde's commitments at the time she utters them (iv. 1415–21), and Troilus casts himself as a humble and faithful lover in urging Criseyde to be true (e.g. iv. 1439, 1477, 1490–1, 1499). In complement to this, Chaucer allows Criseyde a full expression of her intended fidelity: she calls down punishments on herself if she be false to Troilus, cursing herself to the fate of Athamas, mentioned by Dante among the *falsatori* (iv. 1534 ff., 1551–4), insisting that she will be true (iv. 1609–10, 1616–17), and perceiving 'alle trouthe' in Troilus (iv. 1646–9). It is fitting that Chaucer in this way allows his Troilus and Criseyde an added mutual affirmation of fidelity at the close of this interview: Troilus is made to declare that he will never be false (iv. 1654–7); Criseyde sees herself as acting in response to and recognition of the quality of *trouthe* in her lover ('so trewe I have yow founde . . .' (iv. 1665)), although there is poignancy in the way Criseyde can at once dread unfaithfulness in her lover (iv. 1646–9) yet promise an unchanging constancy in herself (iv. 1681–2).

Against the memory of such forcefully expressed commitments and intentions, the failure of *trouthe* in Book V unfolds all the more painfully. Troilus must eventually recognize the limits of loyalty in the object of his own great loyalty in a way that suggests how human love must comprehend and transcend betrayal, in a world where Criseyde's *untrouthe* will seem the prudence of a survivor, and the *trouthe* of Troilus will seem foolishness. The description of Troilus brought into this book by Chaucer's borrowing from Joseph of Exeter sees him as 'Trewe as stiel in ech condicioun' (v. 831), and, in contrast to the constancy—even fixity—of the hero, the faltering *trouthe* of the heroine is only reluctantly admitted. As she herself implies in the Greek camp, her inner constancy is at fatal variance with her circumstances ('I nam but lost, al be myn herte trewe' (v. 706)). It is the loss of her reputation for *trouthe* which Criseyde laments first 'whan that she falsed Troilus' (v. 1053–5), but she resolves instead to be true to

Diomede and praises Troilus for his fidelity ('the gentileste . . . to serven feythfully' (v. 1075–6)). In the letters exchanged between the separated lovers the theme of *trouthe* is also addressed in a way painfully at variance with the march of events, as Troilus calls on Criseyde by their mutual sense of *trouthe* ('Bisechyng hire that sithen he was trewe, | That she wol come ayeyn and holde hire trouthe' (v. 1585–6)), and, when Criseyde's disingenuous letter comes, it still claims to see Troilus as an embodiment of *trouthe* (v. 1616–17). For Troilus, Cassandra's prophecy is 'false' because he cannot accept the truth it tells about Criseyde's *untrouthe*.

The final discovery of Criseyde's *untrouthe* through the brooch on Diomede's 'cote-armure' is marked in Chaucer's poem by the strongest sense of a betrayed faith, and a sense that this specific betrayal is representative of a larger pattern of unfaithfulness in human life. 'Wo bygon ben hertes trewe!' as Pandarus exclaims (iii. 117), and there is throughout the poem a sense of wishing in vain that faithfulness should be rewarded for its sorrows (iii. 1386; v. 1435). Yet perhaps *trouthe* is not really part of this Trojan world, where Helen of Troy can swear with unwitting irony 'If that I may, and alle folk be trewe!' (ii. 1610), and where Troilus calculates the time of Criseyde's return, exclaiming too optimistically, 'I shal be glad, if al the world be trewe!' (v. 651). Long before Criseyde had expressed her distrust of this world ('This false world— allas!—who may it leve?' (ii. 420)), and, once Criseyde has given up her *trouthe*, she fades as a presence in the poem. The *trouthe* of Troilus remains one of his most characterizing qualities, and he is never so unconvincing and so unlike himself as when he must invent a story so as to evade Criseyde's charge that he has deceived her ('And for the lasse harm, he moste feyne' (iii. 1158)). That quality of *trouthe* which sets Chaucer's unworldly Troilus somewhat apart from the world is what eventually earns him translation out of this world and beyond the sphere of change. It is in Chaucer's conclusion to *Troilus* that the thematic patterns of interest in *trouthe* in the poem are fulfilled: the graceful expression of a preference to write about 'Penelopeës trouthe'(v. 1778); the delightful turning round of the theme of the faithless woman, in order to protest that the story has been told 'moost for wommen that bitraised be | Thorough false folk' (v. 1780–1); and, most of all, the recommendation to set aside 'feynede' loves and to love Christ 'For he nyl falsen no wight, dar I seye' (v. 1845). This is that constant and supreme *trouthe*—beyond time and change—against which all others are to be measured.

On aspects of *trouthe*, see Bolton, 'Treason'; Lockhart, 'Semantic, Moral, and Aesthetic Degeneration'; Newman, '"Feynede Loves"'; and Strohm, *Social Chaucer*, 102 ff.

Time and Change

The plot of *Troilus* turns on a change of heart that is set in motion by an exchange of two betrayers, Criseyde for Antenor. The structure of 'double sorwe . . . Fro wo to wele, and after out of joie' (i. 1–4), establishes from the outset an expectation of mutability (cf. ii. 22–3), and provides Chaucer with a subject through which he could explore the themes of mutability, of time and change. Benoît and his successors had interpreted the story as exemplifyng a sad mutability in the fickleness of woman and in changeable fortune; Boccaccio had preceded this story of change with the story of how that love is achieved. Chaucer's poem realizes the potential for a thematic treatment of change in the work of both his predecessors: he charts much more fully and carefully those changes by which the lovers move and develop towards love, and he in turn pays more understanding attention to the processes of how love changes and passes, and this greater sense of moving through change in the poem registers mutability as one of its thematic concerns. That care which Chaucer shows throughout for the time sequence and structure of his poem establishes both an outward and an inward apprehension of time which become increasingly divergent as the force of mutability impinges on the action.

At the outset the very familiarity of the story—of knowing the ending before the beginning—establishes a cyclical sense of time and change. When good fortune and happiness depart, the mutability of the world is lamented, but the characters are aware of mutability in better times as well. It is the achievement of the poem to catch up its reader for a while in the changes in the lovers, and in their establishment of a personal sense of their own time distinct from other time, so that the foreknown changes give new and particular force to the old truths about the mutability of all worldly joy. At the very start, the falling in love of Troilus, the scoffer and scorner at love, is perceived as a process of miraculous change (i. 308). Love alters the moral character of Troilus for the better; Pandarus, by contrast, uses the *carpe diem* argument against Criseyde, that beauty is in continuous change and decline under the impact of time and age (ii. 393 ff.), and she fears the essential changefulness of the lover's life (ii. 756, 772–3, 786–91). Her own transition is signalled through the account of her dream of the white eagle and the exchange of hearts.

Chaucer also insists on those spaces where change cannot be represented within the text. The moment when Criseyde first sees Troilus after being told of his love for her is protested to be no more than a marker at the beginning of a lengthy 'proces' (ii. 666–79); the transition when Criseyde inwardly yields herself to Troilus is stressed to precede—not coincide with—her moment of physical yielding, but it is not presented within the narrative (iii. 1210–11). The reverse process in which Criseyde's affections change away from Troilus is declared to be difficult to set down (v. 1086–90). The bewildering speed of Troilus' 'conversion' in Book I, and the gradual, tentative development of Criseyde towards love in Book II, establish contrasts of pace and process in change which continue throughout the poem. Perception of change as welcome or unwelcome reflects point of view.

Once Troilus and Criseyde are brought to an understanding, they begin to speak in terms of stability and permanence, in a way that reflects their longing for the unchangeable. This is what Antigone's song to Love had foreshadowed, with its coupling of happiness and security in the lover's life (ii. 833) and its praise of the lover as 'stoon of sikernesse' (ii. 843). Criseyde feels Troilus is 'to hire a wal | Of stiel' (iii. 479–80; cf. 982), and the imagery of reassurance and confidence which accompanies the lovers' union suggests the gaining of a security, a centre of rest and peace at the heart of the poem, which is immune to change as the lovers perceive it (iii. 1513, 1678–80). Despite the traditional associations of love with turmoil (e.g. iv. 581), both Troilus and Criseyde look forward to 'resting' their heart upon the other (ii. 760–1, 1325–6); before and after the consummation they hope for rest of heart and soul (iii. 925–6, 966, 1045, 1131, 1279–81, 1518, 1599); Criseyde hails Troilus as 'my pees' (iii. 1309); and the third book is brought to a close with images of Troilus and Criseyde united in rest and peace (iii. 1680, 1819–20).

In their shared private world the lovers think of their feelings as outside normal time and change, to endure unchanged until death:

> And he to ben hire man while he may dure.   (i. 468)
>
>   'for I, whil that my lyf may laste,
> More than myself wol love yow to my laste;'   (i. 536–7)
>
> 'Whos I am al, and shal, tyl that I deye.'   (iii. 1607)
>
> She wol ben his, while that hire lif may laste.   (iv. 677)
>
> 'That I was youre, and shal while I may dure.'   (iv. 1680)

Yet, in a poem preoccupied with enduringness, there is also a series of allusions to origins, birth, and intervening time, so that perspectives in time stretch both backwards and forwards from the present in *Troilus*, creating a sense of process of time and the context of the action within it. The language of Chaucer's lovers expresses value and feeling in terms of time—their speeches are full of 'evere' and 'nevere'—and Troilus prays that other lovers may enjoy a longer love than he could (iv. 326). With the prospect of separation in this life the lovers' minds both leap forward to project an eternity of everlasting lamentation beyond life (iv. 475, 786–7). The rhetoric through which Troilus and Criseyde express their mutual commitment is thus continuously affirming their sense of enduringness as a supreme value, and thereby implicitly denying and resisting the possibility of change.

Chaucer's lovers can both identify with unchanging permanence and duration and be ready to wish the heavens to change from their regular pace and timing in order to serve their love. In her *aubade* Criseyde reproaches night for not remaining with them for three times the length of a normal night (iii. 1427–8); Troilus reproaches the sun for rising (iii. 1469), feels the normal pace of the heavenly bodies has been changed to his detriment (iii. 1705; v. 663–5), and prays that they move differently (v. 655–6). By its sheer extravagance such rhetoric can make its thematic point about mutability, for the union of lovers first made possible by a conjunction of Saturn (traditionally representing Time) with the moon (representing Change) will in due course be destroyed by time and change. The very impossibility of the lovers' wishes to bring the heavens into conformity with their own conception of an unchanging love points to how their love exists within a larger world of inevitable mutability to which it will be subject. This is what Troilus cannot yet see when, in his Boethian song at the close of Book III, he prays that his own love be associated with that larger pattern of permanence and regularity which governs the natural order. Chaucer's Troilus and Criseyde are to learn that the world changes their love rather than being changed by it.

The worldly political deal which is Criseyde's exchange forces upon the lovers an outward change of circumstances, and much of the poem's remaining action is a struggle to preserve an inner unchangingness against outward change. Dame Fortune changes the expression on her countenance, just as the lovers' faces change once they hear of the change in their fortunes (iv. 150, 864–5). The rhetoric of conventional wisdom in the characters' lamentations on the mutability of fortune and of all worldly bliss is offset by their emotional resistance to it in their attempts to defy change. In Book I Pandarus tries to cheer Troilus by arguing that the mutability of Fortune means happiness must succeed unhappiness (i. 851–2), but after the consummation he warns him that 'worldly joie halt nought but by a wir. | That preveth wel, it brest al day so ofte' (iii. 1636–7).

The attempts to rehearse arguments for accepting mutability allow the English Troilus to state all the more forcefully his own determination never to change ('Thow moost me first transmewen in a ston . . .' (iv. 467)), while in her own two paired outbursts on earthly happiness in Books III and IV Criseyde is much preoccupied with the mutability and instability of the 'brotel wele of mannes joie unstable', expressing a keen sense of the change that has occurred between beginnings and endings, and greeting Pandarus as originator of joys 'That now transmewed ben in cruel wo' (iv.

828–30). She sees the mutability of love as part of the larger mutability of all happiness in this world:

> 'Endeth than love in wo? Ye, or men lieth,
> And alle worldly blisse, as thynketh me.'   (iv. 834–5)

There is a poignant distance—eventually to be closed—between the lovers' rhetoric of immutable devotion, or the language of idealization which attempts to step outside time, and the encompassing world of movement, flux, and changing circumstances. The painful process of the fifth book shows that change does come, and that Troilus must at last recognize this. Every sort of change is brought before us: Criseyde's changed countenance (v. 243–5); the deserted palace, now empty of all it once contained (v. 540–53); Troilus' 'chaunged face' (v. 555); Criseyde's joy 'now al torned into galle' (v. 732); Troilus' changed state ('But torned is ... Everich joie or ese in his contrarie' (v. 1378–9)).

After the lovers' parting there is an emphasis on living through memories: the heart's ashes or the brooch as a 'remembraunce' (v. 309–15, 1661–3); the 'ubi sunt' lament over the empty palace (v. 540–53); the lover's haunting of the places of Troy (v. 561–81); Criseyde 'purtraynge' and 'recordynge' in the Greek camp (v. 716–21). But all these are instances of what Pandarus earlier called the worst kind of misfortune, 'A man to han ben in prosperitee, | And it remembren whan it passed is' (iii. 1627–8). Criseyde may try to stand out against the flux of events (v. 762–3), but Troilus must eventually acknowledge the onset of change (v. 1634) and 'hire hertes variaunce' (v. 1670), exclaiming:

> 'Allas, I nevere wolde han wend, er this,
> That ye, Criseyde, koude han chaunged so.'   (v. 1682–3)

It is as a closure to the themes of time and change that the ending of the poem is so satisfying thematically and aesthetically. The very protracted-ness of the sequence through which Troilus comes to accept the mutability of Criseyde's love has held out the possibility that change may not occur, but at the end Troilus in death passes beyond the sublunary world of change into a realm of stability and permanence, beyond the brittleness of this world. Yet, if the hero's spirit is released from the realm of mutability, the story of his love perhaps fittingly remains within it, as the poem worries about the unwelcome corruptive influence of linguistic change but also submits itself to correction and criticism.

See further the studies by Ganim, 'Tone and Time' and *Style and Consciousness*, and
   Mogan, *Chaucer*. On Saturn and the moon as representing time and change, see Mann,
   'Chance and Destiny', 86.

Past, Present, and Future

> 'To late is now to speke of that matere.
> Prudence, allas, oon of thyne eyen thre
> Me lakked alwey, er that I come here!
> On tyme ypassed wel remembred me,
> And present tyme ek koud ich wel ise,
> But future tyme, er I was in the snare,
> Koude I nat sen; that causeth now my care.'   (v. 743–9)

With this anguished acknowledgement Chaucer's Criseyde makes what is the first recorded use of the word *future* in English after that in Chaucer's *Boece* translation, and such innovative diction points to how *Troilus* is concerned to explore human ways of perceiving the past and future in relation to the present. Chaucer's is a treatment of the story of Troilus' love, realized within a sense of the past and of the historical process involved in interpreting old stories: it reconstructs a strong sense of a past but in so doing makes the problem of reconstruction part of the poem and its debate about love.

Troilus consistently shows a historically minded attempt to engage with the 'otherness' of a particular non-Christian past through mention of pagan custom, and by attempting to re-create a sense of the pagan characters' universe of belief. Much detail in *Troilus* is unified by this historical concern for noting pagan practices: Calchas' techniques of divination (i. 64–77; iv. 113–16); Troilus' alibi 'That he was gon to don his sacrifise' (iii. 539) at Apollo's temple on the night of the consummation; his fatalistic soliloquy in Book IV, aptly voiced in a pagan temple; his instructions for his cremation and his belief in augury (v. 295–320). Pagan polytheistic belief becomes part of the texture of the poem through the characters' allusions in oaths and prayers to the ancient gods, especially those particularly applicable to their current situation. Troilus prays to Minerva when about to compose his first letter (ii. 1062), to Venus for his suit to Criseyde (i. 1014), and to Mercury, guide of dead souls, to direct his own soul after death (v. 321–2). Venus is invoked by Pandarus (ii. 1524) and twice by Criseyde (iv. 1216, 1661). Chaucer strives to imagine his way back through Christian forms to how pagans might articulate their beliefs, as when characters pray for the 'grace' of Jupiter or Juno (iv. 1116–17, 1683–4), swear by their hope to see the face of Jove (iv. 1337), or pray to Juno by giving her the Virgin Mary's title of Queen of Heaven (iv. 1594). Yet at other times there is a more ambiguous juxtaposition of Christian and pagan terms:

> 'Immortal god,' quod he, 'that mayst nought deyen,
> Cupide I mene, of this mayst glorifie;'   (iii. 185–6)

'And ech of yow ese otheres sorwes smerte,
For love of God! And Venus, I the herye;
For soone hope I we shul ben alle merye.'   (iii. 950–2)

The characters who seem so 'present' through the vivacity of the narrative are still so 'past' in terms of their paganism.

That the scribes of some *Troilus* manuscripts are moved to add explanatory commentary to such allusions in the manuscript margins is a token of how successfully the poem established historical atmosphere which was felt to need annotation. Often, however, Chaucer tactfully embeds classical allusion within its own gloss. There is that very obliging tendency of Chaucer's characters to swear oaths by classical figures and then helpfully to gloss them in the same breath, or to pray—as it were—with footnotes. We are told that Neptune 'god is of the see' (ii. 443), that satyrs and fauns 'halve goddes ben of wildernesse' (iv. 1545), that the Manes 'goddes ben of peyne' (v. 892), and that Phlegethon is 'the fery flood of helle' (iii. 1600). Troilus directs that at his funeral the ash of his heart be collected 'In a vessell that men clepeth an urne, | Of gold ...' (v. 311–12), and, even when lamenting by herself, Criseyde explains her own reference to 'the feld of pite, out of peyne, | That highte Elisos ...' (iv. 789–90).

Chaucer also seeks to locate the action in the past by making his characters encounter as recent or modern what Chaucer's contemporaries (and ourselves) would regard as ancient. When Pandarus chattily refers Troilus to what Oenone wrote in her despondency ('Yee say the lettre that she wrot, I gesse?' (i. 656)), what to Chaucer's audience was an ancient text—Oenone's epistle to Paris in Ovid's *Heroides*—is mentioned as a contemporary work. Despite the mischievous *trompe-l'œil* in a Trojan's being supposedly able to read the Roman poet Ovid, Chaucer here takes his readers back inside the past by changing the point in time from which the landmark of a classic text is viewed. A similar play with historical perspective is achieved when Pandarus refers to the metamorphosed form of Queen Nyobe as something which can still be inspected (i. 700); or when Chaucer's characters refer to the sufferings of mythological figures as still going on as they speak (like those of Tityus (i. 787)). Many of the mythological figures mentioned in the course of the poem not only evoke a sense of past time but are manifest in the present through a process of metamorphosis which has made them part of the natural world: Procne and Philomela, Myrrha, Ascaphilus, and Scylla, or the loves of the gods which touch the lives of Adonis, Europa, Daphne, or Aglauros. The horizons of the story now extend further, both stretching backwards into a sense of a past that peoples the natural world of birds and stones and trees with unhappy histories of their metamorphosis, but also stretching forward into a pagan hell, where the lovers feel themselves awaited beyond death by Proserpina (iv. 473), by Minos (iv. 1188), and by the Manes (v. 892).

Chaucer gives definition to the historical period of the present action by having his Trojan characters refer to what for them are not so distantly past events, such as the earlier history of Troy, or the history of Thebes (ii. 100–5; iv. 120–6, 300, 547–8, 1538–40; v. 599–602, 936–8). Both past and present are taken up with war, and in developing Cassandra's part Chaucer shows her practising her traditional role as seer through her knowledge as a historian, interpreting Troilus' dream of the boar in the light of Diomede's connection with the boar in 'a fewe of olde stories' (v. 1459) of Theban history, and so bringing into *Troilus* the model and example of a practising historian within Trojan society. There has been considerable debate over the origins of Chaucer's own historical awareness, and its purpose. It has been seen to derive from Chaucer's encounter with the essentially Renaissance attitudes to the past in the Italian literature he had read, as well as from that interest in the pagan past shown by the 'classicizing friars' of his own century. But, however Chaucer's interest arose and was educated, he uses history in *Troilus* as a poet more than as an antiquarian; to explore the double perspective of pagan and Christian, past and present. The second proem is a strategic device which introduces the question of historical change over time, anticipating an audience puzzled over signs of cultural difference and changed mores in its depiction of an ancient love-affair. Yet no such puzzlement would actually have arisen unless invited. The love-affair is presented in what, for Chaucer's contemporaries, would probably have seemed perfectly modern, contemporary terms. The narrative appears—as far as can now be known—to reflect life as led in a medieval courtly society, albeit no doubt an enhanced literary projection of such a life. Here is an account of utmost joy and sorrow in love, but it is the experience long ago of a pagan hero, perhaps to be seen as a virtuous pagan, yet excluded from the understanding of love available to Chaucer's Christian contemporaries. The very otherness or 'alterity' of the Trojan society—the alterity of its ultimate beliefs and values—enables a searching consideration of the essential nature, value, and permanence of love in all times and societies.

In that a more historically aware treatment of the story implies the activity of a historian, the nature and problem of writing history becomes part of the poem's thematic concern with time and change. The second proem's concern with the effect of time on language finds an echo in the comments near the close on the possible effect of the dialectal diversity in English on the future transmission of the poem (v. 1793–8). Where the second proem looks back to past developments that have led up to present diversity, the ending of the poem looks from contemporary diversity towards the future. The second proem raises a 'historical' awareness of cultural difference; the ending sees the very text of the poem itself as subject to time and change in a historical and linguistic process. By so containing the *Troilus* narrative within this historical awareness of change, Chaucer can point towards—without himself claiming—a needed sense of perspective and of objectivity in approaching the interpretation of his story.

As Criseyde's lament suggests ('But future tyme . . . Koude I nat sen . . .' (v. 748–9)), Chaucer's interest in the past is matched by a concern with the possible foreseeing of future events. Prediction and divination become an important element in the poem, in Cassandra and Calchas (a 'gret devyn' (i. 66)), by means of 'calkulynge' (i. 71), 'sort' or divination by the casting of lots (i. 76–7), by oracle from Apollo and by 'astronomye, | By sort, and by augurye' (iv. 115–16). When Criseyde tells Troilus how she will beguile Calchas into allowing her to return to Troy, she emphasizes his methods as a seer: she will delude him despite his skills with oracles or calculations ('For al Appollo . . . Or calkullynge . . .' (iv. 1397–8)); she will persuade him he was wrongly frightened by the Delphic oracle (iv. 1411); she will jog her father's arm, while he tries to make predictions through casting lots (iv. 1401–5).

In constructing this speech of Criseyde's, Chaucer has brought forward materials from the scene in Benoît and Guido where Briseida arrives in the Greek camp and upbraids Calchas for his predictions, which have made him a traitor. Chaucer has turned a retrospective speech in his source into a speech predicting and rehearsing actions in *Troilus*, just as he transmutes the recriminations of Criseida by Troiolo late in *Filostrato* into the uneasy predictions and misgivings of Troilus before Criseyde's departure (iv. 1440 ff.). The same concern for understanding what is to come is explored in the predestination soliloquy of Chaucer's hero, who is so exercised by human lack of foreknowledge and interested in methods of divination: on the night of the consummation—itself closed by an allusion to 'Fortuna major' (iii. 1420), a figure in the art of divination through geomancy— Troilus pretends that he is keeping vigil in the temple 'answered of Apollo for to be' (iii. 541), and later interprets the cry of the screech owl as an omen of his own death (v. 319–20).

It is a sign of the same concern that, where Boccaccio's Cassandra simply makes a family visit to the sick Troiolo as one of his sisters (vii. 84 ff.), in Chaucer's version of the scene Troilus consults Cassandra in her role as a sybil and a great 'devyneresse' (v. 1522). Boccaccio's Troiolo can and does decode his own dream for himself, while Chaucer's Troilus needs to consult the 'devyneresse' for an interpretation of his dream of the boar (v. 1450–4). This is the culmination of a series of emphases in *Troilus* on the value of dreams as portents and predictions of the future: Criseyde hopes there is a good omen in her having dreamed of Pandarus three times during the night before his visit (ii. 89–90); Troilus interprets his dreams after Criseyde has left Troy as predictions of his death (v. 318). Pandarus, by contrast, insists on the impossibility of interpreting dreams or augury as predictions of the future (v. 358–85, 1275–88), declaring 'Ther woot no man aright what dremes mene' (v. 364).

Yet the commonsensical scepticism of Pandarus—and his warnings of the possibility of expounding amiss—are not sufficient to offset the reader's observation that Troilus' dreams of approaching death and of Criseyde's desertion are indeed proved true in the long term, just as the accuracy of Calchas' predictions of the fall of Troy extends credibility to the various means of divination which Chaucer associates with his predictions. Chaucer is able to privilege his characters through his historical hindsight being turned into their predictive foreknowledge, with Pandarus' common sense as the middle ground, reminding the reader of what it is actually like to be living through a temporal process. Chaucer's introduction into *Troilus* of references specifying the methods of divination brings into the poem an awareness of the range of means by which men strive to see the future, and thereby makes both the craving and the feasibility of such prediction part of the thematic preoccupation of *Troilus* with the boundaries of human knowledge and freedom.

See further the studies of Frankis, 'Paganism'; Mayo, 'Trojan Background'; McCall, 'The Trojan Scene' and *Chaucer among the Gods*; Minnis, *Chaucer and Pagan Antiquity*; Spearing, *Medieval to Renaissance*.

Freedom and the Stars

To what extent does the existence in *Troilus* of a structure of astrological reference to the state of the heavens imply a limitation of the characters' freedom? To the pagan Troilus Venus could be both a goddess and a planet, and for the first audience of *Troilus* some sense of the survival of the ancient gods could have been present, through the medieval accommodation of orthodox Christian belief with an acceptance of some planetary influence over human life. The allusions to significant states of the heavens at important junctures in the story allow for the possibility that the characters' actions are predetermined—or at least predisposed—by the movements of the planets.

Chaucer adds several allusions to the horoscopes of the characters at birth, as when Troilus first rides past Criseyde's house 'right as his happy day was' (ii. 621), and the favourable position of Venus both at that moment and at the time of his birth is noted ('she nas not al a foo │ To Troilus in his nativitee ...' (ii. 684–5)). When he is about to enter the bedroom of his beloved, he prays that Venus may intervene with Jupiter to offset any unfortunate features in his horoscope at birth, including unfavourable aspects of Mars or Saturn (iii. 715–18); Criseyde in turn complains that she was 'born in corsed constellacioun' (iv. 745), and Pandarus invokes Jupiter as presiding over birthdays (iii. 150).

Astrological 'elections' are also made at significant turning-points of the poem. An election is the choosing of a favourable time for an action, and the theory of elections was centred on the fortunes of the moon. Thus, at the opening of Book II Chaucer describes how his Pandarus 'caste and knew in good plit was the moone │ To doon viage ...' (ii. 74–5). It is by now 4 May and, upon consulting a moonbook or lunary before he departs on his enterprise, Pandarus would find that the third day of the lunar month is inauspicious, while the fourth day is propitious for new beginnings. Troilus is comparably mindful of the moon when—towards the close of his prayer in the 'stewe'—he beseeches her to look favourably on his coming enterprise (iii. 731–2).

Since these and the many other astrological allusions in the poem have been added by Chaucer to his source, both the astrological structure and any implication in it for the freedom or otherwise of the characters are very much Chaucer's contribution. Chaucer relates particular configurations of the heavens to appropriate actions: he may have carefully calculated an astrologically appropriate moment for Troilus' ride past Criseyde's window (ii. 680–6), where mention of the favourable 'aspectes' of Venus makes it likely that a horoscopic figure is involved; the seventh 'hous of hevene' was believed to be 'of wymmen and of weddinges, of strives sothly, and of participaciouns', and on 4 May 1385 Venus was in conjunction with Jupiter. An invocation of the powerful influence of Venus as planet and as goddess opens the third book, just as Venus is shining brilliantly as an evening star in the western sky on the evening after Diomede's visit to Criseyde. A portentous conjunction is marked by such heavy rain that Criseyde must stay overnight under the same roof as her lover, and, if Troilus that night invokes the planets and makes an election before entering his lady's bedroom for the first time, on 9 June 1385 the planets that matter—Venus, the moon, and Jupiter—were indeed all in the seventh house, while the moon avoids every one of the twelve impediments that were deemed to be so important in an election. The possible allusion to the position of Mars when things go awry in Book IV would also place narrative events in parallel with an unfavourable state of the heavens.

In the standard medieval distinction, planetary influences incline but do not compel, and the medieval Christian view on this point was clear enough. Chaucer's own dismissal of judicial astrology in his *Treatise on the Astrolabe* is at least one instance of the poet conforming emphatically with the conventional view: 'These ben observaunces of judicial matere and rytes of payens, in whiche my spirit hath no feith, ne knowing of her *horoscopum*' (ii. 4). But the characters of *Troilus* are indeed 'payens', who might be predisposed to take a fatalistic view of their subjection to the stars, even if the poem's Christian readers should not. The eloquence of Chaucer's astrological references and their significance and coherence in a structural pattern make space in the poem for a powerful sense of the possibility of planetary influences. There is certainly common sense in a view that questions whether Chaucer would work so painstakingly and elaborately to include an astrological structure which is then to be understood as wholly discounted or subverted by irony. The best available guide is in the sum of attitudes to astrology represented within the poem. To the astrologically minded Pandarus, who checks the position of the moon, or lectures Criseyde on seizing propitious moments (ii. 281–2), the convenient storm that accompanies the aquatic triplicity of the conjunction on the night of his supper party will not have been surprising, however much the narrator exclaims upon it as fateful (iii. 617 ff.), so that the way in which we understand events as influenced by the stars is here suggested to depend on point of view.

The poem presents its astrological structure and lore so as to leave it open to interpretation. It need not be read deterministically, yet the state of the heavens is recurrently kept above the horizon of attention, as when the moon leaves Leo, or when Venus shines over Diomede and Criseyde (v. 1016–22). The very manner of Chaucer's allusions to developments in the heavens as coinciding, in parallel, with developments in the narrative has the effect of preserving a sense of the space between the two planes and with that the essential mysteriousness of the interconnection. Pagans and Christians, narrator and reader, may interpret in different ways the influence of the stars on human actions; *Troilus and Criseyde* is able to suggest within itself the options of acceptance and rejection of planetary determination of human action and can leave these options to define and comment on each other.

On elections and horoscopes, see North, *Chaucer's Universe*, ch. 5 (xi), and pp. 378–80, 382–7. See also McCall, 'Chaucer's May 3'; and, on 4 May, see *Days of the Moon*, in *Works of John Metham*, ed. H. Craig (EETS os 132; London, 1916), 149.

Fortune and Freedom

Chaucer develops his story of the love of Troilus and Criseyde in order to prompt questions in a debate about the interrelated philosophical themes of fortune and the freedom of the individual will. That enquiry into how human freedom is qualified by fallible and partial ability to know and to remember what is known—which is one of the legacies of the *Consolation* to *Troilus*—can also be the mainspring of fictional narrative and character-ization. Such a development is achieved in part by making the characters more disposed (rightly or wrongly) to interpret what happens in their lives and loves in terms of destiny, fortune, and chance. They have the fatalistic opinions that (in the understanding of Chaucer's contemporaries) were typical of pagan belief, and hence distinct from the orthodox Christian view that the agency of fortune functioned within the larger frame of providence, and that Almighty God's omniscience—divinely uncon-strained by our humanly linear conception of time—did not predetermine our free actions. Chaucer's characters discuss and relate their experience to fortune so prominently, or even incongruously, as to alert the reader's attention to the appropriateness of their understanding of fortune, and hence to an element of debate about freedom in the poem.

Most of what is said on the subject has to be weighed in context rather than taken at face value. That Troilus himself is disposed to see whatever happens to him as predestined, and makes strongly fatalistic speeches, does not necessarily mean that the poem as a whole is presenting a sternly predestinarian reading of its story, and so of human life in general. Indeed, it was a paradox of the literary posterity of the *Consolation of Philosophy* that Boethius' vivid account of fortune made more impression and prompted readier imitation than his arguments for the illusoriness of fortune's power. The love of Troilus and Criseyde is discussed in relation to their ideas about fortune, from the moment Troilus declares to himself 'God wolde, | Sith thow most loven thorugh thi destine . . .' (i. 519–20). To Criseyde, Pandarus hints at a 'fair aventure' (ii. 224), and plays on the idea of a fortunate face (ii. 279–80) and a fortunate time (ii. 281–2, 288–91). The idea of a great and mysterious good fortune is stressed to Chaucer's Criseyde before Pandarus reveals that her fortune is a lover, and this association of love and fortune is never to leave the affair. For Criseyde, the language of fortune offers ready commonplaces for complaint and resigna-tion, whether in her own thoughts or in conversation with others (ii. 464, 742). The rigidity with which Troilus thinks of his experience in terms of fate and fortune has a different quality, reflected in the rhetorical high style with which he expresses his thoughts on the subject, as when he reportedly prays to Love's providence (ii. 526–7), sends his first letter to Criseyde to its 'blisful destine' (ii. 1091–2), or thinks to himself of fate as Criseyde is about to step into his sick-room at the close of Book II ('O Lord, right now renneth my sort [fate, lot]' (ii. 1754)). When in the 'stewe' he prays elaborately for help from the Fates, and his tendency to think of himself as predestined is allowed eloquent expression, yet in a context that suggests a failure of proportion and perspective:

> 'O fatal sustren which, er any cloth
> Me shapen was, my destine me sponne,
> So helpeth . . .' (iii. 733–5)

For Pandarus, different commonplaces about fortune suit different contexts, and the contradictions between his advice in various contexts point to a readiness to fit the philosophy to the moment, as with his contradictory remarks on regretting lost happiness (e.g. iii. 1625–8; iv. 393–9, 481–3).

This sense of contradiction reflects the way in which the change in the lovers' circumstances in Book IV places their previous understanding of fortune under strain. Just as the first sorrow in Book I had been analysed through echoes of some of the Boethian speeches in the *Consolation*, so now in the doubled sorrow of Book IV the exchanges between Pandarus and the sorrowing Troilus revert to discussion of the theme of fortune. Yet, with no true equivalent of Lady Philosophy present, there is no framework through which the concept of fortune can be so clearly transcended as in the *Consolation*. The interview between Troilus and Pandarus alludes recurrently to the operation of chance and fortune (iv. 416, 419–20, 491, 499–500); Troilus conceives of himself as having 'fallen' (iv. 271), prays for other lovers on the wheel of Fortune (iv. 323–5), and complains of 'This infortune or this disaventure' (iv. 297). Rather than thinking of fortune as a metaphor for the way of the world, Troilus confuses her with a goddess and, seeing his misfortune as poor recompense for his faithful service, reproaches Fortune for the lack of mercy and of justice in her treatment of him (iv. 260 ff.). The Boethian discussion of free will that follows as his soliloquy has a continuity in theme, and perhaps even in character. There is irony in the depiction of an individual freely arguing away the freedom that he possesses, but stylistically the very awkwardness of the soliloquy aptly expresses the hero's sense of helplessness. As comparison with the *Consolation* reveals (see above, pp. 105–7), the balance has been markedly shifted against free will. This strongly predestinarian monologue—and the failure to include as part of Troilus' awareness anything of the rebuttal of the prisoner Boethius' speech by Dame Philosophy, proving the compatibility of divine prescience and human free will—serves by its very one-sidedness to underline how the characters' declarations about fortune and freedom are to be seen as partial.

The tendency of the characters to see their lives in relation to the workings of fortune is framed by the way in which the narrative—through the fictionalizing of the composition and through its own processes and technique—explores the theme of freedom and predestination. All the play made with sources only adds to the sense that a form of predestination is being illustrated through the narrator's claims to be subject to his sources, yet all the while the narratorial frame and the management of the reader's response is designed to dramatize the coexistence of freedom and constraint. In terms of the narrative, the future of Troilus and Criseyde is foreknown and, as it were, predestined. The end of the old story is already established and in its outward events unalterable. That the reader knows (as past) what is future (and therefore unknown) to the characters effectively puts the reader in God's place. Indeed, it could be argued that the whole poem is a dramatization—or a practical example—of that debate between Philosophy and the prisoner that Troilus uses for his soliloquy, as to whether the author's, or reader's, or onlooker's knowing a thing to be true—whether present, past, or future—necessitates its happening.

The questions about fortune and freedom posed through the characters' comments—and implicitly raised by the narrative processes of the poem— are also pursued by the way in which the narrative contains the story within recurrent and prominent references to fortune. The poem's symmetrical structure represents in itself the rise and fall of Fortune's wheel. At the crucial turning-point in the action, at the opening of Book IV, Chaucer sketches in the image of the spiteful goddess turning her wheel (iv. 6–7), with her blindness, deceptiveness, and treachery all stressed (iv. 1–11), and with Diomede replacing Troilus at the top of the wheel. There is an implicit parallel here between Fortune and 'bright' Criseyde, as Dame Fortune turns away her 'brighte face' from Troilus and transfers her favour to Diomede (iv. 8–11). The fluctuations in the fortunes of the Greeks and Trojans at the siege of Troy are initially related in terms of Fortune's wheel (i. 138–9); the striking of Troilus by Love's arrow is reported in similar terms of ascent and descent ('This Troilus is clomben on the staire, | And litel weneth that he moot descenden' (i. 215–16)). The whole course of the narrative in *Troilus* is punctuated by comments which attribute the sequence and progression of events to the operation of Fortune ('And thus Fortune a tyme ledde in joie | Criseyde and ek this kynges sone of Troie . . .' (iii. 1714–15)), typically linking fortune and the affairs of princes, while hinting at the temporary nature of fortune's gifts. In Book V of *Troilus* the narrative recurrently looks ahead in terms of Fortune, exclaiming as Troilus thinks of the absent Criseyde ('Fortune his howve entended bet to glaze!' (v. 469)), or as Troilus and Pandarus await Criseyde on the walls of Troy ('Fortune hem bothe thenketh for to jape!' (v. 1134)). The whole narrative is brought to its close in this vein as one in which all events fall out according to a predetermined pattern:

But forth hire cours Fortune ay gan to holde . . . (v. 1745)

But natheles, Fortune it naught ne wolde ... (v. 1763)

Important transitions are associated with the operation of fortune so as to bring within the poem careful definitions of how fortune may be understood to work, as in the address to Fortune as 'executrice of wierdes ... under God ye ben oure hierdes' (iii. 617–19), the account of how destiny is committed by Jove to the fates (v. 1–4) or of the relation between providence and fortune that precedes the death of Hector (v. 1541 ff.). Yet such emphasis in the narratorial commentary on a predetermined pattern to events can coexist in *Troilus* with a sense of the mysteriousness of chance and coincidence and a careful attention to that freedom to choose which the characters are shown to exercise.

Chaucer's changes to *Filostrato* increase a sense not only of destinal forces but also of chance. The sighting of Criseyde by Troilus in the temple is said to depend upon chance ('upon cas bifel ...' (i. 271)), but Chaucer matches this by creating the chance first sighting of Troilus by Criseyde as he happens to pass her house just after Pandarus has told her of the prince's love (ii. 610 ff.). The two occasions when Troilus rides past Criseyde's house both seem like chance to her, although the reader knows that the second has been arranged by Pandarus so as to appear like chance. At this and other points the identification of what is chance depends on point of view: the discovery of Troilus languishing by Pandarus seems like a chance event ('com oones in unwar ...' (i. 549)), although Pandarus later describes to Criseyde how his suspicions were already aroused (ii. 505 ff.). Throughout *Troilus* the action is furthered by minor accidents and chance occurrences: Criseyde happens to hear Antigone's song about love, hears the nightingale, and has her dream of the eagle, although her state of mind has predisposed her to take note of them. It is apparently pure luck that Troilus finds a means of getting rid of Helen and Deiphebus from his room before Criseyde enters, by sending them off to read a letter (ii. 1696–7). For the lovers' meeting at his own house Pandarus 'hadde every thyng that herto myght availle │ Forncast' (iii. 520–1), yet events might not have happened as they did if Troilus had not chanced to fall into a faint, providing Pandarus with the opportunity to throw him into Criseyde's bed. The happy union of the lovers is eventually threatened because the Trojans suffer a reversal in the chance of battle and must retrieve some of their captured leaders (iv. 47–56), but the chance of war is compounded by the crucial influence in parliament of the fickle 'peple' who act blindly (iv. 198). In the near-suicide of Troilus after Criseyde's faint both lovers narrowly escape death because Criseyde recovered—'as God wolde'—just in time (iv. 1212 ff.). Chaucer makes Diomede's approaches to Criseyde into something gambled: his second approach is undertaken in a spirit of 'happe how happe may' (v. 796) during a visit rearranged by Chaucer to coincide with the very day that Troilus is expecting Criseyde to return.

The unalterable nature of the ending of the story in betrayal might seem to promise a narrative vehicle expounding a predestinarian philosophy which denies human freedom, but this is not how the poem reads as a whole. The very fullness and richness of Chaucer's characterization—his ability to suggest the force and reality of feelings and experiences in characters—works recurrently and in detail to point to the ways in which characters do possess freedoms and exercise choices. In Book I the implied parallel with the prisoner Boethius uses that powerful and familiar model of a resignation and despair which is famously and rightly overcome to comment by contrast on the wilful passivity of mind in Troilus. The submission of the conduct of the affair to Pandarus continues to comment on the freedom that Troilus has given up, but the real question of freedom for Troilus comes with the dilemma he faces over how to react to the decision to exchange Criseyde. It is this dilemma over whether or not to act which proves the moral pivot of the poem in its exploration of the lovers' freedom. On every level Chaucer shows Troilus declining to exercise the possible choices that are open to him, declining to request Criseyde openly, and submitting to the will of his lady a decision on whether to abduct her and so ambush the exchange. That prowess as a soldier which the poem elsewhere has mentioned suggests that it would be open to Troilus to intervene successfully to prevent the exchange by effective military action. This option he has, but chooses not to take. Chaucer emphasizes the courtly 'observance' of this, so that a freely and deliberately chosen inaction becomes very much the expression of the hero's devotion to the courtly ideal of service in love. Troilus refuses to exercise 'maistrie' or dominion over his lady; his refusal happens to be fatal. In this sense— despite all the fatalistic rhetoric which presents Troilus as helpless before external forces—Chaucer has carefully shown how Troilus has given up choice and freedom rather than never possessing it.

The freedom of Criseyde and its constraints are explored with comparable care in *Troilus*. The whole sequence of action in Book II from the first visit of Pandarus to her entering Troilus' sick-room shows her as intensely concerned with the freedom of action she already possesses and how this may be affected. In both the bedroom interviews with Troilus she is taken advantage of, yet it is also clear that she retains an impressive initiative and control, not least in the implication that she has anticipated more than—for appearances' sake—she would be prepared to admit openly. Criseyde enlists philosophical language to represent her uncle as wholly responsible for what has befallen her ('Pandare first of joies mo than two | Was cause causyng unto me, Criseyde' (iv. 828–9)), yet even in the later books, where she is powerfully constrained by outside factors, Chaucer devotes much space to her determination not to go along with Troilus' plans for escape. Instead she argues for an attempt to evade seeming necessity by stratagem and ingenuity; the tone is sceptical and represents the attempt of human will to overcome apparent necessity, to be lord over fortune (iv. 1587–9). In this way Chaucer—having earlier elaborated the characters' sense of fortune and fate—at the point of crisis and decision gives his Criseyde a confidence in the effectiveness of the human will, which will be set all the more poignantly against the recognition of necessity by both Criseyde and Troilus in the fifth book. In the Greek camp Criseyde's sense of being 'in the snare' of her present predicament through not having been able to foresee the future is immediately followed by her expression of blank determination to return to Troy in the near future 'bityde what bityde' (v. 743–54). With whatever mistaken confidence, Criseyde had been determined to choose her own way of reconciling the demands of love, secrecy, and her reputation, with the outward necessity to go to the Greek camp. This is a willed and chosen acquiescence with the outward necessities of public life; she has reserved a different inner will to act so as to get the better of circumstances, but the collapse of this resolve in her is part of the inevitability of Book V.

See further the studies by Berryman, 'Ironic Design'; Bloomfield, 'Distance and Predestination'; Christmas, '*Troilus and Criseyde*'; Curry, 'Destiny' and *Chaucer*; di Pasquale, '"Sikernesse"'; Elbow, 'Two Boethian Speeches' and *Oppositions*; Howard, 'Philosophies'; Mann, 'Chance and Destiny'; Morgan, 'Freedom'; Owen, 'Problem'; Patch, 'Troilus on Predestination' and 'Troilus on Determinism'; Payne, *Chaucer*, chs. 4 and 5; Shepherd, 'Religion and Philosophy'; Stevens, 'The Winds of Fortune'.

Characterization

For Chaucer the exploration of character was not so much an end in itself as the consequence of his development of theme in the story of Troilus and Criseyde—even though there is much in the presentation of character in *Troilus* that anticipates the novel to an astonishing degree. Such an intensity of psychological interest may not be unique in medieval literature but, by its presentation, to take the reader so near to the very flux and continuum of the characters' existence is the special accomplishment of *Troilus*, and one of the springs of its continuing life. At the same time, it was an old story when the poem was new, and there might be constraints on the freedom of any writer who took it up; but Chaucer actually takes over such constraint as part of his subject, making the relation of his characters to tradition and to type part of the way he develops characterization. They exist within, and are influenced by, the pressures of collective norms of society and of norms of gender, but emerge as larger, more interestingly varied and inconsistent than type figures subordinate to a representative role in the exposition of a theme.

In *Filostrato* the three main characters are in many ways similar and receive similar emphasis: all are of comparable age, all are confident and of a comparably sensual outlook, and as such are not greatly differentiated from Diomede. A comparison of the number of lines spoken by each (see Table 8) is one way to measure the decisive shifts in emphasis that Chaucer has made.

TABLE 8. Troilus: *Lines spoken by characters*

Character	*Filostrato* (5,704 lines)	*Troilus* (8,239 lines)
Troilus	1,706	1,462.5
Pandarus	789	1,804.5
Criseyde	747	1,154
Diomede	98	153

In a much longer poem the amount spoken by the hero grows shorter, while the most dramatic shift is, of course, in the growth in the speaking role of Pandarus. Percentages for the proportion of lines spoken by each character in each of the books underline the dominance of Pandarus and the comparative silence of Troilus in Book II, an almost equal contribution by all three principals in Book III, and a nearly equal contribution by the two lovers in Book IV. Together with these changes of emphasis in the role of the principals, the role of the minor characters in *Troilus* is also expanded, with more significant roles for Hector and Cassandra than in *Filostrato*, and with new roles created for such figures as Antigone, Helen, and Deiphebus.

The effect of all such changes in *Troilus* is that the hero is seen much more in relation to the other figures, who are in turn markedly different in character from him and from each other, and are presented and apprehended differently. This is part of a much larger disposition in Chaucer's poem to set things in juxtaposition, and to explore their relation to each other, in patterns of antithesis and contrast. The way in which the relation between Troilus and Criseyde is presented has been described as contrapuntal rather than harmonic, and the distinction which Chaucer explores between their natures means that something remains mysterious about the heroine for the hero as for the reader. The movement towards union in Book III can therefore involve something of surprise, of discovery, and of unexpected knowledge, just as in Book II there is the impetus and momentum of something discovered about herself in Criseyde's surprised response to love. With Pandarus and Criseyde Chaucer presents a complex tangle of clues and hints which, even while keeping readers guessing, also persuades that there may yet be a possibility to uncover the springs and motives of why they behave as they do. This complexity in turn serves as a foil to the extensively presented consistency and apparent simplicity of Troilus in love, although even about the hero the narrative sometimes points out that its information is incomplete (iii. 447–8); and Chaucer recurrently moves to omit moments in *Filostrato* where characters have the worldly experience to think better of what they are doing, or to foresee disadvantages and criticisms. In a poem that adds a new beginning to an old ending only the aftermath and not the process or moment of decision is included within Chaucer's text.

The Gothic visual arts and the taste and sensibility of the period provide useful parallels for the processes through which Chaucer presents his characters. Contemporary pictures would not be painted from one single point of view, with the perspective dependent on that serving to unify the picture. Different actions at different moments by the same figure might be depicted within the same frame, sharing the same surface, with scenes and subjects set in juxtaposition. Within the same picture might be found flat and stereotypical treatment alongside intensely expressive representation, particularized with realistic detail and a keen sense of pathos. A 'Gothic narrative' like *Troilus* shows closely similar features: the narrative frame, for instance, contains two distinct pictures of how Troilus falls in love, first as he is pierced by the God of Love's arrow (i. 204 ff.), and then when his own sight pierces through the press to hit upon Criseyde (i. 267 ff.). These two moments are separated and made more distinct by the interpolation between them of a lengthy narratorial excursus (i. 211–66), so that the two different pictures are not unified into a single temporal continuum or point of view but are left to stand juxtaposed. Again, the somewhat stiff, formal 'portraits' of Troilus, Criseyde, and Diomede (v. 799–840) are interpolated into a narrative sequence where they seem like a triptych painted in a different style, and on principles of representation by means of stereotypes different from the material alongside them.

To the sense of the narrative as presenting a continuum of the characters' experience the modern reader responds with an admiration which has its standard in the traditional novel, yet Chaucer is not aiming at the unity of a novel. Significant moments occur in unreported gaps in the narrative, as when Criseyde tells Troilus she would not be with him in the bedroom at Pandarus' house if she had not already yielded (iii. 1210–11). The poem simply does not report when Pandarus left the lovers' bedroom, or that he did so, although it does describe him returning next morning.

Nor is there any way of knowing whether the scene with Troilus in the palace garden as Pandarus reports it to Criseyde ever actually happened (ii. 505 ff.).

Together with the exclusions, disjunctions, and juxtapositions, there are inclusions which it can be helpful to see as aspects of a Gothic narrative. This is especially, but not exclusively, true of moments where each of the lovers grows 'philosophical', as in Troilus' predestination soliloquy, or when Criseyde—awakened from sleep in the middle of the night by an extraordinary story of her beloved's suicidal jealousy—launches into a four-stanza speech on the delusive quality of earthly happiness and human apprehension of it (iii. 813–40). Judged by naturalistic or dramatic criteria such inclusions may seem disproportionate in their context, and there has been more tendency to interpret their perceived inappropriateness as funny than to question the appropriateness of naturalistic or dramatic criteria. Both Troilus and Criseyde do passingly use 'philosophical' terms elsewhere (e.g. iv. 829, 1505), but neither is generally given to philosophizing. Their speeches, however, are far from inconsistent with the Boethian cast of the poem as a whole: the characters may sometimes be deployed to give voice to thematic questions, without strict regard for the probability of whether what they say reflects more than they would be likely to understand in the context. This in turn raises a question about whether speech is always to be read as directly expressive of the individual in a highly rhetorical poem. Here the (anachronistic) analogy with opera may be helpful, in which the characters' speeches may be seen to function like operatic arias: responding to events and expatiating upon feelings with a patterned formality of style and structure; vowing, praying, and lamenting at a different pace and extent from spoken utterance. In style and language, as in other aspects, *Troilus* presents a sometimes disconcertingly disjunctive sequence of the stylized, stereotyped, and impersonal, alongside the most vividly idiomatic or passionate.

In opening *Troilus* Chaucer's reader is stepping into the world of a late medieval or Gothic sensibility with its own manners and ceremony, its own emotional idiom and emphases. Such a sensibility has its characteristic emotional colours, its own sense of the shapes and proportions of how a picture should be filled. It is a sensibility disposed to, and identifying with, the most acutely intense sensations: Troilus scorns the very idea that his feelings might be controlled and overcome ('Thow moost me first transmewen in a ston, | And reve me my passiones alle . . .' (iv. 467–8)); he feels that the house left behind by his lady should fall to the ground (v. 545), just as the grieving Richard II was a few years later to order the destruction of the palace in which his queen had died. The capacity to feel is something prized: Pandarus claims to Criseyde that Troilus must be treated exceptionally because he is far beyond any ordinary sensibility ('This is so gentil and so tendre of herte . . .' (iii. 904)), and both lovers fear the effect of sorrow on what they see as the other's especially susceptible nature (iv. 794–5; v. 242–3). The poem's pattern of double sorrow is imbued with an aesthetic of sorrow and lamentation, edged with that 'distresse' so frequently recurring in rhyme like a refrain. When Troilus sits down to write his late letter to Criseyde in Book V he premeditates on 'How he may best discryven hire his wo' (v. 1314), and the hero's striving to achieve the 'best' description of his unhappiness may stand to represent the poem's larger endeavour of art.

Gesture as well as speech informs the characterization in *Troilus*, as when Criseyde wrings her long slender fingers and tears the 'myghty tresses' of her 'ownded' (wavy) hair (iv. 736–8, 816–17). The gentleness and sensitivity of the lovers is signalled through a remarkably detailed attention to outward tokens of feeling, in particular their downcast looks:

> With that she gan hire eighen down to caste, (ii. 253)
>
> With this he stynte, and caste adown the heed, (ii. 407)
>
> How sobrelich he caste down his yën. (ii. 648)
>
> And to the ground his eyen doun he caste. (iv. 522)
>
> And therwithal she caste hire eyen down, (v. 1005)

or their gazes and glances:

> With that she gan hire eyen on hym caste (iii. 155)
>
> For which Criseyde upon hym gan biholde, (iv. 1229)
>
> For pitously ech other gan byholde, (iv. 1249)
>
> And rewfullich his lady gan byholde, (iv. 1691)
>
> Therwith he caste on Pandarus his yë, (v. 554)
>
> And sobreliche on hire he threw his lok. (v. 929)

The lovers' faces so recurrently change and grow pale with anxiousness, distress, or fear that pallor becomes an element in the poem's aesthetic of sorrow and pity:

> That sexti tyme a day he lost his hewe (i. 441)
>
> So wo was hem that chaungen gan hire hewe (iii. 1698)
>
> This Pandarus, ful ded and pale of hewe (iv. 379)
>
> Hire hewe, whilom bright, that tho was pale (iv. 740)
>
> with chaunged dedlich pale face (v. 536)
>
> So pitously and with so ded an hewe. (v. 559)

In reflection and resignation the lovers are much given to sighing (e.g. iii. 801, 812, 972, 1056, 1080–1, 1170–1; v. 58, 715, 738), but Diomede is to put an end to Criseyde's sighs (v. 1034), and Troilus only sighs when he reads Criseyde's strange last letter (v. 1633). Both will tremble or swoon at moments of extreme emotion (e.g. i. 871; ii. 302; iii. 1086–92; iv. 1149–53). Nor is Pandarus altogether outside the realm of such sensibility, for he at least claims that he would swoon to repeat to Criseyde the sorrows of Troilus (ii. 574). The whole action—not simply in the last two books—is accompanied by frequent tears, from Pandarus as well as from the lovers, as part of a larger ritualization of sorrow in lamentation and complaint, given ceremonious form in gestures of supplication between the characters (ii. 974, 1202) and in kneeling (i. 1044–5; iii. 1592).

Through such attention to gesture Chaucer represents his characters' lives in terms of a heightened sensibility in which emotional reactions may be expressed vehemently and extremely: the intense emotions of Troilus are driving him to the verge of madness (e.g. i. 479, 499; ii. 1355; iii. 793–4; iv. 230, 238, 917; v. 206, 1213). Yet such vehemence of feeling is part of the same picture which includes graver tones of studied and deliberate behaviour. When Criseyde has fainted—but seemingly died—there is a scene in which the grieving Troilus laments, weeps, and prays over the 'body' (iv. 1170–4), and the deliberateness through which the strongest feeling is ritualized by a sense of ceremony expresses itself in a disposition—depending on context—to be humble, thankful, and prayerful. In their humility Chaucer's lovers are often anxious and fearful, inclined to accept things passively, in a way that translates the poem's larger thematic concerns with freedom into the texture of individual character. In a long poem the very full narrative attention to the lovers' feelings and thoughts is focused on characters who are other than conventionally 'heroic', and who are not arresting or decided personalities: there are a number of moments where characters are so overcome that they simply do not know what to do. In happiness they are inclined to give thanks—whether to the gods or to each other—while at many points their characteristic reaction is to pray, reflecting the pious, earnest side to their characters.

Such a moral seriousness is also noticeable in the way that Chaucer's characters perceive each other ('she wente ay purtraynge | Of Troilus the grete worthynesse'(v. 716–17)). As soon as he falls in love, Troilus thinks of Criseyde as 'so goodly oon' (i. 373) and 'Good goodly' (i. 458), and Pandarus praises the kindly and active virtue and moral sense in Criseyde:

> 'Ne nevere saugh a more bountevous
> Of hire estat, n'a gladder, ne of speche
> A frendlyer, n'a more gracious

> For to do wel, ne lasse hadde nede to seche
> What for to don ...' (i. 883–7)

Pandarus similarly represents to Criseyde the moral character of Troilus. In answering her polite enquiry about Hector's welfare Pandarus spends more time commending the virtues of Hector's 'fresshe' brother:

> 'The wise, worthi Ector the secounde,
> In whom that alle vertu list habounde,
> As alle trouthe and alle gentilesse,
> Wisdom, honour, fredom, and worthinesse ...' (ii. 158–61)

In this scene is introduced the beginning of a persistent strand in Chaucer's poem, in which Criseyde's love for Troilus is noted to include her response to his moral character. When Troilus rides past Criseyde's house—just after she has learned that he loves her—she is said to be attracted to both his handsomeness of body and his quality of nobility, commending to herself 'His wit, his shap, and ek his gentilesse' (ii. 662). As she considers the opportunity of love, Criseyde's thoughts seem to balance attraction towards Troilus as a man and as a worthy man (ii. 701–2), recalling his good qualities and sensible nature (ii. 723). Having accepted Troilus as her servant, Criseyde feels secure in his wise discretion (iii. 464), and it is in recognition of his good qualities that Criseyde is persuaded to receive the supposedly jealous Troilus in her bedroom (iii. 897–8). After the consummation her pleasure in thinking of her lover is dominated by remembering

> His worthynesse, his lust, his dedes wise,
> His gentilesse, and how she with hym mette ... (iii. 1550–1)

—and when she acknowledges at the close of Book IV the qualities for which she loved Troilus, she praises his 'gentil herte' and his scorn for 'Every thyng that souned into badde' (iv. 1676). Even when she thinks of the lover she is abandoning, she acknowledges that 'I have falsed oon the gentileste | That evere was, and oon the worthieste!' (v. 1056–7).

So much is seen in their experience of love and in each other by Chaucer's lovers that the concept of character is continually shifting between the archetypal and the individual, between pattern and variation, in the exploration of theme. Troilus is both a more typical figure of a lover and yet more of an individual than his prototypes, just as the presentation of Pandarus moves restlessly between the archetypal 'friend' in function and the inimitably individual in expression. Criseyde is recalled by Troilus as essentially and characteristically womanly (v. 244, 473, 577), and has herself a resigned idea of the typical limits of gender and society on what women can effect in the world as it is ('Therto we wrecched wommen nothing konne, | Whan us is wo, but wepe and sitte and thinke' (ii. 782–3)). Yet with both Troilus and Criseyde Chaucer reveals an inner individual far from identical with what outwardly appears to society in arranged behaviour, so as to produce the sense of a dialectic between an inner world of thought and emotion and an outer world of speech and behaviour:

> With sobre chere, although his herte pleyde: (i. 1013)
>
> For which with sobre cheere hire herte lough.   (ii. 1592)

In a poem so concerned to explore the essential mutability of the world and the nature of human knowledge and ignorance, freedom and fallibility, it is the reader's perception and understanding of characters—and the light in which they are seen—that change more than the characters themselves.

On the lines spoken by each character, see Meech, *Design*, 9, and Davenport, *Complaint*, 218. On the 'contrapuntal' relation of Troilus and Criseyde, see Bishop, *Chaucer's 'Troilus and Criseyde'*, 11.

On the minor characters, see Greenfield, 'Role'; Johnson, 'The Medieval Hector'; Knapp, 'Boccaccio and Chaucer'; Knight, *rymyng craftily*; Lumiansky, 'Story'; and Sundwall, 'Deiphobus and Helen'. On processes of characterization, see Gordon, 'Processes'.

Differences from *Filostrato* in characters' understanding and awareness would include the following: Chaucer's first major omission is to cut the moment when Troiolo talks from his own experience in love (*Fil.* i. 23–4); Chaucer retains Troiolo's concern at what lovers will say of him (i. 512) but omits the Italian stanza in which Troiolo knows that other princes will reproach him for wasting time unworthily on love in times of war (*Fil.* i. 52); Chaucer omits Pandaro's admission of the shamefulness in what they are proposing to do with Criseida (*Fil.* ii. 25–6). When Pandaro tells Criseida that a certain man is pining for her, she comes back at him with some direct questions, which do not appear in Chaucer's text ('Are you testing me, or are you telling the truth? ... Who could have his pleasure of me completely, unless he first became my husband?' (*Fil.* ii. 45)). Chaucer omits the stanza (*Fil.* iv. 37) where Troiolo says that, if Criseida's departure had been delayed, he might have grown more used to the idea, and Chaucer also drops the stanza (*Fil.* iv. 59) in which Troiolo declares that, although love can never be driven out, it may well slip away through process of time. In urging Troiolo to abduct Criseida Pandaro argues that it will always be possible to bring her back if things go wrong (*Fil.* iv. 73), but this loophole disappears from Chaucer's text, as does Criseida's argument (*Fil.* iv. 152–3) that, if they could enjoy at will their present furtive passion, it would soon be spent.

On character in the context of the Gothic arts, see Derek Brewer, 'Gothic Chaucer', in Derek Brewer (ed.), *Geoffrey Chaucer* (London, 1974), 1–32, and his *Chaucer* (3rd edn.; London, 1973), ch. 12 ('What kind of a poet is Chaucer?'). See also Brewer, 'Towards a Chaucerian Poetic', and George Henderson, *Gothic* (London, 1967).

On gesture, see Barry Windeatt, 'Gesture in Chaucer', *Medievalia et Humanistica*, 9 (1979), 143–61. See also Hermann, 'Gesture and Seduction', and Robert G. Benson, *Medieval Body Language: A Study of the Use of Gesture in Chaucer's Poetry*, (Anglistica, 21; Copenhagen, 1980). With Chaucer's eye for gesture and the contemporary arts, cf. the figure-painting in the great east window of Gloucester Cathedral (finished by 1349): 'The faces are carefully drawn, the hair and beard are thick and softly curling masses, the hands are delicately rendered with long, slender fingers' (M. Rickert, *Painting in Britain: The Middle Ages* (Harmondsworth, 1954), 160).

On the lovers' passivity, see Crampton, 'Action and Passion', and Manlove, '"Rooteles moot grene soone deye"'; on fear, see Hatcher, 'Chaucer'; on 'gentil' behaviour, see Gaylord, 'Gentilesse'; on the dialectic between an inner world of thought and emotion and an outer world of speech and behaviour, see Mann, '"What is Criseyde worth?"'

Troilus

Chaucer refers to his poem as 'Troilus' and as the 'Book of Troilus', and his Troilus is the central figure who acts as its thematic focus through his feelings and experience. The time span of the poem and its narrative boundaries are determined by two moments of transition in his life: into love and out of life itself. It is the double sorrow of Troilus in love which the poem announces as its subject in its first line—without mentioning the object of that love—so that the form of the poem is coextensive with the life of Troilus as a lover and is organized from his point of view. At the close of each of the five books the focus settles squarely on Troilus. His end is the end of the narrative, which ignores the subsequent lives of other characters: their function is in their contribution to the story of Troilus.

Troilus is the 'hero' of the poem, but one whose nature builds into it an exploration of how the idea of a hero is to be interpreted. He is the key figure in the development of the poem's themes, even if this makes him less interesting as a personality to modern tastes than Criseyde or Pandarus. His emotional life raises the central questions of the poem, and in him the endeavour and venturing associated with the hero is translated into his exceptional capacity to feel. In love for the first time and wholly inexperienced, his power of feeling makes him single-mindedly committed but anxiously full of 'drede', and given to every kind of idealization of what is happening to him. He has no competitor or companion in his strength of feeling. His characteristic expression is through apostrophe, song, complaint, or prayer, rather than the interchange of dialogue. For Troilus love is transcendent; he proves—as Pandarus predicts—to have the zeal of the convert, devoted in love and no longer the scoffer in the temple (who had scoffed from ignorance rather than disillusionment). In his new devotion Troilus proves—as he vowed—unchangingly loyal until death.

Yet Chaucer makes a point of turning his Troilus into rather more of a romance hero than had usually been encountered. Other heroes of romance are prostrated by love or dependent on an intermediary, but the effect in *Troilus* is intensified by the focus and scale of narrative attention to the hero's reactions. Chaucer's identification of his Troilus with such lover figures of romance tradition acknowledges what is good and well intentioned in their idealizing attitudes and aspirations, while also turning that very conventionality of attitude and conduct into a comment against itself. The effect is that Troilus' experience as a lover comes across as both 'fresshe' and archetypal, at once individual and awkward, yet also universal, because of the intensity with which his experience is conveyed and the patterns into which it is ordered.

   Troilus is a hero by virtue of being in love, but he is also an ancient pagan of moral worth, not wise in the ways of the world, but acting generously to others and doing no mean thing himself. He sees his experience of love—both its happiness and sadness—in some metaphysical dimension. In earlier texts—except *Filostrato*—Troilus had been a hero by dint of his heroism in feats of arms, and the traditional character of Troilus is of a very young and spirited warrior, impulsively brave to the point of rashness. Chaucer only mentions Troilus' youth when he appears as a warrior more than a lover (ii. 636; v. 830), and the portrait of Troilus inset from Joseph of Exeter (v. 827–40) brings again into *Troilus* at a late point the traditional picture of youth and unsurpassable valour which Criseyde saw from her window. What receives more emphasis in Chaucer's narrative is the 'feere' and 'drede' of the hero in love, fearful of confessing his love (i. 875), and fearful of his lady's responses to him and his love (i. 1019–21; ii. 1080; iii. 92–4, 706–7, 1647). As the heroic stature in the field of 'this fierse and proude knyght' (i. 225) is understood but scarcely shown, the status of Troilus as the hero of this narrative must be newly defined in terms of a 'domestic' experience in which his independent initiative and eventual success are qualified. (Criseyde dreams of a *white* eagle, suggesting that she sees purity as well as royalty and power in her lover.) The game of love is no game for this lover, for whom all is taken literally and in earnest. His reliance on the game being conducted through Pandarus leaves him in a state removed from direct action, risk, and responsibility which fits with the passive disposition of his character.

   The excursus which follows the moment when Troilus is hit by Cupid's arrow—emphasizing love's humbling of pride and the wisdom of acquiescing in the irresistible designs of love—anticipates that passive disposition which Chaucer's Troilus is to show both in the winning of Criseyde and in his response to the moves to exchange her. The whole conduct of the affair through a go-between, the supposedly 'accidental' meetings of Troilus with Criseyde, the hero's entering his lady's bed by being thrown into it in an unconscious state—all allow Chaucer's Troilus to be brought together with his lady with the minimum of active initiative or direct responsibility on his part: he is characteristically 'allone abedde ... in a traunce | Bitwixen hope and derk disesperaunce' (ii. 1305–7); he places himself under Criseyde's governance (iii. 136–46); he declares himself ready to do as Pandarus wishes (i. 1029; ii. 945), even to serve as his slave (iii. 390–2), until he surrenders himself to Criseyde's wishes (iii. 1176). For Troilus the role of the petitioner and supplicant in love lends ritual form to a naturally passive disposition, although the lover's passivity in *Troilus* becomes exceptional and eccentric to the point of questioning the conventions from which it derives.

Chaucer's development of this disposition of his Troilus in the earlier 'courtship' books prepares for his presentation in the traditional story of the lovers' separation, in which he does not intervene to prevent the departure of Criseyde. In what seems a natural extension of his character all along, he is now submissive in the face of what he sees as the inevitability of the exchange (iv. 287, 568). In submission to his lady's wishes he cannot act to forestall her exchange (iv. 458–9). It is for such a devotedly humble servant of his lady's wishes that Chaucer introduces the soliloquy in which Troilus argues for the predetermined nature of his misfortune ('He seyde he nas but lorn, weylaway!' (iv. 957)), the culmination of the emphasis on Troilus' recurrent concern to seek out larger patterns of meaning in his personal experience.

Throughout the poem—at important junctures, transitions, and crises—Troilus is inclined to pray: his first words on seeing Criseyde in the temple are 'O mercy, God' (i. 276). He prays before revealing his secret to Pandarus (i. 597–8), as he returns to battle (i. 1046–7), against despair (ii. 526–32), when writing to Criseyde (ii. 1059–62), and when he receives a reply (ii. 1317–19). In Book III he is especially full of prayers and thankfulness, most notably in the 'stewe' (iii. 705 ff.), hailing Love with an adapted prayer from Dante (iii. 1261–7), uttering prayers or prayerful thanks for subsequent meetings with Criseyde (iii. 1526, 1672), and praying to Love in his hymn ('Bynd this acord . . .' (iii. 1750)). In his last reported letter to his lady, Chaucer's reverent hero is still gravely praying that all be as Criseyde wishes (v. 1362–3, 1410–11).

Such an exploration of behaviour which is not conventionally masculine raises questions about the values of traditional sexual stereotypes. There is no uncertainty in the outward, public role of Troilus as a soldier: it is 'his manhod and his pyne' (ii. 676) which have their effect on Criseyde's heart. But in the private world of his 'pyne' Troilus is acted upon by his feelings in ways that do not conform to 'manhod'. His fainting is reproved both by Pandarus ('O thef, is this a mannes herte?' (iii. 1098)) and by Criseyde ('Is this a mannes game? | What, Troilus, wol ye do thus for shame?' (iii. 1126–7)). The prince who at his first appearance in the poem is accustomed to 'gide' a group of young men as they confidently eye up women in the temple (i. 183 ff.) is by the midpoint of the poem being chastened and admonished by his lady as if he were a child (iii. 1169, 1180). Sexual possession comes only after the man's moment of deepest humiliation and humility in the face of the woman's suspicions (iii. 1072 ff.). Yet cumulatively the poem establishes a respect for the way Troilus is himself and for the restraint that he shows. For Pandarus, 'manly' self-assertion can shape the world to one's wishes (iv. 538, 622); but Troilus' restraint is the more difficult course than conventional conformity to male self-assertion. His humble submission to Criseyde is called 'his manly sorwe' (iii. 113), and his 'manhod' is associated with discretion and self-control (iii. 428–9; v. 30). Criseyde declares that she loved Troilus both for his 'gentil herte' and his 'manhod' (iv. 1674). As the narrative progresses, the ways in which Troilus restrains himself from solving his problems 'so lik a man of armes and a knyght' in order to submit to his lady's wishes represent a kind of 'manhod' which may not be understood by Pandarus but which the poem seems designed to invite us to value. The exceptional understanding in *Troilus* of the nature of Criseyde as a woman is reflected and complemented by the way that her lover is understood to be a better man because of approximating in his 'manhod' to some of the qualities of 'womanhede'.

Troilus' outward humble gentleness of manner has its inward counterpart in a persistent idealism. There is a heroism of the heart in the intensity of Troilus' sorrows in Books IV and V which is not simply cancelled out by the possibility that from some viewpoints the very seriousness of such feeling may appear comic. Indeed, the 'weakness' of Troilus as presented by the poem may be seen as a moral strength. The contrast between his brutally abrupt end and the privilege of his soul's ascent beyond death suggests a worth in the spirit of Troilus—that spirit which this narrative has created as opposed to the warrior figure of epic and history—that has merited an altered level of perception after death. It is this sense of a virtue in the character of Troilus that gives a larger worth to his experience of the sorrow of loss which—when the poem is seen entire—can stand to represent the failure of any earthly attachment or ideal.

Yet, despite all this Troilus may still seem—by comparison with Criseyde or Pandarus—a relatively simple character, in the sense that there is nothing withheld or unknowable about him. The interest lies rather in the possibility of different interpretations. He is recurrently praised as second only to Hector, but such praise is always inseparable from an implicit qualification. He is given the rhetoric and the lyrical artistry to invest a value in human love which experience will test, and in this he may seem heroic or misguided, or both. Seen in one light Troilus' capacity for suffering will seem little more than ludicrous; in another it emerges as heroism transposed into an unexpected key. In one view Troilus is a character of some moral worth, a virtuous pagan, both philosopher and lover; in another view, he may be seen as given to wilful self-enslavement, irrational self-delusion, and subjection to the senses. His capacity for idealization may seem either sincerely and admirably high-minded, or simply naïve and impractical. His humble passivity may seem a reflection of gentleness and modesty, but it may appear either comically or tiresomely ineffectual. His decision not to abduct Criseyde may be seen as moral responsibility or as weakness. His unchanging loyalty may be read as consistency and steadfastness, or as rigidity and obsession. Chaucer's presentation of Troilus includes both archetypal and naturalistic features, occurring in close proximity but with no presumption that they need to be integrated from a single point of view. Such a juxtaposition of the archetypal and individual in the figure of Troilus accepts that the coming of love is unique yet part of a universal pattern. The narrative enables its reader to be fully aware of all the ways of looking at Troilus' experience, for the conventionality of Troilus as a lover—like his standing in relation to Hector—has more than one side to it, and its quality of the superlative and hyperbolical can catch the light in different ways.

On Troilus, see especially Aers, *Community*; Crampton, 'Action and Passion'; David, 'Hero'; Meech, *Design*, 402–10; Stanley, 'About Troilus'; Utley, 'Chaucer's Troilus'; and Wenzel, 'Chaucer's Troilus'. On the age of Troilus, see Brewer, 'Ages', and Steadman, 'Age'. In Muscatine's influential interpretation: 'Depending on perspective Troilus can be viewed as an ideal hero of romance or as an ancestor of Don Quixote. . . . Troilus is *too* perfect a courtly lover. In him convention has taken on the superior purity that is only possible in nostalgic retrospect' (*Chaucer*, 137). For Lambert, 'in Chaucer's narrative statement, Troilus is the subject, Criseyde the object' ('*Troilus*', 105).

Instances where Chaucer gives his hero a more passive disposition than in *Filostrato* would include the following: where Troiolo is enflamed by Love (*Fil.* i. 40), Troilus is not only enflamed but held in thrall (i. 439–40); Troiolo simply cannot think of a means to communicate his feelings to Criseida (*Fil.* i. 49), but Troilus would not dare to do so (i. 503–4); Troilus feels himself destroyed ('fordon' (i. 525)), and uniquely isolated and rejected ('refus of every creature' (i. 570)); Troiolo tells Pandaro to leave him to fight against his distress (*Fil.* ii. 8), while Troilus declares 'but be thow in gladnesse, | And lat me sterve, unknowe, of my destresse' (i. 615–16); Troilus recurrently expresses himself or behaves with humility (e.g. i. 433; ii. 1069, 1257–9; iii. 96, 141, 1487; iv. 1499; v. 1320, 1354), and resigns his destiny into Pandarus' hands ('My lif, my deth, hol in thyn hond I leye' (i. 1053)).

See also Mann (*Geoffrey Chaucer*, ch. 5 ('The Feminised Hero')), who comments how, when Chaucer mentions Troilus' manhood, 'he habitually pairs it with the "feminised" characteristics—capacity for feeling or suffering, "gentilesse"—that cleanse it of aggression: "his manhod and his pyne", "manly sorwe", "gentil herte and manhod" (ii. 676; iii. 113; iv. 1674)', and notes how 'the story in which he is set casts him in a feminine role in that it assimilates him to the women of Ovid's *Heroides*—abandoned and betrayed by his lover, immobilised, frustrated of action and movement, finding relief only in memory, lamentation and fruitless letter-writing' (p. 168).

Criseyde

Chaucer's poem announces itself to be telling of the sorrows of Troilus, but it becomes the story of Criseyde as well. The two stories modify each other, but, in that Criseyde's role is not anticipated as the subject of the poem, the significance of her role has to emerge as the narrative progresses. The fascination of her character challenges but eludes any single explanation of her nature, and so makes the difficulty of interpreting and accounting for character one of the themes of *Troilus and Criseyde*.

The question of interpretation is made acute because Criseyde is a character who comes with a history, already a type-figure identified with an act of infidelity. Indeed, this infidelity is her only action, and the reason for her being written about, in the works of Benoît and Guido. When Boccaccio had in *Filostrato* invented backwards from the traditional ending to imagine what preceded it, his invention of an earlier life and character for Criseyde fits all too predictably with the ending. For most of *Troilus* Criseyde's past treachery is still to come, and the challenge taken up by Chaucer's text is to persuade its audience to reopen the case. Here is a problem in the interpretation of character which must work backwards from the disillusioning evidence of one unchangeable act at the end of the traditional story, and so raises questions about the development, consistency, and predictability of character. Was Criseyde's change of heart a part of her character all along? Boccaccio had provided one solution, yet Chaucer's version differs not only in the sympathetic fullness and delicacy of his realization of Criseyde—which tempts readers to feel they know her—but also in a continual suggestion of the impossibility of fully understanding her. The cumulative effect is to entrance but baffle the reader. The text presents information but warns, whether explicitly or implicitly, of its uncertainties, and her ultimate mysteriousness is one clue to her continuing life and the attraction she exercises. Criseyde is shown, first changing to accept Troilus, and then changing away from him in a painful reversal; she shares in the mutability of Fortune. The late description of her as 'Tendre-herted, slydynge of corage' (v. 825) when she is near to changing her affections from Troilus only highlights those elements in her nature which allowed her to accept him as a lover in the first place. In his own way the *Troilus* narrator is presented as in love with his Criseyde. But this is a dotingly sentimental and tenderly protective affection, always indulgent and defensive about his heroine, reluctant to see other than the best in her, despite the well-known end of her story. Is there calculation or innocence in Criseyde's character, or an ambiguous mixture of both?

In the narrative of a secret love-affair presented with such narratorial chivalry about a lady much concerned with her honour, the extent of Criseyde's knowledge and desires, and hence her responsibility, remain tantalizingly beyond our power to determine. Is this the narrative of a seduction or of a pretence of being seduced? To pose alternatives about

Criseyde is usually to see that the narrative allows no exclusive interpretation as one thing or another. When Criseyde watches Troilus riding past
her house, the narrating voice intervenes to expostulate against any
imputation that she fell in love too hastily (ii. 666–79); when she asks
Pandarus whether Troilus will be at his house on the evening of the supper
and Pandarus assures her that he will not be, the narrating voice intervenes
to insist that his source does not divulge whether Criseyde thought
Pandarus was telling the truth (iii. 575–81), although the narrative goes
straight on to record how she besought him to be careful about gossip 'and
wel avyse hym whom he broughte there' (iii. 586). By such interventions
the narrative raises questions about her state of mind which might well not
have occurred to most readers. But the latter question in particular is so
evidently important that it demands an answer, and the narrative's
declared inability to provide one only reminds the reader of how much
remains to be surmised. Are we to suppose that Criseyde has forgotten how
Pandarus tricked her before at Deiphebus' house? (Although we are told
beforehand that she was 'al innocent of Pandarus entente' (ii. 1723), we
never hear what she said to Pandarus about it afterwards). Are we to
suppose that she has altogether forgotten the moment when Pandarus so
forgot himself as to look forward to the time 'Whan ye ben his al hool' (ii.
587)?

The narrative of *Troilus*, which at points takes us so close to Criseyde's
consciousness, moves us outside her at critical moments, leaving such
questions open. Criseyde emerges as an intelligent woman: are we to accept
that she believes the cock-and-bull story about her lover's jealousy of
Horaste? The narrative declares loyally that she took the story at face value,
and insists that she always did everything for the best (iii. 918–24). What of
her promise to return to Troy while nevertheless insisting on going to the
Greek camp? The sources are invoked to insist on her sincerity (iv.
1415–21), just as they will later be invoked to insist on the slowness and
painfulness of her change to Diomede, even though the narrative contradicts this (v. 1086–92, 842, 1023–36). Her final infidelity is reluctantly
reported as hearsay ('Men seyn—I not [ne wot]—that she yaf hym hire
herte' (v. 1050)), even though two stanzas later the narrative also reports
Criseyde's very own words accepting her change of heart ('To Diomede
algate I wol be trewe' (v. 1071)).

A narrative method that underlines the complexities in our response and
assessment is not necessarily presenting a complex character, but our sense
of knowing Criseyde is always accompanied by a sense of what we do not
know. How old is Criseyde? It is not until a curiously belated afterthought
(v. 826) that we are told that such information is unavailable. In discussion
with Troilus, Pandarus refers to her 'youthe' (i. 982), and she thinks of

herself as 'right yong' (ii. 752). Yet she dismisses dancing as something for maidens and 'yonge wyves' (ii. 119), and after the lovers' first meeting thinks of love as having come late to her (iii. 468–9). Criseyde has already had a husband; has she had children? Again, such information is unavailable (i. 132–3). When Troilus urges Criseyde to yield to him, she retorts that, if she had not already yielded, she would not be there (iii. 1210–11); but this decisive moment has never actually been depicted. This is in keeping with the presentation of Criseyde in Book II, which approaches so closely to her very consciousness yet describes no moment of decision: she first sits in debate with herself on the pros and cons of love, then listens to her niece singing a lady's song in praise of love and her lover, and finally sleeps and dreams of hearts exchanged.

So intimate a narrative creates the very texture, the ebb and flow of mood, in Criseyde's inner self ('Now hoot, now cold; but thus, bitwixen tweye . . .' (ii. 811)). But there is a contrast between the inner and outer worlds as we both witness Criseyde's private feelings and see her from the outside, in a contrast that sometimes contains a disconcertingly double perception. The cool self-possession of her inward assessment of the news of Troilus' love (ii. 387, 462) contrasts with the outburst with which she then curiously belatedly reacts to that news with indignation (ii. 409–27). Which is Criseyde's real response? The careful, inward self-possession, or the outward moral outrage? Or is neither intended to exclude the other? Chaucer's text presents both, without comment and without relating them to each other, so as to suggest that both are equally valid expressions of Criseyde's reaction.

By preserving some mystery about what Criseyde knows, Chaucer preserves for her an autonomy, an independence both of other characters in the poem and of the poet and reader. We know, and she ostensibly does not, how so much that happens to her in Books II and III is manipulated. It would be possible to draw the conclusion that Criseyde realizes exactly what is afoot and is happy to let herself be manœuvred into what she really wants, because all the manipulation preserves appearances, allowing her to seem innocent of what is happening. Such an interpretation would be plausible—she shows every sign both of intelligence and of a kind of shrewd prudence. In saying to herself 'It nedeth me ful sleighly for to pleie' (ii. 462), Criseyde realizes she must outplay a clever player at a game of wits, and the poem does not suggest it is an unequal match. Yet her self-possession coexists intriguingly with other signs of a nature easily frightened but also easily soothed, a mixture of heightened emotions and of a certain calculation. As she appears in the temple in Book I she is 'ay undre shames *drede*' (i. 180), yet also 'with ful *assured* lokyng and manere' (i. 182), and with a look that, however modestly, shows some conviction of her own worth (i. 292). No wonder these early descriptions are limited to noticing what 'men myght in hire gesse':

> But alle hire lymes so wel answerynge
> Weren to wommanhod, that creature
> Was nevere lasse mannyssh in semynge;
> And ek the pure wise of hire mevynge
> Shewed wel that men myght in hire gesse
> Honour, estat, and wommanly noblesse.   (i. 282–7)

What her outward appearance is taken to reveal is insistently related to a quintessentially female nature and attributes.

Such a studiedly enigmatic characterization of Criseyde in *Troilus* is the distinctive product of that interventive and combinative process of translation and transformation which is Chaucer's response to his sources. At the first mention of Criseyde in the narrative, Chaucer doubles the single stanza of description he found in *Filostrato*, and in so doing introduces some of the characteristics that will be part of his Criseyde throughout the poem:

> Avea Calcàs lasciato in tanto male,
> sanza niente farlene sapere,
> una sua figlia vedova, la quale
> sì bella e sì angelica a vedere
> era, che non parea cosa mortale:
> Criseida nomata, al mio parere,
> accorta, onesta, savia e costumata
> quant'altra che in Troia fosse nata   (*Fil.* i. 11)

In such a difficult situation, without giving her any warning, had Calchas left his widowed daughter, who was so beautiful and so like an angel to behold that she did not seem to be anything mortal—she was named Criseida, as I believe, and was as prudent, honourable, wise, and well-mannered as any woman born in Troy.

> Now hadde Calkas left in this meschaunce,
> Al unwist of this false and wikked dede,
> His doughter, which that was in gret penaunce,
> For of hire lif she was ful sore in drede,
> As she that nyste what was best to rede;
> For both a widewe was she and allone
> Of any frend to whom she dorste hir mone.
>
> Criseyde was this lady name al right.
> As to my doom, in al Troies cite
> Nas non so fair, forpassynge every wight,

So aungelik was hir natif beaute,
That lik a thing inmortal semed she,
As doth an hevenyssh perfit creature,
That down were sent in scornynge of nature.   (i. 92–105)

Chaucer introduces the idea of Criseyde's fearfulness and isolation, her not knowing what to do for the best, yet emphasizes her heavenly perfection of appearance. The tendency shown here to cast her as the heroine of a romance, a beautiful lady in need of gallant protection, invites a view of her not always easy to square with the perceptive, shrewdly knowing, and self-assured woman that Criseyde also shows herself to be in other parts of the text. The Criseyde who is later shown consulting on business with a fond uncle is not really without 'any frend' after all, although she justifies her unfaithfulness by reference to needing friends even when she is with her father (v. 1026–7). The Criseyde who is so frightened of weapons that she must promise suicide by fasting to death (iv. 771–7) shortly afterwards tells her lover that she would have killed herself with his sword rather than survive him (iv. 1240–1). Her first action in the poem is to fall at Hector's feet to ask for protection (i. 106–12), and this opening picture of the female suppliant before the lord is a striking emblem of Criseyde which remains in the mind's eye as the frontispiece to her ensuing story.

The focus on womanly nature becomes very much part of the poem's distinctive tone and meaning. The long conversations between Pandarus and Criseyde in Book II, and her subsequent moments alone with her women, are all couched with alertness to the nature and position of women. The uncle's flattery, his arguments to persuade his niece, and his unspoken assumptions about the way such 'tendre wittes' as women's minds work (ii. 267–73), all speak volumes about the ways men think of women. Once she knows about Troilus, Criseyde's thoughts recur to her own position as a woman and that of women in general ('I am myn owene womman . . .' (ii. 750); 'we wrecched wommen nothing konne' (ii. 782); 'The tresoun that to wommen hath ben do!' (ii. 793)). In the garden scene an audience of women listen to a song composed by a Trojan lady about the qualities of her lover, for in *Troilus* both sexes are given opportunities to contemplate the behaviour of the other. In Book III Troilus twice addresses his lady as 'wommanliche wif' (iii. 106, 1296), and at the conclusion of the book he makes 'swich a feste and swich a proces . . . of Criseyde, and of hire womanhede' (iii. 1739–40). By having his Troilus describe the heroine in these terms, Chaucer sets his conception of Criseyde's defining 'woman-hede' alongside such rather different approaches to the character of women as are expressed in the manipulations of Pandarus or in the narrative commentary when Criseyde submits to Troilus ('For love of God, take every womman heede | To werken thus, if it comth to the neede' (iii. 1224–5)).

Fear has sometimes been taken as the key to understanding Criseyde. She is frightened by her father's defection (i. 108), the dangerous predicament of Troy (ii. 124), and the threats in Pandarus' persuasions. She is described as 'the ferfulleste wight | That myghte be ...' (ii. 450–1); her response to the news of love and a lover is recurrently described in terms of her overcoming fear (ii. 605–6, 770, 901). Her fearfulness expresses itself in a disposition to see all the drawbacks, difficulties, and even mere inconveniences in any new turn of events, and governs the way she perceives and values the qualities of Troilus. In her uncertain position in Troy she is both flattered by the attentions of a prince and mindful of possible consequences in refusing them—her late disclaimer (iv. 1667) that she was influenced by Troilus' rank seems a passage of bravura and yet another intriguing contradiction. Both Pandarus' stratagems play in turn on Criseyde's fears, first of the lawsuit and then of what Troilus might do in jealous despair; at the consummation a beautiful image describes Criseyde as a bird overcoming its fear (iii. 1233–9). Fearfulness is an understandable reaction from a woman who is without power in a world she cannot control and can only propitiate. Her soliloquy on false worldly happiness (iii. 820–33) distinguishes two character types—one instinctively trusting of worldly joys and one contrastingly distrustful—which parallel some of the distinctions between Troilus and Criseyde, just as there is a distinction between Troilus' swooning in Book III at the thought of not attaining what he has sought and Criseyde's swoon in Book IV at the thought of losing what she has possessed. To pursue the theme of fear into the later books would be to explain Criseyde's decision to stay in the camp as swayed by her fear of the dangerous return journey (v. 704), by her fear of isolation (v. 1026–7), and by her fear, played upon by Diomede, that Troy will indeed fall (v. 883–917, 1025).

Pagan religion was thought to be founded on fear, as Criseyde herself acknowledges: 'Drede fond first goddes, I suppose' (iv. 1408). Yet this same remark is part of a larger scepticism and confidence in Criseyde who—whatever she may say—does not demonstrate Troilus' inclination to think of the self as acted upon by greater forces, and who behaves as if she thinks herself mistress of her actions. Chaucer stresses that her pivotal decision in the poem—her plan to comply with her exchange and then return—stems from an obstinate self-confidence in her power to deceive the seer Calchas, in the most pointed moment in the poem (and the most fatal), where she shows a limited self-knowledge. Fearfulness does not explain this crucial moment, but to highlight any single quality as the key to Criseyde's character proves deceptive, because to isolate one characteristic—however important—is contrary to the mixture of attributes which the text accumulates. Timidity is but one part of an impressionable side to Criseyde's character which enables Chaucer to show her responding to so many suggestions and possibilities without also revealing the depth and effect of all the impressions she receives. Chaucer's version of Criseyde's character is as elusive as it is sympathetically full, and it is a play between this amplitude and the difficulty of interpretation which the narrative presentation develops.

To precede the story of Criseyde as an unfaithful lover Chaucer provides as 'preface' a narrative—longer than the narrative of her infidelity—in which the gradualness of her succumbing to love is presented. On the one hand there is always precariousness in Criseyde's position as the daughter of a traitor to Troy, but she is presented as being of much higher social status than her prototypes, mistress of a palatial household, and enjoying the very highest social connections. With Criseyde's respectable social station—and her knowledge that it is vulnerable to detractors—comes a keen concern for what society will think and say of her. Criseyde remains an essentially social creature: at home in the social and domestic realities of everyday life, she is shaped, limited, and changed by her circumstances, and her inner self is never wholly independent of her environment. Pandarus in vain urges her to cast off the respectable attire and the self-restraint of which it is an emblem and outward social sign (ii. 110, 222). But Criseyde has no taste for adventures or heroics outside the pages of books and finds Pandarus' behaviour somewhat 'wylde' (ii. 116). That widowhood which for Boccaccio's Criseida was an opportunity for sensuality is for Criseyde a vulnerable estate, and she is understandably more concerned with *sikernesse* and with avoiding *jupartie*. As a medieval lady she has been schooled to wish to please, and Criseyde is characteristically concerned to conform with whatever the person she is with expects her to be, although, by trying to please everyone, she eventually loses everything. There is poignancy in the way Chaucer shows Criseyde internalizing and accepting the values and norms of the world made by men, such that she can cry out 'What is Criseyde worth, from Troilus?' (iv. 766).

With Criseyde's social position come also charm, a gracious bearing, and a graceful wit. The refinement is an outward sign of her sensitivity, while there is a quickness in her wit which betokens intelligence. Criseyde's first words in the poem are of banter with Pandarus, and Book II establishes her as a character well able to tease her uncle and to parry words with him in sport. The contrast in character between Troilus and Pandarus is balanced by the way in which the characters of Criseyde and Pandarus complement and understand each other, joined in confidentialities and private jokes which the text does not lay completely open to the reader (ii. 213–19,

1180–3). To her uncle and to her lover Criseyde speaks in a different idiom. With Troilus she is measured and without levity. In Criseyde's conversations with her uncle there is often an engaging sprightliness, yet her manifest perceptiveness seems greater than would normally be taken in by Pandarus. If Criseyde is never shown to act in an unconsidered way, her circumstances in Troy give her ample cause, but there is a sense in which caution brings a touch of calculation to Calchas' daughter. When it is too late, Criseyde blames herself for having lacked prudence in going to the Greek camp (v. 743–9), although more generally she may rather have been too prudent: her attraction to Troilus seems conditioned by her cautious sense of her circumstances until she finds sufficient assurance in the consummation scene. Even here—at least by comparison with Troilus— there seems some restraint in Criseyde, something held in reserve. There is little suggestion that she is a person of strong imaginative life, or that love will possess her and transform her spiritually as it utterly transforms the life of Troilus. It is Criseyde who gently returns high-flown moments to earth (iii. 1306–7; iv. 1242–4), and the earlier presentation of the divided-ness of her feelings makes a stronger impression than any succeeding avowal of wholeheartedness. Indeed, there is more attention overall to how Criseyde inspires the love of Troilus than to how she loves herself. The element of the unknowable in Criseyde includes the extent and degree of her love for Troilus. Yet Criseyde thinks of her lover's sufferings as well as her own (iv. 757, 795–8), for the poem's 'double sorwe' takes on a further dimension as a deep, mutual sympathy between the lovers in their shared affliction:

> 'Allas, for me hath he swich hevynesse?
> Kan he for me so pitously compleyne?
> Iwis, his sorwe doubleth al my peyne.'   (iv. 901–3)

And where Troilus had imagined for himself an eternity of solitary complaint in hell (iv. 473–6), Criseyde imagines her spirit united with that of Troilus in the afterlife, never to be parted (iv. 785–91).

In his account of the more traditional story of her infidelity, Chaucer has given Criseyde proportionately much more attention than his sources and more belated self-knowledge: she acknowledges her self-delusion in coming to the Greek camp (v. 689–707); she laments the necessities driving her to be unfaithful (v. 1023 ff.); her accommodation to the idea of abandoning Troilus for Diomede is presented fully, yet at some unspecified future time in relation to the narrative. The one stanza which does summarize the influences 'in hire soule'—

> Retornyng in hire soule ay up and down
> The wordes of this sodeyn Diomede,

> His grete estat, and perel of the town,
> And that she was allone and hadde nede
> Of frendes help; and thus bygan to brede
> The cause whi, the sothe for to telle,
> That she took fully purpos for to dwelle   (v. 1023–9)

—neither relates those influences to each other nor suggests their relative force, and confines itself to early developments in causation. The infinitely saddening effect is that the charming, graceful figure who once seemed so near slips ever further out of focus. Criseyde can now be known only through a disingenuous letter (v. 1590–1631), which we see Troilus reading but not Criseyde writing, and which in its empty flourishes, evasions, and insinuations of bad faith in Troilus painfully mirrors a disintegrating personality. The span and process of the narrative engages the reader in an increasing and then bewilderingly diminishing knowledge of Criseyde which accompanies that of her lover. Neither the hero nor the reader can altogether understand or explain what has happened to Criseyde, and the reader's baffled sense of 'losing' Criseyde in the narrative shares in the bafflement and pain of Troilus himself.

On Criseyde, see especially Aers, 'Criseyde' and *Chaucer, Langland, and the Creative Imagination*; Burnley, 'Criseyde's Heart'; David, 'Chaucerian Comedy'; Dinshaw, *Chaucer's Sexual Poetics*; Donaldson, 'Ending', 'Criseide', 'Chaucer and the Elusion of Clarity', and 'Briseis'; Fries, '"Slydynge of Corage"'; Knapp, 'Nature'; Lambert, '*Troilus*'; Mann, *Geoffrey Chaucer*; Martin, *Chaucer's Women*; McAlpine, *Genre*, ch. 6; Meech, *Design*, 395–402; Mizener, 'Character'; Muscatine, *Chaucer*; Pearsall, 'Criseyde's Choices'; Saintonge, 'Defense'. Donaldson comments: 'If the poet were trying to make her motivation psychologically clear, he failed badly, but this was not his purpose' (*Chaucer's Poetry*, 968).

For readings of Criseyde, and Troilus, in the context of Chaucer's treatment of gender, see particularly Dinshaw, *Chaucer's Sexual Poetics*; and Martin, *Chaucer's Women*; also Mann, who discusses how, by an act of imaginative retrieval, Chaucer rescues the betrayal from an antifeminist meaning, showing human change in its fully tragic dimensions, and exploring how the lovers' relationship represents a fusion of two wills into one, 'the spontaneous moulding of oneself to the other, so that it is no longer possible to say whose will dominates and whose is subjected' (*Geoffrey Chaucer*, 21–31, 101–11).

Pandarus

> Standosi in cotal guisa un dì soletto
> nella camera sua Troiol pensoso,
> vi sopravvenne un troian giovinetto
> d'alto legnaggio e molto coraggioso;
> il qual veggendo lui sopra il suo letto
> giacer disteso e tutto lacrimoso,
> 'Che è questo,' gridò, 'amico caro?'   (*Fil.* ii. i)

While Troiolo was thus one day in his room alone with his thoughts a youthful Trojan of noble descent and great spirit unexpectedly appeared there, and when he saw him stretched out on his bed all in tears, cried out: 'What's the matter, my dear friend?'

So steps into the story Boccaccio's Pandaro, introduced as the contemporary of the youthful prince Troilus, and presumably not far in age from Criseida, who is shortly revealed to be his cousin. The Italian Pandaro is not presented as a person of any greater experience of life than the lovers, nor does he present himself as possessing any greater practical wisdom than they; he does not show any particular sense of humour; and he remains without especially distinctive personality, a relatively undeveloped and functional figure.

Such a sketch of his prototype shows at once how Chaucer has extended the role and complicated the personality of his Pandarus. The greater passivity of Troilus and the greater resistance of Criseyde mean that he can be given more space, not only initially as the go-between, but also later as the ingenious architect of opportunities for the lovers to meet. Not for nothing does Troilus cry 'But, thow wis, thow woost, thow maist, thow art al!' (i. 1052). With this more active role comes a great extension of the amount of speech: he becomes the most talkative character in *Troilus* (see above, Table 8). For Pandarus, speech is a means to an end, and it is emblematic that the very first words he speaks in the poem do not mean what they say but are uttered for the occasion and for a purpose in teasing play (i. 561).

Chaucer draws on other antecedents to supplement the prototype from *Filostrato* and produce an extraordinarily original and lively character. In many romances sympathetic go-betweens—often persons of refinement and rank, and related to one of the lovers—with delicacy and tact help bring the lovers to a mutually desired meeting, but in the active manipulation of events by Pandarus towards the lovers' physical fulfilment there is an outline resemblance to the function of some bawd-figures and procurers of very different literary tradition. Here the hired go-between helps bring about a sexual union by trickery or surprise of the woman (see above, pp. 169–72). Chaucer's Pandarus is an unnerving combination of these functions and associations, while in his mastery of language and argument he also recalls some of the preceptors and counsellors of love, such as the 'Ami' of the *Roman de la rose*.

Yet Chaucer catches the essence of Pandarus' nature as a friend: he is simply superb company, and to open *Troilus* is to be in that company. In so dramatizing the compelling attractiveness of Pandarus as a companion, Chaucer re-creates what draws Troilus and Criseyde to their friend. Chaucer's Pandarus is a man of inexhaustible wit, vivacity, and charm, full of gaiety and sly good humour, affably ready to help and to sympathize, a loyal and devoted friend who wants the happiness of his friends. Such is his exuberance that the whole pulse of the action quickens whenever he makes his entrance. Of all the vivid figures Chaucer's poems summon up to the imagination, Pandarus and the Wife of Bath are surely his most memorably vivid creations, and both attain to a kind of autonomous life. The intelligence of Chaucer's Pandarus is of a wholly different order from that of other characters in the text. He takes advantage of life as it passes and has an exceptional power to extemporize. He is the essence of initiative and resourcefulness; he generates action and shapes it. It is a performance tuned to the highest pitch yet also gracefully urbane and seemingly effortless. A figure with such control shows an inventiveness in making things happen which parallels (even rivals) the inventive power of the narrative that is presenting him, and the motives of Pandarus as a character merge with his role as the 'fictile power' of the romance.

Very different analyses of Pandarus have been made by emphasizing different aspects of the role Chaucer gives him in different parts of the poem: not for nothing is the god Janus invoked to guide him (ii. 77). Like his conversation, Pandarus is protean, mobile, shifting—and his consistency is in so being. Pandarus does so much, but why he does it—other than because of a ready sympathy for his friend—is never explained. He is revealed—characteristically—to have 'connections', to be on good terms with the royal family, and to spend whole days in the presence of the king (v. 284). But what does he look like? The poem offers no description of Pandarus' physical appearance, although it presents full portraits of the other three main characters. He becomes Criseyde's uncle rather than her cousin, which may imply but does not confirm that he is in a different generation from Troilus. A generational difference might explain the inclination of Pandarus to give worldly advice and to argue sententiously to Troilus, although it evidently has no effect on his extraordinary vitality and relish for life, nor on his own interest in love. He is presented as an unsuccessful lover (i. 622), and this is a subject of banter on his visits to Criseyde (ii. 97–8), where arch comments establish that he is seen as something of a ladies' man (ii. 211–12). Like Troilus' successful career as a warrior, Pandarus' unsuccessful life as a lover is something off-stage to this poem, and the explanation for his lack of success is never enlarged upon. The resulting irony is that Pandarus is the architect and stage-manager for others of a love-affair which, granted the character of both lovers, could scarcely have begun without him. This gap between personal experience and assumption of authority—between personal failure in love and professed belief in its worth—may indicate a fault line of something fallible or incoherent in the character from the start, but at his very first appearance Chaucer's Pandarus—with a flurry of sententious examples—ingeniously argues his own lack of success in love into an advantage in advising Troilus (i. 624 ff.).

The role Chaucer creates for his Pandarus hence requires him to be both a highly practical fixer and a theoretician about love—and by extension about life in general. He is both a pragmatist and something of an idealist in the art and craft of love, and his lack of success dramatizes the disparity between theory and practice. Interest in love notwithstanding, Pandarus is the antithesis of all that tends to the spiritual and idealizing in the lovers. He is consistently this-worldly, where Troilus would be 'celestial' (and ends in the heavens). It is through the conversation of Pandarus that so much of the down-to-earth lore, proverbs, adages, and colloquial expression are introduced into *Troilus*. His inventive speech always has the potential to surprise and dare ('What? who wol demen, though he se a man | To temple go, that he th'ymages eteth?' (ii. 372–3)), and his conversation also alludes to a range of matters from geometry to dream theory and philosophy, so that his talk holds together and seems the outcome of a range of observation and reading, from folk wisdom to bookish allusion. Against belief in dreams he can argue in an intellectual manner, shrewdly summarizing the contradictions in dream theory, before dismissing all such superstition as beneath the dignity of man (v. 358–85). Small wonder that Pandarus has often seemed like some partial projection or self-portrait of Chaucer himself: an outsider to love, in middle life, a diplomat and negotiator, moving in courts and among princes, and something of a jester, whose sophistication and humour express an understanding of human nature. Of the characters, it is Pandarus who has views on books and writing, who refers to his reading, who cites books as authorities, and who picks up a book to pass the time (iii. 980).

Pandarus is also an opportunist: for him, fortune helps those who help themselves (iv. 600–2). Pandarus always tries to seize the time, but in seizing it he serves it too. The short term may be ingeniously contrived; the longer term is unconsidered—his argument is always bent to a local and specific aim. A character who is full of arguments for doing things is not shown in any kind of reflection. He has no lyrics, monologues, or introspective debates. This is one of the disquieting undertones in Chaucer's presentation of his pragmatic Pandarus: such engaging bustle and busy-ness, such a restless readiness for activity. But does he believe in anything?

Pandarus' characteristic activity is a foil to both Troilus and Criseyde, and in his dealings with Troilus he is recurrently given the comic character's privilege of voicing an audience's censored exasperation. Urgent, 'undignified' physical actions are the role of this friend: shaking

Troilus or stripping him, poking Criseyde or thrusting a letter into her bosom, whispering urgently into the ears of both the lovers. Pandarus is active and in perpetual motion in a way that complements the passivity of Chaucer's Troilus: 'He shof ay on, he to and fro was sent' (iii. 487). Pandarus rushes (iv. 350), leaps about (ii. 939, 1637), runs and sweats (ii. 1464–5), while Troilus rests inert. He goes out and talks to people while Troilus shuts himself away. Yet the directness of his outward movements ('And Pandarus, as faste as he may dryve, | To Troilus tho com, as lyne right' (iii. 227–8)) only emphasizes by contrast how indirectly he prefers to move events. Pandarus is recurrently impatient of the slothfulness of not trying ('Now help thiself, and leve it nought for slouthe!' (ii. 1008)), and for him it is evidently part of being alive to believe that something can usually be done to affect a situation, and that it is always worth trying or risking. In this he is a necessary foil to both his friend and his niece, for Chaucer's Criseyde is by nature cautious. Not that Pandarus is rash, for his is a boldness seasoned by the experience of a calculating spirit, confident that by deft stage-management the lives of his two friends can be directed to a goal that he evidently feels they desire, even though neither he nor they will actually say so: 'Fox that ye ben!', as Criseyde greets him on the morning after the consummation (iii. 1565).

This Pandarus is wont to see the conduct of life and love as play and game, and he is a cunning player whose moves stay exultantly far ahead of others in a game that involves the manipulation of appearances both to Criseyde and to society in general (ii. 1400). The two stratagems by which Pandarus brings the lovers together, first at the house of Deiphebus and then in his own home, are in his terms triumphs of manipulated events (ii. 1743). But they also involve so many apparently unneeded risks as to amount to a kind of virtuoso performance in the art of delusion, and materials for his manipulation are the intimate lives of his closest friend and his niece. Is there perhaps something unsavoury or suspicious in the nature of such manipulation, however much it is prompted by the inexperience of Chaucer's lovers? His own lack of success in love and enforced celibacy make Pandarus something of a priest of Venus — celebrating the union of the lovers, which he has accomplished with almost a magician's sleight of hand — and an androgynous element may also be detected in his nature. He seems to anticipate the gladness of three people instead of two ('And so we may ben gladed alle thre' (i. 994)), and his proximity to the lovers threatens to betray a voyeuristic interest in their sexual fulfilment. Does the phallic gesture of Pandarus — in thrusting his arm under Criseyde's bedclothes on the morning after the consummation (iii. 1572–5) — imply that the uncle then has intercourse with his niece ('Pandarus hath fully his entente' (iii. 1582))? It is not impossible to think so, especially when not actually inside the process of reading the narrative; but baseness remains an implication, whilst benevolent sympathy and humour are everywhere explicit. Is the poem's form suggesting the difficulty in disentangling the more and less pure in human actions rather than failing to distinguish them for us? At his first appearance Pandarus is moved with what proves an unflaggingly sympathetic desire to help his friend, but a confession of love is for Pandarus the beginning of 'game' (i. 868), and his enthusiasm is so infectious that his own feelings and motives can barely be seen separately from his zest for the game. Pandarus will apparently go to any lengths to make friends happy, without pondering the nature of happiness, declaring that he would not discourage Troilus from pursuing his love, even if it were incestuous and adulterous (i. 676–9). Although it is possible to feel such sympathy is undiscriminatingly devoted to gaining pleasure and reveals the amoral shallowness of a mind which lives only in and for the present, yet Chaucer's narrative leaves it to the reader to define an ulterior motive for Pandarus, just as it leaves us to decide the question why, if Pandarus is a voyeur, we are not equally so ourselves.

It is only necessary to imagine how grotesque or tasteless a comic characterization of the role and function of Pandarus in the poem might have been to realize how an unerringly tactful use of humour allows the character of Pandarus to contribute its full potential to the poem without unbalancing it. The Pandarus who at one point echoes the words of a court jester (ii. 400–6) is a jester who is also ready 'of hymself to jape' (iii. 555), and has a special capacity for raillery that is his own in the poem. He is the only character who can be tongue-in-cheek about the solemnities and excesses of lovers and who mocks at the conventions of love which he also observes (ii. 1096 ff.); his innuendo and risqué remarks (rings and rubies (ii. 585)) express a kind of bodily humour which acknowledges the demands of the body which he accepts and serves. The juxtaposition of idealization with practicality and opportunistic efficiency points to the limited reliability of either. So concerned with the means to an end as it is, the activity of Pandarus may ultimately ask questions about both means and ends which he does not ask himself.

Throughout the poem Pandarus' attitudes serve as a foil to the dispositions of Chaucer's Troilus and Criseyde, not least once their fortunes have changed. In the first three books Chaucer's Pandarus seems the very embodiment of a genial humour which casts its spell as long as the lovers' happiness lasts. Pandarus has enabled the lovers to have what he thinks they need and want, but he is not well placed to help them once their desires move out of line with what is practically possible after Fortune's wheel has turned. In the opportunistic outlook of Pandarus the practical alternatives for Troilus are either to abduct Criseyde before she can be exchanged or to accept that a change of circumstance will bring a change of love and a new lady. The assumption of Pandarus is that human love cannot survive separation and absence, that it is inseparable from the presence of the body and cannot survive in the mind through its memory and hope. New experience causes former experience to be forgotten, and to Pandarus it would seem sensible for Troilus to accept the inevitable and find a new love. Ever ready to accept change, he is too much of a realist to hope that anything can be done to avert the consequences for Troilus and Criseyde of a political agreement between the Greeks and Trojans.

In the later books he can no longer make any effective initiatives, and there is no point in his saying what he thinks, since Troilus will not hear. The characterizing speech and bustle of Pandarus lose their earlier relevance and influence, although even now he is still instinctively framing schemes, tossing out a suggestion that Criseyde go to the Greeks and then return (iv. 935), and it is these words of her uncle's which Criseyde duly repeats to Troilus as her own plan (iv. 1261–3, 1275–9). His later conversations with Troilus become either increasingly vain remonstrance or palliative comments. His speech is no longer initiating and shaping actions but responding to events, and his last speech opens by accepting that 'I may do the namore' (v. 1731). After his own earlier machinations his final response to his niece's betrayal of Troilus is vehement ('I hate, ywys, Cryseyde' (v. 1732)). It is the aptest end to this role as talker and fixer that his last words in the poem are an admission of speechlessness when confronted by the evidence of Criseyde's infidelity: 'I kan namore seye' (v. 1743).

On Pandarus, see especially Braddy, 'Chaucer's Playful Pandarus'; Carton, 'Complicity and Responsibility'; Fyler, 'Fabrications'; Howard, 'Philosophies'; Meech, *Design*, 412–19; Muscatine, *Chaucer*; Rowland, 'Pandarus'; Slocum, 'How Old is Chaucer's Pandarus?'
On Pandarus as 'fictile power', see Wetherbee, *Chaucer*, 63; as time-server and voyeur, see Spearing, *Chaucer: 'Troilus and Criseyde'*, 43–7; as voicing an audience's censored exasperation, see Lambert, 'Telling the Story', 68.

Diomede

Although Diomede appears only in the last book of *Troilus*, his significance is out of proportion to the number of lines devoted to his speeches, thoughts, and actions. In presenting the successful rival in love to Troilus, Chaucer draws together materials from at least three earlier texts—by Joseph of Exeter, Benoît de Sainte-Maure, and Boccaccio—and dramatizes three crucial occasions on which Diomede woos Criseyde. In *Filostrato*, only one visit by Diomede causes a change in the heroine's affections; by contrast, in the *Roman de Troie* Benoît devotes more space to the interplay between Diomede and Briseida than to her earlier affair with Troilus and presents Diomede as a man thrown into helpless and exquisite suffering by a passionate love. In his own characterization of Diomede, Chaucer always seeks to distinguish him from the other three main characters, but especially as a disturbingly distorted and debased replica of Troilus as lover of Criseyde. Chaucer's Diomede is positively created to be the least complicated of the principal characters in *Troilus*. He is the only significant character in the poem whose intentions are unambiguous, while in pursuing them with Criseyde he is consistently disingenuous. There is no mystery about what motivates Diomede: he is always completely self-serving ('He is a fool that wol foryete hymselve' (v. 98)), and wholly self-confident. Where Chaucer allows us into Diomede's thoughts (v. 92–105, 771–98) he is planning what we then see him putting successfully into effect as a characteristically skilful operator, and the visit on the tenth day is shown to follow an extended period of premeditation beforehand. The first thing noticed about Diomede in the narrative is his perceptiveness about how things stand between Troilus and Criseyde (v. 88). Diomede is an adventurer, a cynical opportunist who feels he may as well try to exploit his opportunity with Criseyde, to try the door and see if it is actually locked ('For he that naught n'asaieth naught n'acheveth' (v. 784)).

A key difference between the Diomede of *Troilus* and his prototypes is that Chaucer has replaced all evidence in his sources that he is motivated by love; his thoughts are wholly given over to calculation and he is not shown to act under any impress of feeling. The contrast with Troilus could scarcely be more marked. The impression is that Diomede is motivated above all by the desire to gain, to win, to conquer. Throughout his speeches there is a sustained series of references to the comparative merits of the Greeks and Trojans as lovers (v. 118–19, 124–6, 169–71, 860–1, 918–21, 931), which reflects an individual male competitiveness as well as identification of the winning of Criseyde from her Trojan lover with the larger war over a woman between the Greeks and Trojans. He thinks that the man who wins Criseyde from her lover 'myghte seyn he were a conquerour' (v. 794), presumably a conqueror twice over: of the lady and of another man, her present lover.

Diomede's ideal image of himself is hence aggressively competitive and assertive, and Chaucer throughout establishes an impression of physical and temperamental masculinity. He is stressed to be characteristically bold (v. 795, 802), but, in approaching Criseyde, Diomede's is a nicely calculated boldness and effrontery. He may be 'this sodeyn Diomede' (v. 1024), but in fact he shows an excellent sense of pace. He is said to have the 'sleghte' or cunning necessary to the successful huntsman or fisherman, interested in netting or hooking his prey (v. 775–7), a calculated will to constrain another creature not altogether inseparable from cruelty. There is a symbolism in Diomede's taking of the rein and bridle of Criseyde's horse to lead her away from Troy and from Troilus (v. 90, 92), and he evidently sees it that way when later referring back to the act as if it represents a prior claim (v. 873). This 'sodeyn' Diomede sets the pace by forcing it: he takes Criseyde's hand when they have only just met (v. 152), and he is described as taking rather than being given Criseyde's glove (v. 1013). In the way that Chaucer summarizes events from the *Roman de Troie* the emphasis is on Criseyde's gifts to Diomede: her giving back of the horse presented to her by Diomede, her gifts of a brooch, her sleeve, and eventually her heart (v. 1037–50). There is no answering sense that Diomede gives anything of himself in return, and Troilus' dream of Criseyde lavishing her kisses on a sleeping and unresponsive boar is an image of the imbalance between give and take in the union of Criseyde and Diomede (v. 1238–41).

The principal image of Diomede in the poem is his 'portrait' in the set of three taken from Joseph of Exeter and positioned just before his visit to Criseyde on the tenth day. It is the aptest moment for a portrait of Diomede because this is his hour in the story, and the lateness of Troilus' portrait and the timeliness of his rival's tell against each other. Chaucer limits himself to describing tersely the powerful frame and forceful qualities of a warrior:

> This Diomede, as bokes us declare,
> Was in his nedes prest and corageous,
> With sterne vois and myghty lymes square,
> Hardy, testif, strong, and chivalrous
> Of dedes, lik his fader Tideus.
> And som men seyn he was of tonge large;
> And heir he was of Calydoigne and Arge.   (v. 799–805)

This is a study in power, a power which will express itself in the will to sexual possession for its own sake.

The ambiguous comment, itself reported hearsay, that Diomede was 'of tonge large'—suggesting that he was rather free of speech—forms the larger context for his use of language in his scenes with Criseyde, where he does almost all the talking. His cynicism extends to the language he uses so well as a persuasive means: it is all just breath, just words, as he several times reminds himself (v. 96, 798) before both his principal conversations with Criseyde in the poem. The reader's prior knowledge of Diomede's real attitude makes his speeches to Criseyde a study in insincerity, in the debased and dishonest application of that love language we have already heard so sincerely applied by Troilus. The impression formed of Diomede is almost inseparable from his role as the duplication in the poem of both Criseyde's and the reader's experience of the speeches of a lover. That impression is further compounded by the sense—contrary to his protestations—that lover's language is for Diomede a repetition of speeches he has already applied many times before. His prompt intuition of the love between Troilus and Criseyde is attributed to experience ('As he that koude more than the crede | In swich a craft ...' (v. 89–90)), and Diomede's claim to Criseyde that he is yet to learn how to 'compleyne aright' as a lover (v. 160–1) is belied by the evidently practised facility with which he has just spoken and is to speak again (v. 1033).

Between Diomede's thinking and speaking Chaucer pursues a chilling distinction in style as in substance which makes its own comment on the character. To himself Diomede speaks in simple, forceful language, quite without subtlety, and often with a monosyllabic briskness. The thoughts that he so expresses are tritely and meanly commonplace, clinched with well-worn maxims and proverbial phrases (v. 97–8, 784, 791, 796), and suggest more of efficient calculation than of individuality. In contrast, his speeches to Criseyde rehearse all the now familiar rhetoric of the lover, but this has already been signalled as directness pretending to indirectness ('I shal fynde a meene | That she naught wite as yet shal what I mene' (v. 104–5)). The effrontery of his prompt suit is camouflaged by a diversionary reference to the old romance tradition of love for a lady as yet unseen (v. 162–5) and by claims that Criseyde is his first love (v. 157, 940). Diomede's insincere rehearsal of the language of feudal service (v. 113–16, 132, 143, 146–7, 172–5, 921–3, 939) is revealed to the reader as a feigned humility, in which the supposed will to serve is actually in the service of a will to master, and the traditional claim to be helplessly in the power of love (v. 166–8) is belied by the very effective manner in which the claim is pressed. Diomede's cynically insincere duplication of the language of courtship at once undermines the sense of timelessness and uniqueness it carried in Troilus, inviting impatience and disgust. Diomede's hope in his first speech that Criseyde will find a Greek *as* true and kind as any Trojan (v. 124–6) has become by his second speech the boast that Criseyde will find a *more* kind and perfect lover among the Greeks (v. 918–20). After this boast the lover's moment of blushing, quavering voice and downcast look (v. 925–31)—which comes through from *Filostrato*—adds to that doubleness in Diomede that Chaucer has earlier characterized.

It is not the character of the seducer who holds the attention in this story, however, but his victim—not his techniques in courtship but their effect on Criseyde and the implications for her character. Given that the insincerity of Diomede's seduction is made so palpable to the reader, there is a sadness in seeing his gambits received by Criseyde with a meek,

apparently unsuspecting politeness, which is then to merge with accom-
modatingness, and eventually with concessions. When Criseyde receives
Diomede on the tenth day with a gracefully easy social manner—

> And forth they speke of this and that yfeere,
> As frendes don, of which som shal ye heere   (v. 853–4)

—the reader may not fault the heroine's good manners in themselves but
regrets that it is Diomede who receives such a welcome. When Criseyde
tells Diomede—in her first reference in the poem to her dead husband—
that she once had a lord she loved (v. 974–6) but has had no other love
since his death (v. 977–8), loyal readers are placed in the position of
trusting that their heroine is telling one lie and hoping that she is telling
two, but the effect of this speech—even if a piece of enforced, tactical
dissembling—is to deny and exclude Troilus in the presence of Diomede.
By the end of this visit Diomede's menacing talk of the doom of Troy (v.
990–4) has done its work: it is Criseyde's initiative to invite Diomede back
the next day, with a despairing equivocation:

> 'If that I sholde of any Grek han routhe,
> It sholde be youreselven, by my trouthe!
>
> 'I say nat therfore that I wol yow love,
> N'y say nat nay; but in conclusioun,
> I mene wel, by God that sit above!'   (v. 1000–4)

There is a studiedly vague sense of something completed, yet without
definition or limit, in the accounts of what Diomede has achieved by the
ends of these two successive days:

> And finaly, whan it was woxen eve
> And al was wel, he roos and tok his leve.   (v. 1014–15)
>
> And finaly, the sothe for to seyne,
> He refte hire of the grete of al hire peyne.   (v. 1035–6)

The text need not enter into detail to convey that Diomede has already won
what he set out to win.

On the character of Diomede, see Knight, *rymyng craftily*, and Meech, *Design*, 410–12.
   Donaldson describes him as 'that blunt, aggressive, unillusioned mind' (*Chaucer's
   Poetry*, 977).

Ending(s)

> 'But he that departed is in everi place
> Is nowher hol . . .'   (i. 960–1)

As both the preceding sections of this chapter and the earlier chapters on Sources and Genres have suggested, Chaucer has included a rich variety of materials in *Troilus and Criseyde*, which have the effect of extending and deepening the implications of his version of the Troilus story. It may well be asked not only whether Chaucer has integrated all he has borrowed and invented, but also whether he needed or aimed to do so. The very act of accumulating so many possibilities for interpretation works against seeing the resulting poem from any single, unifying perspective, and, although there is always a strong impetus to celebrate the unity of any literary work, the design of *Troilus* can educate its reader to see a wholeness in the poem distinctively different from the expectations of unity brought to a text by a modern reader, a unity more through inclusion than exclusion, more in process than in product.

Chaucer's exceptional achievement in shaping the structure of *Troilus* is a clue to his ambition for the unity of his work, for the elaborate pattern of bilateral or 'doubled' symmetry builds into the poem's form a developing perception of the story's double aspect, at first looking forward with hope, and then looking back from the far side through memory ('Remembryng hir, fro heven into which helle | She fallen was . . .' (iv. 712–13)). When loss and separation follow gain and union, a celebratory spirit and a generous sense of spaciousness and possibility is succeeded by contraction and foreclosure, and it is possible to see a process of prior illusion 'corrected' and disabused in hero, narrator, and reader. The poem's narrative form as a re-creation of experience matches with the poem's wisdom that all human experience—for Troilus' love comes to represent any worldly commitment—is conditioned and modified by the process of time and change. The lesson of such wisdom is realized through the way the poem's narrative progressively educates its reader into accepting that there can be no single fixed point of view in this world. Such a juxtaposition of contrasts, a structure of antitheses in a continuing dialectical process, typifies the structure and poetic texture of *Troilus* to the last. In juxtaposition one element modifies and is seen in relation to another. By reaching out to contain polarities within its vision, the poem promotes a kind of multiconsciousness in which any interpretation of its text must necessarily be offset by its opposite or contain its contrary within itself. This is not another way of saying that the poem succeeds in making no sense of its subject: rather, that the bewilderment and pain of seeking the sense in human experience is re-created through each reader's experience of the poem.

'By his contrarie is every thyng declared' (i. 637), as Pandarus remarks to Troilus, and, in continuing 'Ech set by other, more for other semeth' (i. 643), he states what may be seen as one of the conjoining principles of contrast and interrelation in *Troilus* as a whole, in which at last 'of two contraries is o lore' (i. 645). Not least because of Pandarus, the impetus of the action often lies in dialectic. Alternatives are ever in debate, shaping conceptions of action and character, as when Pandarus imagines Troilus imagining an inward debate in the mind of Criseyde between 'Kynde' and 'Daunger', that is, between natural inclination and disdainful reserve (ii. 1373–6), or when 'with hire riche beaute evere more | Strof love in hire ay, which of hem was more' in the portrait of Criseyde (v. 818–19). As the poem moves towards its centre, perceived patterns of contraries in experience are fulfilled:

> 'For how myghte evere swetnesse han ben knowe
> To him that nevere tasted bitternesse?' (i. 638–9)

> And now swetnesse semeth more swete,
> That bitternesse assaied was byforn . . . (iii. 1219–20)

As the action moves beyond the consummation, unhappiness is expressed in perceptions of a movement into a contrary state ('As he that nevere yet swich hevynesse | Assayed hadde, out of so gret gladnesse' (iii. 1446–7)), and in the retrospective of the latter half of the poem it is this sense of antithesis that pervasively structures the way characters relate their present to their past:

> 'My good, in harm; myn ese ek woxen helle is;
> My joie, in wo; I kan sey yow naught ellis,
> But torned is—for which my lif I warie [curse]—
> Everich joie or ese in his contrarie.' (v. 1376–9)

Near the centre of the poem is the idea of an achieved equipoise in which past unhappiness is counterbalanced by present joy ('And passed wo with joie contrepeise' (iii. 1407)). Although that counterbalancing is soon upset, there are a number of references in *Troilus* to how a character's feelings or problems stand balanced between two forces or states, and this notion of 'betweenness' as a token of qualification and interrelation is a characteristic aspect of the poem's presentation of its events: in Book II Troilus finds himself 'bitwixen hope and derk disesperaunce' (ii. 1307), and in Book V he is twice described 'bitwixen hope and drede' (v. 630, 1207). At the consummation the lovers

> That nyght, bitwixen drede and sikernesse,
> Felten in love the grete worthynesse. (iii. 1315–16)

When Pandarus describes himself as having become 'Bitwixen game and ernest, swich a meene | As maken wommen unto men to comen' (iii. 254–5) his locating of his activity between 'game' and 'ernest' points to a wider ambiguity by which almost everything in *Troilus* may be taken as both serious and comic, in an equivocal interrelation that questions what indeed is 'game' or 'ernest' but leaves all evaluation to the reader.

Yet does there come a point towards the end of such a structure that serves to reconcile what is contrary and to resolve what is ambiguous? The ending Chaucer designs for *Troilus* acknowledges such a sense of need, even as it reminds the reader that endings are artificial. If 'th'ende is every tales strengthe' (ii. 260) as Pandarus declares, echoing a commonplace, the particular 'strengthe' of the ending of *Troilus* lies in its continuing capacity to invite questions about itself and so about the ends and design of the poem as a whole. A conclusion that includes rejection of the pagan gods and their world, and that urges its audience to turn to the love of God, in a seeming renunciation of the love of Troilus, has prompted almost as many interpretations as there are essays about *Troilus*. One view has always been to accept that, at its conclusion, *Troilus* 'reverts' to a more conventionally 'medieval' and orthodox approach to the pagan past and to the enjoyment of earthly love than prevails in the body of the narrative. A sudden shift of perspective, a transference of allegiance, may be seen to convey the exaltation of a perception of religious truth and with it a resurgence of faith and renewed devotion: an uncompromising tone befits a final vision that has passed beyond compromise, with all the momentum of a return journey, an act of repentance, and a reaffirmation. For other readers, the very conventionality of sentiment (after a poem that has explored the limits of convention), and an assertiveness at variance with the narrative technique earlier in the poem, allow such an ending to be disregarded. Another view, by contrast, has insisted on a unity in *Troilus* that overrides apparent disjunction between the poem and its ending, arguing that the other-worldly dimension to the conclusion has been implicit, by means of irony, throughout the presentation of a love-affair that inscribes its end in its beginning when it opens by announcing how it will end. In such a structure of ironies the ending has a key role, for the anticipations of changing fortune, the naïveties and self-delusion of an unreliable and incompetent narratorial persona, and the (mis)applications of a religious language to the earthly experience of love, are among elements that act as ironic portents, keeping before the reader the memory of the story's end, which thus comes as the fulfilment of a pattern of expectation.

Views of the ending as either disjunctive or implicit are more complementary than contradictory. The ending is implicit through a complex structure of ironies in the narrative, yet when it does come it retains a power to surprise with the unforeseen. However conventional many of its materials and sentiments, it makes no conventional effect. It is in the nature of the story of Troilus and Criseyde that it offers no single event to serve as a climactic ending: the precise moment when Criseyde's fidelity to Troilus comes to an end is obscured by a chivalrous narrative; the end of Troilus is not synchronized with his betrayal by Criseyde, while his death in battle, taken at a disadvantage and slaughtered by Achilles, is not in itself a climactic event; nor did tradition as yet include the end of Criseyde (although Henryson would later invent it). While the story of Troilus and Criseyde remained a series of episodes interwoven in a larger narrative, its intrinsically anticlimactic nature posed little problem. In *Filostrato*—the only free-standing version of the story prior to *Troilus*—the end of a disillusioned Troilus in battle can lead into some antifeminist warnings, before the poem is sent on its way, ostensibly to Boccaccio's mistress, with a graceful envoy. As he exhausted the material for his fifth book of *Troilus* and neared the point of Troilus' death, Chaucer faced the problem of finding a conclusion not only to a story that lacked a climactic end, but also to all that abundance of implication with which he had invested his version of the story in his poem. In *Troilus* the endings of the story and of the poem become distinct, and through the performative quality of the ending Chaucer is able to dramatize the problematic process of making an ending.

If the *story* had no climax, could there be a climax to the *poem*, a climax of interpretation, of derived meaning? The ending contrived by Chaucer dramatizes a climactic awareness of difficulty, to form the framework that contains within itself an array of diverse closural devices and formulas. One such closural feature is the narrator's setting aside of the fictional world (giving up the earlier fiction that the narrative is not a fiction). With this comes a return from the imagined world of a pagan past into the Christian present shared with the poem's projected audience, a return from pagan incompleteness to Christian completeness, and with it a transition from Troilus' last, loveless solitariness to the community of those who share in the love of God. Like many medieval poems at their conclusions, the ending of *Troilus* is recurrently prayerful and ends in fervent prayer so that, recalling the first proem, the whole poem is bracketed within the act of prayer. Perhaps prompted by Philosophy's idea of the flight of thought into the heavens in the *Consolation*, the sky-journey of Troilus' soul after death also brings intimations of a corrective and final progression from blindness to perception, from ignorance to knowledge, and from folly to wisdom, even if Troilus' own learning and progress are left largely to implication.

The ending of *Troilus* gives the impression—to be analysed shortly—of deploying as many closing devices and formulas as can be thought of, with shifts and turns, in a way that dramatizes narratorial nervousness and anxiety. The very number and variety of forms of ending employed draws attention to the process of making an ending, and to its necessary artificiality as a made conclusion to the narrative: a series of closing devices that represents both an attempt at comprehensiveness and an uncertain accumulation of alternatives. The impression is given of a succession of different endings, of ending repeatedly in a way that postpones any single ending, and this works against seeing the concluding phase of the poem as a homogeneous single unit in itself, or as the single point from which the poem is to be explained as a whole. 'Ne jompre ek no discordant thyng yfeere' (ii. 1037) had been part of Pandarus' advice to Troilus on writing, yet the distinctive jumbling of matter and tone in the ending reflects the strain of confronting a difficult task. Flurries of renunciation and rejection of what has gone before are presented so as to be seen as part of a particular process of dramatized reaction (perhaps over-reaction), rather than as some measured judgement handed down authoritatively to the reader. Although cumulatively the end of Chaucer's fifth book does act as 'an' ending to *Troilus*, it is essentially pluriform, and as such it may be seen to take its place as part of the poem's debate about love 'of kynde' and love 'celestial'. As Chaucer had translated from Boethius' *Consolation* in his *Boece*: 'al that evere is iknowe, it is rather comprehendid and knowen, nat aftir his strengthe and his nature, but aftir the faculte (that is to seyn, the power and the nature) of hem that knowen' (v, pr. 4, 137–41).

The difficulty of seeing the poem from any single point of view has in some modern interpretation of *Troilus* been seen as the culminating difficulty of a narratorial character completely and consistently distinguishable from the author at every point in the preceding poem. Yet it is not necessary to accept the implications for the whole poem of this somewhat anachronistic emphasis on 'character' in order to appreciate the role in the ending of a self-dramatizing narrating voice. In the ending such a voice may be heard to confront the sad but foreknown conclusion of his long past story, which the spell of his narrative has caused both his readers and himself more than half to forget while the story lasts, a narrative forgetfulness in identifying with the present and immediate that is mimetic of the lovers' own experience. In closing, the poem includes within itself a momentary sense of its own art as an illusion and a futility, along with the love that art has celebrated and made to live again, and the sense of strain in the ending mimes an acute awareness of change and difference between the beginning, middle, and end of the whole narrative that challenges 'the faculte (that is to seyn, the power and the nature) of hem that knowen'. A promise of lasting joy in human life and sexuality was not the pattern in Chaucer's borrowed story, and the wisdom of his version of that story lies in realizing both the climax and the ensuing anticlimax with such intensity that each reader will still feel challenged to relate them, and may not succeed. The exhaustion of earthly hopes and prospects for a pagan hero and pagan lover creates a vacuum that, for the medieval poet and his Christian audience, may be filled by an ending that suggests not so much a renunciation and rejection as a transference and redirection of that longing and striving for love instilled in the human heart ('And of youre herte up casteth the visage' (v. 1838)). Only in a love for God—truth and reality itself, beyond time and change—can the aspirations of the human power to love be fulfilled. Yet the poem does not map a development in which there is some charted progression from the one love to the other, although it presents an extended sequence of experience in which each love successively fills the picture.

The rest and peace found in God beyond the narrative of the characters' lives makes 'love celestial' both spiritually and aesthetically an aptly unifying point of resolution and renewal after a tale whose anticlimactic end realizes precisely that shattered unity that Troilus had earlier described as the antithesis of love ('And lost were al that Love halt now tohepe' (iii. 1764)). Yet a resolution achieved only after the end of the story for the reader, and only after death for the pagan hero, is pointedly outside and apart from the depicted experience of human love. The removal of the story to a pagan era allows the apparent joy and potential in human love to be celebrated to the full, while the poem's ending can look back on that love's limitations in a Christian perspective for an implied Christian audience without needing to press too closely the application of such an interpretation or pass judgement on the story's pagan lovers. In translating Petrarch's Sonnet 132 into Troilus' first Canticus (i. 400–20), Chaucer does not take over for his hero the lines in which Petrarch describes himself:

> 'sì lieve di saver, d'error sì carca
> ch'i' medesmo non so quel ch'io mi voglio'.

'so light of wisdom, so laden with error, that I myself do not know what I want'.

For Troilus to think of himself as lacking in wisdom, and in error, and so early in the poem, would be contrary to Chaucer's development of a story in which to discover the knowledge of what one should want becomes part of the accumulated implication of the whole. To open up spiritual perspectives without condemnation of individuals: such is the openness of approach, and charity of spirit, that the ending achieves.

To understand the ending(s) as a process it can be helpful to analyse the concluding stanzas of *Troilus* as a succession of endings, and to this we now turn. Where does Chaucer's *Troilus* begin to end? The death of Troilus and an account of the war is still to come, but there is a sense of the poem's main emotional action coming to a close in the stanza where Pandarus utters his last words (v. 1737–43), the last reported of any character in the poem, praying God to remove Criseyde soon from the world—although, despite his prayer, it will in fact be Troilus who is the first to be delivered from this world.

The next stanza (v. 1744–50) shifts to a wider narrative focus on behaviour and events over a period, to which Chaucer adds three lines of reflection on the world and a prayer, first of a number of prayers in the ending of *Troilus*:

> Swich is this world, whoso it kan byholde:
> In ech estat is litel hertes reste.
> God leve us for to take it for the beste! (v. 1748–50)

There follows a stanza (v. 1751–7) emphasizing the nobility, might, and knighthood of Troilus, but also his wrath, paid for 'ful cruwely' by the Greeks 'in many cruel bataille'.

This is the cue for referring the reader elsewhere ('Rede Dares ...') for a full account of the hero's military career, and there follow five more stanzas (v. 1765–99) before the single stanza briefly describing the death of Troilus in battle. It is perhaps these five stanzas that contribute most to the impression of jerking transitions and unforeseen shifts of subject left by the ending of *Troilus* as a whole. They contain in effect three different forms of address: to readers interested in Troilus' larger career, to women, and to the book of *Troilus* itself. The stanza directing readers to 'Dares' itself draws attention to the fact that the present work deals with only one side of the hero's life, different aspects of which may be read about in different books and other genres, and no one book necessarily matches with the whole of his life. This theme of books and lives forms an underlying connection to the next two stanzas (v. 1772–85), which open 'Bysechyng every lady bright of hewe, | And every gentil womman ...', although the address to women at this point comes unforeseen and reads like a symptom of narratorial anxiety. The emphasis here is that other books all contain the same story of Criseyde, the story of her wrongdoing ('Ye may hire gilt in other bokes se' (v. 1776)). It is the existence of these comparable accounts of Criseyde's guilt that exonerates the present writer and his work. A book on a different aspect of women's behaviour must be a book about different women, or so it is implied ('And gladlier I wol write, yif yow leste, | Penolopeës trouthe and good Alceste' (v. 1777–8)). This curiously belated preference to write about another subject is one of a number of points in the ending of *Troilus* that dramatize a wish not to have entered on the present work and now to be elsewhere, variously a sense of embarrassment ('Be nat wroth with me' (v. 1775)), or of weariness and resignation ('But for that I to writen first bigan | Of his love, I have seyd as I kan ...' (v. 1768–9)).

The second stanza of the address to women (v. 1779–85) is an artfully controlled dramatization of ingenuousness, opening with the claim 'N'y sey nat this al oonly for this men, | But moost for wommen that bitraised be | Thorugh false folk . . .' (v. 1779–81). To identify with women who are betrayed comes both unexpectedly—in the context of traditional antifeminist views of Criseyde—and yet by no means inappropriately in the aftermath to this particular story of Criseyde as the prey first of Pandarus, then of Diomede. To this identification with 'wommen that bitraised be' follows a prayer (v. 1781–3), which serves as an apt point of departure for the stanza's final line, presented as if blurted under the impress of strong feeling: 'and in effect yow alle I preye, | Beth war of men, and herkneth what I seye!' (v. 1784–5). That, at the close of the story of Criseyde's infidelity to Troilus, women in the audience should be begged to beware of men, is a dramatization of an author's unease with the traditional implication of the material he has reworked, of thinking that he cannot risk leaving his version of the story to speak for itself.

Such writerly concerns are the bridge to the final two stanzas of the five (v. 1786–99), beginning 'Go, litel bok . . .' Here is the belated description of *Troilus* as 'litel myn tragedye'. Here, too, is the prayer to be allowed 'to make in som comedye', whether this means to 'match' tragedy with comedy, or more simply to write comedy, either in some subsequent opus or imminently, before the conclusion of *Troilus*. The whole stanza is an ambivalent mixture of tones, ostensibly humble, but less so in implication. The 'litel' book is instructed not to vie with 'makyng' or verse in English, but 'subgit be to alle poesye', by which Chaucer may have meant poetry seen in its classical and European dimension (perhaps with Dante in mind; see above, pp. 131–4). Yet even to imagine the poem as following in the wake of the five classical poets named in the text—Virgil, Ovid, Homer, Lucan, and Statius—is to make an implicit claim (and a grand one) for the standing of *Troilus* in relation to the classics, which could be made for no other contemporary English poem.

The sense of the poem's worth includes the specialness of such a poetic accomplishment in the English language, and this leads on to the stanza (v. 1793–9) praying that the text be not miswritten nor rendered unmetrical through the process of transcription and 'That thow be understonde, God I biseche!' (v. 1798). Even before the death of the hero is at last described, the poem is worrying about future distortions of its form and meaning, and, by fearing possible misunderstanding, it implicitly raises the problems of interpretation. The order and proportion of material implies that reporting the end of Troilus is a formality overtaken in importance by the concerns of concluding the poem. Nothing conveys how the continuum has shifted to narratorial procedure as much as the holding back of one very obvious and natural end to any narrative—the death of its hero—by the interpolation of reflections which do not need to precede that event and might more fittingly follow it.

Coming as it does in this context, the version in *Troilus* of what was already a designedly terse account in *Filostrato* of the hero's murder by Achilles must read even more abruptly because of Chaucer's shaping of the stanza's structure. First comes a report of how the wrath of Troilus causes the deaths in battle of thousands of Greeks, then one last rehearsal of the poem's refrain of qualified praise:

> As he that was withouten any peere,
> Save Ector, in his tyme, as I kan heere   (v. 1803–4)

and then an expression of pity, counterbalanced with pious resignation ('But—weilawey, save only Goddes wille' (v. 1805)). And finally—aptly pressed against the limit of the verse-form—one last line of a stanza suffices to report the merciless death of Troilus, a central figure in the previous eight thousand lines and more ('Despitously hym slough the fierse Achille' (v. 1806)). A single line in a single stanza to contain the hero's end matches in a truncation of form the cutting down of Troilus in the field.

The ensuing sense of release, of a widening vista and extent of space achieved by movement and ascent beyond this life, makes a striking contrast, and enacts within the poem the frequent prayer at the end of romances that the hero's soul be brought to bliss. In taking over for Troilus the flight of Arcita's soul in the *Teseida*, Chaucer avails himself of all the resonance of a tradition of such soul journeys and apotheoses, although there has been much dispute as to the insight and reward implied by associating this tradition with the hero of Chaucer's poem. Any such insight comes with translation to the skies and an accompanying shift of perspective, rather than being a lesson learnt from and within earthly experience. There is no mention of love during the flight of Troilus' soul, and no attempt explicitly to relate the hero's previous role as a lover to his perceptions after death. There is, however, an emphasis on the achievement of perception, both in the whole image of the flight and downward gaze, and also in the note that Troilus 'saugh with ful avysement' (v. 1811). That the hero's soul ascends 'ful blisfully' and that he has the satisfaction of hearing the harmony of the music of the spheres—'ful of hevenyssh melodie' and so transcending all previous mention of music and song in the poem—seem to be signals that the flight is some endorsement of value in the hero, even though his more permanent abode remains uncertain. It is characteristic of Chaucer's *Troilus* to endow its hero with a final vision, yet not to make that the whole and ultimate vision of the poem. What Troilus does achieve—and the reader with him—is a vivid sense of comparison, contrast, and proportion. From being engrossed in the pleasures and pains of worldly life, he sees the world from beyond death in a context that

evidently diminishes it. There is still an affectionate embrace for the world
left behind in the way that Troilus now notices its smallness ('This litel
spot of erthe that with the se | Embraced is . . .' (v. 1815–6)). Yet this
lingering affection is immediately followed by lines in a different key ('and
fully gan despise | This wrecched world' (v. 1816–17)) and by an insistence
on the contrast between this world and heaven ('and held al vanite | To
respect of the pleyn felicite | That is in hevene above . . .' (v. 1817–19)). It
is the combination of the soul's flight with the inward laugh of Troilus—he
who has so rarely laughed or smiled—that makes this passage such a
striking representation of transition and relinquishing. The grief of those
that mourn a death is an object of laughter to the dead person who can
watch them mourning from beyond death, and the living reader for a
moment shares the viewpoint of the dead hero. Troilus perceives one last
contrast,

> And dampned al oure werk that foloweth so
> The blynde lust, the which that may nat laste,
> And sholden al oure herte on heven caste;   (v. 1823–5)

where the plural verb *sholden* implies the subject *we* and so extends the
hero's thought to include both narrator and reader, before Mercury leads
Troilus to a dwelling place undisclosed by the text.

Now, and for the last of so many times in his interaction with
Boccaccio's poem in the making of *Troilus*, Chaucer turns back to base his
next two stanzas (v. 1828–41) on two stanzas in *Filostrato* (viii. 28–9),
although their wording and tone are significantly altered. In the first of
these stanzas the repeated pattern of 'Swich fyn hath . . .' develops
Boccaccio's phrase into a much more insistent and plangent rhetorical
pattern. The *Troilus* passage uses the formula to regret the end of
admirable or desirable things ('worthynesse', 'estat real', 'noblesse') as well
as 'false worldes brotelnesse', whereas Boccaccio's passage exclaims on the
end of an ill-conceived love, unequalled sorrow, and vain hope. Chaucer
mentions only the beginning of love and pointedly does not include its
ending ('And thus bigan his lovyng of Criseyde | As I have told, and in this
wise he deyde' (v. 1833–4)).

The following stanza in *Filostrato* (viii. 29) is a warning address to young
men, and the moral transformation this has undergone in being rendered
into English is an eloquent token of the warmth and depth in that
consideration of love in Chaucer's poem that is completed by the ending of
Troilus:

> O giovinetti, ne' quai con l'etate
> surgendo vien l'amoroso disio,
> per Dio vi priego che voi raffreniate

i pronti passi all'appetito rio,
e nell'amor di Troiol vi specchiate,
il qual dimostra suso il verso mio;
per che, se ben col cuor gli leggerete
non di leggieri a tutte crederete.   (*Fil.* viii. 29)

O you young men, in whom amorous desire springs as you grow up, in God's name I pray you to check your eager steps towards evil passion and see yourselves mirrored in the love of Troiolo which my verse has set forth. For if you take it to heart as you read you will not lightly put your trust in all women.

O yonge, fresshe folkes, he or she,
In which that love up groweth with youre age,
Repeyreth hom fro worldly vanyte,
And of youre herte up casteth the visage
To thilke God that after his ymage
Yow made, and thynketh al nys but a faire,
This world that passeth soone as floures faire.   (v. 1835–41)

The added idea of 'fresshe' suggests the unsullied glad intent of the young, and the broadening of address to 'he or she' witnesses that the poem is talking about love seen from both sides, a mutual human love, and acknowledges that a growing capacity for love is inseparable from the process of growing up itself. *Filostrato*'s injunction to 'see yourselves mirrored' in the love of Troilus has in turn prompted a marvellous change into the injunction in *Troilus* 'up casteth the visage | To thilke God that after his ymage | Yow made ...' (v. 1838–40). The implications for earthly love—like that of Troilus—of the other-worldly perspectives of the close are largely left to emerge from a juxtaposition between the ending and the poem, although there is a double aspect even to the language of other-worldliness:

and thynketh al nys but a faire,
This world that passeth soone as floures faire.   (v. 1840–1)

Here—in a doubling built into the rhyme itself—the lines advise detachment from the transience of the world, yet convey an acknowledgement that the instances of its transience are among its loveliest things. The ending of *Troilus* maintains the wisest and most generous view of human love, including more than excluding. Love 'of kynde'—as represented in a story of mutable love between pagans, and allowed its fullest celebration in the narrative—is not as such rejected in the ending, although the accents of simple rejection are rehearsed as one possibility, and an implicit contrast with 'love celestial' emerges, a contrast that readers are left to interpret for themselves.

The increasingly other-worldly focus recommended in the course of Chaucer's previous stanza finds its natural development in a commendation to love Christ (v. 1842–8), in the first of four last stanzas—themselves a typically Gothic mixture—that Chaucer has written to bring his poem to a close. Even now, no explicit condemnation of earthly love is included, yet to commend a love for Christ as a love that will never be betrayed ('For he nyl falsen no wight . . .' (v. 1845)) offers implicitly—to those who wish to see it—a contrast with earthly love in general and the previous narrative in particular. In a way representative of the poem's whole debate on love— open and without exhortation—the recommendation to love Christ ends simply with a question:

> And syn he best to love is, and most meke,
> What nedeth feynede loves for to seke?   (v. 1847–8)

From this sense of the primacy of love for Christ there is some inevitability in the momentum to reject those pagan gods, practices, and poetry that the earlier poem has so carefully studied (v. 1849–55), and the rhetorical example of the repeated 'Swich fyn . . .' is applied again, but in a less open-minded spirit ('Lo here, of payens corsed olde rites!' (v. 1849)). The last six stanzas hence fall into a pattern of two triads of stanzas, each opening with renunciation, and each ending with religious affirmation, before a last exclamation on 'olde clerkis speche | In poetrie . . .' provides the impetus for the next stanza to open with a contrasting dedication of *Troilus* to the contemporary writers, Gower and Strode.

The dedication implicitly asserts that the poem's treatment of its subject befits and merits the attention of those who are themselves 'moral' and 'philosophical' in cast of mind, yet the artist's sense of worth in his creation is in turn followed and contrasted with the humble stance of worship and of prayers for mercy in the poem's final movement (v. 1860–9). Here contemplation of the majesty of the Trinity is complemented by invocations of aspects of Jesus and his mother—'sothfast', merciful, and 'benigne'—that have been at the heart of what the poem has valued in human love and human character. As Chaucer had himself translated into English from Boethius' *Consolation of Philosophy*, and as is suggested by the openness of *Troilus* in its ending: 'of thinges that han ende may ben maked comparysoun, but of thynges that ben withouten ende to thynges that han ende may be makid no comparysoun' (*Boece*, ii, pr. 7, 106–9).

Brewer sets response to the ending of *Troilus* in the historical context of developing responses to Chaucer by remarking that 'It is not surprising that about the same time as *Troilus* is first described as a novel, early in the nineteenth century, so critics also start complaining about the ending of *Troilus*, which is non-organic, unconcerned with plot and character, non-illusionist as the poet changes the plane of fiction' (*Chaucer* (3rd edn.), 167).

For early critical response to the ending of *Troilus*—concerned with whether to term the last stanzas an 'epilogue', or even a 'palinode', and with whether or not the ending is implicit and prepared for in the poem—see especially Curry, 'Destiny'; Kittredge, *Chaucer and his Poetry*; and Tatlock, 'Epilog'. To Curry the ending is 'dramatically a sorry performance . . . not a part of the whole and . . . detachable at will' and represents 'the complete separation of the pure artist from the religious man' (pp. 66–9). To Kittredge the ending is 'an utter abandonment of the attitude so long sustained, and therein lies its irresistible appeal' (p. 144). To Tatlock the ending 'is in no way foreshadowed at the beginning or elsewhere; it does not illuminate or modify; it contradicts . . . He tells the story in one mood and ends in another' (p. 636).

Lewis, in a justly famous passage, gives a context to the religious effect of the poem's ending: 'We hear the bell clang; and the children, suddenly hushed and grave, and a little frightened, troop back to their master' (*The Allegory of Love*, 43), and a sense of the cultural inevitability of a prayerful closure has been explored from very different points of view.

Robertson ('Chaucerian Tragedy') sees renunciation as the only appropriate conclusion to a story that rehearses the fall of Adam; and that the ending declares explicitly what is suggested and anticipated implicitly through the poem's narrative has become a common element in most modern interpretations, however different these may be in other ways.

Waswo points out that, while most critics distance themselves from Robertson's critique, most actually 'accept its basic terms and set themselves the task of demonstrating that the primary technique by which the narrator of the *Troilus* evokes unstated values is irony' ('Narrator', 3), although the outcome of their efforts only demonstrates the poem's seemingly inexhaustible capacity to generate in different readers different perceptions of what is ironical.

Muscatine's influential interpretation sees the ending as the 'third view' which follows on to the poem's characterizing dualism and tension: 'To present secular idealism as a beautiful but flawed thing, and to present practical wisdom as an admirable but incomplete thing, to present them, indeed, as antithetical and incongruous to each other, is by implication to present a third view, higher and more complete than either. This philosophical third view hovers over every important sequence in *Troilus* and is made explicit in the ending' (*Chaucer*, 132). It follows from this that 'The moral of the epilogue is inherent in the poem from its beginning', for 'That the poem is a criticism as well as a celebration of secular life is announced in its very first line' (p. 162). The unity of the implicit and the explicit, the poem and its ending, has repeatedly been located in conceptions of a narratorial 'character', and for Muscatine 'the ending takes on the nature of a dramatic climax to his participation in the poem' (p. 161), in which some of the declamatory earnestness can be related to the narrator's psychology.

Donaldson develops eloquent interpretations of a narratorial character subject to divisions of allegiance and feeling. In the ending's form and its relation to the poem 'the simultaneous awareness of the real validity of human values—and hence our need to commit ourselves to them—and of their inevitable transitoriness—and hence our need to remain uncommitted—represents a complex, mature, truly tragic vision of mankind' (*Chaucer's Poetry*, 980). The sense of strain in the ending shows how 'an extraordinary feeling of tension, even of dislocation, develops from the strife in the narrator's mind between what should be and what was. . . . The meaning of the poem is not the moral, but a complex qualification of the moral' ('Ending', 91–2). Such an understanding of the narrator allows earlier biographical criticism of the ending to be left behind: 'So skilfully has Chaucer mirrored his narrator's internal warfare—a kind of nervous breakdown in poetry—that many a critic has concluded that Chaucer himself was bewildered by his poem' (p. 91).

Salter argues that the relation between the ending and the poem expresses an unresolved struggle within a poet 'whose gradually changing purposes involve him in greater and greater difficulty with his sources' (*'Troilus and Criseyde*: A Reconsideration', 88), in a poetic text that very immediately reflects the processes of its composition. The outcome is that, 'like many of Chaucer's answers to complicated problems, the final answers given in *Troilus* do not match the intelligence and energy of the questions asked' (p. 106).

David expresses a comparable view with comparable conviction ('Chaucer in composing *Troilus* did *not* fully know what he was about and was a true Poet *because of*, not in spite of this fact' (*The Strumpet Muse*, 29–30)), and he questions the view of the ending as expounding at last a moral hitherto implicit in the poem: 'In a hundred ways we have been prepared to accept the moral. And when it comes we find—at least I find it so—that it is unacceptable. And ultimately I think it is also unacceptable to Chaucer the poet' (p. 33). For David the ending represents a struggle between what Chaucer's 'intellect as a medieval moralist tells him ideally should be and what his feelings as a poet tell him actually is true' (p. 30).

Gaylord ('Lesson'), by contrast, sees such notions of the poet in resistance to the moral impetus of his material as smacking more of the Romantic poets than of that 'chastisement and correction' of its narrator and audience that is the poem's purpose. The lesson in the poem has been the subject of a succession of learned interpretations of *Troilus* and its ending—e.g. Gordon, *Double Sorrow*; Kean, *Chaucer*—especially in terms of a larger reading of the poem in relation to its traditions and sources: in Steadman, *Disembodied Laughter*; in the final chapters of McAlpine, *Genre*, Taylor, *Chaucer Reads*, and Wetherbee, *Chaucer*; and particularly in relation to Dante (Morgan, 'Ending'; Wheeler, 'Dante').

Rowe (*O Love, O Charite!*) also interprets the poem and its ending in the context of tradition as a pattern of contraries harmonized into a unified conception and a unified work of art, while David points out that *Troilus* 'retains its wonderful doubleness yet still remains all of a piece' ('Chaucerian Comedy', 91), for modern arguments for the poem's thematic unity are often prompted by a need to account for a perceived challenge to unity in the nature of the ending.

Brewer—arguing that the unity of *Troilus* should be understood by analogy with ideas of unity and form in the Gothic arts—proposes that Chaucer 'at the debatable end of *Troilus* . . . seems to juxtapose irreconcilable conditions rather than organically to unify them. . . . It is as if the poet himself refrained from a preferred sense, leaving an autonomy with the words, and with the reader, that can sustain or reject many ambivalences or frank incompatibles' (*Chaucer* (3rd edn.), 170).

Lowes originally noted the 'tumultuous hitherings and thitherings of mood and matter' in the ending (*Geoffrey Chaucer*, 153), and the kaleidoscopic shifts and reconstitutions of form and emphasis may be related to a rhetorical structure for the ending in a succession of apostrophes which transfer responsibilities to the audience (Evans, '"Making Strange"'), and hence contribute to the openness of the ending, along with its deployment of such a variety of closural formulas and devices (Dean, 'Chaucer's *Troilus*'; Spearing, *Readings*; Tatlock, 'Epilog').

Bishop compares the ending's self-consciousness about processes, conventions, and forms with the effects achieved by actors' 'unmasking' and setting aside their assumed identity, or by 'turning up the house lights' somewhat before the close of a theatrical or operatic performance (*Chaucer's 'Troilus and Criseyde'*, 94), although for Chaucer's audience such unmasking does not return to the 'reality' in which they partake in the world, but depends on an 'intellectual scheme of objective value which related material appearances in the end very low and set true reality beyond them' (Brewer, *Chaucer* (3rd edn.), 167). In the end, 'whether we share Chaucer's Christian conformity or not, we should recognize what an admirable source of detachment it is to him' (Bayley, *Characters*, 72).

Spearing—who notes of *Troilus* that 'almost no interpretative statement can be made about it that does not require correction by its opposite'—interprets the poem's ending as offering 'a final but not a total view of the meaning of its story', and concludes: 'The essence of the work lies in movement, change; and the reader must move through it again and again, realizing it as a shape changing in time, and himself changing with it' (*Chaucer: 'Troilus and Criseyde'*, 3, 63).

Style

The Play of Style

Language and style, and their nature and use, are central to the poem's exploration of its subject. Style is of the essence both for the rhetoric of hopeful love, and for the rhetoric of elegiac lamentation and disappointment. To his treatment of this story Chaucer brings a sheer multiplicity of styles and an abundance of language that contribute to the ambitious scope and variety of implication of the poem.

Throughout *Troilus* the narrating voice of the poem is highly conscious of language, prompting readers to awareness of the choices represented by the styles and diction of the poem, reminding by various devices that the language being used remains open to question. Part of the second proem addresses itself as if to a reader criticizing the unfamiliar mode and idiom of love depicted in the poem, and such criticism is countered by a reminder that language and custom change over time. Yet, since the language and manners of courtship portrayed are apparently contemporary, the imagined criticism and the defence aim more to attract attention to the language used than to address a real problem about archaic terminology. By emphasizing that the narrator writes from no personal experience of his subject-matter of love—that he therefore has no more discrimination in the idiom than a blind man has in judging colours (ii. 19–21)—this proem serves to establish at the outset of the courtship that the narrator's judgement is no more authoritative in questions of language than in other matters. The choice of language is being made by a narrator who reminds his reader, by means of an analogy with sculpture, that there is always a choice of mediums and materials for expression ('Ek som men grave in tree, some in ston wal' (ii. 47)). Indeed, at the height of the consummation scene the narrator interposes to submit the language he has used to correction or adjustment by his readership in the light of their own experience of such love (iii. 1328–36). In the context of a narratorial fiction so ostentatiously humble about his command of language, conventional assertions of the indescribability of some surpassing joy or grief take on a fresh force (e.g. v. 267–73). The poem is ostensibly being narrated by someone who casts himself as unable to match scenes of high feeling with adequate language:

> But, as for me, my litel tonge,
> If I discryven wolde hire hevynesse,

It sholde make hire sorwe seme lesse
Than that it was, and childisshly deface
Hire heigh compleynte, and therfore ich it pace.   (iv. 801–5)

In power of expression, as in experience, he is as far from the lovers' passionate emotions as a child.

The narrative makes a show of describing its own rhetorical processes in specialized terms ('It were a long digression' (i. 143); 'But fle we now prolixitee ...' (ii. 1564)), and Pandarus too has a capacity to apply rhetorical definitions to his speeches while he makes them:

'And nere it that I wilne as now t'abregge
Diffusioun of speche, I koude almoost
A thousand olde stories the allegge ...'   (iii. 295–7)

When in Book II Pandarus visits his niece to inform her of Troilus' love, he prefaces his introduction of the subject with an elaborate disclaimer of prefaces, disparaging those who are prone 'with subtyl art hire tales for to endite' and exclaiming 'What sholde I peynte or drawen it on lengthe?' (ii. 257, 262), although his private reason for plainness is an unflattering estimate of Criseyde's ability to understand 'if I my tale endite | Aught harde, or make a proces any whyle' (ii. 267–8). As a disclaimer of rhetoric it is highly rhetorical, and in a symmetrically placed parallel instance in Book IV Pandarus is to make a comparable allusion to literary style and structure when visiting Criseyde in her grief, promising to be brief because she 'May to no long prologe as now entende ...' (iv. 893). The writing of Troilus' first letter is prefaced by Pandarus' advice on the style and diction to be used; he stresses stylistic decorum, and puts more emphasis on what to avoid than to include, drawing analogies between composition and music or painting. There should be avoidance not only of repetition and monotony, which palls even from the finest performer and the finest instrument (ii. 1030–6), but also of discordant registers and inappropriate diction:

'Ne jompre ek no discordant thyng yfeere,
As thus, to usen termes of phisik
In loves termes; hold of thi matere
The forme alwey, and do that it be lik;
For if a peyntour wolde peynte a pyk
With asses feet, and hedde it as an ape,
It cordeth naught, so were it but a jape.'   (ii. 1037–43)

Although the substance of such advice is traditional, its inclusion draws attention to the concealed craft of composition, for Pandarus is emphatic that the product of this labour of style and choice of language should not smell of the lamp ('Ne scryvenyssh or craftyly thow it write' (ii. 1026)).

The process of courtship gives rise to a much more extensive and

elaborate use of rhetoric than in *Filostrato*. Criseyde's first question to
Pandarus about this prospective lover is:

> 'Kan he wel speke of love?' quod she; 'I preye
> Tel me, for I the bet me shal purveye,'   (ii. 503–4),

and an important part of the remainder of Book II is taken up with the
letters, in which such speaking of love is put into practice, and judged by
the lady. (Pandarus anxiously asks Criseyde's opinion of Troilus' ability in
the writing of love-letters: 'Kan he theron? For, by my trouthe, I noot' (ii.
1197)). In fact, Troilus' first letter is reported, in summary, as the
quintessence in style and substance of such courtly epistles (ii. 1065–85;
see above, p. 227)). Criseyde in turn takes up her pen to write what is—she
declares—the first letter she has ever written (ii. 1213–14), a detail which
serves to underline the specialness of the letters as the expression of one of
the lovers to the other. To her uncle (who has already teased her that he
will 'write', i.e. transcribe, any letter, provided she compose the contents
(ii. 1159–62)) Criseyde claims not to know what she can write in reply to
Troilus (ii. 1206). But she goes along with Pandarus' request at least to
send some expression of thanks (ii. 1208), and her own letter is examined
intently by Troilus just as she had examined his (ii. 1321–30, 1177–8). A
correspondence between the lovers now ensues, reported by the narrative
in only the briefest outline, yet also with emphasis that such letter-writing
was one among the 'observaunces | That til a lovere longeth in this cas . . .'
(ii. 1345–6).

Troilus as a lover is therefore Troilus as a stylist: newly in love, he
polishes (*borneth*) his language (i. 327); he later employs language so well as
to win Criseyde's complete approval (iii. 471–2); he delivers to Pandarus
such an elaborate 'proces' in celebration of Criseyde (iii. 1739–42) that 'It
was an hevene his wordes for to here', and discourses so well about matters
of feeling that all lovers approve what he says (iii. 1796–9). The anxious
rhetorical question in the very last line of Book II ('O myghty God, what
shal he seye?' (ii. 1757)) emphasizes the dramatic transition from letter-
writing to speech as the lovers are about to meet and speak to each other for
the first time. Leaving nothing to the moment, both lovers have already
inwardly composed their prepared speeches. Criseyde has 'avysed wel hire
wordes and hire cheere' (ii. 1726), while the diction of what he shall say in
his 'complaint', and how he will comport himself in uttering it, have
already been painstakingly prepared by Troilus:

> Recordyng his lesson in this manere:
> 'Mafay,' thoughte he, 'thus wol I sey, and thus;
> Thus wol I pleyne unto my lady dere;
> That word is good, and this shal be my cheere . . .'   (iii. 51–4)

Yet, even though Troilus' memorized entreaty goes clean out of his mind, the lovers do still address eloquent, stylistically elaborate, first speeches to each other, which serve to establish the terms of their understanding as lady and servant in love. This first meeting of the lovers represents a union in understanding through the medium of language, and the amount of time the lovers spend on words here moves Pandarus to jest that, when they next meet at his house, they will have the leisure to compete for a prize in speaking of love ('And lat se which of yow shal bere the belle | To speke of love aright!' (iii. 198–9)). Pandarus' innuendo is that, on such a future occasion, they will have other things to do than talk, yet the physical consummation does indeed take place in the course of a scene of heightened rhetorical exchanges between the lovers. Love-talking in *Troilus* is not so much superseded by love-making but fittingly stands to represent it in a poem where love is so bound up with rhetoric, and where 'wordes' have great power not only to console (iii. 1133–4; iv. 1434–5) but to represent one absent lover to another (iv. 857–8). It is in such a context that Troilus fears Criseyde's father will so praise potential Greek suitors that he will 'yow glose | To ben a wif, and . . . ravysshen he shal yow with his speche' (iv. 1471–4), while after his separation from Criseyde Troilus laments anxiously 'Who speketh for me right now in myn absence?' (v. 236). In a contrast that says everything, Diomede feels that—at the worst—talking of love will help shorten the ride from Troy (v. 96), and later that 'I shal namore lesen but my speche' (v. 798).

Such attention to the act of composition and the use of language alerts the reader both to the accuracy and appropriateness of language used in the poem, and also to its range and variety. When Pandarus toys with the technical language of penance to the languishing Troilus ('remors of conscience . . . attricioun' (i. 554, 557)), the narrative makes clear that this is a deliberate manipulation of language for a particular purpose (i, 561), so that Pandarus at his very first appearance in the poem is immediately identified with the ability to use language to say something other than what it ostensibly means. In Pandarus' mastery of styles the language of 'holynesse' is one of the many available registers, along with his character-istic sententiousness: when in Book IV Pandarus deploys a stream of proverbs to encourage the anguished Troilus the narrative comments 'Thise wordes seyde he for the nones alle . . . He roughte nought what unthrift that he seyde' (iv. 428–31), again confirming his capacity to manipulate words for effect. When persuading Criseyde to receive her lover in her bedroom, Pandarus assures her that, if Troilus were of a different temperament, he would fob him off with some specious fine talk, 'a fewe wordes white' (iii. 901). The phrase returns on the morning after the consummation, when Criseyde accuses Pandarus of having 'caused al this fare ... for al youre wordes white' (iii. 1566–7), so that the delusive role of language is implicated in the accomplishment of the lovers' union.

Chaucer's characters in some contexts show concern to use language for proper definition, for naming, while in other places the action depends on not naming and euphemism. A momentary qualm drives Pandarus into an uncharacteristically coy circumlocution to refer to his own role as a go-between (iii. 253–6), but Troilus insists on calling it 'gentilesse, | Compassioun, and felawship, and trist' (iii. 402–3), on the grounds that a proper distinction must be made between apparently similar things:

> 'Departe it so, for wyde-wher is wist
> How that ther is diversite required
> Bytwixen thynges like ...'   (iii. 404–6)

Some of the discursive moments in *Troilus* focus upon such questions of definition, as in the exchange between Troilus and Pandarus in Book I ('The oghte not to clepe it hap, but grace' (i. 896)), Pandarus' attempts to define the term 'dulcarnoun' or dilemma (iii. 932–3), and Criseyde's distinction between various kinds of jealousy,

> 'If it be likkere love, or hate, or grame;
> And after that, it oughte bere his name',   (iii. 1028–9)

concluding that Troilus' own 'passioun' is 'but illusioun' (iii. 1040–1). In such moments the relationship is made explicit between language and what it is being used to describe, to reveal, or to conceal.

The characters are similarly alert to the implications of different levels of style. They assume a formal tone, or react against formality in others. Pandarus, for instance, advises Troilus not to tangle his love letter with 'argumentes tough' (ii. 1025), while Troilus expresses impatience with the rhetorically elaborate consolation offered by Pandarus (iv. 496–7); Criseyde at first implores Pandarus not to speak to her in such a 'fremde manere speche' (ii. 248), but then attempts to secure her position by means of formal utterance ('And here I make a protestacioun' (ii. 484). When Pandarus calls on Deiphebus to enlist the prince's support for Criseyde, he introduces his subject very politely ('Lo, sire, I have a lady in this town, | That is my nece ... ' (ii. 1416–17)), but to this Deiphebus courteously responds by setting aside Pandarus' formal and distant style, explicitly describing it as such and redefining Criseyde in his own terms, although as a prince Deiphebus addresses Pandarus as 'thow' ('O, is nat this, | That thow spekest of to me thus straungely, | Criseyda, my frend ... ?' (ii. 1422–4).

The juxtaposition of styles, the difficulty of determining tone in many contexts, builds in the course of the poem towards a sense of continual interplay between styles. One instance occurs when Criseyde accepts her uncle's invitation to stay overnight at his house:

> She thoughte, 'As good chep may I dwellen here,
> And graunte it gladly with a frendes chere,
> And have a thonk, as grucche and thanne abide;
> For hom to gon, it may nought wel bitide.'

> 'I wol,' quod she, 'myn uncle lief and deere;
> Syn that yow list, it skile is to be so.
> I am right glad with yow to dwellen here;
> I seyde but a-game I wolde go.'   (iii. 641–8)

By taking the reader into Criseyde's thoughts at this point, the narrative underlines a striking difference between the clipped, brisk diction and syntax in which Criseyde reasons inwardly to herself that she has little choice but to stay, and the speech of measured courtly politeness, itself rewriting her earlier refusal, with which she outwardly accepts his invitation. In response Pandarus equivocates about accepting Criseyde's excuse that she had been joking, and the doubt as to whether the speech was 'a game' (iii. 650) may stand to represent an uncertainty that plays about much of the characters' speech and the style of the narrative. During Criseyde's evening walk in her garden in Book II a highly rhetorical description of the sunset is suddenly interrupted by the narrating voice in a way that draws attention with comic self-consciousness to an elaborate style:

> The dayes honour, and the hevenes yë,
> The nyghtes foo—al this clepe I the sonne—
> Gan westren faste . . .   (ii. 904–6)

Such a tonal ambivalence often links the style of the narrative with that of Pandarus. The facility with which Pandarus can slip into many different types of language indeed helps bring multiplicity and play of language into *Troilus*, beyond that fundamental stylistic distinction between the speech of Troilus and Pandarus which runs through the poem, two different styles which Criseyde moves between in her own speech. Pandarus' mode of addressing Troilus in private can range from 'thow fol' (i. 618) to 'Sire' (ii. 957), or 'Myn alderlevest lord, and brother deere' (iii. 239). In his first address to Troilus Pandarus uses the polite second person plural form 'yow' (i. 553), but in the very next line—the first of a new stanza—he switches to the familiar 'thow', and in private they continue to address each other in the singular form throughout the poem, apart from a few grave moments of conjuration (i. 682–4; iv. 596–7; v. 496). In the presence of others, and upon introducing Criseyde into the room at the lovers' first meeting, Pandarus addresses Troilus formally as 'yow' (ii. 1688–9; iii. 62–3), yet by the end of the same meeting Pandarus is addressing both Troilus and Criseyde in the familiar form (iii. 193–4), in an exception to the prevailing pattern of address between Pandarus and Criseyde. The fond uncle and niece almost always use the polite second person plural to each other, although in a few moments of intense feeling Pandarus lapses into the second person singular to Criseyde (ii. 352–7, 395–6; iv. 848–54). To each other Troilus and Criseyde almost invariably use the second person plural, apart from a handful of occasions involving oaths or apostrophe (iii. 1512; iv. 1209; v. 734–5, 1258).

With the earnest Troilus Pandarus at times rises to match and endorse the lover's seriousness, at times comically deflating such moments with some impatient or disrespectful colloquialism:

> Quod Pandarus, 'Thow hast a ful gret care
> Lest that the cherl may falle out of the moone!'   (i. 1023–4)

From his first conversation with Troilus Pandarus imports into the poem the language of proverb and a larger tendency to the sententious phrase and aphorism; he can contrive to sound proverbial without citing recognizable proverbs and can meet any occasion with a pithy expression. Yet he can also strike a high note of moral abstraction in sketching to Troilus the virtues of Criseyde's character:

> 'For of good name and wisdom and manere
> She hath ynough, and ek of gentilesse.
> If she be fayr, thow woost thyself, I gesse.'   (i. 880–2)

That final flicker of irony is, however, characteristic of the way in which Pandarus' facility makes any style he takes up the strategy and disguise of the moment.

In Pandarus' exchanges with Criseyde there is even more interplay of style, and he assumes various grave tones for effect. There is a manner of epic style for recalling Troilus' feats of bravery on the battlefield:

> 'He was hire deth, and sheld and lif for us,
> That, as that day, ther dorste non withstonde
> Whil that he held his blody swerd in honde.'   (ii. 201–3).

Yet, even though Troilus is indeed a valiant warrior, the histrionic element in Pandarus, and a sense of context, opens up his attempt at epic style to the possibility of comic overstatement. The style of oath which Pandarus swears to Criseyde likewise seems too vehemently serious to be taken seriously:

> 'O cruel god, O dispitouse Marte,
> O Furies thre of helle, on yow I crye!

> So lat me nevere out of this hous departe,
> If I mente harm or any vilenye!' (ii. 435–8)

No wonder Criseyde can appear to take this seriously, but a reader may also notice the signs of a bravura performance, in which the high style of extreme emotionalism and pathetic sentiment is combined with bluntness and managed with a faultless control of pace. To his niece Pandarus can also employ a (too?) highly rhetorical style of moral injunction—

> 'Wo worth the faire gemme vertulees!
> Wo worth that herbe also that dooth no boote!
> Wo worth that beaute that is routheles!
> Wo worth that wight that tret ech undir foote!' (ii. 344–7)

—or the idiom of (over?) refined and ceremonial conduct of love:

> '... certein, best is
> That ye hym love ayeyn for his lovynge,
> As love for love is skilful guerdonynge.' (ii. 390–2)

Yet Pandarus can also mock at the conventional style of love language, glancing humorously at its oxymorons ('I have a joly wo, a lusty sorwe' (ii. 1099)) or at its borrowed religious and metaphysical notions ('That certeinly namore harde grace | May sitte on me, for-why ther is no space' (i. 713–14)). Into this jesting about the style of love-talk Criseyde is able to enter spiritedly as a knowing player ('Tel us youre joly wo and youre penaunce. | How ferforth be ye put in loves daunce?' (ii. 1105–6)), just as she is able to match Pandarus in the language of urgent moral injunction ('This false world—allas!—who may it leve?' (ii. 420)).

Troilus is a poem supremely of talk, of conversation, and of argument and persuasion. Chaucer's accomplishment is that *Troilus* is at once his masterpiece of artificial form—a poem highly wrought and perfected—yet a work filled with the accents of natural English speech. Within the poem's compass there can be the loftiest style, as in the crafted suspense and syntactical delay of the opening lines:

> The double sorwe of Troilus to tellen,
> That was the kyng Priamus sone of Troye,
> In lovynge, how his aventures fellen
> Fro wo to wele, and after out of joie,
> My purpos is ... (i. 1–5)

Yet elsewhere there can also be the quickening pulse, the authentic accents, of gossip and sheer chat:

> And seyde, 'Nece, who hath araied thus
> The yonder hous, that stant aforyeyn us?'

> 'Which hous?' quod she, and gan for to byholde,
> And knew it wel, and whos it was hym tolde ...   (ii. 1187–90)

There can be stanzas almost entirely constituted of dialogue, with the interchange of speakers left implicit or very briefly signalled:

> 'What, nat as bisyly,' quod Pandarus,
> 'As though myn owene lyf lay on this nede?'
> 'No, certes, brother,' quod this Troilus,
> 'And whi?'—'For that thow scholdest nevere spede.'
> 'Wostow that wel?'—'Ye, that is out of drede,'
> Quod Troilus; 'for al that evere ye konne,
> She nyl to noon swich wrecche as I ben wonne.'   (i. 771–7)

Troilus so catches within itself the momentum of talk that there are sequences—at once seen and heard—when speech and gesture or movement seem in motion together:

> And with that word he gan hym for to shake,
> And seyde, 'Thef, thow shalt hyre name telle.'   (i. 869–70)

Nor does this inclusive poem of talk evade moments when utterance leaves words behind, as when Criseyde 'gan to homme' in embarrassment (ii. 1199), shies away from Pandarus' innuendo with a nervous laugh (ii. 589), or has her exclamation of 'I! what?' mimicked by Pandarus as they stand by Troilus' bed (iii. 120, 122). In other contexts a sense of strain or incoherence in the syntactical movement of speech or reported thoughts may serve to suggest the confusion and strain in the predicament of a character:

> This accident so pitous was to here,
> And ek so like a sooth at prime face,
> And Troilus hire knyght to hir so deere,
> His prive comyng, and the siker place,
> That though that she did hym as thanne a grace,
> Considered alle thynges as they stoode,
> No wonder is, syn she did al for goode ...   (iii. 918–24)

In a way significant for the reading of character in the poem, there can also be some of the imprecision and slackness of ordinary speech. Loose ends may be left mysteriously trailing, as in Criseyde's words to Pandarus and his response, after he has invited her to supper and assured her Troilus will not be there:

> And seyde hym, 'Em, syn I moste on yow triste,
> Loke al be wel, and do now as yow liste.'
>
> He swor hire yis, by stokkes and by stones,
> And by the goddes that in hevene dwelle,

> Or elles were hym levere, soule and bones,
> With Pluto kyng as depe ben in helle
> As Tantalus—what sholde I more telle? (iii. 587–93)

Is Pandarus swearing that he will see that all is well, or that he will do as he likes? What exactly is the connection in Criseyde's quiet request between her uncle's seeing that 'al be wel' and doing as he pleases? (It is not even altogether clear whether Criseyde is saying that she 'must' trust her uncle or that she trusts him 'most'.) The vehemence with which Pandarus swears 'yes' in reply—and even the narrating voice wearies of that—does not serve to make the connection any clearer, in this early sample of free indirect speech that catches a character's tone while suggesting a mere summary of a much longer outburst. In a passage where narrative statement and narratorial implication are so much at odds, the spaciousness of the verse-form and the syntax allow Chaucer to take away with one hand what he gives with the other. Even the smallest words (like 'yet') may gain in context a poignant irony:

> 'Now God, thow woost, in thought ne dede untrewe
> To Troilus was nevere yet Criseyde' (ii. 1053–4)

and the stock phrases of conversation may come to seem evasions.

Such informal aspects of the poem's style are not compartmentalized: much of the interplay of style is suggested by juxtaposition, overlapping and superimposition of various stylistic levels and layers, so that conversation can move quickly in and out of different registers, with learned and specialized terms occurring in perhaps surprising contexts. 'Tel me plat [bluntly]', begins Pandarus, as he tries to draw from Troilus the reason for his sorrow, yet within the same sentence his own speech is soon far from blunt:

> 'And tel me plat what is th'enchesoun [reason]
> And final cause of wo that ye endure;
> For douteth nothyng, myn entencioun
> Nis nat to yow of reprehencioun [reproach] ...' (i. 681–4)

The initial stages of the affair are accompanied by the language of disputation ('I sette the worste ...' (ii. 367); 'Now sette a caas ...' (ii. 729); 'As thus: I pose ...' (iii. 310)). There is also word-play and punning: 'this Calkas knew by calkulynge' (i. 71); 'that Troye sholde | Destroyed ben' (i. 76–7). Even the end of the story in Chaucer's version is associated with some ending of rhetoric, in Criseyde's evident failure to accomplish the persuasion of her father as she predicted ('I shal hym so enchaunten with my sawes' (iv. 1395)), with the consequence that the later part of the poem is given over to the rhetoric of complaint and lamentation.

The range of styles and richness of language is registered by some of the commentary in the manuscript margins, as also by those kinds of commentary which the poem carries in itself, alerting and educating its reader or hearer to the variety of language being used, and both these different types of commentary can offer pointers to what may have seemed special in the poem's style to early readers. There is evidence in the marginal glosses to some *Troilus* manuscripts of a response to the rhetorical form of the poem, with passages signalled, for example, as 'invocacio' (ii. 8; Dg, H1, J, S2), 'comendacio' (i. 99; H5, R), 'narracio' (ii. 316; H4), 'declaracio' (iv. 145; H4), 'recapitulacio' (iii. 911; H4), 'oratio & laudacio' (iii. 185–6; H4), 'requisicio' (iii. 131; H4), or 'peticio' (iv. 68; v. 594; H4). Behind such labelling is a suggestion of early recognition and attempted analysis of the poem's stylistic variety. The exceptional quality of the language in *Troilus* is also the object of glossing by some scribes, and in a passage like that on the death of Hector—

> Among al this, the fyn of the parodie   *duracio parodie*
> Of Ector gan aprochen wonder blyve.
> *decorporare*
> *destine fate* The fate wolde his soule sholde unbodye,
> *achilles*
> And shapen hadde a mene it out to dryve,
> Ayeyns which fate hym helpeth nat to stryve;   (v. 1548–52)

—the provision of glosses registers a sense that classical references ('fate'), difficult diction ('parodie'), or unusual formations ('unbodye') may need explanation, while the oblique manner of referring to Achilles in this elevated passage also calls for clarification. (The glosses are those of H4; three other MSS—Cp, H1, S1—gloss 'parodie'.) Yet much unusual diction is placed within the text alongside a more familiar synonym in a 'doublet' which unobtrusively serves to gloss it, just as in the poem's 'self-footnoting' technique with classical and other learned references (see pp. 255–6 above); and sometimes it seems to comment on its own high style ('By day, he was in Martes heigh servyse, | This is to seyn, in armes as a knyght . . .' (iii. 437–8)).

Figurative language is also recurrently found accompanied by a commentary within the text. Pandarus immediately explains 'The preie ich eft . . . That privete go with us in this cas' with 'That is to seyn, that thow us nevere wreye' (iii. 282–4), just as Criseyde glosses ' "Of gilt misericorde!" | That is to seyn, that I foryeve al this . . .' (iii. 1177–8). Such care to use the figurative but to clarify it marks Chaucer's practice throughout the poem: the medicine of love is unobtrusively explained in the next breath ('This lechecraft, or heeled thus to be . . .' (iv. 436)); Troilus attempts to persuade Criseyde to elope with him by making a learned allusion ('And thynk that folie is, whan man may chese, | For accident his substaunce ay to lese' (iv. 1504–5)), but must then twice add 'I mene thus ...' (iv. 1506) and 'Thus mene I ...' (iv. 1511). There is a comparable tendency when proverbial expressions are accompanied with a commentary: in talking to Criseyde Pandarus scorns by means of a homely expression the practice of playing a man along, but must then explain ('And maken hym an howve above a calle, | I meene, as love another in this while ...' (iii. 775–6)); in her description to Troilus of how she will deceive Calchas Criseyde invokes what might well seem a proverb of self-evident meaning ('men seyn that hard it is | The wolf ful and the wether hool to have' (iv. 1373–4)), but she then proceeds to draw out the sense and application of her proverbial figure ('This is to seyn ... And how I mene, I shal it yow devyse ...' (iv. 1375–9)). It is in a speech of Diomede to Criseyde that this commenting style is taken to its extreme, when the text contains a word so unfamiliar that two further lines must be devoted to glossing what it means:

'And but if Calkas lede us with ambages—
That is to seyn, with double wordes slye,
Swiche as men clepen a word with two visages ...'   (v. 897–9)

It is fitting that Diomede—falsifier of the language of love that the whole earlier poem has developed—should use and explain a term which means a sly, 'double' word, a word which more largely can stand to represent the Janus-like qualities of language. It is these aspects of language—as it were two-faced, or facing in several directions—and the ambiguous relation of language to *trouthe*, which are revealed in the course of a poem pointedly set in a world where 'goddes speken in amphibologies, | And for o soth they tellen twenty lyes' (iv. 1406–7). It is also very fitting that such 'ambages' should be suspected of a soothsayer who is a traitor. As an apt acknow-ledgement of the possible ambiguity of words, Diomede's speech is a token of that exploration of the potential and limits of language which the multiplicity of style and language in *Troilus* invites. As an acknowledge-ment of the need to gloss words Diomede's speech points to a concern with interpretation of language which Chaucer makes central to his deepened and extended treatment of the love-story of Troilus and Criseyde. Taking cues from the glosses offered by the poem itself and by its early readers, the following sections explore the figurative, the proverbial, and the unfamil-iar, as further aspects of the style of a poem in which language conveys both the formality and the intensity through which the experience and aspirations of love are expressed.

See Derek Brewer, 'Chaucer's Poetic Style', in Boitani and Mann (eds.), *The Cambridge Chaucer Companion*, 227–42, and also his three studies, 'The Relationship of Chaucer to the English and European Traditions', 'Some Metonymic Relationships in Chaucer's Poetry', and 'Towards a Chaucerian Poetic', all reprinted in his *Chaucer: The Poet as Storyteller* (London, 1984).

On Chaucer's style in *Troilus*, see Payne, *Key*, and for some particular studies, see Burnley, *Chaucer's Language*; Lanham, 'Opaque Style'; Lockhart, 'Semantic, Moral, and Aesthetic Degeneration'; McKinnell, 'Letters'; Muscatine, *Chaucer*; Reichl, 'Chaucer's *Troilus*'; Stokes, 'Recurring Rhymes' and '"Wordes White"'; Taylor, 'Terms'; Wallace, 'Chaucer's "Ambages"' and *Chaucer and the Early Writings of Boccaccio*. See also Siegfried Wenzel, 'Chaucer and the Language of Contemporary Preaching', *Studies in Philology*, 73 (1976), 138–61.

On the use of the second person singular and plural, see Johnston, 'The Pronoun of Address'; Schmidt, 'Das Anredepronomen'; and Walcutt, 'The Pronoun of Address'.

On ellipsis, see Elliott, *Chaucer's English*, 71–2; for some instances of ellipses of the subject, see *Troilus*, iii. 691; v. 1825; and for the verb, see i. 126; ii. 1108; iii. 548.

'In science so expert'

To achieve the play of style and the range of references brought together in *Troilus* demanded exceptional resources of language, yet to what work in English could Chaucer look for a predecessor? To the popular English romances he was indebted for some of the elements of style in all his narrative poems, and the task of translating into the English prose of his *Boece* the mixture of Latin prose and poetry in the *Consolation of Philosophy* has in turn had a pervasive influence on the poetic language of *Troilus*. Yet the posterity of *Troilus* in English poetry is easier to discern than its antecedents. The responses of scribes in manuscripts of *Troilus*, and the influence of *Troilus* on poetic diction in English during succeeding generations, suggest that the poem was admired for an exceptional richness and elevation of style, a perfected versification of its kind, a minted and polished diction of erudition, range, and depth, and the dignity and resonance of Latinity in interplay with a syntax that moves between the highly wrought and the limpidly natural.

In such stylistic accomplishment *Troilus*, along with Chaucer's other poems, benefits from the developing status of English in the later fourteenth century, as it came to be increasingly used for writing, no longer the least prestigious of languages in a trilingual culture, along with the Anglo-Norman form of French and with Latin. The unrepeatable circumstances of the English language during his lifetime allowed Chaucer to create—in *Troilus* perhaps supremely—poems with all the poise of manner and depth of background in traditions and contexts familiar from a wider European culture, yet at the same time with the freshness of a vernacular almost unused for any equivalent secular writing. With access to Chaucer's sources it is now possible to see how the accomplished model of diction and stanzaic narrative in *Filostrato*—with its fluent match of Italian syntax to the demands of stanza and rhyme—was a potent influence in shaping the language and form of Chaucer's verse paragraphs in English (see above, pp. 50–3). This was supplemented by the resources of language recollected from his recent work on the text of Boethius. No wonder that, in narrating his story, Chaucer was interested by the cultural differences between ancients and moderns. At the period when he was composing *Troilus* Chaucer's activities as a writer are inseparable from the task of translating, mediating between cultures, and re-expressing faithfully or freely in his mother tongue what already existed in another language.

If Chaucer was composing in English under the impress of his reading in Italian, French, and Latin, it is unsurprising that many words of romance origin are recorded in *Troilus* for the first, or indeed the only, time in English, though the nature and effect of such newness has to be defined in the context of a culture where the recognizably equivalent words may well be already known through French or Latin, and where an audience may hence have a more flexible sense than modern readers of what is and is not English. In addition, the accident of what survives as the earliest written instance of a word may not always be an accurate reflection of how unusual that word was. Despite all such qualifications, none the less, there is a resonant, polysyllabic Latinity to many lines in *Troilus* and many of the instances of words first recorded in the poem occur in contexts involving some of Chaucer's characteristic thematic and formal interests, not least style itself: that Troilus *borneth*, or polishes, his 'speche' (i. 327); *digression* (i. 143); *declamed* (ii. 1247); or 'I wilne as now t'abregge | *Diffusioun* of speche' (iii. 295–6). To Pandarus Chaucer appropriately gives the only recorded instance of *proverb* as a verb (iii. 293), or the nonce word *jompre* (i.e. jumble) in his advice on the style of writing love-letters (ii. 1037).

The poem's endeavour to present an ancient world of pagan belief and custom leads to the first recorded occurrences of *augurye* (iv. 116; v. 380); the *marcial* tournament (iv. 1669); the *palestral* games (v. 304); and such aspects of pagan myth and belief as *fate* (v. 209, 1550–2), or the *Manes* (v. 892). The Fates, the Muses, nymphs, 'satiry and fawny' (iv. 1544), and the pagan *funeral* ceremony (v. 302), appear earlier only in Chaucer's own works, *Boece* or the *Knight's Tale*.

Other of the first recorded usages come from learned and scientific terminology: that Troilus 'in teris gan *distille*, | As licour out of a *lambyc*' (iv. 519–20); cosmological terms such as *influence* (iii. 618), *convers* (v. 1810), *erratik* (v. 1812), and the verbs *cerclen* (iii. 1767) and *circumscrive* (v. 1865; from Dante). Moral or religious terms such as *bestialite* (i. 735) or *attricioun* (i. 557) are recorded for the first time in an English text, although familiar from Latin or French, as would also be the case with *comporte* (v. 1397), *mocioun* (iv. 1291), or *mansuete* (v. 194). In a poem co-dedicated to the lawyer and philosopher Ralph Strode, a legal term such as *advocacies*

(ii. 1469) makes its earliest recorded appearance, as do the adjectives *casuel* (iv. 419) or *philosophical* itself (v. 1857).

It is from the *Consolation of Philosophy* as he translated it that Chaucer carries forward and uses again in *Troilus* much vocabulary first recorded in these two texts. In part such an overlap occurs, as might be expected, in the more philosophical sections of *Troilus*, where *mutabilite* (i. 851) and the *muable* (iii. 822) quality of things is felt, where the problem of *future* time is considered (v. 748), and where ideas of order and harmony are invoked as opposed to the *discordant* (ii. 1037). Yet, by extension, originally Boethian terms also become applied in such romance contexts as how the lovers *entrechaungeden* their rings (iii. 1368), how the coming of dawn made *disseveraunce* of them (iii. 1424), how Pandarus is to *dispone* of the lover's effects (v. 300), and how various influences will *commeve* the behaviour of individuals (iii. 17; v. 1386, 1783). The occurrence of related words first recorded in *Troilus* also reflects some of the patterns of theme and imagery in the poem: Fortune as *adverse* (iv. 1192); the *constreinte*, or distress, of love (ii. 776; iv. 741); a concern for the *natif*, or inborn (i. 102) and hence an interest in the *natal* attributes of Jove, the god who presides over birthdays.

In the coining of terms, too, Chaucer gives his poem what may well have struck contemporaries as a special, indeed 'minted' quality of diction. This is not to say that *Troilus* would have seemed bristling with strange neologisms: Chaucer's technique is rather to achieve some fresh combination of already-known elements, as when Troilus in his misery calls himself a 'combre-world, that may of nothyng serve' (iv. 279). There is a large category of diction in *Troilus* where by the use of a prefix or suffix Chaucer creates what the *Middle English Dictionary* cites as the first (or only) recorded instance of a word, and, while this does not prove that Chaucer's early audiences would never have met every such word before, the sheer number of them does suggest a distinctive characteristic of Chaucer's style.

This is especially striking in the case of words prefixed with *un-*, particularly adjectives and past participles, but also verbs, adverbs, and some nouns. The relieved Troilus 'Weex of his wo, as who seith, untormented' (i. 1011), where that brief parenthesis of preparatory hesitation presents and draws attention to the adoption of an unusual form. It was a form with which Chaucer was evidently experimenting at the period when he was engaged on *Boece* and *Troilus*, for between them these two works contain about one hundred unique nonce *un-* words, compared with less than fifty for all Chaucer's other works. Nouns prefixed by *un-* are relatively infrequent, although the narrator's profession of 'unliklynesse' (i. 16) suggests by its unusualness his peculiar mixture of ability and disability, while Troilus' reflection that, in committing suicide, he would 'bothe don unmanhod and a synne' (i. 824) conveys the reversal of values that suicide would involve. It is in the use of verbs prefixed by *un-* that Chaucer achieves some striking effects of language: Criseyde must *unfettre* her heart (ii. 1216); Troilus bids his spirit to *unneste* from his heart (iv. 305); Criseyde vows to *unshethe* the soul from her breast (iv. 776), and nothing can *unbynde* love from her heart (iv. 675); the lovers' hearts *unswelle* through weeping (iv. 1145–6); fate wishes to *unbodye* the soul of Hector (v. 1550); Troilus cannot *unloven* Criseyde (v. 1698); or, with participles, Pandarus' crisp advice 'Unknowe, unkist, and lost that is unsought' (i. 809).

There is a comparable use of words prefixed with *mis-* and *dis-*: unique usages such as *misacounted* (v. 1185), *mysforyaf* (iv. 1426), *myswrite* and *mysmetre* (v. 1795–6); *disblameth* (ii. 17); the earliest recorded usages of *mysconstruwe* (i. 346), *myslyved* (iv. 330), *mystrusted* (ii. 431), *mysspak* (i. 934), *disavaunce* (ii. 511), *dysaventure* (ii. 415; iv. 297, 755; v. 1448), and *disturne* (iii. 718). In a comparable area, the injunction to Troilus to *biblotte* (ii. 1027) his first love-letter with his tears is the only recorded instance, and when he *byreyned* (iv. 1172) the unconscious Criseyde with his tears this is an early instance. Chaucer's imaginative sense of what can be conveyed by prefixes also prompts him to the earliest instances of *entrecomunen* (iv. 1354), *entreparten* (i. 592), *forknowynge* (i. 79), *forseynge* (iv. 989), *in-knette* (iii. 1088), and *over-haste* (i. 972). Such experimentation is complemented by attention to suffixes: *knotteles* (v. 769), *heleles* (v. 1593), and *lightles* (iii. 550); and to feminine forms such as *devyneresse* (v. 1522), *herdesse* (i. 653), or *executrice* (iii. 617).

The peculiar effectiveness of such deployment of prefixes and suffixes is that it achieves concision through an experimentation which is transparent and immediately open to understanding. A sense of technical accomplishment and control of form is promoted which goes together with extension and enhancement of meaning, and draws attention to the very process of creating meaning through language.

On Chaucer's language generally, see David Burnley, *The Language of Chaucer* (London, 1989); Norman Davis, 'Chaucer and Fourteenth-Century English', in Derek Brewer (ed.), *Geoffrey Chaucer* (London, 1974), 58–84; and Elliot, *Chaucer's English*. For reference, see Kittredge, *Observations*, and Joseph Mersand, *Chaucer's Romance Vocabulary* (2nd edn., New York, 1939).

On words prefixed with *un-*, see Elliott, *Chaucer's English*, 162.

On *Troilus* and the language of *Filostrato*, see the excellent account in Wallace, *Chaucer and the Early Writings of Boccaccio*, ch. 6.

'This paynted proces'

In response to Pandarus' disclaimer of a style of speech painted with the 'colours' of rhetoric, Criseyde exclaims 'Is al this paynted proces seyd—

allas!—| Right for this fyn?' (ii. 424–5). The exchange serves stylishly to highlight the use of rhetoric it disclaims and bring into focus the rich density of poetic texture in Chaucer's language, particularly as this expresses the inner lives of his characters and their aspirations, and a vivid sense of process. It is diction distinctively vehement in expression, characterized by the use of hyperbole and superlative, by the language of solemn declaration and of sententiousness, as well as by a vigorous figurativeness.

Elements of this intensifying process may be seen in the differences between a single stanza of *Filostrato* and of *Troilus* in which Troilus is responding to Pandarus' entreaty to divulge the name of the lady he loves:

> 'Amore, incontro al qual chi si difende
> più tosto pere ed adopera invano,
> d'un piacer vago tanto il cor m'accende,
> ch'io n'ho per quel da me fatto lontano
> ciascheduno altro, e questo sì m'offende,
> come tu puoi veder, che la mia mano
> appena mille volte ho temperata,
> ch'ella non m'abbia la vita levata.' (*Fil.* ii. 7)

'Love—against which he who defends himself is soonest taken and strives in vain—with a fond delight so kindles my heart that for its sake I have put far from me all other pleasures; and this so afflicts me, as you may see, that a thousand times I have barely held my hand back from taking my life'.

> 'Love, ayeins the which whoso defendeth
> Hymselven most, hym alderlest avaylleth,
> With disespeyr so sorwfulli me offendeth,
> That streight unto the deth myn herte sailleth.
> Therto desir so brennyngly me assailleth,
> That to ben slayn it were a gretter joie
> To me than kyng of Grece ben and Troye.' (i. 603–9)

As so often in his translating, Chaucer starts his own stanza by responding closely to the syntactical movement and rhyme-word of his Italian model, before moving away through freer translation towards a different kind of resolution at the close. As early as the second line Chaucer is sharpening *Filostrato*'s paradoxes, so that whoever defends himself *most* benefits *least of all* (i. 604). The language in which the hero analyses his emotions has been extended to contain some more abstract moral vocabulary, while at the same time an intensified sense of despair is expressed in vividly figurative terms, as Chaucer's Troilus imagines his heart sailing towards death, feels himself (in a more strongly stated version of *Filostrato*) burningly assailed by desire, and gives a more extravagant expression of feeling in hyperbolical, superlative language ('to ben slayn it were a gretter joie . . . than kyng of Grece ben and Troye' (i. 608–9)).

In a brief compass Chaucer's adaptation of this stanza draws together and may serve to introduce a number of larger stylistic patterns in the 'paynted proces' of the poem: a language marked by use of superlative forms and hyperbole; the language of solemn declaration, oaths, and acclamation; and a distinctive language of the heart.

The use of superlatives by Troilus and by the poem's narrating voice is one very characteristic feature of the style of *Troilus*. Pandarus and Criseyde use them more sparingly. For Troilus the superlative expresses his idealizing and aspiring character: most aspects of his experience are described as the very best, or the absolute, or the most enduring. No man was ever caught so fast by love (i. 534), and no man was ever so beholden to the gods, although the least deserving (iii. 1259, 1268), for Love has bestowed him 'in so heigh a place | That thilke boundes may no blisse pace' (iii. 1271–2); he feels that his experience in love is unique ('he demed, as I gesse, | That ther nys lovere in this world at ese | So wel as he . . .' (iii. 1727–9); his vision of Criseyde is a characteristically superlative one, for she is 'the faireste and the beste | That evere I say' (iii. 1280–1), and surpasses all comparisons ('make no comparisoun | To creature yformed here by kynde!' (iv. 450–1)); to impugn her is like traducing Alcestis, 'of creatures . . . kyndest and the beste' (v. 1528–9). It follows that, when his fortunes change, Troilus casts himself as superlatively miserable, 'the wofulleste wyght | That evere was . . .' (iv. 516–17; cf. iv. 303–4). He uses the same distinctive pattern of expression when he calls Pandarus 'frend of frendes the alderbeste | That evere was . . .' (iii. 1597–8), describes himself as 'wrecche of wrecches' (iv. 271), or addresses Criseyde's deserted palace as 'hous of houses whilom best ihight' (v. 541). After Criseyde's departure Troilus is described as 'In sorwe aboven alle sorwes smerte' (v. 198).

There is a comparable inclination to the superlative in the narrative voice. The power of love in overcoming Troilus is presented as a movement between extremes ('moost in pride above . . . moost subgit unto love' (i. 230–1)); of Criseyde at her first appearance 'Nas nevere yet seyn thyng to ben preysed derre' (i. 174); than Troilus in Book III 'gladder was ther nevere man in Troie' (iii. 357), assisted by a friend who cannot be bettered (iii. 489–90). When change comes, the sorrow of Criseyde at leaving Troy is stressed to be incomparable (v. 20), as it is later when she betrays Troilus ('Ther made nevere womman moore wo' (v. 1052)). To the last it is loyally insisted that 'Ther myghte ben no fairer creature' (v. 808), but she is also 'the ferfulleste wight | That myghte be' (ii. 450–1). There is evidently no match for the experience and characters of the story (iii. 476, 781–2; v. 839–40); and it is against this established pattern that Criseyde's final echoes of these superlative terms seem so poignant, when she admits to herself that she has betrayed 'oon the gentileste | That evere was, and oon the worthieste!' (v. 1056–7).

A distinctive feature of the superlative and hyperbolical is that actions and reactions are recurrently declared to occur hundreds or indeed thousands of times: in Troilus' first languishings for love (i. 457, 531); in his kissing his letter to Criseyde a thousand times (ii. 1089–90); in his thousand 'esy sykes' (or hundred, in some manuscripts) at the consummation (iii. 1360). Here style strives to match the spirit of the scene more than to report numbers feasibly. Criseyde (iv. 753) calls upon death a thousand times, and after she has left Troy Troilus re-reads her letters a hundred times (v. 472).

This hyperbolical emphasis already occurs in Chaucer's Italian source, but is actually increased and extended in *Troilus*. Numbers in *Filostrato* become multiplied in *Troilus*: when Troiolo feels Criseida exceeds what he expected (*Fil.* iii. 53) this becomes Troilus' view that Criseyde 'A thousand fold was worth more than he wende' (iii. 1540). Only Chaucer has his hero implore his lady when faced with her departure 'A thousand tymes mercy I yow preye' (iv. 1500), or includes the vow by a thousand deaths (iii. 388–9, 573–4). Such numerical hyperbole is also a feature of those parts of his poem where Chaucer freely invents. At the centre of the lovers' union Troilus kisses Criseyde a thousand times (iii. 1252), and Criseyde is so happy that 'Twenty thousand tymes, er she lette, | She thonked God that evere she with hym mette' (iii. 473–4). Such aspiring language cannot remain true to the realities of human experience, and all the superlative language of earthly love in the poem is at last to be set against those superlatives which are aptly applied to Christ ('And syn he best to love is, and most meke . . .' (v. 1847)).

It is through the medium of oaths and other solemn declarations that a sense of God and another world is most sustainedly introduced into the language of the poem, and the love-affair of Troilus and Criseyde is surrounded with oaths at every stage,

> Yet lasse thyng than othes may suffise
> In many a cas, for every wyght, I gesse,
> That loveth wel, meneth but gentilesse.   (iii. 1146–8)

Chaucer has introduced much more asseverative language, much more swearing of oaths, than he found in his sources, so that the actions and events of *Troilus* are recurrently set alongside the expression of commitment and aspiration. The added oaths cut across the received story-line with the invocation of greater commitments and different understanding of the affair at every step: in the early undertakings of Pandarus to both

Troilus and Criseyde, in the lovers' meetings and union, and in the prospect and the persistence of their separation.

'Yet lasse thyng than othes may suffise . . .' Do Chaucer's characters indeed protest too much? As with other aspects of Chaucer's stylistic elaboration of his story, the effects are cumulative, and more positive than negative: asseveration and oath reflect the deeper, more urgent significance that his characters see in their experience, even if the traditional end of the story will bring their greater aspirations to a keener disappointment.

There is evident design in the use of oaths in *Troilus*, in their nature, incidence, and distribution among the speakers. There is most swearing in Books II and III, and Pandarus swears more than other characters: he swears over eighty times, with Criseyde uttering over sixty oaths, Troilus about forty, and the narrating voice about twenty more, not counting the recurrent use of 'pardee', 'benedicitee', and 'God woot'. The majority of such oaths involve mention of 'God', bringing constantly within the poem, in however formulaic a way, some sense of a religious dimension to the context where they occur. While there is often some Christian echo in such oaths (e.g. ii. 381; iii. 1501–3), there is enough consistent reference to the classical gods to present the characters as at once Trojan ancestors and as contemporaries of Chaucer's audience. Troilus uses the French oath 'Mafay' (iii. 52), and both he and Criseyde utter the oaths 'pardieux' (i. 197; ii. 759) and 'depardieux' (ii. 1058, 1212). Oaths by Pandarus (and by Criseyde to some degree) tend to be more vigorously expressed than those of Troilus, often with some colloquial edge.

In Book I most of the oaths concern Pandarus' commitment of himself to the affair, with a typical vehemence: swearing that he will never reveal Troilus' love even if tortured (i. 674–5), or that he may be hung, drawn, and quartered if he does not cure his friend's troubles (i. 833). Moreover, it is in a solemn exchange of oaths that Chaucer dramatizes the agreement of Pandarus to help Troilus at the climax and close of the first book. There is a markedly asseverative tone to the exchanges between Pandarus and Criseyde in the second book (e.g. ii. 299, 352–3, 1145–8), and, when Pandarus goes from Criseyde to Troilus, he swears that his niece will love him 'And therto hath she leyd hire feyth to borwe' (ii. 963). Yet there are many ambivalent oaths in the world of the poem: Pandarus swears to the truthfulness of his various exaggerated representations to Criseyde of the desperate plight of Troilus (ii. 1126–7); both Helen of Troy and Troilus swear to defend Criseyde against troubles that we may suspect to be fictitious.

In Book III the union of the lovers is set around with solemn language of declaration and oath, though there is still variety of tone and level in the use of such language. Pandarus can swear oaths banteringly to his niece (iii.

566), and many oaths are part of the common coin of that easy conversation which Chaucer so develops in the poem. Such variety emphasizes the solemnity with which declarations are made in the exchanges between the lovers at their meetings, first at the house of Deiphebus ('God woot, for I have ... Ben youres al, God so my soule save' (iii. 100–2)), and more extendedly in the central consummation scene. The action of this book is made to turn on assurances: the oaths that reassure Criseyde that Troilus is out of town (iii. 570); the oaths that dispel misunderstanding between the lovers; the repeated oaths of fidelity by Criseyde (iii. 999–1002, 1045–9, 1053–4, 1111, 1492 ff.). The language of asseveration gives formal expression to the concern with fidelity and constancy in *Troilus*, and the juxtaposition of ambiguous, disingenuous oaths only highlights the aspiration in those genuinely felt.

In the last books oaths are comparably used to give expression formally and urgently to the sense of commitment in the lovers, in a pointed contrast with the inevitable processes and shape of the story. When Troilus and then Criseyde hear of the exchange, they commit themselves solemnly to lament their separation, and when they meet for the last time Chaucer has them express their devotion with a sequence of oaths (e.g. iv. 1236–9, 1444–6). Their exchanges on their last night together are supplemented with a whole sequence of terrible oaths by which Criseyde swears that she will return to Troilus (iv. 1534–54), and Troilus too swears his fidelity in a resounding oath (iv. 1654–7). It is against the memory of these idealistic intentions that the disappointment and disillusionment of Book V can be more poignantly contrasted ('Who shal now trowe on any othes mo?' (v. 1263)), culminating in the particular significance of the brooch Troilus had given Criseyde, that now becomes a token of her pledge to Diomede ('And she hym leyde ayeyn hire feith to borwe | To kepe it ay!' (v. 1664–5)).

The lovers' elevated mode of address, the way they hail each other with acclamations that acknowledge and praise what is seen in the other person, whether uttered directly or in soliloquy, brings into the poem much resonant expression of value and esteem. This comes with a certain formality, just as the lovers rarely address each other in anything but the polite second person plural. Such poetry of acknowledgement and recognition, acclamatory in tone and form, is especially prominent in Book III. At their first meeting Troilus addresses Criseyde in terms which recognize and affirm what he sees in her ('O goodly fresshe free ... my lady right and chief resort' (iii. 127–47)). Accord and acceptance is given the outward rhetorical form of acclamation. Criseyde hails Troilus with a recognition of the qualities she sees in him ('myn owen hertes list, | My ground of ese ...' (iii. 1303–4)) until the moment of surrender is conveyed rhetorically through the ceremonious formality of greeting and acclamation ('Welcome, my knyght, my pees, my suffisaunce!' (iii. 1309)). At their parting Criseyde hails Troilus as 'Myn hertes lif, my trist, al my plesaunce' (iii. 1422), and in response he acclaims her as 'My lady right, and of my wele or wo | The welle and roote, O goodly myn Criseyde' (iii. 1472–3). In the later books this poetry of address is still prominent, but, once the lovers are separated, its continuing use is turned against itself by events, as Troilus apostroph-izes his absent lady ('O herte myn, Criseyde, O swete fo! | O lady myn . . .' (v. 228–9)).

Throughout the poem Chaucer floods the lovers' speech with the rhetoric of tender endearment, the language of the dear and sweet heart. This may often have been simply convenient to Chaucer for purposes of rhyme, but it is none the less a pervasive feature of the style and aesthetic of *Troilus*, everywhere in the poem and remarkably uniform, even limited. Criseyde is almost invariably Troilus' 'swete' or 'dere' heart or lady, and also his 'owne' or 'righte'. This is a rhetoric that affirms love and belonging insistently and repetitiously, and in his fifth book Chaucer creates two particularly poignant moments when Troilus remembers and repeats Criseyde's affectionate words to him in the past:

> 'And yonder ones to me gan she seye,
> "Now goode swete, love me wel, I preye . . ."' (v. 571–2)

> 'She seyde, "I shal ben here, if that I may,
> Er that the moone, O deere herte swete,
> The Leoun passe, out of this Ariete . . ."' (v. 1188–90)

The effects of such a language of endearment are cumulative, for, if the style remains the same, the context sadly alters, with Troilus continuing to the last to be at one with his rhetoric of love.

The place of the heart in a rhetoric of endearment and acclamation is only a part of what may be called a language of the heart that characterizes the style of *Troilus*. The word *herte* is one of the most frequently used in the poem, and the recurrence of the word is one stylistic aspect of the poem's focus on the inward experience, the thinking and feeling of the characters. The heart is by tradition the seat and centre of emotion and perception, but in its sense of process the poem frequently goes strikingly beyond conventional usage ('His herte, which that is his brestez yë, | Was ay on hire . . .' (i. 453–4)). Such descriptions of emotional processes are not aiming to represent scientifically or mimetically what Chaucer's contem-poraries might have understood as the physiology of inner responses. Such scientific awareness is indeed present in the poem—as in the careful account of how Troilus' faint was brought on ('Therwith the sorwe so his herte shette . . .' (iii. 1086 ff.))—but more characteristic of *Troilus* are passages in which the heart is the setting or agent in some imagined process through which the intensity of a response of joy, fear or sorrow may be figuratively conveyed. At Troilus' first sight of Criseyde 'his herte gan to sprede and rise' (i. 278; cf. iv. 1422) and at the first signs of success he exclaims that his heart is swelling such that it will burst (ii. 979–80), but more often the lovers' hearts are quaking or heaving with fear (ii. 809, 1321; iii 57), or about to break with sorrow.

Perception is recurrently suggested to be an experience of forceful physical impact or invasion, as in the account of Troilus' first reactions to the sight of Criseyde and his memory of that:

> That in his herte botme gan to stiken
> Of hir his fixe and depe impressioun ... (i. 297–8)
>
> Right with hire look thorugh-shoten and thorugh-darted. (i. 325)

The heart of Criseyde as a lover is in turn imprinted (ii. 900), penetrated (ii. 650, 902), mined (ii. 677), and engraved (ii. 1241; iii. 1499). Through devotion and memory one lover is 'shut' within the heart of the other (iii. 1549), who also has her heart fixed 'fast' upon him (v. 954). Lovers' hearts become so firmly 'set' upon each other (iii. 1488; iv. 673–7), so intergrown (ii. 872), that any disturbance or separation can only be a wrenching break; feelings are recurrently associated with the breaking of the heart, and in sorrow the lovers' hearts are gnawed (iv. 621; v. 36), twisted and pressed (iv. 254, 1129), squeezed by constraining cramps (iii. 1069–71), or swollen by sorrow (v. 201; cf. iv. 1146). So deeply inward are the heart's associations that strong emotions must leave it through being torn up by the roots (v. 954; cf. iii. 1015). The soul is imagined as nested (iv. 305), or sheathed (iv. 776), within the heart, and sorrow rends the soul from Troilus' breast (iv. 1493) and rends the soul out of his heart (iv. 1700).

The emotional responses of the lovers are also conveyed by projecting them as imagined processes within the heart involving melting, weeping, and bleeding, notions that are brought together when Troilus, about to part from Criseyde after their first night together, feels 'The blody teris from his herte melte' (iii. 1445). By extension, the heart may come to represent the emotional character of individuals and hence themselves (v. 596–7, 988), and from this the heart becomes in courtly discourse and ritual an emblem of feeling: when Troilus plans his funeral, he wishes the ashes of his heart to be sent to Criseyde in a golden urn 'for a remembraunce' (v. 309–15), and the courtly idea of an exchange of hearts which symbolizes the establishment of mutual love lies behind Criseyde's dream of an eagle replacing her heart with its own and flying away 'with herte left for herte' (ii. 931). In a more deliberate, ritual moment on the night of the lovers' first union (iii. 1370–2), Criseyde pins on the 'sherte' of

Troilus a brooch set with a heart-shaped ruby: it is a ceremonious token of her giving him her heart, the emblem of an inward process which is part of the larger pattern of Chaucer's imagery in *Troilus and Criseyde*.

Examples of 'hundred' and 'thousand' retained from *Filostrato* would include: i. 457, 531; ii. 977, 1089; iii. 1360; iv. 337, 753; v. 472. Chaucer's additions of 'hundred' and 'thousand' include: iii. 252, 297, 389, 574, 1595, 1684; iv. 826, 1500.

On oaths in Chaucer, see Elliott, *Chaucer's English*, ch. 5; on the heart in *Troilus*, see Clark and Wasserman, 'Heart'; Leyerle, 'The Heart and the Chain'; Windeatt, 'The "Paynted Proces"'.

For the heart as weeping, see ii. 567; as bleeding: i. 502; ii. 950; iii. 1524; v. 17, 1200.

'Verray signal of martire'

When Criseyde gives Troilus that heart-shaped ruby brooch, the lovers also exchange rings, but of these the poem makes a point of declaring that it can 'tellen no scripture' (iii. 1369), either being ostentatiously mysterious—that it has no information—or more particularly declaring that it has nothing to divulge about the inscriptions or mottoes that may have been on the lovers' rings. The world inhabited by the characters of *Troilus* is full of objects, phenomena, or events that may be read and interpreted as signs: the rubies mentioned in the poem could be interpreted with reference to the 'lapidaries', books which expounded the qualities and properties of precious stones; Criseyde's joined eyebrows (v. 813)—not a good sign, alas—might be referred to the works which analysed physiognomy; allusions to birds and other animals might be read in terms of traditional beast lore, and Pandarus refers to the pagan practice of augury (v. 380); references to the heavenly bodies and the signs of the zodiac are part of a complex study of signs; and the images presented by dreams were similarly the object of interpretation as signs, a study itself reviewed, like astrology, within *Troilus*.

As Chaucer extends and enriches the figurative language he found in his sources, some recurrent patterns of imagery emerge, and these are the subject of this section, especially imagery of heat and cold, light and darkness, sight and blindness, navigation, binding and constraint, and a range of animal and hunting imagery.

Chaucer associates the feelings of his lovers with fewer images of heat than in *Filostrato*, but adds many images of cold to balance a sense of the fire of love with the chill of misery and unhappiness. The main effects of these changes in his imagery of heat are, in the case of Troilus, to emphasize less the sensuality of his feelings than how they conform to conventional patterns and, in the case of Criseyde, to remove that heat of sensuality which enflames the Italian Criseida. The excursus after Troilus has fallen in love refers to the setting on fire of his heart ('Yet with a look his herte wex a-fere' (i. 229)), but concludes with a promise to include the contrary extremes of Troilus' experience ('Bothe of his joie and of his cares colde' (i. 264)). The contrary sensations of heat and cold, felt with death-like intensity, are part of that experience of the lover which Chaucer imports into the poem for his Troilus through his borrowing of the Petrarch sonnet (i. 420). The notion of the fire of love is so conventional that Troilus tries to hide his symptoms with a pretended illness ('lest men of hym wende | That the hote fir of love hym brende' (i. 489–90)), but he feels such fire as a pain worse than death (i. 607–9), and this is typical of how imagery of heat and fire is used to suggest great intensity of feeling in Chaucer's hero rather than sensual delight. In a curious development in *Troilus* Chaucer prefers to associate heat with Criseyde's feelings much later in his own poem, noting how after hearing of her exchange 'she brenneth both in love and drede' (iv. 678), or how in the Greek camp 'she sette hire woful herte afire | Thorugh remembraunce of that she gan desire' (v. 720–1). Having preserved his heroine from the associations of a heated sensuality before the consummation, Chaucer instead uses fire imagery to convey the intensity of her grief and sorrow.

The contrasting imagery of coldness is very much Chaucer's addition. The traditional phrase 'cold care' is introduced recurrently (e.g. iii. 1202, 1260; iv. 1692; v. 1342–3, 1747). Coldness is associated with not loving (iii. 1769) or with death (iv. 511), where warmth is identified with security (iii. 1630) and with hopefulness (ii. 967–73), and in the later books Chaucer develops his distinct aesthetic sense of the chill of misery. Troilus feels his heart grow cold as frost at the sight of Criseyde's deserted house, and he longs to kiss its 'colde dores' (v. 535, 551–2); and his heart grows suddenly chill as he recognizes the brooch on the captured 'cote-armure' of Diomede (v. 1658–9).

Chaucer develops imagery of light and darkness throughout *Troilus*, both in the narrative and in the proems that frame and articulate his poem. The central book of *Troilus* opens with a ringing address to the 'blisful light of which the bemes clere | Adorneth al the thridde heven faire!' (iii. 1–2), and the other proems can be seen to group around this central address to light. In the first proem the narrator expresses his distance from the experience of love as 'derknesse' (i. 18); at the opening of Book II, as the focus moves from Troilus' sorrows to a more hopeful subject, the imagery is that of leaving darkness and black seas behind for clear weather (ii. 1–2). At the end of the central third book, complementing the proem, is an address to Venus as 'Thow lady bryght . . .' (iii. 1807), probably recollected in part from a passage in Dante which acclaims the light of Venus in her epicycle. The fourth proem again marks the change in terms of light and darkness: Fortune now averts from Troilus her previously 'brighte face' and aid is invoked for the narrative from the Furies, identified as the three daughters of the night (iv. 8, 22).

Within the *Troilus* narrative there is also a marked use of imagery of light and darkness which is not in Chaucer's source. Criseyde's widow's black is complemented by a suggestion of inward light and brightness ('Nor under cloude blak so bright a sterre | As was Criseyde . . .' (i. 175–6)), and the association of Criseyde with light gives point to all the other imagery of light in the poem. Imagery of sunshine and cloud is introduced to evoke Criseyde's alternating mood as she ponders the possibility of love (ii. 781); Antigone's song associates love with the sun (ii. 862–5); and Troilus lies near 'derk disesperaunce' as he awaits Criseyde's first letter (ii. 1307). The timing of the consummation at the centre of the poem 'whan lightles is the world' (iii. 550) and on a night of 'smoky reyn' (iii. 628) only highlights by contrast the associations of light. The Boethian hymn of Troilus acclaims love as the force which governs the regular alternation of day and night (iii. 1755–6), so that, when he bestows upon Criseyde the conventional epithet of 'bright' in the consummation scene and recurrently afterwards, there is a special significance in the association of the lady with light, just as there is in Criseyde's oath that the sun will first fall from its sphere before she is unfaithful (iii. 1495).

In the later books these patterns of light and darkness are used with cumulative effect to emphasize the sorrows of the lovers. In two arresting images of the stripping away of happiness, Chaucer likens the grieving Troilus to a tree bereft of its leaves in winter and pent 'in the blake bark of care' (iv. 229), and later visualizes Troy being plucked of its 'fetheres brighte' (v. 1541–7). The 'sonnyssh' hair of Criseyde (iv. 736, 816) and her 'snowissh' throat (iii. 1250) are only outward aspects of the radiance of that lady who is the 'hertes day' of her lover (v. 1405). In Troilus' three-stanza address to her empty house, her absence becomes darkness:

'O thow lanterne of which queynt is the light,
O paleys, whilom day, that now art nyght . . .' (v. 543–4),

and Troilus feels his eyes will weep away their sight 'Syn she is queynt that wont was yow to lighte' (iv. 313), imagining himself as another blinded Oedipus 'in derknesse' (iv. 300).

Chaucer also makes extended use of the figurative associations of sight and blindness. At the very start of the love-story Troilus mocks at lovers as 'veray fooles, nyce and blynde be ye!' (i. 202), just before he is struck by Love's arrow in the form of seeing Criseyde and 'the subtile stremes of hir yen' (i. 305). This is accompanied by the exclamation 'O blynde world, O blynde entencioun!' (i. 211), so that the hero's very moment of seeing and loving is identified with blindness as he is shot by the god of love, 'Thy blynde and wynged sone ek, daun Cupide' (iii. 1808).

The process by which Chaucer's lovers come together allows for a further development of the figurative associations of sight and blindness. Through Pandarus and through the poem's narrating voice the progress of the affair is advanced by those who associate themselves with blindness: Pandarus in commenting on his ability to help Troilus despite his own lack of success in love ('I have myself ek seyn a blynd man goo | Ther as he fel that couthe loken wide' (i. 628–9)), and the narrator on his personal inexperience of what he is describing ('A blynd man kan nat juggen wel in hewis' (ii. 21)). In addition, Chaucer's version of his source includes more consciously significant moments of actual seeing—as in the two rides past of Troilus—while, in the more formal progress of the lovers' understand-ing, the act of seeing represents a certain stage in relations: Troilus requests 'That with the stremes of youre eyen cleere | Ye wolde somtyme frendly on me see' (iii. 129–30), and there Criseyde professedly wishes matters to remain (ii. 1295). The union of the lovers is only achieved when their watchful society is 'blynd | In this matere' (iii. 528–9; cf. ii. 1743), and at dawn Troilus imagines every chink of sunlight in their bedroom as so many eyes peering in (iii. 1453).

Once the lovers are separated out of sight of each other the associations of seeing become especially poignant: Troilus gazes at the moon (v. 647–9), Criseyde gazes back at Troy from the Greek camp (v. 709–11), and Troilus looks out from Troy so yearningly that he mistakes a cart for Criseyde (v. 1162). This is accompanied by the developing realization of the lovers' figurative blindness. Chaucer's Criseyde admits to herself her lack of foresight by alluding to the three eyes with which Prudence sees past, present, and future (v. 744–5), and into the account of Troilus' anxious gazing from the walls of Troy is inserted the quiet comment 'But al for nought; his hope alwey hym blente' (v. 1195). The clouded earthly vision of Troilus is overcome by the ascent of his spirit after death, where true understanding is dramatized as a form of seeing ('And ther he saugh . . .' (v. 1811)) as Troilus damns the 'blynde lust' and as the narrative commends the young to turn towards God the countenance of their hearts (v. 1838), representing a proper understanding of love as a form of looking and seeing.

It is a view from where he may contemplate the 'erratik sterres' (v. 1812) that puts in proportion Troilus' earlier view of Criseyde as a 'sterre, of which I lost have al the light' (v. 638) and without whom he is navigating blindly and in darkness. Such imagery of sailing and navigating is given force and direction by the Cantici Troili in Books I and V, and the second proem likening composition to a sea-voyage. From the Petrarch sonnet that becomes Troilus' first Canticus comes the likening of the hero's quandary to being in a rudderless boat at sea between contrary winds, though Chaucer adds that their contrariness is everlasting (i. 415–18). In a symmetry of imagery Troilus' final Canticus is also figured in comparable terms of a boat at sea ('Toward my deth with wynd in steere I saille' (v. 641)). The dominant image here is of Criseyde as the guiding star, which echoes other references in the last book to her as the pole-star for her lover (v. 232, 1392). The idea of the lover's experience as a voyage is repeated at other moments as the narrative develops. Troilus complains that love so assails him 'That streight unto the deth myn herte sailleth' (i. 606), and wishes that he were arrived at the port of death (i. 526–7); Pandarus encourages Troilus with 'Stond faste, for to good port hastow rowed' (i. 969), and Criseyde asks Pandarus 'What manere wyndes gydeth yow now here?' (ii. 1104). Images of navigation, of seas and winds, were also defined for Chaucer by their appearance in the *Consolation*, and its echoes are to be heard when Troilus laments to Fortune that 'Nought roughte I whider-ward thow woldest me steere' (iv. 282).

The distinctive Boethian imagery of binding order and control lies behind many of the poem's allusions to both actual and metaphorical binding, entrapment, and imprisonment. In the background to this imagery lie other traditions, including the literary convention of love as a snare and religious imagery of the bondage of the flesh and the devil. Such a rich background of traditions enables the notion of love, at first spoken of in terms of 'capture' and snare, to become associated with that cosmic binding force which Boethius calls love. Each of Chaucer's three main characters draws attention to the figurative nature of the language about constraint and ensnarement: Pandarus congratulates his niece with the news of Troilus' love, declaring it is she 'that han swich oon ykaught withouten net!' (ii. 583); Troilus, as he embraces Criseyde on their first night together, asks her 'How koude ye withouten bond me bynde?' (iii. 1358); and Criseyde describes her plans to catch out Calchas 'Withouten net' (iv. 1371).

The figurative associations of binding first appear in the assertions of the excursus after Troilus has been hit by love's arrow: 'Love is he that alle thing may bynde' (i. 237; cf. i. 255–6). The conventional notion of love as a capture ('kaught is proud . . .') is here reasoned—however speciously—into a submission of one's freedom to an irresistibly stronger and constraining force. Troilus vehemently expresses his sense of love as an ensnarement ('O fool, now artow in the snare . . . now gnaw thin owen cheyne!' (i. 507–9)); complaining how Fortune 'pleyeth with free and bonde' (i. 840), he anticipates his own fate; and, when Pandarus swears an oath by one figure in Hades ('To Cerberus yn helle ay be I bounde' (i. 859)) this will resonate in the closing stages of the poem where Troilus in his misery is likened to Ixion bound to his wheel in hell (v. 212).

The hopes and fears in love of both Troilus and Criseyde are in part presented through such imagery of constraint and binding. Troilus exclaims 'al brosten ben my bondes' at the first encouraging signs from Criseyde (ii. 976); his recovery from his faint is likened to release 'fro bittre bondes' (iii. 1116); his union with Criseyde is seen as rescue and escape (iii. 1242); and his heart remains so tightly enmeshed within Criseyde's 'net' that no other lady's beauty can begin to untie its knots (iii. 1730–6). Criseyde's approach to love, by contrast, is traced through her fear of being bound. In a striking image of an animal free to graze, Chaucer's Criseyde thinks of herself as 'unteyd in lusty leese' (ii. 752); resolves that Troilus 'shal me nevere bynde in swich a clause' (ii. 728); and writes that she will not 'make hireselven bonde | In love' (ii. 1223–4). She is, however, content that love 'of alle joie hadde opned hire the yate' (iii. 469). Troilus is to her 'a wal | Of stiel and sheld' (iii. 479–80), while he is happy that Criseyde 'of his herte berth the keye' (v. 460); for a related aspect of the imagery of binding is the characters' longing for permanence, in which rootedness is a value (i. 964–6) and rootlessness is misery and the prospect of death (iv. 770).

It is in Book III that Chaucer's Troilus associates his own love with a Boethian sense of love as the binding force of the universe, addressing love in the consummation scene as 'thow holy bond of thynges' (iii. 1261), and at the close of the book singing his Boethian hymn to love which makes the elements 'holden a bond perpetuely durynge', that 'constreyneth' the tides, and without whose 'bridel', 'al that now loveth' should fly apart. His purpose is to pray Love to 'bind' the accord that he has with his Criseyde (iii. 1750), and the remainder of the poem sets the bonds of human love a test which will suggest they are not to be identified with the bonds of that love which holds the universe together. In vain did Troilus pray that night be forever bound to his hemisphere (iii. 1439), for his own hymn to love as the bond of cosmic order acknowledges the ordered succession of time. Now Chaucer's imagery of binding may comment poignantly on the impermanence of the love and commitments it describes (as in saying of Criseyde 'that al this world ne myghte hire love unbynde' (iv. 675)). There is once again the constriction of sorrowful feelings (iv. 710–11, 229). In jaded rehearsal of the old idiom of love as ensnarement, Diomede schemes how 'Into his net Criseydes herte brynge' (v. 775), after he has taken the bridle of her horse and led her from Troy. It was perhaps from the *Consolation* that Chaucer derived a hint for the loosing of the soul of Troilus from the bonds of this world ('And yif the soule, whiche that hath in itself science of gode werkes, unbownden fro the prysone of the erthe, weendeth frely to the hevene, despiseth it nat thanne al erthly ocupacioun; and usynge hevene rejoyseth that it is exempt fro alle erthly thynges?' (*Boece*, ii, pr. 7, 152–7)).

Such imagery is further developed through the imagery of hunting — hawking, fishing, and the netting and snaring of prey. Association with some form of hunt also unites much, although by no means all, of the animal imagery in the poem. Throughout the earlier books of *Troilus* the pursuit of love is recurrently likened to sporting pursuits, from the moment when Troilus short-sightedly mocks lovers with 'youre prey is lost' (i. 201), and is taken by love just as birds are taken by the liming of twigs (i. 353). His desire is described as breeding inward arguments like young deer (i. 465), or being kept like a falcon 'in muwe' (i. 381); and, as he considers whether to reveal his love to Pandarus, Troilus thinks 'Ek som tyme it is a craft to seme fle | Fro thyng whych in effect men hunte faste' (i. 747–8). Of his own lack of success in love Pandarus says 'I have no cause, I woot wel, for to sore | As doth an hauk that listeth for to pleye' (i. 670–1); he uses the terminology of a hunt for river birds when arguing that there is still a diversity of ladies with different talents for Troilus to choose from ('Both heroner and faucoun for ryvere' (iv. 413)); and he imagines Troilus as the one hunted, when the gaining of Criseyde's 'love of friendship' is held to have lamed the sorrow that is pursuing Troilus (ii. 964). The implication of such language overall, however, is that, in the venery of love, Criseyde is the quarry of the hunt: when planning to bring Criseyde to visit Troilus in his feigned illness Pandarus tells the lover to conceal himself at his 'triste' (or hunting 'station'), towards which Pandarus will drive the deer:

> 'Lo, hold the at thi triste cloos, and I
> Shal wel the deer unto thi bowe dryve.'   (ii. 1534–5)

Such hunting language in *Troilus* has more effect for not being applied too often and directly to the progress of the love-affair, and this makes all the more striking the association of predatory language with Diomede's approach to Criseyde (but a boar hunt figures in his ancestry, as Cassandra will recall (v. 1464–84)). When Diomede contrives how he may best 'Into his net Criseydes herte brynge' (v. 775), this process is linked with preparing to catch a fish with hook and line (v. 777). In the background to this image lies the traditional medieval etymology which derived the Latin word for love (*amor*) from the word for a fishhook (*hamus*), and in *Troilus* this is the culmination of various associations of fishing with affairs of love, as when the proem to Book III likens the chance of falling in love with the taking of fish in traps at weirs (iii. 33–5).

The whole movement towards, and fulfilment of, the lovers' union in Book III is accompanied by allusions to hunting and by a sequence of bird images. Criseyde's losing her fears like the 'newe abaysed' nightingale (iii. 1233) looks back to the nightingale's song preceding her dream of the eagle who exchanged hearts with her (ii. 918–31), and this idea of an overmastering bird is taken up again as the consummation nears ('What myghte or may the sely larke seye, | Whan that the sperhauk hath it in his foot?' (iii. 1191–2)). At the dawn parting Criseyde invokes the various impossibilities that will occur before Troilus is out of her heart, including 'And everich egle ben the dowves feere' (iii. 1496), a modification of the aggressive stereotype which is explored more widely in the poem's associations of the hero not only with the lion (i. 1074) but also with a snail (i. 300), an ass (i. 731), a mouse (iii. 736), a gnat (iv. 595), a roosting bird (v. 409), or a horse that has to accept that he is harnessed, 'in the trays' (i. 222). The whole action is accompanied by references to bird-lore and to the role of birds in signalling time. The lovers' meetings are safe from gossip of 'goosissh' people (iii. 584) because 'it cler was in the wynd | Of every pie' (iii. 526–7), while at significant junctures in the action birds function as important signs: the archetypal role of the cockerel is suggested by designating him as the 'comune astrologer' (iii. 1415); the cry of the screech owl is interpreted as an ominous sign (v. 319–20, 380–2); dawn on the longed-for day of Criseyde's expected return to Troy is signalled by the song of that bird into which the treacherous daughter of King Nysus was metamorphosed (v. 1110); and the image of Troilus as a plucked peacock near the opening of the first book (i. 210) is mirrored symmetrically near the close of the last book by an image of Troy itself as a bird being plucked of its bright plumage (v. 1546). These last images also draw on the resources of that language of proverb so characteristic of Chaucer's style throughout *Troilus*.

On figurative associations in the language of *Troilus*, in comparison with *Filostrato*, see Meech, *Design*, ch. III. On imagery of binding and constraint, see especially Barney, 'Troilus Bound'; Sundwall, 'Criseyde's Rein'; and Van, 'Imprisoning and Ensnarement'; for some further image patterns, see Schuman, 'Circle', and Stevens, 'The Winds of Fortune'.

In imagery of heat and fire, various patterns distinguish *Troilus* from *Filostrato*. Chaucer avoids any sense that the heroine burns with desire and that she kindles a flame in her lover that she has the power to quench. Troiolo thinks his fervour could be quenched by Criseida (*Fil.* i. 41), but Chaucer omits this (cf. i. 435–48), and, whereas Troiolo's first letter addresses Criseida as 'The only one who can quench the flame of my love' (*Fil.* ii. 88), this idea disappears from *Troilus*. Troiolo talks of the pleasure in being kindled with the fire of love (*Fil.* ii. 7), but his prayer to Love to kindle flames beneath Criseida's dark garments (*Fil.* ii. 60) survives only as Troilus' belief that his own love burns more intensely because hidden (ii. 538–9). After hearing of Troiolo's love, Criseida urges herself to 'satisfy his burning desire' (*Fil.* ii. 74); after reading his letter she thinks 'Now I must find a time and means to quench this flame' (*Fil.* ii. 115), and in her reply describes herself as burned by cruel flames (*Fil.* ii. 127). Nothing of this survives into Criseyde, and Chaucer's different plot excludes the scene of Criseida with torch in hand greeting Troiolo, 'both burning with equal passion' (*Fil.* iii. 30). In the ensuing scene Troiolo addresses Criseida's 'amorous eyes', declaring 'You sent the fiery shafts of love into my heart and set me all aflame' (*Fil.* iii. 36), but none of this fiery imagery remains in *Troilus* (cf. iii. 1353–4), and Chaucer also omits *Filostrato*'s account of the lovers' second night together, when Criseida tells Troiolo of 'the fierce desire that you have kindled in my breast' and speaks of her inextinguishable passion that blazes up more than before (*Fil.* iii. 67–8).

In the prose proem to *Filostrato* the poet, separated from his lady and longing to see her, acknowledges that sight of the beloved gives much greater pleasure than talking or thinking about her, and Boccaccio's ensuing narrative has a series of moments of amorous seeing that become part of the reference to sight in *Troilus*: Troiolo's sight of Criseida in the temple (*Fil.* i. 25 ff.); his later lament about that moment, reported by Pandaro (*Fil.* ii. 58); the exchange of glances between the lovers (*Fil.* ii. 82); Troiolo's address to his lady's eyes at the consummation (*Fil.* iii. 36); his feeling that his eyes will become useless in the absence of Criseida (*Fil.* iv. 35); his memory of how Criseida's eyes first caught him (*Fil.* v. 55).

'Proverbes may me naught availle'

Troilus contains a higher proportion of proverbs than any other of Chaucer's works. This gives a reflective and discursive dimension to the poem, whereby characters comment on their experience, interpreting and seeing patterns within it, and the very universality of proverbial statement makes some movement away from the particulars of the narrative towards the universal. Yet Chaucer also makes use of the way that proverbs offer a typically 'limited' wisdom, cautious, pragmatic, unidealizing. Proverbial language thus serves a double purpose: while it is in the nature of proverb to invite assent and the recognition of a familiar pattern of accepted wisdom, the use of proverbs in *Troilus* achieves a more ambiguous effect in which—as context governs—recognition may not necessarily be followed by consent.

Proverbial expressions may be found scattered throughout the poem, but there are some particularly thick clusterings at key points: the excursus in Book I after Troilus has been struck by love's arrow; the arguments of Pandarus to Troilus in Book I, and to Criseyde in Book II; Criseyde's thoughts to herself in Book II after Pandarus' departure; the discussion of discretion and secrecy early in Book III; Pandarus' comforting of Troilus early in Book IV; Criseyde's arguments to Troilus for her going to the Greek camp in Book IV.

The following tables present instances of proverbs, sentences, and proverbial phrases in the order in which they occur in the poem, citing the line references to *Troilus* and matching these with the entry in B. J. Whiting's standard dictionary of medieval English proverbs (where, however, instances cited from *Troilus* are sometimes accompanied only by

few and later instances, possibly influenced by Chaucer). The tables set out to chart how the development of the Troilus story in Chaucer's version is accompanied by these pieces of traditional wisdom. (The speaker of each proverb is indicated by [T], [C], [P], [D], with [N] for the narrating voice, [A] for Antigone, and [Ck] for Calchas.)

BOOK I

After Troilus mocks lovers by echoing a proverb ('O veray foles . . .'), and is himself smitten by love, most of the remaining proverbs in the book are cited about him or quoted at him. His 'fall' can be associated with proverbs about pride, with proverbial comparisons with animals, and with proverbs about the power of love. With the entry of Pandarus the incidence of proverbs becomes much more frequent.

202–3	[T]	One fool cannot beware by another (F449)
210	[N]	As proud as a peacock (P280)
214–16	[N]	Pride will have a fall (P393)
217	[N]	Oft fails the fool's thought (F448)
218	[N]	Bold as blind Bayard (B71)
219	[N]	His corn pricks him (C430)
237	[N]	Love binds (L497)
241–4	[N]	No man can escape love (M224)
257–8	[N]	Better bow than break (B484)
300	[N]	To draw in one's horns (H491)
406	[T]	The more some drink the more they wax dry (D403)
630	[P]	A fool may give a wise man counsel (F404)
631–2	[P]	A whetstone makes sharp carving tools (W217)
635	[P]	'Often wise men ben war by foolys' (M340)
637	[P]	'By his contrarie is every thyng declared' (T110)
638–9	[P]	He knows not sweetness that never tasted bitterness (S943)
640–1	[P]	No man knows bliss that never endured woe (M236)
642	[P]	White seems more by black (W231)
645	[P]	'Of two contraries is o lore' (C415)
662	[P]	Leech heal thyself (L171)
694–5	[P]	Woe to him who is alone (Eccles. 4:10)
708–9	[P]	'To wrecche is consolacioun To have another felawe in hys peyne' (W715)
731	[P]	Like an ass to the harp (A227)
740–2	[N]	To make a rod for oneself (S652)
747–8	[T]	Sometimes it is a craft to flee from thing which in effect one hunts (C526)

809	[P]	Unknown, unkissed (U5)
837	[T]	Fortune is my foe (F529)
857–8	[P]	A leech may not heal a wound unless the sick man shows it to him (L173)
946–9	[P]	The ground that bears wicked weeds bears wholesome herbs (G478)
950	[P]	Next the valley is the hill (V2)
951	[P]	Next the dark night the glad morrow (N108)
956	[P]	He hastes well that wisely can abide (H171)
964–6	[P]	A tree set in divers places will not bring forth fruit (T474)
976–8	[P]	No man can escape love (M224)
1064	[N]	To have time and place (cf. ii. 1367–70) (T328)
1074	[N]	To play the lion (L353)

BOOK II

Here there is a sustained use of proverbs by Pandarus in his visit to his niece, and by Criseyde in her responses to him and then in her thoughts alone, which crystallize her perception at successive points. In the latter half of the book proverbs and proverbial phrases continue to mark Pandarus' arguments for action to both Troilus and Criseyde.

21	[N]	The blind man cannot judge in colours (M50)
28	[N]	'In sondry londes, sondry ben usages' (T63)
42	[N]	'Forthi men seyn, ecch contree hath his lawes' (cf. ii. 28)
193	[P]	As thick as bees; to swarm like bees (B167, B177)
260	[P]	The last word binds the tale (W598)
328	[P]	To fish fairly (F240)
343	[P]	Advisement is good (A62)
398	[P]	Too late aware (B155)
470	[C]	Of two harms choose the least (E193)
483	[C]	Let the cause cease and the effect will cease (C121)
538–9	[P]	Wry [cover] the gleed [coal] and hotter is the fire (G154)
600	[N]	As still as stone (S772)
671	[N]	Everything has a beginning (E164)
715	[C]	There is measure in all things (M464)
754	[C]	To say checkmate (C169)
789	[C]	Harm done is done, whoso may rue it (H134)
791	[C]	A wonderful beginning often gives way at the end (B201)
798	[C]	Nothing has its being of nought (N151)
807–8	[C]	He that undertakes nothing achieves nothing (N146)

811	[C]	Now (this) now (that) (N179)
861	[A]	Many men speak of Robin Hood that never bent his bow (R156)
926	[N]	As white as bone (B443)
989	[P]	All thing has time (T88)
1022	[P]	To have one's ears glow (E12)
1033	[P]	To harp on one string (S839)
1083	[N]	Without ho (H398)
1104	[C]	What wind guides you here? (W343)
1256	[N]	Red as (a, any, the) rose (R199)
1276	[N]	Smite while the iron is hot (I60)
1320	[P]	In black and white (B328)
1335	[N]	From a small shoot comes a great tree (O7, G418, K12)
1380–3	[P]	An oak falls all at once (O8)
1381	[P]	An oak is feeble that falls at the first stroke (T471)
1387–9	[P]	The reed that bows will arise when the wind ceases (R71)
1461	[N]	As straight as a line (L301)
1615	[N]	To ring like a bell (B234)
1681	[N]	To affile one's tongue (T378)
1745	[P]	Wagging (movement) of a straw (rush) (W4)

BOOK III

Although Book III has a richness of proverbial phrases and comparisons, there is a predictable shift from previous books in the relative absence of passages of a sustained monitory nature. The early exchanges between Pandarus and Troilus on discretion, Pandarus' persuasions of Criseyde to receive Troilus, and the late discussion between Pandarus and Troilus on preserving the happiness that has been won, are the main contexts in which proverbs are associated with conduct.

88	[N]	To sing a fool a mass (F458)
114	[N]	A heart of stone would melt (H277)
115	[N]	To weep as one would turn to water (W81)
198	[P]	To bear the bell (B230)
228	[N]	As right as any line (L300)
294	[P]	The first virtue is to keep the tongue (V41)
309	[P]	An avaunter and a liar are all one (A244)
329	[P]	The wise are chastised by the harm of fools (cf. C161)
404–6	[T]	Diversity is required between like things (D270)
425	[N]	To burn like the fire (F200)

526	[N]	To be clear in the wind (W314)
615	[N]	All things have an end (T87)
695	[N]	Love's old dance (L535)
733–4	[T]	One's destiny was shaped before one's shirt (D106)
764	[P]	It's not good to wake a sleeping hound (H569)
813–15	[C]	Every joy is mingled with pain (J59, S516)
832	[C]	Not to set a mite (M608)
853	[P]	Peril is drawn in with delay (P145; cf. D157)
854	[P]	Not worth a haw (H193)
855	[P]	All thing has time (cf. ii. 989) (T88)
861	[P]	Farewell fieldfare (F130)
885	[C]	Blue is true (B384)
896	[P]	Lost time cannot be recovered (T307)
901	[P]	White words (W627)
931	[C]	To be at one's wit's end (W412)
936	[P]	Not worth two vetches (V27)
1024	[C]	Jealousy is love (J22)
1060–1	[N]	'of a ful misty morwe Folowen ful ofte a myrie someris day' (M693, 686)
1062	[N]	After winter follows May (W372)
1137	[P]	Light is not good for sick folk's eyes (L260)
1192	[N]	To flee as the lark does the sparrowhawk (L84)
1194	[N]	As bitter as soot (S480, 482)
1200	[N]	To quake like an aspen leaf (A216)
1212–15	[N]	A bitter drink heals a fever (D393)
1263	[T]	To fly without wings (W363)
1282	[T]	Mercy passes right (M508)
1577	[N]	God forgave his death (G205)
1625–8	[P]	The worst infortune is to remember past prosperity (I41)
1630	[P]	To sit warm (W48)
1633	[P]	As red as (any, the) fire (F171)
1634	[P]	As great a craft is keep well as win (C518)
1636–7	[P]	The world's bliss lasts but a while (W671)
1665	[N]	Span new (S547)
1784	[N]	Fresh as falcon (F25)

BOOK IV

Pandarus continues to call on the authority of proverb in advising first Troilus and then Criseyde. The most concentrated use, however, comes in the long closing interview of Troilus and Criseyde, where Criseyde argues by frequent recourse to proverb for the necessity of complying with the exchange agreement, and in reacting to her arguments Troilus makes more use of proverbs than he does elsewhere.

3–5	[N]	Fortune seems truest when she will beguile (F535)
101	[Ck]	Now or never (N178)
183	[N]	'Vox populi vox Dei' (V54)
337	[N]	As hot as any gleed (G142)
367	[N]	To melt like the snow (S445)
423–4	[P]	Seldom seen soon forgotten (S130)
434	[N]	In at one ear and out at the other (E4)
461	[T]	In dock, out nettle (D288)
461	[T]	Now this, now that (N179)
588	[P]	A wonder lasts but nine nights (W555)
600–2	[P]	Fortune helps hardy man (F519)
622	[P]	To set (all) at six and seven (S359)
626	[P]	To lie (in field, etc.) like a dog (D329)
684	[N]	Dear enough a mite (M596)
765	[C]	Like a fish out of water (F233)
770	[C]	Rootless green must soon die (G453)
834–6	[C]	The end of bliss is sorrow (E80)
927	[P]	'flat than egge' (cf. A30)
936	[P]	Women are wise in short advisement (cf. iv. 1261–3) (W531)
1137	[N]	As bitter as gall (G8)
1166	[N]	One's song is welawey (S469)
1283	[C]	Lost time cannot be recovered (T307)
1369	[C]	Covetousness is natural to the old (C490)
1398	[C]	Not avail three haws (H189)
1408	[C]	Dread first found [invented] gods (D385)
1453–4	[T]	The bear thinks one thing, but his leader thinks another (B101)
1456	[T]	Men may the old (wise) outrun but not outwit (O29)
1457–8	[T]	It is hard halting before a cripple (H50)
1459	[T]	Argus and his eyes (A180)
1568	[C]	Hasty man never wants [lacks] care (M97)
1584	[C]	'The suffrant overcomith' (S865)
1585	[C]	'Whoso wol han lief, he lief moot lete' (L233)
1586	[C]	To make a virtue of necessity (V43)
1628	[C]	'Who may holde a thing that wol awey?' (H413)
1645	[C]	Love is full of dread (L517)

BOOK V

Here the citation of proverbs becomes less sustained, and the 1869 lines of Book V contain fewer proverbs than the 1092 lines of Book I. Towards the close of the poem proverbial sentiments of resignation and detachment are invoked.

343	[P]	Friends may not always be together	(F639)
347	[P]	Fair and soft	(F17)
350	[P]	As time hurts time cures	(T300)
358 ff.	[P]	Dreams are false (cf. 1277); (D387)	
362	[P]	A straw for ...	(S816)
425	[T]	As glad as any bird of day	(B292]
469	[N]	To make one a houve [hood] of glass	(H624)
741–2	[C]	'To late comth the letuarie [remedy]'	(L168)
757–9	[C]	Whoso will heed every word shall never thrive	(W629)
783–4	[D]	He that undertakes nothing achieves nothing	(N146)
830	[N]	Hardy as lion	(L314)
831	[N]	True as steel	(S709)
1062	[C]	To have one's bell rung	(B233)
1265	[T]	To be gospel	(G401)
1266–7	[T]	A familiar enemy is the worst	(E97)
1277	[P]	Dreams are false	(D387)
1433	[N]	Pipe in an ivy leaf	(I72)
1434	[N]	Thus goes the world	(W665)
1639–40	[N]	Sooth shall be seen	(S491)
1748–9	[N]	The world's bliss lasts but a while	(W671)
1840	[N]	This world is but a fair	(W662)
1841	[N]	To pass like (the, a) flower	(F326)

In *Troilus* Chaucer achieves a distinct ambiguity by means of proverbial expression. The sheer intelligence and perceptiveness of the poem may embolden the reader to wonder whether the lives of Troilus and Criseyde are really to be interpreted solely in terms of such patterns. Context does govern, for there may be some tension between the self-contained, encapsulated wisdom of the proverb—a contextless piece of generally applicable wisdom—and the particular context and rhetorical purpose where it occurs in *Troilus*. In the Book I excursus on the need to submit to love, the comparison of Troilus with the horse Bayard serves to emphasize how the hero's experience has brought him down to the same level as the common experience of man (i. 218 ff.), so setting Troilus in relation to the commonplace, the generally valid, although the very attempt to do this is enough to prompt questions and qualifications. The narrating voice is dramatized as being eager to establish with his audience a common acceptance of proverbial wisdom:

> For ay the ner the fir, the hotter is—
> This, trowe I, knoweth al this compaignye;   (i. 449–50)

and his disposition to proverb and set-phrases invites but does not justify immediate acceptance of what he says.

Pandarus is most associated with proverbial language, his technique being to overcome the brevity of individual proverbs by creating a flurry of similar sentiments in an accumulating series: the thrust of the argument is single-minded, the sense of a variety and range of evidence is largely rhetorical. Many of the individual points—considered in the abstract—express what would be good sense in many predicaments, and that Pandarus should have such ready recourse to such a corpus of lore becomes part of his characteristic knowledgeableness in the poem. His advice is expressed with a typical combination of bookish authority and homely proverbial emphasis (e.g. iii. 852–4), yet his flow of instances is checked by Troilus in a way that bluntly questions their relevance to his case ('For thi proverbes may me naught availle . . . Lat be thyne olde ensaumples, I the preye' (i. 756, 760)). The relating of events in the poem to proverbial patterns is not allowed to proceed without dispute, and, although Pandarus is always ready to talk in terms of sententious instances (once remarking to Troilus 'Proverbes kanst thiself ynowe' (iii. 299)) this is to be seen as expressing one among many attempts at interpretation in the poem rather than a single monopoly of wisdom which commands acceptance. Pandarus argues to Troilus that, although a great oak takes time to fell, its fall is irreversible, whereas a reed leans over and grows upright again depending on the wind (ii. 1380–9); yet, in view of how the story will end, the truth in Pandarus' example may appear to be the opposite of how he applies it, for Troilus is the tree that falls, while Criseyde is the reed and 'Ful lightly, cesse wynd, it wol aryse' (ii. 1388).

Pandarus' sententiousness brings within *Troilus* a rhetorical and thematic variety which it might otherwise lack. When, with a stanza of commonplaces, Pandarus encourages Troilus to be hopeful ('Next the foule netle, rough and thikke, | The rose waxeth swoote and smothe and softe' (i. 948–9)), or when he urges Criseyde to use time and beauty before they pass, Chaucer uses such passages to set the action of his narrative in relation to archetypal patterns, in a way that gives some resonance beyond the limits of the particular proverbs. New departures and transitions in the narrative, persuasions and decisions, are often set against the recollection of some traditional wisdom, but Pandarus' facility in deploying proverbs to suit his argument comes to subvert their claims to a universal truth. He gives the impression of being able to manufacture proverbially sounding comments to suit his persuasive purpose, and a division often suggests itself between the claim to a general truth, and an evident ulterior motive in a particular context. The sententious cast of mind brings into the poem some reflections of great beauty ('Wo worth the faire gemme vertulees!' (ii. 344 ff.)), but these are often subverted by the motive of the speaker. Through the proverbs of Pandarus the lovers are offered distinctly practical and realistic advice which seeks to find the norm and mean, avoiding extremes, favouring survival, taking advantage of opportunity while not forsaking prudence. He is all for accepting time and change, and for him all behaviour can be explained through truisms that assume a common pattern in life. Everything has happened and been noticed before; the present is always a repetition, confirming by conforming to an established pattern of commonplaces. For Troilus, by contrast, the present is not a repetition, and is not experienced and felt like one. The present has a unique reality of experience without precedent, and Troilus' experience of love begins precisely as he loses his previous ability to stand aloof and observe patterns in lovers' behaviour ('O veray fooles . . . Ther nys nat oon kan war by other be!' (i. 202–3)). In this light it is Pandarus who is most given to using proverbs and Troilus who uses them least: Pandarus who tends to see life fulfilling commonplace patterns; Troilus whose experience seems to him to stand alone.

In Criseyde, it is perhaps a typical caution and reliance on others' opinions that Chaucer hopes to suggest, a family resemblance to her uncle, and to the worldly pragmatism and self-interest of Pandarus and Diomede, rather than to Troilus' single-minded idealism. Criseyde is much disposed to proverbial expressions when Pandarus visits her in Book II: there is a pretext for concession to Troilus ('Of harmes two, the lesse is for to chese' (ii. 470)), as also in the familiar argument against extremes ('In every thyng, I woot, ther lith mesure' (ii. 715)). It is by means of proverbial wisdom that Criseyde persuades herself to allow the affair to develop ('He which that nothing undertaketh, | Nothyng n'acheveth . . .' (ii. 807–8)), a proverb shared with Diomede (v. 784), and her response to later misfortune is similarly phrased in terms of commonplaces. It is with a self-consciously sententious phrase that she represents to herself her separation from Troilus ('For which ful ofte a by-word here I seye, | That "rooteles moot grene soone deye"' (iv. 769–70)). Her fondness for 'by-words' or proverbs is to prove sadly ironic, in view of the way that she was to become a by-word in herself. Her unwise insistence on going to the Greek camp is marked by its dependence on proverbial wisdom, as if Chaucer wishes to draw attention in this crucial speech to the way she thinks by means of proverbs and commonplace patterns.

As the tables demonstrate, proverbs cluster at important points of persuasion and self-persuasion in *Troilus*. Such frequent recourse to proverbs cumulatively imports into the poem an awareness both of the limitations of prudential wisdom and of its value, for there is value in the wisdom of proverb which generations have acknowledged, and to view all instances of the proverbial in *Troilus* as uniformly ironic in their context is only to be uniformly anachronistic. Yet, as used in *Troilus*, proverbs remain perennially open to interpretation. That a proverb's meaning may vary according to context tends to subvert that stability promised by its fixed form, and this interplay between an immutable form and modification-by-context contributes to the poem's concern with change. In this process the worldly wisdom that proverbial language and comparison represent may come to symbolize some compromise of what had begun as an ideal, although it is ironically the prudential wisdom of Criseyde which is eventually to bring the emotional action of the poem to paralysis. With the coming of disappointment and frustration the prudential wisdom of proverbs comes more than ever to seem contradictory, and their flexibility and potential for universal application become even more disturbingly slippery, in a way that contributes to a larger concern with the reliability of language in *Troilus and Criseyde*.

On the incidence of proverbs in *Troilus*, see B. J. Whiting, *Chaucer's Use of Proverbs* (Harvard Studies in Comparative Literature, 11; Cambridge, Mass., 1934), 49. For studies of the use of proverbs in *Troilus*, see Lumiansky, 'Function'; Frank, '*Troilus and Criseyde*'; and Taylor, 'Proverbs' and *Chaucer Reads*.

For the citations of proverbial wisdom in *Troilus* in this section, and for reference, see B. J. and H. W. Whiting, *Proverbs, Sentences, and Proverbial Phrases from English Writings Mainly Before 1500* (Cambridge, Mass., 1968).

'*To ryme wel this book til I have do*'

> O lady myn, that called art Cleo,
> Thow be my speed fro this forth, and my Muse,
> To ryme wel this book til I have do;
> Me nedeth here noon other art to use. (ii. 8–11)

To emphasize a complete dependence on sources for the substance of *Troilus and Criseyde* is at the same time to claim credit merely for putting the material into rhyme, even if there is some quiet claim in the versifier's wish 'to ryme wel this book'. Although it was once part of what made *Troilus* so much admired in the first centuries after its composition, Chaucer's versification does not always attract much attention nowadays. Yet that *Troilus* is a narrative poem in seven-line stanzas is a fundamental and characterizing aspect of its style and structure as a poem, and to have carried through a stanzaic narrative on such a scale is in itself a remarkable accomplishment. Through Chaucer's poetic craftsmanship the stanza becomes an adaptable medium for narrative, in which an unfalteringly (but unobtrusively) constant and patterned rhyme structure provides a frame-work within which the full variousness of Chaucer's style may be given expression yet also conformed to a larger unity of form and style in *Troilus* as a whole.

Although his *Complaint of Venus* lamented 'Syth rym in Englissh hath such skarsete' (80), Chaucer's achievement in maintaining the 'rhyme-royal' pattern of his stanzas in *Troilus* is manifest throughout his 1,177-stanza poem and needs no extended illustration in itself. (The exception, the closural 'Go, litel bok' stanza (v. 1786–92) with its special rhyming together of the stanza's first five lines, only proves the rule.) In turn, a strong presumption establishes itself that in the majority of lines there will be ten syllables arranged in a broadly alternating pattern of unstressed and stressed syllables, albeit with possibilities for flexibility and variety. Such is the syllabic and stress pattern in undisputed lines (i.e. for which there are no manuscript variants), and it would appear to be the pattern held to by the scribes of earlier manuscripts of *Troilus*, in the face of that fast pace of contemporary linguistic change in English that Chaucer correctly foresaw would be detrimental to the metre of his lines as they were repeatedly transcribed in times to come (v. 1793–6). These manuscripts maintain most consistently the contracted verb forms necessary for such a metre and sometimes found only in *Troilus* among Chaucer's poems:

> 'Hym *tit* [= tideth] as often harm thereof as prow.'   (i. 333)
> '*Ret* [= redeth] me to love, and sholde it me defende?'   (ii. 413)
> 'Lo, yond he *rit*! [= rideth]' Quod she, 'Ye, so he doth!'   (ii. 1284)
> 'I write, as he that sorwe *drifth* to write . . .'   (v. 1332)

This highly crafted stanza-form has a patterning and ordering effect on the poem more largely. The stanza unit is too unvarying and too brief to be altogether comparable in form and effect to a verse paragraph; instead of its altering to match changes of material, material different in subject, mood, and pace is narrated within the same, unchanging verse-form. (The very first stanza of the poem, with its syntactical control and its unusual rhyming of 'fro ye' with 'joie' and 'Troye', establishes at the outset expectations of stanza-form and rhyme-scheme that are to be maintained.) In the ABABBCC scheme of the rhyme-royal stanza the first half of the structure works through alternation and interlinking, while the BBCC couplets tend to build a closural effect into the second half, so that every stanza has the potential to remind the reader of the poem's bilateral symmetry and becomes an emblem in miniature of the formal and thematic structure of *Troilus*. The effect of closure in the stanza's final couplet is compounded in numerous instances by a tendency to devote the last line of a stanza to general or sententious reflections, exclamations, oaths, prayers, and wishes, of which the following are simply a small selection:

> Lo, here his lif, and from the deth his cure!　(i. 469)
>
> 'A ha!' quod Pandare; 'Here bygynneth game.'　(i. 868)
>
> 'As love for love is skilful guerdonynge.'　(ii. 392)
>
> 'And Elde daunteth Daunger at the laste.'　(ii. 399)
>
> 'Swouneth nought now, lest more folk arise!'　(iii. 1190)
>
> Thus sondry peynes bryngen folk in hevene.　(iii. 1204)
>
> Wher hym was wo, God and hymself it wiste!　(iv. 1162)
>
> 'This Diomede is inne, and thow art oute.'　(v. 1519)

The last line or two of the stanza is often also the place for addresses to the audience, promising a transition to new material (e.g. iii. 238; iv. 238, 1127; v. 952). Given that in *Troilus and Criseyde* the caesura comes almost always between the second and third stress, usually before the fifth syllable, it has also been suggested that the second metrically stressed syllable of the seventh line often comes to bear a particular significance and so—because of the emphasis upon the stanza's last line—becomes pivotal to the movement and sense of the whole stanza.

In *Troilus* a stanza often contains and so frames a single action or reaction, a single short speech or reply. The movement in and out of a stanza can be an effective way of marking and matching the entry into a character's private inner thoughts and then out again into speech, as in the representation of Criseyde's thinking and speaking during the first visit of Pandarus (ii. 456, 463). The boundaries of the stanzas and the breaks between them may also be used to mark transitions or completion in action and speech, as when the stanza's final line coincides with a step forward in the plot (iii. 742, 1099), or when the close of an action or process is synchronized with the close of a stanza (ii. 1085). In the context of such a pattern exceptions accordingly make their mark, where a speech or process of thought or action is not completed within the stanza but continues to the beginning of the next.

The number of occasions when a sentence unmistakably runs on from one stanza to end in another is not large in proportion to the length of the poem, but such exceptions may make striking effects, whether of some special emphasis (as at Criseyde's first appearance):

> Nas nevere yet seyn thyng to ben preysed derre,
> Nor under cloude blak so bright a sterre
>
> As was Criseyde, as folk seyde everichone ...　(i. 174–6)

or to convey the impetus of a continuous action (as Pandarus' secret plot thickens and quickens):

> Whan al was wel, he roos and took his leve,
> And she to soper com, whan it was eve,
>
> With a certein of hire owen men,
> And with hire faire nece Antigone,
> And other of hire wommen nyne or ten.
> But who was glad now, who, as trowe ye,
> But Troilus, that stood and myght it se
> Thorughout a litel wyndow in a stewe,
> Ther he bishet syn mydnyght was in mewe,
>
> Unwist of every wight but of Pandare?
> But to the point ... (iii. 594–604)

Continuity may also be created by the concatenation of stanzas interlinked through the repetition of a word or words from the last line of the preceding stanza near the opening of the next:

> And ay the ner he was, the more he brende.
>
> For ay the ner the fir, the hotter is—   (i. 448–9)
>
> 'Of kynde non avauntour is to leve.
>
> 'Avauntour and a lyere, al is on ...'   (iii. 308–9)

In some instances a continuous sentence may run through three stanzas, as in the case of Troilus' courtly speech of petition to Criseyde early in Book III (127–47), where the syntactical unity of the stanzas, and their style and diction, may have had the effect of setting such stanzas somewhat apart in the manner of a lyric unit within the narrative (cf. Criseyde's later response to Troilus (iii. 988–94)). The syntactical structure of the stanza can also be used to represent confusion, as with Troilus' learned but then forgotten 'lesson', where the stanza's movement shadows the flustered state of the nervous lover, reduced at last to a final blurted entreaty (iii. 92–8). Chaucer's capacity to catch the accents and rhythms of speech also works through the whole stanza unit, despite the apparent distance of his stanza-form from speech. When Pandarus first arrives to visit his niece, their conversation (ii. 113–40) is an example of how unconstrained by stanzaic form Chaucer may be in constructing dialogue that effortlessly satisfies the formal requirements of versification, yet has the backwards and forwards movement of a sprightly exchange, full of banter and exclamation:

> 'And I youre borugh, ne nevere shal, for me,
> This thyng be told to yow, as mote I thryve!'
> 'And whi so, uncle myn? Whi so?' quod she.
> 'By God,' quod he, 'that wol I telle as blyve!

> For proudder womman is ther noon on lyve,
> And ye it wiste, in al the town of Troye.
> I jape nought, as evere have I joye!'   (ii. 134–40)

Another moment when Pandarus warms to his theme must serve to represent the many telling instances of *enjambement* within stanzas, where the sense runs on past the rhymed line end:

> 'And nere it that I wilne as now t'abregge
> Diffusioun of speche, I koude almoost
> A thousand olde stories the allegge
> Of wommen lost through fals and foles bost . . .'   (iii. 295–8)

Although Chaucer's versification rarely gives any impression of straining after effect, and despite his complaints about the scarcity of rhyme in the English language, Masui's catalogue of the rhyme-words in *Troilus* and their frequency shows that most rhymes in the poem are used relatively sparingly. Yet some of the most frequently listed rhyme-words confirm some of the generic, thematic, and stylistic emphases already noticed in *Troilus*, and the following is a selection of rhyme-words with the number of their appearances in rhyme (the number is the total including variant spellings):

destresse (34); drede (43); d(e)ye (45); peyne (44); smerte (26); sorwe (24); sterve (16); wo (29);

che(e)re (38); de(e)re (130); herte (41); knyght (28); service (23); swe(e)te (22);

biseche/bisoughte (22); compleyne(d) (9), compleynte (2), compleynynge (2); pleyne (16), pleynte (3); preye (24), preyde (6); rede (25); seyde (47), seye (69); speche (27); telle (30);

dwelle (26); ende (23); laste (21); reste (29); wende (30).

There are also some telling cases where the same words will tend to rhyme together in *Troilus*. The collocation of *joie* with *Troye* 31 times (23 times in the final couplet of the stanza) develops a particular resonance, ironic and poignant, as the narrative takes its course towards the unhappy end of Troy. The collocation of *routhe* and *trouthe* (19 times, 14 in the final couplet), as also of *rewe* and *trewe*, begins in the courtly context of Criseyde's pity on the faithful integrity or *trouthe* of Troilus as a lover and develops through a sad transformation as Criseyde loses her own *trouthe*. The collocations of *serve* with *deserve* (and with *sterve*), as also of *entente* with *mente*, allow the relation between these terms to be explored in a way that contributes to the poem's analysis of love and the lovers. The association of ideas and terms that is sounded and reiterated by the chime of rhyming is tested when circumstances and contexts shift during the narrative. Not for nothing does the narrative declare of Troilus:

> His resons, as I may my rymes holde,
> I yow wol telle, as techen bokes olde. (iii. 90–1)

On aspects of Chaucer's versification in *Troilus*, see Michio Masui, *The Structure of Chaucer's Rime Words* (Tokyo, 1964); Stanley, 'Stanza and Ictus'; Stokes, 'Recurring Rhymes'; and Windeatt (ed.), *Troilus and Criseyde*, 55–64.

As examples of sentences continuing across the break between stanzas, see i. 49–50, 175–6; ii. 56–7, 70–1, 546–7, 1267–8, 1694–5; iii. 161–2, 511–12, 539–40, 1120–1; iv. 154–5, 189 f., 1288–9; v. 259–60, 952–3, 1022–3, 1414–15, 1498–9, 1659–60. With iii. 595–6 and 602–3, and perhaps with iv. 84–5 and 91–2, three stanzas are linked. Where a stanza begins with 'And seyde', or some equivalent, without subject (e.g. i. 729; ii. 1156, 1359) there is room for doubt as to whether this forms part of the last sentence of the preceding stanza, or is simply an introductory phrase for the speech (Stanley, 'Stanza and Ictus', 124).

For the *Troye/joie* rhyme, see i. 2–4, 118–19, 608–9; ii. 139–40, 643–4, 748–9, 881–2; iii. 356–7, 790–1, 874–5, 1441–2, 1450–2, 1714–15; iv. 55–6, 90–1, 274–6, 335–6, 1306–7, 1441–2, 1630–1; v. 27–8, 118–19, 393–5, 426–7, 608–9, 615–16, 729–31, 779–81, 930–1, 1380–2, 1546–7.

For rhymes on *trouthe/routhe* and *trewe/rewe*, see i. 582–4, 769–70; ii. 349–50, 489–90, 664–5, 786–9, 1138–9, 1280–1, 1502–3, 1609–10; iii. 114–17, 120–2, 1511–2, 1770–1; iv. 1476–7, 1490–1, 1528–31, 1609–10, 1672–3; v. 706–7, 1000–1, 1070–1, 1098–9, 1364–5, 1385–6, 1586–7, 1686–7.

For rhymes on *serve/disserve*, see i. 818–19, 1058–60; iii. 174–5, 387–90, 440–1, 1290–3; v. 146–7, 972–3, 1387–9, 1721–2. For *serve/sterve*, see i. 15–17, 426–7; ii. 1150–2; iii. 153–4, 389–90, 713–14; iv. 279–80, 321–2, 447–8, 517–18; v. 174–5, 312–13.

For rhymes on *entente/mente*, see ii. 1219–21, 1560–1; iii. 125–6, 1185–8; iv. 172–3, 1416–18; v. 867–8, 1693–4.

For further discussion of rhyme patterns, see Stanley, 'Stanza and Ictus', and Stokes, 'Recurring Rhymes'.

Imitation and Allusion, c.1385–1700

> And red wherso thow be, or elles songe,
> That thow be understonde, God I biseche!   (v. 1797–8)

Future response to the matter and form of *Troilus* is anticipated in the conclusion of Chaucer's poem (v. 1786–98). Such response is then made part of the fiction in what was probably Chaucer's very next poem—the *Prologue* to the *Legend of Good Women*—where the poet-figure is rebuked for his translations by the God of Love, who mentions specifically the *Romaunt of the Rose* and then *Troilus*, the reference to the poem being revised between the two versions of the *Prologue*:

> 'And of Creseyde thou hast seyd as the lyste,
> That maketh men to wommen lasse triste,
> That ben as trewe as ever was any steel.'   (F 332–4)

> 'Hast thow nat mad in Englysh ek the bok
> How that Crisseyde Trylus forsok
> In shewynge how that wemen han don mis?'   (G 264–6)

Alcestis defends the poet on the grounds that he did not know what he was translating (G 344–5), and the poet eventually defends himself by pointing to how his poem may be interpreted:

> 'what so myn auctour mente,
> Algate, God wot, it was myn entente
> To forthere trouthe in love and it cheryce,
> And to be war fro falsnesse and fro vice
> By swich ensaumple; this was my menynge . . .'   (G 460–4)

Although we cannot know what response *Troilus* actually did receive in the courtly circles to which Chaucer was close, in his next poem Chaucer chooses in this way to fictionalize a poet-figure encountering reactions to *Troilus*, and attempting to respond to those responses.

In his manner of bidding farewell to his poem at the conclusion of *Troilus* Chaucer seems to know very well what he has done: despite all the conventional professions of modesty he evidently foresees that his poem will have a posterity, even as he prays that it be correctly understood. Chaucer foresaw the future quite correctly. It would be hard to overestimate the esteem and influence enjoyed by Chaucer's work in England in the first centuries after his death, and not least by *Troilus*, his greatest single accomplishment. Such esteem, however, is only occasionally expressed through imitation of any substantial part of the *Troilus* narrative by subsequent writers. Between the composition of *Troilus* and Dryden's play of *Troilus and Cressida* there is a rich and varied tradition of references to the story of Troilus and Criseyde, although it is not always easy to tell whether these constitute direct references to the text of Chaucer's poem about that story. Yet, as a model of poetic language and technique and as a treasury of poetic forms—lyric, complaint, testament, and epistle—*Troilus and Criseyde* continues to exert a discernible influence for several centuries after Chaucer's death.

Evidence for responses to *Troilus* by other writers during Chaucer's lifetime is confined to the work of John Gower and Thomas Usk, and perhaps the French poet Eustache Deschamps. In his *Confessio amantis* Gower includes reading and listening to 'Troilus' among a list of courtly pursuits, and otherwise refers to the story of Troilus and Criseyde a number of times in his poems, although it is difficult to discern in these references any particular influence from the poem co-dedicated to him by Chaucer at its close (see above, pp. 95–6). By contrast, Usk's prose allegory *The Testament of Love*—a dialogue between the prisoner and Love—reflects a minute familiarity with the texture of *Troilus*; he is evidently steeped in its diction and phrasing. The *Testament* is written out of the influence of both *Troilus* and Chaucer's *Boece*, and it is to *Troilus* that Usk has the figure of Love refer the lover for definitive resolution of the issue of free will:

> 'Myne owne trewe servaunt, the noble philosophical poete in Englissh, whiche evermore him besieth and travayleth right sore my name to encrese (wherfor al that willen me good owe to do him worship and reverence bothe; trewly, his better ne his pere in scole of my rules coude I never fynde)—he (quod she), in a tretis that he made of my servant Troilus, hath this mater touched, and at the ful this question assoyled. ... In the boke of Troilus, the answere to thy question mayst thou lerne.' (III. iv)

It is pre-eminently *Troilus*, among his works written by the 1380s, that determines such a characterization of Chaucer as love poet and 'philosophical poete'. This esteem for the 'philosophical' *Troilus* receives support from a professional philosopher towards the end of the fifteenth century, when John of Ireland concludes a scholastic discussion of divine prescience addressed to James IV of Scotland: 'And for þe instruccioune of þi hieness, I haue now tretit part of þis mater, for it pertenis mare to my crafft þan to Chauceire, þat has tretit þis mater in þe buk of Troylus, & richt excedandly for a temporale man and clerk nocht greit in theologie.'

For other treatments of the story of Troilus and Criseyde, see Benson, 'True Troilus'; Mieszkowski, 'Reputation'; Miskimin, *The Renaissance Chaucer*; and Rollins, 'The Troilus–Cressida Story'. For reference, see Brewer, *Critical Heritage*, and Spurgeon, *Five Hundred Years*.

Deschamps's *balade* honouring Chaucer—traditionally dated *c*.1385—contains the lines

qu'i as
Semé les fleurs et planté le rosier
Aux ignorans de la langue Pandras,
Grant translateur, noble Geoffrey Chaucier!

[you] who have sown the flowers and planted the rose-tree, a Pandarus for those who do not
know the [French] language, great translator, noble Geoffrey Chaucer! (Deschamps,
Œuvres, ed. le marquis de Saint-Hilaire and G. Raynaud, (Societé des Anciens Textes
Français; 11 vols.; Paris, 1878–1904), ii. 138–9); for a text and different translation, see
also Brewer, *Critical Heritage*, i. 39–42.
'La langue Pandras' has usually been taken to allude to Pandrasus in Wace's *Brut*, who
fought Brutus, the founder of Britain, and so to mean 'the French language'. But
Mieszkowski, '"Pandras"', argues for the translation above, so identifying the go-
between role of Pandarus with Chaucer's role as a cultural intermediary in his translating
from French to English in some of his work.
Thomas Usk, *The Testament of Love*, in W. W. Skeat (ed.), *Chaucerian and Other Pieces*
(Oxford, 1897), 123.
John of Ireland, *The Meroure of Wyssdome*, ed. C. Macpherson, (STS NS. 19; Edinburgh,
1926), 74. Cf. J. A. W. Bennett, 'Those Scotch Copies of Chaucer', *Review of English
Studies*, NS 32 (1981), 294–6.

The later Middle Ages saw a number of renderings into English of Guido's
Historia, some of which show Chaucer's influence. The author of *The Laud
Troy Book* (*c*.1400) apparently did not know *Troilus*; he certainly shows
little of Chaucer's sympathy for the lovers. After describing the sorrows of
Troilus and Criseyde, the author of the alliterative *Gest Hystoriale of the
Destruction of Troy* tells his reader 'Who-so wilnes to wit of þaire wo fir
[further] | Turne hym to Troilus, & talke þere ynoghe!' (8053–4). When he
describes Troilus gazing out from the walls of Troy ('Tristly may Troiell
tote [gaze] ouer the walle, | And loke vpon lenght, er his loue come!'
(8178–9)), he apparently recalls Book V of *Troilus*, while his sympathetic
attitude towards 'true' Troilus may also have been influenced by a reading
of *Troilus and Criseyde*.

The version of Guido in John Lydgate's *Troy Book* is written with self-
conscious reference to the existence of Chaucer's *Troilus*. Lydgate broadly
follows Guido's narrative of the story (which corresponds to Chaucer's
Books IV and V), but he interpolates in that narrative so as to refer to the
whole story in Chaucer's version. In Lydgate's list of 'portraits' the task of
describing Criseyde prompts him to an excursus apologizing for describing
what Chaucer has already described and praising Chaucer's achievements
for poetry in English (ii. 4698–9). The excursus (ii. 4677–735) is over twice
as long as the description of Criseyde, a typical Lydgatean 'de luxe' version
of Chaucer's portrait (ii. 4736–62), but no such apology for doing what
Chaucer has already done is prompted by the long description of Troilus
(ii. 4861–95), itself a tissue of phrases from Chaucer's poem. There is also a

fluent narrative outlining the sorrows of Troilus and then Criseyde at news of her exchange, their doleful meeting, and the near suicide of Troilus, a summary narrative corresponding to the most moving scenes of Chaucer's fourth book and selecting descriptive detail and phrases from *Troilus*. The lovers' sorrowful last night together prompts Lydgate to refer his reader gracefully to *Troilus* for a full account ('Sith my maister Chauncer her-a-forn | In þis mater so wel hath hym born' (iii. 4197–8)). Here Lydgate inserts a retrospective summary of how the love-affair started, which of course was absent from Guido: how Troilus falls in love in the temple and is helped by Pandarus until Fortune frowns upon him ('Lo! here þe fyn of false felicite, | Lo! here þe ende of worldly brotilnes' (iii. 4224–5)). It is recollection of *Troilus* which then prompts Lydgate's most fulsome praise of Chaucer's achievements (iii. 4234–63).

Lydgate's handling of Criseyde's change of heart represents a curious compromise between the account in the *Historia* and a reading of *Troilus*, which leaves his two authorities in unexplained and inconsistent juxtaposition. Having followed Diomede and Criseyde to the Greek camp, Lydgate refers his reader to *Troilus* (iii. 4432–4), but he then goes straight on to report Guido's denigratory and antifeminist account—quite contrary to Chaucer's—of how Criseyde abandoned Troilus the same evening that she left Troy ('But Guydo seith, longe or it was nyȝt, | How Cryseyde for-soke hir owne knyȝt' (iii. 4435–6)). For the remainder of the story—Diomede's courtship of Criseyde, her disloyalty to Troilus, Troilus' death at the hands of Achilles—Lydgate follows the narrative in Guido, although with various phrasal echoes of *Troilus*. Later English redactions of Guido—such as Caxton's *The Recuyell of the Historyes of Troy* (*c.*1474) or Thomas Heywood's *Troia Britanica* (1609), a reworking of Caxton—show no substantial influence from Chaucer's *Troilus*.

The *Troy Book* contains Lydgate's most sustained treatment of the Troilus and Criseyde story, but his influential and voluminous writings contain various other references to Troilus and Criseyde. His allusions to the hero lie near the beginning of a long tradition of Troilus as the stereotype of the faithful lover, while some of his references to Criseyde are startlingly dismissive by comparison with the account of her in the *Troy Book*. Such differing assessment of Criseyde within the work of one writer anticipates the range of emphases with which her character will be presented and interpreted by subsequent writers.

The Laud Troy Book, ed. J. Ernst Wülfing (EETS os 121, 122; London, 1902–3). See especially, 13553–64, including the comment on Briseida and Diomede: 'For to him sche ȝaff al hir talent, | For he hadde mechel on hir y-spent' (13561–2). See also R. M. Lumiansky, 'The Story of Troilus and Briseida in the *Laud Troy Book*', *Modern Language Quarterly*, 18 (1957), 238–46.

The 'Gest Hystoriale' of the Destruction of Troy, eds. G. A. Panton and D. Donaldson (EETS os 39, 56; 1869, 1874). See McKay Sundwall, 'The *Destruction of Troy*, Chaucer's *Troilus and Criseyde*, and Lydgate's *Troy Book*', *Review of English Studies*, NS 26 (1975), 313–17.

John Lydgate, *Troy Book*, ed. H. Bergen (EETS ES 97, 103, 106, 126; London, 1906–20). For the possible influence of *Troilus* on another English version of Guido, see C. David Benson, 'Chaucer's Influence on the Prose "Sege of Troy"', *Notes and Queries*, 216 (1971), 127–30.

Caxton's *Recuyell*, which is a translation of Raoul Lefevre's compilation, *Recueil des Hystoires Troyennes*, refers its reader to Chaucer for the 'storye hooll' of Troilus: 'who that lyste to here of alle theyr love, late hym rede the booke of Troyllus that Chawcer made; wherin he shall fynde the storye hooll, whiche were to longe to wryte here', *The Recuyell of the Historyes of Troye*, ed. H. Oskar Sommer (2 vols.; London, 1894), 604. Heywood similarly notes 'The passages of love betwixt Troylus and Cressida, the reverent poet Chaucer hath sufficiently discourst, to whom I wholly refer you.'

In the background of the traditional character of Troilus are his qualities as a knight, which Lydgate acknowledges in Chaucerian idiom in his *Fall of Princes* ('in knyhthod so manli eek was founde, | That he was named Ector the secounde' (i. 5935–6)). The brevity of Lydgate's other allusions suggests how readily and absolutely the figure of Troilus is identified with that of a loyal lover, and other fifteenth-century writers similarly allude to the figure of Troilus in ways which seem influenced by a reading of Chaucer. 'The Lover's Mass' acknowledges 'The grete trouthe of Troylus | perseuerant to hys lyves ende'. A poem of Charles d'Orléans, his Rondeau CCII, assumes that sympathetic reading about Troilus may be part of courtly life, and the Troilus figure resembles Chaucer's:

> Lire vous voy fais merencolieux
> De Troïlus, plains de compassion;
> D'Amour martir fu en sa nascion:
> Laissez l'en paix, il n'en est plus de tieux!

I see you—full of compassion—reading of the suffering of Troilus. He was a martyr of love in his land. Leave him in peace. There are no more like him!

The speaker of the lyric 'To his Mistress, Flower of Womanhood' can cast himself as a new Troilus in the constancy of his love:

> Go, litill bill, with all humblis
> vnto my lady, of womanhede þe floure,
> And saie hire howe newe troiles lithe in distreȝ,
> All-onely for hire sake and in mortall langoure;
> And if sche wot nat whoo it is, bute stonde in erore,
> Say it is hire olde louer þat loueth hire so fre, trewe,
> hir louynge a-lone—not schanginge for no newe.

If Chaucer's sympathetic account of Troilus helped to fix him as the figure of 'true Troilus', that stereotyping in turn has the effect of confirming some of the stereotyping of Criseyde, and in ways which often ignore Chaucer's sympathetic account of her. When Gavin Douglas at the end of the fifteenth century refers in his *Palice of Honour* to 'Trew Troilus, vnfaithfull Cressida' (565), he is echoing a commonplace already intrinsic to Chaucer's sources and which *Troilus* has not succeeded in completely replacing. Yet the tradition of the Criseyde character is by no means monolithic: in addition to hostile allusions to her promiscuity and unfaithfulness, there are apparently neutral references to her as one of the great beauties of antiquity, and even some evidently approving references to her as a lover. A partial Middle English translation of Boccaccio's *De claris mulieribus* includes Criseyde in a list of beautiful women, with the moral *memento mori* ('Evys bewte, wytt and womanhede | Excedyd Dido, Cryseyde and Heleyne' (359–60)); Charles d'Orléans lists Criseyde in his Ballade LX with the refrain 'Ce monde n'est que chose vaine'; and in his poem 'That Now Is Hay Some-Tyme Was Grase', Lydgate also includes Criseyde in a list of beauties.

Both the 'English Chaucerians' Lydgate and Stephen Hawes illustrate the way Criseyde can be referred to by the same writer in different contexts with differing degrees of censure. In his *Troy Book* Lydgate comes closest to a Chaucerian understanding of her difficulties, but in other poems he shows no such recollection of Chaucer's Criseyde. In 'Amor Vincit Omnia Mentiris Quod Pecunia' he presents Criseyde as an example of women who sell themselves:

> Remembre Troye, of Troylus and Cres[e]ide,
>   Eche in theyr tyme furtherd to plesaunce;
> But what fille after longe or Troylus deyde?
>   A false serpent of chaunge and variaunce
>   Withouten any lengger attendaunce
> Put out Troylus, and set in Dyomede.
>   What shal I say or conclude in substaunce?
> Love was set bakke, gold went afore, and mede.

In his *Fall of Princes* Lydgate refers readers to his *Troy Book* for the story of Criseyde, but suggests that she would have left both Diomede and Troilus for some third lover:

> Off hir nature a quarel thus she took,
> Tassaie bothe, yiff neede eek wer, to feyne
> To take the thridde, & leue hem bothe tweyne.   (i. 6018–20)

Such a view of Criseyde figures in an earlier fifteenth-century text, the anonymous 'Chance of the Dice' (1440), which is the rhymed text for a fortune-telling game. The stanzas are numbered, and the game is to throw dice and tell fortunes by matching the throws with the stanza numbers. A promiscuous fortune is represented by Criseyde (379–85), yet even a poem which presents such a stereotype can elsewhere use the love of Troilus and Criseyde as a type instance of joy (139–40).

There is a comparable range in the allusions to Criseyde's story by Hawes and by Skelton. In a passage in *The Pastime of Pleasure* (1509) Hawes tells how in *Troilus* Chaucer laments the hero's sorrows and the doubleness of Criseyde (1331–4); she is cited elsewhere as an instance of women's changeability ('Yf by fortune there come an other newe | The fyrst shall be clene out of her fauoure | Recorde of Creseyd ...' (3565–7)); but Hawes also cites the story as an example of happiness in love:

> Remembre ye / that in olde antyquyte
> How worthy Troylus / the myghty champyon
> What payne he suffred / by grete extremyte
> Of feruent loue / by a grete longe season
> For his lady Cresyde / by grete trybulacyon
> After his sorowe / had not he grete Ioye
> Of his lady / the fayrest of all Troye.   (1807–13)

In his poem 'Phyllyp Sparowe' (?1505) John Skelton refers to Criseyde in terms of her transgression ('For she dyd but fayne; | The story telleth playne' (695–6)), and her deserved disgrace ('She was moch to blame; | Disparaged is her fame | And blemysshed is her name, | In maner half with shame' (710–13)). But just as Hawes includes Criseyde in a list of beauties (*Pastime*, 1758–9), so Skelton in his *Garlande of Laurell* (1523) can identify 'my Lady Elisabeth Howarde' with:

> Goodly Creisseid, fayrer than Polexene,
> For to envyve [enliven] Pandarus appetite;
> Troilus, I trowe, if that he had you sene,
> In you he wolde have set his hole delight.
> Of all your bewte I suffyce not to wryght:
> But, as I sayd, your florisshinge tender age
> Is lusty to loke on, plesaunt, demure, and sage.   (871–7)

Lydgate's Fall of Princes, ed. H. Bergen (EETS ES 121–4; London, 1924–7). Cf. the portrait of Troilus in *Troy Book*, ii. 4879–83, and see also the references to Troilus in 'On Gloucester's Approaching Marriage' and 'A Wicked Tunge Wille Sey Amys', and to Criseyde in 'That Now Is Hay Some-Tyme Was Grase' and 'Amor Vincit Omnia Mentiris Quod Pecunia', in *John Lydgate, The Minor Poems*, ed. H. N. MacCracken (EETS OS 192; London, 1934), 601–8, 842, 810, 745.
'The Lover's Mass': in *English Verse between Chaucer and Surrey*, ed. E. P. Hammond (Durham, NC, 1927), 213.
Charles d'Orléans: Poésies, ed. P. Champion (CFMA; Paris, 1923), i. 85 (Ballade LX), ii. 406 (Rondeau CCII).

'To his Mistress, Flower of Womanhood': in *Secular Lyrics of the XIVth and XVth Centuries*, ed. R. H. Robbins (Oxford, 1952), 190–1.

The Palice of Honour: in *The Shorter Poems of Gavin Douglas*, ed. P. J. Bawcutt (STS 4th ser., 3; Edinburgh, 1967).

For the English version of *De claris mulieribus*, see *Die mittelenglische Umdichtung von Boccaccios De claris mulieribus*, ed. G. Schleich (Leipzig, 1924), especially lines 358–62.

'The Chance of the Dice', ed. E. P. Hammond, *Englische Studien*, 59 (1926), 1–16.

Stephen Hawes, The Pastime of Pleasure, ed. W. E. Mead (EETS os 173; London, 1928).

John Skelton: The Complete English Poems, ed. J. Scattergood (Harmondsworth, 1983).

A variety of evidence—in such diverse forms as inventories, excerpts, quotings, or echoes of the poem—points to the reading of *Troilus* during the fifteenth century. One copy of *Troilus*, now lost, is listed in the inventory of a library in Burgundy; another is included in an inventory (*c.* 1475–9) of the books of Sir John Paston II, although it was lent to a neighbour ten years earlier, who in turn lent it to someone else. *Troilus and Criseyde* is for many fifteenth-century writers a source book for diction and versification, and echoes are everywhere. The moments of formal self-consciousness in Chaucer's poem become formulas employed recurrently by subsequent poets. The English version of the romance of *Partonope of Blois*, for instance, reflects the influence of Chaucer's proems and has learned a lesson from the example of a narratorial persona in *Troilus*. The 'Go, litel bok' formula is taken over by Lydgate, Hoccleve, James I, Sir Richard Roos, Hawes, Skelton, and various anonymous writers. *Troilus* becomes a great formulary and lexicon, and, in the English poems of Charles d'Orléans Chaucer's poem can be a model of both sentiment and expression, a role it also fulfils for much less gifted poets, such as the author of 'The Lufaris Complaynt'. In his *Amoryus and Cleopes* (1449) John Metham tries (unsuccessfully) to emulate *Troilus* in form as a stanzaic romance and dimly echoes something of its narrative. Other 'Chaucerian' writers—such as the translator of *Partonope* or the poet of *The Isle of Ladies*—apparently have in mind as a model the handling of love-scenes in *Troilus*, although their own love-scenes end up removed from Chaucer in texture and tone.

For the Burgundian allusion, see E. P. Hammond, 'A Burgundian Copy of Chaucer's *Troilus*,' *Modern Language Notes*, 26 (1911), 32, and for the Paston inventory, see *The Paston Letters and Papers of the Fifteenth Century*, ed. N. Davis (Oxford, 1971), i. 516–18, and pl. VII.

The Middle English Versions of Partonope of Blois, ed. A. T. Bodtker (EETS es 109; London, 1912). Cf. Barry Windeatt, 'Chaucer and Fifteenth-Century Romance: *Partonope of Blois*', in Ruth Morse and Barry Windeatt (eds.), *Chaucer Traditions: Studies in Honour of Derek Brewer* (Cambridge, 1990), 62–80.

'The Lufaris Complaynt' is a poem surviving in a late-fifteenth-century manuscript which contains, among other things, a copy of *Troilus* and the unique text of *The Kingis Quair* (Bodleian Library, MS Arch. Selden B.24). The 'Complaynt' draws freely on the idiom of love's pain in *Troilus*, which is also its model for references to Fortune and to the classical deities. For a text, see K. G. Wilson, '*The Lay of Sorrow* and *The Lufaris Complaynt*: An Edition', *Speculum*, 29 (1954), 708–26.

The Works of John Metham, ed. H. Craig (EETS os 132; London, 1916).

The Isle of Ladies, or The Ile of Pleasaunce, ed. A. Jenkins (New York, 1980).

In the excerpting and quoting of *Troilus* in manuscripts of other works there also emerge various signs of response to Chaucer's poem, reflecting especially appreciation of the poem's sententious passages and love lyrics. Pandarus' stanza beginning 'A wheston is no kervyng instrument . . .' and concluding 'By his contrarie is every thyng declared' (i. 631–7) is copied separately in two manuscripts, as are his three stanzas of advice to Troilus against boasting (iii. 302–22), in one case incorporated into a larger composite poem called 'The Tongue'. The first stanza of the first Canticus Troili (i. 400–6) is excerpted three times to serve as a free-standing poem, while the whole song and the two following stanzas (i. 400–34) appear in the Bannatyne Manuscript (1568). The first stanza of the song is also quoted—by way of illustrating the 'furst token of carnal love'—in a discussion of the features of earthly love in a treatise for nuns entitled *Disce morus*. Extracts in several other manuscripts may well reflect another way Chaucer's text was put to use. In one manuscript (Rawlinson MS. C.813, fo. 48ᵛ) nine separate stanzas from various books of *Troilus* have been selected and rearranged into a 'poem' in the form of a lover's epistle. Such an instance of Chaucer's text cannibalized into a 'new' poem may record how his *Troilus* was drawn upon more widely as the source and model for love-letters, which could be in this fashion 'written out of Chaucer'. In the Devonshire Manuscript (*c*.1532–41) one poem is made up entirely of excerpts from Book IV of *Troilus* rearranged as a continuous new 'poem': three stanzas of Troilus' lament at Criseyde's departure and his subsequent address to lovers are introduced by two lines from the fourth proem (so that the 'poem' comprises iv. 13–14, 288–308, 323–9). The same Devonshire Manuscript contains six other free-standing excerpts from *Troilus* along with poems cannibalized from Chaucer's *Anelida*, from Hoccleve's *Letter of Cupid*, and the *Belle dame sans merci* of Sir Richard Roos.

Troilus, i. 631–7: in Trinity College, Cambridge, MS R.3.20, fo. 361ʳ, printed in F. J. Furnivall (ed.), *Odd Texts of Chaucer's Minor Poems* (Chaucer Society; London, 1868); and in Huntington Library, EL 26.A.13, cf. H. N. MacCracken, *Modern Language Notes*, 25 (1910), 126–7.

Troilus iii. 302–22: in Trinity College, Cambridge, MS R.4.20, fo. 171ᵛ, and in 'The Tongue', in Cambridge University Library, MS Ff.1.6, fo. 150ʳ (see Furnivall (ed.), *Odd Texts*).

Troilus, i. 400–6: in Cambridge University Library, MS Gg.4.12, fo. 105ᵛ; Huntington Library, EL 26.A.13; and British Library, MS Cotton Otho A XVIII, on which see G. B. Pace, *Speculum*, 26 (1951), 307–8. *Troilus*, i. 400–34: in Advocates Library, Edinburgh,

MS 1.1.6, fo. 230^r; see *The Bannatyne Manuscript*, ed. W. T. Ritchie, iii (STS 23; Edinburgh, 1928), 304–5.

Disce morus: in Bodleian Library, MS Laud Misc. 99, fo. 252^r, and Jesus College, Oxford, MS 39, fo. 311^r. See Lee Patterson, 'Ambiguity and Interpretation: A Fifteenth-Century Reading of *Troilus and Criseyde*', *Speculum*, 54 (1979), 297–330, repr. in his *Negotiating the Past* (Madison, Wis., 1987).

On MS Rawlinson C.813, see F. M. Padelford, *Anglia*, 31 (1908), 362–3, and *Anglia*, 34 (1911), 284–7.

R. Southall, 'The Devonshire Manuscript Collection of Early Tudor Poetry, 1532–41', *Review of English Studies*, NS 15 (1964), 142–50.

As a courtly manuscript miscellany, containing poems by Sir Thomas Wyatt and others along with its excerpts from Chaucer, the Devonshire Manuscript is an important token of early Tudor taste and of the place of Chaucer's *Troilus* within that taste. It postdates and reflects the influence of the 1532 edition of Chaucer's works by Thynne, which was to have a signal influence on responses to *Troilus*. The *Troilus* had already been printed on its own by Caxton (*c*.1483) and by Wynkyn de Worde (1517), and had been included with the *Canterbury Tales* and other Chaucer poems in Pynson's collection (1526). It is in Thynne's text that Chaucer's poem was first made widely available in a conscientiously edited form where some care was taken to present a metrical text. Here, too, is the earliest extant text of Robert Henryson's *Testament of Cresseid* (written *c*.1475), printed without attribution so as to follow immediately after the conclusion of Chaucer's *Troilus*. At the opening of the *Testament* the narrator refers to reading a book about Troilus and Criseyde 'Writtin be worthie Chaucer glorious', so that it would always have been clear to any careful reader that the *Testament* was written by someone other than Chaucer. Yet the unattributed inclusion of the *Testament* as a pendant or sequel to *Troilus* in an influential edition of Chaucer's works—and that memorable ending to Criseyde's life which Chaucer's narrative had not included—meant that the *Testament* came to be regarded as the sixth book of Chaucer's poem. It was reprinted in all editions of Chaucer down to Urry's (1721), and could still be included as Chaucer's in Chalmers' *Works of the English Poets* (1810). In his *Animadversions* on Speght's 1598 edition of Chaucer Francis Thynne had commented: 'yt wolde be good that Chaucer's proper woorkes were distinguyshed from the adulterat, and suche as were not his, as the Testamente of Cressyde'. But in his 1602 edition Speght inserted a new introductory heading to the text of *Troilus* which included the narrative of Cresseid's end in the *Testament* as part of Chaucer's poem: 'In this excellent Booke is shewed the feruent loue of Troylus to Creiseid, whome hee enioyed for a time: and her great vntruth to him againe in giuing her selfe to Diomedes, who in the end did so cast her off, that she came to great miserie. In which discourse Chaucer liberally treateth of the diuine purueiaunce.' The first known acknowledgement of Henryson's authorship of the *Testament* is found in the comments on *Troilus* by Sir Francis Kynaston.

The *Testament of Cresseid* is a strikingly beautiful and original poem which can stand entirely free of *Troilus and Criseyde*, although it gains in poignancy and resonance when read with Chaucer's poem in mind. In presenting the fate of Criseyde after she has betrayed Troilus and been eventually abandoned by Diomede, the *Testament* is not exactly a sequel to *Troilus*, since in Henryson's version of events Criseyde dies before Troilus. Rather, the *Testament* responds to that narrative exclusion in *Troilus* by which Chaucer's poem passes in silence over what becomes of Criseyde. In outline, the poem tells how Cresseid—by now reduced to prostitution—reproaches the gods for her abandoned fate. For her blasphemy she is divinely punished with leprosy, in a marvellous set-piece of word-painting in which the gods appear while she lies in a swoon. The poem then tells of her retirement to a leper house and, after her doleful lament, finds its climax in the graphic moment when Troilus rides past the lepers begging for alms and—reminded of his love by something in Cresseid's now unrecognizably disfigured face—tosses a purse of gold and jewels into her lap before riding on. When Cresseid discovers the identity of the knight she laments ('O fals Cresseid, and trew knicht Troilus!'), and before dying makes her 'testament', in which she sends news of her end to Troilus, who arranges for her tomb and epitaph.

That the *Testament* presents events outside the narrative of *Troilus* does not prevent its being a form of commentary on Chaucer's poem, especially its fifth book. Its opening, with the narrator telling of reading Chaucer's poem and then taking up another unspecified book to read of Criseyde's end, prepares for this in a way that develops the lessons of Chaucer's narratorial technique for its own ends. There is also some debt to the language of emotion in *Troilus*, and to its sensibility of distress and of swoons. But most importantly, Henryson has confronted and transcended any sense that Chaucer's account is too charitable in extenuating wrongdoing and omitting any punishment for Criseyde. In the gruesomeness of leprosy—usually diagnosed in the Middle Ages as a venereal disease—Henryson can show Cresseid punished for her sexual infidelity terribly and specifically, so as to release a sympathy that is in turn enhanced by the dignity with which Cresseid accepts guilt and affliction ('Nane but myself as now I will accuse' (574)). Her final laments, and her pathetic regret that she cannot return to Troilus the gifts of his that now are Diomede's, recall the Criseyde of Chaucer's Book V, and in their consistency with Chaucer's heroine they achieve a special poignancy in showing how Criseyde's sorrows in *Troilus* were only the beginning of her tragedy.

The Poems of Robert Henryson, ed. Denton Fox (Oxford, 1981), pp. xciv ff. On aspects of
Troilus and the *Testament*, see Douglas Gray, *Robert Henryson* (Leiden, 1979); M. P.
McDiarmid, *Robert Henryson* (Edinburgh, 1981); A. C. Spearing, *Medieval to Renaissance in English Poetry* (Cambridge, 1985), ch. 5.
For Thynne's *Animadversions*, see Spurgeon, *Five Hundred Years*, i. 154–5.

The association of Henryson's *Testament* with *Troilus* has a profound
influence on the subsequent perception of Criseyde's character. In addition
to the interpretation of her as a type of the faithless woman, familiar since
Benoît and Guido, there now develops a tradition in which she is typecast
as a prostitute and a leper. In *Henry V* Pistol calls Doll Tearsheet a 'lazar
kit of Cressid's kind' (II, i, 76), and in many a moralizing context the
memory of Cressid's end is invoked. In Thomas Howell's 'The Britlenesse
of Thinges Mortall' (*c.* 1570), revised in 1581 to 'Ruine the rewarde of
Vice', the leprous punishment of Cressid is gruesomely lingered over as an
awful warning, and the moral is then pressed home in an echo of *Troilus*
('Lo here the ende of wanton wicked life . . . | Lo here of vice the right
rewarde and knife . . .'). In his *Rocke of Regard* (1576) George Whetstone
includes a poem 'Cressids Complaint', in which Cressid admits she was a
prostitute even while involved with Troilus, and the complaint ends 'Loe!
here the fruits of lust and lawlesse love'. The prose 'Argument' to this
poem shows that Whetstone sees the story through Henryson's version
('her lothsome leprosie after lively beautie . . .') and uses Cressid's example
in preaching to his own times ('And for as much as Cressids heires in every
corner live, yea, more cunning then Cressid her selfe in wanton exercises,
toyes and inticements, to forewarne all men of such filthes . . .'). Cressid's
end as a leper is also a frequent motif in Elizabethan verse. Turberville,
who is fond of characterizing his mistress as Criseyde, himself as true
Troilus, concedes of his lady: 'I pray she may have better hap | Than beg
her bread with dish and clap | As she, the sielie miser, did, | When Troilus
by the spittle rid.'

Yet the example provided by the story as a whole was too various to be
wholly determined by the non-Chaucerian ending which Chaucer's poem
had now acquired, and many sixteenth-century allusions to both the lovers
and their story do not acknowledge Henryson's conclusion. In courtly
lyrics and in ballads there are references to the story as one of happily
fulfilled love. The poem in *Tottel's Miscellany* (1557) entitled 'A comparison of his loue wyth the faithfull and painful loue of Troylus to Creside'
begins by referring to reading the story, urges the poet's lady to see the
parallels between his state and that of Troilus, and begs the lady to grant
her grace 'As Creside then did Troylus to'. In his *Epitaphes* (1567)
Turberville—whose poems are full of allusions to *Troilus*—also has a poem
entitled 'The Lover in utter dispaire of his Ladies returne, in eche respect

compares his estate with Troylus'. In William Fulwood's *The Enimie of Idlenesse* (1568) there is a model verse-letter in which 'A constant Lover doth expresse His gripyng griefes' and which concludes 'Therfore graunt grace, as Cressida, did unto Troylus true: | For as he had hir love by right, so thine to me is due.' As late as 1612 Richard Johnson compares himself to Troilus and mentions Criseyde favourably in a 'Lovers Song in Praise of His Mistress' in *A Crowne-Garland of Goulden Roses*.

Thomas Howell, 'The Britlenesse of Thinges Mortall', in *New Sonets, and Pretie Pamphlets*, in A. B. Grosart (ed.), *Occasional Issues of Unique or Very Rare Books* (privately printed, 1879), 121–2; 'Ruine the Rewarde of Vice', in *Howell's Devises*, ed. W. Raleigh (Oxford, 1906).

Cressid is mentioned in such works and collections as: *A Handful of Pleasant Delights* (1584), *Paradyse of daynty deuises* (1576), *Gorgious Gallery of gallant Inuentions* (1578), *A poore Knight his Pallace of priuate pleasures* (1579), W. A.'s *Speciall Remedie against the furious force of lawlesse Loue* (1579), and *Willobie His Avisa* (1594).

George Turberville, *Epitaphes, Epigrams, Songs and Sonets* (1567), fos. 139ʳ–140ᵛ; *Tragical Tales* (1587), fo. 164ᵛ. On the brother, nephews, and cousins of Turberville named Troilus, Rollins, 'Troilus–Cressida Story', 403 n., cites Hutchins's *History and Antiquities of Dorset* (3rd edn.), i. 139, 201. On the name Troilus, see Spurgeon, *Five Hundred Years*, iv. 53.

'A comparison of his loue . . .' in *Tottel's Miscellany*, ed. Hyder E. Rollins (rev. edn.; Cambridge, Mass., 1966), i. 183–5.

A number of sixteenth-century ballads take the love-affair of Troilus and Criseyde as their theme. A ballad by William Elderton, 'The panges of Loue and louers fittes' (1559–60), has the lines:

> Knowe ye not, how Troylus
> Languished and lost his joye,
> With fittes and fevers mervailous
> For Cresseda that dwelt in Troye;
> Tyll pytie planted in her brest,
> Ladie! ladie!
> To slepe with him, and graunt him rest,
> My dear ladie.

A ballad in the Percy Folio begins:

> Cressus: was the ffairest of Troye,
> whom Troylus did loue!
> the Knight was kind, & shee was coy,
> no words nor worthes cold moue,
> till Pindaurus soe play his part
> that the Knight obtained her hart
> the Ladyes rose destroyes:
> [They] held a sweet warr a winters night
> till the enuyous day gaue light;
> which darkness louers ioyes.

A jovial ballad of *c*.1565 ('When Troylus dwelt in Troy towne') follows more of the narrative line of Chaucer's first three books, including the Book I temple scene:

> Tyll at the last he cam to churche
> Where Cressyd sat and prayed a,
> Whose lookes gave Troylus such a lurche,
> Hys hart was all dysmayde a . . .

and the exchange between Pandarus and Criseyde on the morning after the lovers' union:

> 'In faythe, old unkell,' then quoth she,
> 'Yow are a frend to trust a,'
> Then Troylus lawghed, and wat you why?
> For he had what he lust a . . .

In a poem in the Bannatyne Manuscript the speaker on his morning stroll meets someone who turns out to be Pandarus ('I sperit his name and he said panderus | That sumtyme servit the gud knycht troyelus'). The speaker wishes to discuss 'questionis of luve' and asks Pandarus 'Quhen ladeis to thair luvaris salbe leill . . .' The whole tone of the piece, and its oblique encounter with the theme of constancy in Chaucer's poem, is sprightly, wry, but uncensorious.

'The panges of Loue . . .' is reprinted in H. L. Collman (ed.), *Ballads and Broadsides* (Roxburghe Club; London, 1912), 111.
'Cressus was the fairest of Troye' is reprinted in *Bishop Percy's Folio Manuscript*, eds. J. W. Hales and F. J. Furnivall (London, 1868), iii. 301–2.
'When Troylus dwelt in Troy towne' is reprinted in T. Wright (ed.), *Songs and Ballads, with Other Short Poems; Chiefly of the Reign of Philip and Mary* (Roxburghe Club; London, 1860), 195–7.
'Furth ouer the mold at morow as I ment', is in *The Bannatyne Manuscript*, ed. W. T. Ritchie (STS NS 26; Edinburgh, 1930), iv. 40–2.

In his pains and sorrows as a lover Troilus continues to be the stock figure of a suffering lover, as in the *Handful of Pleasant Delights* ('And *Troylus* eke . . . Constrained by loue did neuer mone | As I my deer for thee haue done' (1620–2)). In a ballad beginning 'Whan Cressyde came from Troye' the speaker concludes with the promise that he will be 'Trew Troylus to my grave'. That Troilus is 'true' is now proverbial, as in the poem beginning 'In winters just returne' by Henry Howard, Earl of Surrey, whose poems also reflect a general familiarity with *Troilus*. In the background of true Troilus there is the heroic figure of the Trojan War, a figure less indebted to Chaucer than to works such as Caxton's *Recuyell* ('Ther was not in alle the Royaume a more strong ne more hardy young man' (ii. 543)). This is also the Troilus Thomas Heywood summons up in his *Apologie for Actors* (1612) when describing a dramatic re-enactment of the Trojan War. Even references which have the Troilus of Chaucer's poem in mind can sometimes be more light-hearted. In his *Tale of Troy* Peele briefly refers to the love-affair ('So hardie was the true knight Troylus, | And all for love of the unconstant Cresid') but declares he will say no more about Troilus

> Whose passions for the raunging Cressida,
> Read as fair England's Chaucer doth unfold,
> Would tears exhale from eyes of iron mould.   (283–5)

In 'Dan Bartholomew his first Triumphe' in his *Posies* (1575), Gascoigne paints a fond but gently ridiculous picture of Troilus as the wronged and unsuccessful lover, and his reference to Lollius shows Gascoigne had been reading Chaucer's *Troilus*:

> Thy brother Troylus eke, that gemme of gentle deedes,
> To thinke how he abused was, alas, my heart it bleedes!
> He bet about the bushe, whiles other caught the birds,
> Whome crafty Gresside mockt to muche, yet fede him still with wordes;
> And God he knoweth, not I, who pluckt hir first sprong rose,
> Since Lollius and Chaucer both make doubt upon the glose.   (i. 101)

True Troilus had become too true, as a passage in Lyly's *Euphues* (1578) makes plain ('Though Aeneas were to fickle to Dido, yet Troylus was to faithfull to Craessida'). Indeed, some of Shakespeare's brief allusions scattered through the plays suggest the variousness of associations to the Troilus story and to Chaucer's poem. In *The Merchant of Venice* the poetry of longing in Chaucer's romance is summoned up by a reference to Book V of *Troilus* as Lorenzo says to Jessica:

> 'in such a night
> Troilus methinks mounted the Trojan walls,
> And sigh'd his soul toward the Grecian tents
> Where Cressid lay that night.'   (V. i. 3–6)

By contrast, when Rosalind in *As You Like It* ridicules the notion that men have ever actually died for love, she archly cites Troilus as her first example: 'Troilus had his brains dashed out with a Grecian club, yet he did what he could to die before, and he is one of the patterns of love' (IV, i, 99–100). And in *The Taming of the Shrew* Petruchio has named his spaniel Troilus (IV, i. 150).

   If Troilus is 'O parfyte Troylus ... The moste treuest louer / that euer lady hadde' (as he is described in some stanzas of commentary appended after Book V in Wynkyn de Worde's 1517 edition), then the reputation of the Criseyde who betrays him is depressed in proportion. The Wynkyn edition commentary—which goes on to mention women's undoing of Aristotle, Virgil, and Samson—begins by stressing the wrong done to Troilus by Criseyde:

> And here an ende / of Troylus heuynesse
> As touchy*n*ge Cresyde / to hy*m* ryght vnky*n*de
> Falsly forsworn*n* / defloury*n*g his worthynes
> For his treue loue / she hath hy*m* made bly*n*de
> Of feminine gendre / ye woma*n* most vnky*n*de
> Dyomede on her whele / she hathe set on hye
> The faythe of a woman / by her now maye you se.

Such references usually reveal little if any individual response to Chaucer's text. Yet it is possible to find allusions like the summary of *Troilus* by 'ornate Chaucer' in *La conusaunce damours* (?1525), where the story and both the lovers are presented without censure. In a poem like 'Ane dissuasion from vaine lust' in the *Gude and Godlie Ballatis* (1567) the story of Troilus' love can be used as a warning ('Sic plesoure bringis miserie') without explicit criticism of Criseyde. In *Common Conditions* (before 1576) there is sympathy for Criseyde, caught between her two lovers, and even reproof for Troilus ('Though *Troilus* therin was iust yet was hee found vntrewe'). And a ballad in the *Paradise of dayntie deuises* (1580), (K3ᵛ–K4ᵛ), beginning 'If *Cressed* in her gadding moode | Had not gone to the greekish hoste', also has a spirited 'Replye' spoken by Criseyde, which blames Troilus for failing to keep his lady with him in Troy by marrying her ('The seede of shame had not bine sowne | If knightly prowes his minde had lead | By rightfull force to keepe his owne'). In a Scottish poem probably by William Fowler, 'The Laste Epistle of Creseyd to Troyalus' (*c*.1604)—drawn from a reading of both *Troilus* and the *Testament*—the dying Criseyde sends a last letter to Troilus, in which she tries to excuse her unfaithfulness, but acknowledges her guilt and asks to be forgiven.

A Handful of Pleasant Delights (1584), ed. Hyder E. Rollins (Cambridge, Mass., 1924), 56.

The 'Trew Troylus to my grave' ballad is in British Library, Additional MS 28635. For a text, see *Tottel's Miscellany*, ed. Hyder E. Rollins (rev. edn.; Cambridge, Mass., 1966), ii. 294–5. Cf. the ballad entitled 'A new Dialogue betweene Troylus and Cressida', included in Thomas Deloney's *Strange Histories* (1612), K2ᵛ–L1ᵛ.

For 'In winters just returne', see *Henry Howard, Earl of Surrey: Poems*, ed. Emrys Jones (Oxford, 1964), 12–14, and 117 ('Surrey was clearly well acquainted with Chaucer's poem'). The poem is printed in *Tottel's Miscellany* as the 'Complaint of a diyng louer'. Poem 15 in Jones's edition of Surrey and poem 11 ('The sonne hath twyse brought forthe the tender grene') both reflect Surrey's knowledge of *Troilus and Criseyde*.

An Apology for Actors (Shakespeare Society Reprint; London, 1841), 20.

The Tale of Troy (lines 279–85), in *Life and Minor Works of George Peele*, ed. D. H. Horne (New Haven, Conn., 1952).

The Posies: The Complete Works of George Gascoigne, ed. J. W. Cunliffe (Cambridge, 1907).

John Lyly, *Euphues, The Descent of Euphues*, ed. J. Winny (Cambridge, 1957), 35.

For a transcription of the Wynkyn stanzas, see C. David Benson and David Rollman, 'Wynkyn de Worde and the Ending of Chaucer's *Troilus and Criseyde*', *Modern Philology*, 78 (1981), 275–9.

'Ane dissuasion from vaine lust', in A. F. Mitchell (ed.), *A Compendious Book of Godly and Spiritual Songs* (STS 1st ser., 39; Edinburgh, 1897), 213.

Common Conditions, ed. Tucker Brooke (Elizabethan Club Reprints, 1; New Haven, Conn., 1915).

The Works of William Fowler, ed. H. W. Meikle (STS NS 6; Edinburgh, 1914), 379–87.

It is against these traditions of interpretation that Shakespeare writes his play *Troilus and Cressida* (1602). There had already been a number of plays on the subject: on Twelfth Night 1515/16 at Eltham, Cornish and the Children of the Chapel Royal acted the *Story of Troylous and Pandor*, with 'Kryssyd imparylled lyke a wedow of onour, in blake sarsenet and other abelements for seche mater'; Bale records that Nicholas Grimald wrote a Latin comedy *Troilus ex Chaucero*, which does not survive. In *The Rare Triumphes of Love and Fortune* (1582) there is a 'show' of Troilus and Cressida as an exemplum of Love's power, but no dialogue or description exists except the comment of Mercury ('Beholde, how Troylus and Cresseda | Cryes out on Love, that framed their decay'), followed by that of Vulcan ('That was like the old wife when her ale would not come | Thrust a fire brand in the groute, and scratcht her bum'). In Thomas Heywood's *The Iron Age* (1596) the story of Troilus and Criseyde is included in the play's account of the Trojan War. In Part II Sinon proves his point to Diomedes that Cressida is 'neither constant, wise, nor beautifull'. As Diomedes looks on from one side Sinon persuades Cressida in the space of fifty lines to change her affections from Diomedes to himself, after which Diomedes breaks with Cressida and leaves her lamenting. A lost play of Troilus and Cressida by Dekker and Chettle (1599) is known only through notes in Henslowe's papers (including the note: 'Enter Cressida with Beggars, pigg Stephen, mr Jones his boy & mutes to them Troylus, & Deiphobus & proctor exeunt'). In Marston's *Histriomastix* (1599) a scene is put on from a Troilus play:

> TROY. Come *Cressida* my Cresset light
>   Thy face doth shine both day and night,
>   Behold, behold, thy garter blue,
>   Thy knight his valiant elboe weares,
>   That When he shakes his furious Speare,
>   The foe in shivering fearefull sort,
>   May lay him downe in death to snort.
> CRES. O knight with vallour in thy face,
>   Here take my skreene, wear it for grace,
>   Within thy Helmet put the same,
>   Therewith to make thine enemies lame.

The scene is described within the play as 'Lame stuffe indeed the like was never heard', and there is a similarly mocking allusion in Samuel Rowland's *The Letting of Hvmours Blood in the Head-Vane* of 1600 ('Be thou the Lady *Cresset-light* to mee, | Sir Trollelollie I will prooue to thee').

In Chapman's *Sir Giles Goosecap* (1602) plot, dialogue, ideas, and images from the first three books of *Troilus* are translated into the terms of an Elizabethan comedy. Troilus becomes a priggish Renaissance gentleman-scholar, a Neoplatonic philosopher, while Cressida is also something of a scholar. Her amiable uncle, Lord Momford, is a sympathetic character whose main aim is the lovers' marriage. The plot includes the lover's confession of love in a letter, the lady's reply, the uncle's inviting of his niece to supper at his house, and the lover's feigned illness, while the structure of dialogue in the uncle's visits to woo his niece for his young friend is especially indebted to Pandarus' scenes with Criseyde in Book II of *Troilus*.

For the *Story of Troylous and Pandor*, see Rollins, 'Troilus–Cressida Story', 388.

For Grimald's *Troilus ex Chaucero*, see Spurgeon, *Five Hundred Years*, i. 95.

The Iron Age: in *The Dramatic Works of Thomas Heywood* (London, 1874), iii.

On the Dekker and Chettle play, see G. Bullough, 'The Lost *Troilus and Cressida*', *Essays and Studies*, 17 (1964), 24–40.

John Marston: Plays, ed. H. H. Wood (Edinburgh, 1934–9), 265.

Chapman: The Comedies, ed. T. M. Parrot (London, 1912). For a discussion of *Sir Giles Goosecap*, see Thompson, who comments that 'This is perhaps the only play in the whole period for which we can say definitely that its author must have had a copy of Chaucer open beside him as he wrote' (*Shakespeare's Chaucer*, 43).

The relation to Chaucer's poem of Shakespeare's play *Troilus and Cressida* is altogether more elusive, and close verbal parallels between the two texts are hard to find. Yet it would have been surprising if Shakespeare did not know the poem of Chaucer's most admired by the Elizabethans, and the influence of *Troilus and Criseyde* may be discerned in the treatment of the lovers in *Romeo and Juliet*. To see the obvious divergences of the play from the poem is easier but less interesting than to allow for the possible ways in which Shakespeare's imagination may have formed aspects of his play in reaction to the delicate balance of sympathetic ironies in Chaucer's narrative, and the presence of Chaucer's poem has been detected in Shakespeare's treatment of the love-story element in his play, its characterization, and dramatic structure. Act I, Scene i and Scene ii, of the play may be seen to correspond to Chaucer's first and second books, and Act III, scene ii, corresponds to Book III of *Troilus*, with some recollection of Book II. In the first scene there are possibly some reminiscences of the passivity of Chaucer's Troilus and his dependence on Pandarus. The initial presentation of the self-possessed, calculating heroine is far removed from Chaucer, but gains in force if seen as a contrast to Chaucer's Criseyde. In Act III, Scene ii, Shakespeare compresses the lovers' first meeting and the consummation into a single occasion, but Chaucer's account of Deiphebus' party, the excuses, the use of Troilus' brothers as unwitting accomplices, the role of Helen, all find echoes in the play. In both play and poem Troilus is wholly dependent on Pandarus, indulges in rhetoric before the crucial point, but is then speechless with shyness. However, Shakespeare's Cressida has already revealed that she knows what is happening to her, and Pandarus' cynical comments to the audience at the end of this scene, and also at the close of the play, may be a parody of the narratorial comments at the close of Chaucer's third book. In the declining action Shakespeare creates a deliberately coarsened and more explicit version of Pandarus' visit to Criseyde on the morning after the consummation, and the structure of Chaucer's Book IV may influence the shaping of the lovers' separate responses to the news of their impending separation. Echoes of Chaucer's poem in the affecting moments of the lovers' sorrows and their exchange of tokens coexist with Shakespeare's very different handling of Diomede, or of Cressida's arrival in the Greek camp. That equivocal balance between sympathy and irony possible in a narrative poem is evidently neither feasible nor intended in a play that takes a view of the story so different from the defensive, extenuating approach in *Troilus*, and does not imitate Chaucer's allusion to a corrective frame of reference. Yet it may be that the philosophical concerns of *Troilus* with time, knowledge, and will, worked out through its dealings with character, lie in part behind the markedly questioning and philosophical dimension in Shakespeare's play of the same love-story.

For two excellent and complementary studies of the relation between Chaucer's poem and Shakespeare's *Troilus and Cressida*, see Donaldson, *The Swan at the Well*, 74–118, 148–57, and Thompson, *Shakespeare's Chaucer*, 111–65. See also Bradbrook, 'What Shakespeare did', and Mann, '"What is Criseyde worth?"'

The influence and example of Chaucer's *Troilus* in the sixteenth century are much greater than the number of specific references to the poem and quotations from its text. When such early sixteenth-century figures as Tyndale and Sir Thomas Elyot use *Troilus* as an example of that secular reading of which they disapprove, they confirm the contemporary standing of Chaucer's poem. Among literary men towards the end of the century there is evident knowledge and appreciation of *Troilus*. In his *Apologie for Poetrie* Sir Philip Sidney eloquently praises Chaucer's achievement: 'Chaucer, undoubtedly, did excellently in his *Troilus and Criseyde*; of whom, truly, I know not whether to marvel more, either that he in that misty time could see so clearly, or that we in this clear age walk so stumblingly after him. Yet had he great wants, fit to be forgiven in so reverent antiquity.' Gabriel Harvey noted in his copy of Speght's edition of Chaucer that *Troilus* is 'A peece of braue, fine, & sweet poetrie. One of Astrophils cordials,' and in his preface to the Speght edition Francis Beaumont commends the five-book *Troilus* (i.e. without the *Testament*) as 'so sententious, as there bee fewe staues in that Booke, which are not concluded with some principall sentence: most excellently imitating *Homer* and *Virgil*, and borrowing often of them, and of *Horace* also, and other the rarest both Oratours and Poets that haue written'. Indeed, the verse-form of *Troilus* was taken as a kind of model. When a character in the self-consciously literary Cambridge play *The Return from Parnassus* (1600) is employed to imitate the verse of Chaucer, Gower, Spenser, and Shake-speare, the Chaucerian examples are imitations of passages in Book II of *Troilus*. In his *Short Treatise on Verse* (1584) the future James I had recommended 'For tragicall materis, complaintis, or testamentis, use this kynde of verse following, callit Troilus verse—', and in the *Arte of English Poesie* (1589) Puttenham writes: 'His meetre Heroicall of *Troilus* and *Cresseid* is very graue and stately, keeping the staffe of seuen, and the verse of ten, his other verses of the Canterbury tales be but riding ryme.' Puttenham also comments on the origin of *Troilus* in translation, as do Peacham and Dryden, though such acknowledgement does not diminish esteem for Chaucer's achievement: for Dryden in his Preface to the *Fables* the story is 'much amplified by our *English* Translatour, as well as beautified'.

   The publication by Dryden in his *Fables Ancient and Modern* of verse translations of some Canterbury Tales represents a watershed in response to Chaucer and also to *Troilus*. In his Preface Dryden accepts that the increasing difficulty experienced in reading Chaucer's language makes modernization necessary. Indeed, earlier in the seventeenth century *Troilus* had twice been the object of more extensive attempts at keeping Chaucer's text accessible to readers. The first three books were 'Translated into our Moderne English' *c.*1630 by 'J.S.', presumed to be one Jonathan Sidnam. The title-page of the manuscript describes what follows as 'a paraphrase', directed towards those 'Who either cannot, or will not, take ye paines to vnderstand | The Excellent Authors | Farr more Exquisite, and significant Expressions | Though now growen obsolete, and | out of vse'. Sidnam rephrases obsolete diction and syntax, with some consequent recasting of the stanzas and their rhymes, and his text concludes with two stanzas of his own, in which he declines to pursue the story into Criseyde's infidelity.

A more scholarly enterprise to preserve the text of *Troilus* from the depredations of time and change in the English language was that of Sir Francis Kynaston, courtier, scholar, and antiquary, who translated it into the timeless medium of Latin. Not that Kynaston intended his version to supersede Chaucer's original: his project was to publish Chaucer's poem in parallel with a translation into Latin verse. In 1635 the *Amorum Troili et Creseidae libri duo priores Anglico-Latini* was published at Oxford, and presents a reprint of the 1598 Speght text of Books I and II in parallel with Kynaston's Latin, which even attempts to reproduce Chaucer's original rhyme-scheme. His preface looked forward to the eventual publication of a complete text with facing translation, together with a commentary of explanatory notes. These were completed in manuscript by 1639 and licensed for the press on 2 June 1640, but Kynaston died before his work could be published and the whole project never appeared. His commentary shows a scholarly and erudite interest in explicating the detail of Chaucer's text that anticipates the concerns of a modern edition. The history of subsequent response to *Troilus* might well have been different indeed if Kynaston's whole project had been published, for it would have set a standard for scholarly commentary much higher than anything then available for Chaucer's other poems.

But instead, both Sidnam's paraphrase and most of Kynaston's efforts remained in manuscript. Neither was available to alter the perception of Chaucer as an ancient author increasingly difficult to read, nor could they influence that shift in estimation of Chaucer in which the *Canterbury Tales* come to be seen as his most characteristic work, to the relative neglect of his other poems. This is a major shift of emphasis in the appreciation of Chaucer's poems, in which Dryden's discussion of Chaucer's genius in the Preface to the *Fables* plays an important part, for since they contained his modernizations of the Knight's, Wife's, and Nun's Priest's Tales, his Preface discusses Chaucer largely in terms of the *Tales*. Although Dryden could justifiably declare that 'no Man ever had, or can have, a greater Veneration for *Chaucer* than myself', his generous praise of Chaucer makes little acknowledgement of *Troilus and Criseyde*, and his rewriting of Shakespeare's play into his own play, *Troilus and Cressida, or Truth Found Too Late* (1679), does not suggest that he had recourse to Chaucer's poem.

By the end of the seventeenth century Chaucer has become an ancient poet and a classic, but his work is no longer part of a living tradition. The perception of *Troilus* changes within such a larger change, and there are no longer signs of that readership and influence which the poem had enjoyed as late as the Elizabethan period. There is to be a scattering of references to *Troilus* across the years by scholars and admirers of Chaucer. William Godwin includes a critical discussion of the poem in his *Life of Chaucer*

(1803); Keats in *Endymion* alludes to 'The close of Troilus and Cressid sweet' (ii. 13); and Wordsworth 'modernizes' a lyrical section of Book V, Troilus' sorrow in Troy after his return from Sarpedon (v. 519–686). But it was not to be until the later twentieth century that *Troilus and Criseyde* would once again be fully recognized as Chaucer's finest single achievement, his masterpiece.

William Tyndale, in his *The Obedience of a Christian Man* (1528), deplores the reading of 'Robyn hode & bevise of hampton | hercules | hector and troylus with a tousande histories & fables of love & wantones & of rybaudry as fylthy as herte can thinke | to corrupte ye myndes of youth with all' (fo. xx); in Sir Thomas Elyot's dialogue *Pasquill the Playne* (1533) the contrast is deplored between a New Testament held in the hand and a copy of *Troilus* hidden in the bosom ('Lorde what discord is bytwene these two bokes . . .'). Cf. Brewer, *Critical Heritage*, i. 87, 90.

Sir Philip Sidney, *An Apology for Poetry, or The Defence of Poesy*, ed. Geoffrey Shepherd (London, 1965), 133.

For Gabriel Harvey's annotations, see Brewer, *Critical Heritage*, i. 123.

For Francis Beaumont's preface, see Brewer, *Critical Heritage*, i. 136–9.

The Return from Parnassus: in *The Parnassus Plays*, ed. J. B. Leishman (London, 1949). Cf. lines 1146–52 with *Troilus*, ii. 967–73; lines 1153–8 with ii. 1026–7, 1091–2; lines 1159–65 with ii. 1041–3.

King James VI of Scotland, *Short Treatise on Verse* (1584), reprinted in *Elizabethan Critical Essays*, ed. G. G. Smith (Oxford, 1904), i. 222.

For extracts from Puttenham's *Arte of English Poesie*, see Brewer, *Critical Heritage*, i. 126–7.

For Dryden's Preface to the *Fables*, see *The Poems of John Dryden*, ed. J. Kinsley (Oxford, 1958), iv. For Henry Peacham's *The Compleat Gentleman* (1622), see Brewer, *Critical Heritage*, i. 149.

The version of *Troilus* by 'J.S.' is in British Library, Additional MS 29494. See *A Seventeenth-Century Modernisation of the First Three Books of Chaucer's 'Troilus and Criseyde'*, ed. H. G. Wright (Bern, 1960).

On Sir Francis Kynaston, see Beadle, 'The Virtuoso's *Troilus*'.

For Dryden's *Troilus and Cressida, or Truth Found Too Late*, see *The Works of John Dryden*, ed. M. E. Novack and G. R. Guffey, xiii (Berkeley, Calif., Los Angeles, and London, 1984).

William Godwin's remarks on *Troilus* in his *Life of Chaucer*—in effect, the earliest critical essay on the poem—include both criticism and ample praise, of which the following is a selection:

It is certainly much greater in extent of stanzas and pages, than the substratum and basis of the story can authorise.

It is also considerably barren of incident. There is not enough in it of matter generating visible images in the reader, and exciting his imagination with pictures of nature and life. There is not enough in it of vicissitudes of fortune, awakening curiosity and holding expectation in suspense.

Add to which, the catastrophe is unsatisfactory and offensive . . .

But, when all these deductions have been made from the claims of the Troilus and Creseide upon our approbation, it will still remain a work interspersed with many beautiful passages . . . of exquisite tenderness, of great delicacy, and of a nice and refined observation of the workings of human sensibility. . . . His personages always feel, and we confess the truth of their feelings; what passes in their minds, or falls from their tongues, has the clear and decisive character which proclaims it human, together with the vividness, subtleness and delicacy, which few authors in the most enlightened ages have been equally fortunate in seizing. (298 ff.)

BIBLIOGRAPHY

ADAMS, JOHN F., 'Irony in Troilus' Apostrophe to the Vacant House of Criseyde', *Modern Language Quarterly*, 24 (1963), 61–5.

ADAMSON, JANE, 'The Unity of *Troilus and Criseyde*', *Critical Review*, 14 (1971), 17–37.

AERS, DAVID, 'Criseyde: Woman in Medieval Society', *Chaucer Review*, 13 (1979), 177–200.

—— *Chaucer, Langland, and the Creative Imagination* (London, 1980), ch. 5 ('Chaucer's Criseyde: Woman in Society, Woman in Love').

—— *Chaucer* (Brighton, 1986), ch. 4 ('Chaucer's Representations of Marriage and Sexual Relations').

—— *Community, Gender and Individual Identity* (London, 1988), ch. 3 ('Masculine Identity in the Courtly Community: The Self Loving in *Troilus and Criseyde*').

ANDERSON, DAVID, 'Theban History in Chaucer's *Troilus*', *Studies in the Age of Chaucer*, 4 (1982), 109–33.

ANTONELLI, ROBERTO, 'The Birth of Criseyde—An Exemplary Triangle: "Classical" Troilus and the Question of Love at the Anglo-Norman Court', in Piero Boitani (ed.), *The European Tragedy of Troilus* (Oxford, 1989), 21–48.

APROBERTS, ROBERT P., 'The Central Episode in Chaucer's *Troilus*', *PMLA* 77 (1962), 373–85.

—— 'Criseyde's Infidelity and the Moral of the *Troilus*', *Speculum*, 44 (1969), 383–402.

—— 'The Boethian God and the Audience of the *Troilus*', *JEGP* 69 (1970), 425–36.

—— 'Love in the *Filostrato*', *Chaucer Review*, 7 (1972), 1–26.

—— 'The Growth of Criseyde's Love', in W.-D. Bald and H. Weinstock (eds.), *Medieval Studies Conference Aachen 1983* (Frankfurt, 1984), 131–41.

—— and SELDIS, ANNA BRUNI (trans.), *Giovanni Boccaccio: Il Filostrato* (New York, 1986), 'Introduction'.

ARN, MARY-JO, 'Three Ovidian Women in Chaucer's *Troilus*: Medea, Helen, Oënone', *Chaucer Review*, 15 (1980), 1–10.

ASTELL, ANN W., 'Orpheus, Euridice and the "Double Sorwe" of Chaucer's *Troilus*', *Chaucer Review*, 23 (1989), 282–99.

BAILEY, SUSAN E., 'Controlled Partial Confusion: Concentrated Imagery in *Troilus and Criseyde*', *Chaucer Review*, 20 (1985), 83–9.

BARNEY, STEPHEN A., 'Troilus Bound', *Speculum*, 47 (1972), 445–58.

—— (ed.), *Chaucer's 'Troilus': Essays in Criticism* (London, 1980).

BARON, F. XAVIER, 'Chaucer's Troilus and Self-Renunciation in Love', *Papers on Language and Literature*, 10 (1974), 5–14.

BASWELL, C. C., and TAYLOR, P. B., 'The *Faire Queene Eleyne* in Chaucer's *Troilus*', *Speculum*, 63 (1988), 293–311.

BAYLEY, JOHN, *The Characters of Love* (London, 1960), ch. 2 ('Love and the Code: *Troilus and Criseyde*').

BEADLE, RICHARD, 'The Virtuoso's *Troilus*', in Ruth Morse and Barry Windeatt (eds.), *Chaucer Traditions: Studies in Honour of Derek Brewer* (Cambridge, 1990), 213–33.

BENNETT, J. A. W., 'Chaucer, Dante and Boccaccio', in Piero Boitani (ed.), *Chaucer and the Italian Trecento* (Cambridge, 1983), 89–113.

BENSON, C. DAVID, 'A Chaucerian Allusion and the Date of the Alliterative *Destruction of Troy*', *Notes and Queries*, 21 (1974), 206–7.

—— 'King Thoas and the Ominous Letter in Chaucer's *Troilus*', *Philological Quarterly*, 58 (1979), 264–7.

—— ' "O Nyce World": What Chaucer Really Found in Guido delle Colonne's History of Troy', *Chaucer Review*, 13 (1979), 308–15.

—— *The History of Troy in Middle English Literature* (Cambridge, 1980).

—— 'True Troilus and False Cresseid: The Descent from Tragedy', in Piero Boitani (ed.), *The European Tragedy of Troilus* (Oxford, 1989) 153–70.

—— *Chaucer's 'Troilus and Criseyde'* (London, 1990).

—— and WINDEATT, BARRY, 'The Manuscript Glosses to Chaucer's *Troilus and Criseyde*', *Chaucer Review*, 25 (1990), 33–53.

BENSON, LARRY D. (ed.), *The Riverside Chaucer* (3rd edn., Boston, 1987; Oxford, 1988).

BERRY, GREGORY L., 'Chaucer's Mnemonic Verses and the Siege of Thebes in *Troilus and Criseyde*', *English Language Notes*, 17 (1979), 90–3.

BERRYMAN, CHARLES, 'The Ironic Design of Fortune in *Troilus and Criseyde*', *Chaucer Review*, 2 (1967), 1–7.

BESSENT, BENJAMIN R., 'The Puzzling Chronology of Chaucer's *Troilus*', *Studia neophilologica*, 41 (1969), 99–111.

BESTUL, THOMAS H., 'Chaucer's *Troilus and Criseyde*: The Passionate Epic and its Narrator', *Chaucer Review*, 14 (1980), 366–78.

BIANCIOTTO, GABRIEL, 'Édition critique et commentée du "Roman de Troyle", traduction française du xve siècle du *Filostrato* de Boccace', Dissertation (Paris, 1977).

BIE, WENDY A., 'Dramatic Chronology in *Troilus and Criseyde*', *English Language Notes*, 14 (1976), 9–13.

BISHOP, IAN, *Chaucer's 'Troilus and Criseyde': A Critical Study* (Bristol, 1981).

—— '*Troilus and Criseyde* and the *Knight's Tale*', in *The New Pelican Guide to English Literature*, i. *Medieval Literature: Chaucer and the Alliterative Tradition*, ed. B. Ford (Harmondsworth, 1982), 174–87.

BLAMIRES, ALCUIN, 'The "Religion of Love" in Chaucer's *Troilus and Criseyde* and Medieval Visual Art', in K. J. Höltgen, P. M. Daly, and W. Lottes (eds.), *Word and Visual Imagination: Studies in the Interaction of English Literature and the Visual Arts* (Erlangen, 1988), 11–31.

BLOOMFIELD, MORTON W., 'Chaucer's Sense of History', *Journal of English and Germanic Philology*, 51 (1952), 301–13.

—— 'Distance and Predestination in *Troilus and Criseyde*', *PMLA* 72 (1957), 14–26; repr. in Richard J. Schoeck and Jerome Taylor (eds.), *Chaucer Criticism*, ii. *Troilus and Criseyde and the Minor Poems* (Notre Dame, Ind., 1961), 196–210.

—— 'The Eighth Sphere: A Note on Chaucer's *Troilus and Criseyde*, v, 1809', *Modern Language Review*, 53 (1958), 408–10.

—— 'Troilus' Paraclausithyron and its Setting: *Troilus and Criseyde V*, 519–602', *Neuphilologische Mitteilungen*, 73 (1972), 15–24.

BOITANI, PIERO, *English Medieval Narrative in the 13th and 14th Centuries* (Cambridge, 1982), ch. 6 ('Chaucer: The Romance').

—— (ed.), *Chaucer and the Italian Trecento* (Cambridge, 1983).

—— (ed.), *The European Tragedy of Troilus* (Oxford, 1989).

—— *The Tragic and the Sublime in Medieval Literature* (Cambridge, 1989).

—— and MANN, JILL (eds.), *The Cambridge Chaucer Companion* (Cambridge, 1986).

BOLTON, W. F., 'Treason in Troilus', *Archiv*, 203 (1966), 255–62.

BORTHWICK, SISTER MARY CHARLOTTE, 'Antigone's Song as "Mirour" in Chaucer's *Troilus and Criseyde*', *Modern Language Quarterly*, 22 (1961), 227–35.

BOUGHNER, DANIEL C., 'Elements of Epic Grandeur in the *Troilus*', *English Literary History*, 6 (1939), 200–10; repr. in Richard J. Schoeck and Jerome Taylor (eds.), *Chaucer Criticism*, ii. *Troilus and Criseyde and the Minor Poems* (Notre Dame, Ind., 1961), 186–95.

BOWERS, JOHN M., 'How Criseyde Falls in Love', in Nathaniel B. Smith and Joseph T. Snow (eds.), *The Expansion and Transformation of Courtly Literature* (Athens, Ga., 1980), 141–55.

BOWERS, R. H., 'The "suttel and dissayvabull" world of Chaucer's *Troilus*', *Notes and Queries*, 202 (1957), 278–9.

BRADBROOK, MURIEL C., 'What Shakespeare did to Chaucer's *Troilus and Criseyde*', *Shakespeare Quarterly*, 9 (1958), 311–19, repr. in her *The Artist and Society in Shakespeare's England* (Brighton, 1982), 133–43.

BRADDY, HALDEEN, 'Chaucer's Playful Pandarus', *Southern Folklore Quarterly*, 34 (1970), 71–81.

BRENNER, GERRY, 'Narrative Structure in Chaucer's *Troilus and Criseyde*', *Annuale mediaevale*, 6 (1965), 5–18, repr. in Stephen A. Barney (ed.), *Chaucer's 'Troilus': Essays in Criticism* (London, 1980), 131–44.

BREWER, DEREK S., 'Love and Marriage in Chaucer's Poetry', *Modern Language Review*, 49 (1954), 461–4; repr. in his *Tradition and Innovation in Chaucer* (London, 1982), 22–6.

—— '*Troilus and Criseyde*' in W. F. Bolton (ed.), *The Middle Ages* (London, 1970), 195–228.

—— 'The Ages of Troilus, Criseyde and Pandarus', *Studies in English Literature* (Tokyo), (1972), 3–13; repr. in his *Tradition and Innovation in Chaucer* (London, 1982), 80–88.

—— *Chaucer* (3rd suppl. edn.; London, 1973).

BREWER, DEREK S., 'Honour in Chaucer', *Essays and Studies*, 26 (1973), 1–19; repr. in *Tradition and Innovation in Chaucer* (London, 1982), 89–109.

—— 'Towards a Chaucerian Poetic', *Proceedings of the British Academy*, 60 (1974), 219–52; repr. in his *Chaucer: The Poet as Storyteller* (London, 1984), 54–79.

—— (ed.), *Chaucer: The Critical Heritage* (2 vols.; London, 1978).

—— 'Root's Account of the Text of *Troilus*', *Poetica* (Tokyo), 12 (1981), 36–44.

—— 'Observations on the Text of *Troilus*', in P. L. Heyworth (ed.), *Medieval Studies for J. A. W. Bennett*, (Oxford, 1981), 121–38.

—— *An Introduction to Chaucer* (London, 1984), ch. 10 ('A Philosophical Romance: *Troilus and Criseyde*').

—— 'Comedy and Tragedy in *Troilus and Criseyde*', in Piero Boitani (ed.), *The European Tragedy of Troilus* (Oxford, 1989), 95–109.

—— 'The History of a Shady Character: The Narrator of *Troilus and Criseyde*', in R. M. Nischik and B. Korte (eds.), *Narrative Modes* (Würzburg, 1990), 166–78.

BROOKHOUSE, CHRISTOPHER, 'Chaucer's *Impossibilia*', *Medium Ævum*, 34 (1965), 40–2.

BROWN, WILLIAM H., jun., 'A Separate Peace: Chaucer and the Troilus of Tradition', *JEGP* 83 (1984), 492–508.

BRÜCKMANN, PATRICIA, '*Troilus and Criseyde*, III, 1226–1232: A Clandestine Topos', *English Language Notes*, 18 (1981), 166–70.

BURLIN, R. B., *Chaucerian Fiction* (Princeton, 1977), ch. 6 ('*Troilus and Criseyde*: "Love in Dede"').

BURNLEY, J. D., 'Proude Bayard: *Troilus and Criseyde*, I.218', *Notes and Queries*, 23 (1976), 148–52.

—— *Chaucer's Language and the Philosophers' Tradition* (Cambridge, 1979).

—— 'Criseyde's Heart and the Weakness of Women: An Essay in Lexical Interpretation', *Studia neophilologica*, 54 (1982), 25–38.

BURJORJEE, D. M., 'The Pilgrimage of Troilus's Sailing Heart in Chaucer's *Troilus and Criseyde*', *Annuale mediaevale*, 13 (1972), 14–31.

BURROW, J. A., *Medieval Writers and their Work* (Oxford, 1982).

CARTON, EVAN, 'Complicity and Responsibility in Pandarus' Bed and Chaucer's Art', *PMLA* 94 (1979), 47–61.

CHRISTMAS, PETER, '*Troilus and Criseyde*: The Problems of Love and Necessity', *Chaucer Review*, 9 (1975), 285–96.

CLARK, JOHN W., 'Dante and the Epilogue of the *Troilus*', *JEGP* 50 (1951), 1–10.

CLARK, S. L., and WASSERMAN, JULIAN N., 'The Heart in *Troilus and Criseyde*: The Eye of the Breast, the Mirror of the Mind, the Jewel and its Setting', *Chaucer Review*, 18 (1984), 316–28.

CLAYTON, MARGARET, 'A Virgilian Source for Chaucer's "White Bole"', *Notes and Queries*, 26 (1979), 103–4.

CLOGAN, PAUL M., 'Chaucer and the *Thebaid* Scholia', *Studies in Philology*, 61 (1964), 599–615.

—— 'Chaucer's Use of the *Thebaid*', *English Miscellany*, 18 (1967), 9–31.

—— 'Two Verse Commentaries on the Ending of Boccaccio's *Filostrato*', *Medievalia et Humanistica*, NS 7 (1976), 147–52.

—— 'The Theban Scene in Chaucer's *Troilus*', *Medievalia et Humanistica*, NS 12 (1984), 167–85.

CLOUGH, ANDREA, 'Medieval Tragedy and the Genre of *Troilus and Criseyde*', *Medievalia et Humanistica*, 11 (1982), 211–27.

CONLEE, JOHN W., 'The Meaning of Troilus' Ascension to the Eighth Sphere', *Chaucer Review*, 7 (1972), 27–36.

COOK, DANIEL, 'The Revision of Chaucer's *Troilus*: The Beta Text', *Chaucer Review*, 9 (1974–5), 51–62.

COOK, ROBERT G., 'Chaucer's Pandarus and the Medieval Idea of Friendship', *JEGP* 69 (1970), 407–24.

COPE, JACKSON I., 'Chaucer, Venus and the "Seventhe Spere"', *Modern Language Notes*, 67 (1952), 245–6.

CORMICAN, JOHN D., 'Motivation of Pandarus in *Troilus and Criseyde*', *Language Quarterly*, 18: 3–4 (1980), 43–8.

CORSA, HELEN S., 'Dreams in *Troilus and Criseyde*', *American Imago*, 27 (1970), 52–65.

—— 'Is This a Mannes Herte?', *Literature and Psychology*, 16 (1966), 184–91.

COTTON, MICHAEL E., 'The Artistic Integrity of Chaucer's *Troilus and Criseyde*', *Chaucer Review*, 7 (1972), 37–43.

COVELLA, SISTER FRANCIS DOLORES, 'Audience as Determinant of Meaning in the *Troilus*', *Chaucer Review*, 2 (1968), 235–45.

CRAMPTON, GEORGIA R., 'Action and Passion in Chaucer's *Troilus*', *Medium Ævum*, 43 (1974), 22–36.

CUMMINGS, HUBERTIS M., *The Indebtedness of Chaucer's Works to the Italian Works of Boccaccio* (Cincinnati, Ohio, 1916).

CURETON, KEVIN K., 'Chaucer's Revision of *Troilus and Criseyde*', *Studies in Bibliography*, 42 (1989), 153–84.

CURRY, WALTER CLYDE, 'Destiny in Chaucer's *Troilus*', *PMLA* 45 (1930), 129–68; repr. in Richard Schoeck and Jerome Taylor (eds.), *Chaucer Criticism*, ii. *Troilus and Criseyde and the Minor Poems* (Notre Dame, Ind. 1961), 39–70.

—— *Chaucer and the Mediaeval Sciences* (2nd edn.; Oxford, 1960).

DAHLBERG, CHARLES, 'The Narrator's Frame for *Troilus*', *Chaucer Review*, 15 (1980), 85–100.

—— *The Literature of Unlikeness* (University of New England Press, 1988), ch. 6.

DAVENPORT, W. A., *Chaucer: Complaint and Narrative* (Cambridge, 1988), ch. 6.

DAVID, ALFRED, 'The Hero of the *Troilus*', *Speculum*, 37 (1962), 566–81.

—— *The Strumpet Muse: Art and Morals in Chaucer's Poetry* (London, 1976), ch. 2 ('The Theme of Love in *Troilus*').

—— 'Chaucerian Comedy and Criseyde', in Mary Salu (ed.), *Essays on 'Troilus and Criseyde'* (Cambridge, 1979), 90–104.

DAVIS, NORMAN, 'The Litera Troili and English Letters', *Review of English Studies*, 16 (1965), 233–44.

DEAN, CHRISTOPHER, 'Chaucer's Play on the Word *Beere* in *Troilus and Criseyde*', *Chaucer Review*, 15 (1981), 224–6.

DEAN, JAMES, 'Chaucer's *Troilus*, Boccaccio's *Filostrato*, and the Poetics of Closure', *Philological Quarterly*, 64 (1985), 175–84.

DELANY, SHEILA, 'Techniques of Alienation in *Troilus and Criseyde*', in A. P. Foulkes (ed.), *The Uses of Criticism* (Frankfurt, 1976), 77–95.

DENOMY, A. J., 'The Two Moralities of Chaucer's *Troilus and Criseyde*', *Transactions of the Royal Society of Canada*, 44 (1950), 35–46; repr in Richard J. Schoeck and Jerome Taylor (eds.), *Chaucer Criticism*, ii. *Troilus and Criseyde and the Minor Poems* (Notre Dame, Ind., 1961), 147–59.

DEVEREUX, JAMES A., SJ, 'A Note on *Troilus and Criseyde*, Book III, Line 1309', *Philological Quarterly*, 44 (1965), 550–2.

DE VRIES, F. C., '*Troilus and Criseyde*, Book III, Stanza 251, and Boethius', *English Studies*, 52 (1971), 502–7.

—— 'Notes of Troilus's Song to Love', *Parergon*, 3 (1972), 17–19.

DIAMOND, ARLYN, '*Troilus and Criseyde*: The Politics of Love', in Julian N. Wasserman and Robert J. Blanch (eds.), *Chaucer in the Eighties* (Syracuse, 1986), 93–103.

DINSHAW, CAROLYN, *Chaucer's Sexual Poetics* (Madison, 1989), ch. 1 ('Reading Like a Man . . .').

DI PASQUALE, PASQUALE, jun., ' "Sikernesse" and Fortune in *Troilus and Criseyde*', *Philological Quarterly*, 49 (1970), 152–63.

DONALDSON, E. TALBOT (ed.), *Chaucer's Poetry: An Anthology for the Modern Reader* (New York, 1958), 'Commentary', pp. 965–80.

—— 'The Ending of *Troilus*', in his *Speaking of Chaucer* (London 1970), 84–101, repr. from Arthur Brown and Peter Foote (eds.), *Early English and Norse Studies Presented to Hugh Smith* (London, 1963).

—— 'Criseide and her Narrator', in his *Speaking of Chaucer* (London, 1970), 65–83.

—— 'The Masculine Narrator and Four Women of Style', in his *Speaking of Chaucer* (London, 1970), 46–64.

—— 'Chaucer and the Elusion of Clarity', *Essays and Studies*, NS 25 (1972), 23–44.

—— 'Chaucer's Three Ps: Pandarus, Pardoner, and Poet', *Michigan Quarterly Review*, 14 (1975), 282–301.

—— 'Briseis, Briseida, Criseyde, Cresseid, Cressid: Progress of a Heroine', in Edward Vasta and Zacharias P. Thundy (eds.), *Chaucerian Problems and Perspectives* (Notre Dame, Ind., 1979), 3–12.

—— *The Swan at the Well: Shakespeare Reading Chaucer* (New Haven, Conn., and London, 1985).

DOOB, PENELOPE B. R., 'Chaucer's "Corones Tweyne" and the Lapidaries', *Chaucer Review*, 7 (1972), 85–96.

DRAKE, GERTRUDE C., 'The Moon and Venus: Troilus's Havens in Eternity', *Papers on Language and Literature*, 11 (1975), 3–17.

DRONKE, PETER, 'The Conclusion of *Troilus and Criseyde*', *Medium Ævum*, 33 (1964), 47–52.

DUNNING, T. P., 'God and Man in *Troilus and Criseyde*', in N. Davis and C. L. Wrenn (eds.), *English and Medieval Studies Presented to J. R. R. Tolkien* (London, 1962), 164–82.

DURHAM, LONNIE J., 'Love and Death in *Troilus and Criseyde*', *Chaucer Review*, 3 (1968), 1–11.

DURLING, ROBERT M., *The Figure of the Poet in the Renaissance Epic* (Cambridge, Mass., 1965), ch. 2, pp. 44–66.

EBEL, JULIA, 'Troilus and Oedipus: The Genealogy of an Image', *English Studies*, 55 (1974), 15–21.

ELBOW, PETER, 'Two Boethian Speeches in *Troilus and Criseyde* and Chaucerian Irony', in P. Damon (ed.), *Literary Criticism and Historical Understanding: Selected Papers from the English Institute* (New York, 1967), 85–107.

—— *Oppositions in Chaucer* (Middleton, Conn. 1975).

ELDREDGE, LAURENCE, 'Boethian Epistemology and Chaucer's *Troilus* in the Light of Fourteenth-Century Thought', *Mediaevalia*, 2 (1976), 50–75.

ELIASON, N. E., *The Language of Chaucer's Poetry* (Anglistica, 17; Copenhagen, 1971).

ELLIOTT, R. W. V., *Chaucer's English* (London, 1974).

ELLIS, DEBORAH, ' "Calle It Gentilesse": A Comparative Study of Two Medieval Go-Betweens', *Comitatus*, 8 (1977), 1–13.

ERZGRÄBER, WILLI, 'Tragik und Komik in Chaucers *Troilus and Criseyde*', in Dieter Riesner and Helmut Gneuss (eds.), *Festschrift für Walter Hübner* (Berlin, 1964), 139–63.

—— 'Zu Chaucers *Troilus and Criseyde*, Buch IV', in Manfred Bambeck and Hans Helmut Christmann (eds.), *Philologica Romanica: Erhard Lommatzsch gewidmet* (Munich, 1975), 97–117.

EVANS, MURRAY J., ' "Making strange": The Narrator(?), The Ending(?), and Chaucer's "Troilus" ', *Neuphilologische Mitteilungen*, 87 (1986), 218–28.

EVERETT, DOROTHY, *Essays on Middle English Literature*, ed. P. Kean (Oxford, 1955), ch. V.

FALKE, ANNE, 'The Comic Function of the Narrator in *Troilus and Criseyde*', *Neophilologus*, 68 (1984), 131–41.

FANSLER, D. S., *Chaucer and the Roman de la Rose* (New York, 1914).

FARNHAM, ANTHONY E., 'Chaucerian Irony and the Ending of the *Troilus*', *Chaucer Review*, 1 (1967), 207–16.

FISH, VARDA, 'The Origin and Original Object of *Troilus and Criseyde*', *Chaucer Review*, 18 (1984), 304–15.

FISHER, JOHN H., 'The Intended Illustrations in MS Corpus Christi 61 of Chaucer's *Troylus and Criseyde*', in J. B. Bessinger, Jr. and R. R. Raymor (eds.), *Medieval Studies in Honor of Lillian Herlands Hornstein* (New York, 1976), 111–19.

FLEMING, JOHN V., 'Deiphoebus Betrayed: Virgilian Decorum, Chaucerian Feminism', *Chaucer Review*, 21 (1986), 182–99.

—— *Classical Imitation and Interpretation in Chaucer's 'Troilus'* (Lincoln, Nebr., 1990).

FRANK, ROBERT W., jun.,'*Troilus and Criseyde*: The Art of Amplification', in Jerome Mandel and Bruce A. Rosenberg (eds.), *Medieval Literature and Folklore Studies: Essays in Honor of Francis Lee Utley* (New Brunswick, NJ, 1970), 155–71.

FRANKIS, JOHN, 'Paganism and Pagan Love in *Troilus and Criseyde*', in Mary Salu (ed.), *Essays on 'Troilus and Criseyde'* (Cambridge, 1979), 57–72.

FRANTZEN, ALLEN J., 'The "Joie and Tene" of Dreams in *Troilus and Criseyde*', in Julian N. Wasserman and Robert J. Blanch (eds.), *Chaucer in the Eighties* (Syracuse, 1986), 105–19.

FREIWALD, LEAH RIEBER, 'Swych Love of Frendes: Pandarus and Troilus', *Chaucer Review*, 6 (1971), 120–9.

FRIEDMAN, JOHN B., 'Pandarus' Cushion and the "pluma Sardanapalli"', *JEGP* 75 (1976), 41–55.

FRIES, MAUREEN, ' "Slydynge of Corage": Chaucer's Criseyde as Feminist and Victim', in Arlyn Diamond and Lee R. Edwards (eds.), *The Authority of Experience: Essays in Feminist Criticism* (Amherst, 1977), 45–59.

FROST, MICHAEL H., 'Narrative Devices in Chaucer's *Troilus and Criseyde*', *Thoth*, 14 (1974), 29–38.

FROST, WILLIAM, 'A Chaucerian Crux', *Yale Review*, 66 (1977), 551–61.

—— 'A Chaucer–Virgil Link in *Aeneid* XI and *Troilus and Criseyde* V', *Notes and Queries*, 26 (1979), 104–5.

FRY, DONALD K., 'Chaucer's Zanzis and a Possible Source for *Troilus and Criseyde*, IV, 407–413', *English Language Notes*, 9 (1971), 81–5.

FYLER, JOHN M., *Chaucer and Ovid* (New Haven, 1979).

—— 'The Fabrications of Pandarus', *Modern Language Quarterly*, 41 (1980), 115–30.

—— '*Auctoritee* and Allusion in *Troilus and Criseyde*', *Res publica litterarum*, 7 (1984), 73–92.

GALLAGHER, JOSEPH E., 'Theology and Intention in Chaucer's *Troilus*', *Chaucer Review*, 7 (1972), 44–66.

—— 'The Double Sorrow of Troilus', *Medium Ævum*, 41 (1972), 27–31.

—— 'Criseyde's Dream of the Eagle: Love and War in *Troilus and Criseyde*', *Modern Language Quarterly*, 36 (1975), 115–32.

GALWAY, MARGARET, 'The *Troilus* Frontispiece', *Modern Language Review*, 44 (1949), 161–77.

GANIM, JOHN M., 'Tone and Time in Chaucer's *Troilus*', *English Literary History*, 43 (1976), 141–53.

—— *Style and Consciousness in Middle English Narrative* (Princeton, NJ, 1983).

GARBÁTY, THOMAS, 'The *Pamphilus* Tradition in Ruiz and Chaucer', *Philological Quarterly*, 46 (1967), 457–70.

—— '*Troilus* V, 1786–92 and V, 1807–27: An Example of Poetic Process', *Chaucer Review*, 11 (1977), 299–305.

GAYLORD, ALAN, 'Uncle Pandarus as Lady Philosophy', *Papers of the Michigan Academy of Science, Arts and Letters*, 46 (1961), 571–95.

—— 'Chaucer's Tender Trap: The *Troilus* and the "Yonge, Fresshe Folkes"', *English Miscellany*, 15 (1964), 25–42.

—— 'Gentilesse in Chaucer's *Troilus*', *Studies in Philology*, 61 (1964), 19–34.

—— 'Friendship in Chaucer's *Troilus*', *Chaucer Review*, 3 (1969), 239–64.

—— 'The Lesson of the *Troilus*: Chastisement and Correction', in Mary Salu (ed.), *Essays on 'Troilus and Criseyde'* (Cambridge, 1979), 23–42.

GILL, SISTER ANNE BARBARA, *Paradoxical Patterns in Chaucer's Troilus: An Explanation of the Palinode* (Washington, DC, 1960).

GILLMEISTER, HEINER, 'Chaucer's *Kan Ke Dort (Troilus* II, 1752), and the "Sleeping Dogs" of the Trouvères', *English Studies*, 59 (1978), 310–23.

GLEASON, MARK J., 'Nicholas Trevet, Boethius, Boccaccio: Contexts of Cosmic Love in *Troilus*, Book III', *Medievalia et Humanistica*, 15 (1987), 161–88.

GORDON, IDA L., 'The Narrative Function of Irony in Chaucer's *Troilus and Criseyde*', in F. Whitehead, A. H. Diverres, and F. E. Sutcliffe (eds.), *Medieval Miscellany Presented to Eugene Vinaver by Pupils, Colleagues, and Friends* (Manchester, 1965), 146–56.

—— *The Double Sorrow of Troilus: A Study of Ambiguities in Troilus and Criseyde* (Oxford, 1970).

—— 'Processes of Characterization in Chaucer's *Troilus*', in W. Rothwell, W. R. J. Barron, David Blamires, and Lewis Thorpe (eds.), *Studies in Medieval Literature and Languages in Memory of Frederick Whitehead* (Manchester, 1973), 117–31.

GORDON, R. K., (ed. and trans.), *The Story of Troilus* (London, 1934; repr. Toronto, 1978).

GRANSDEN, K. W., 'Lente currite, noctis equi: Chaucer, *Troilus and Criseyde* 3. 1422–70, Donne, "The Sun Rising", and Ovid, *Amores*, i. 13,' in David West and Tony Woodman (eds.), *Creative Imitation and Latin Literature* (Cambridge, 1979), 157–71.

GREEN, D. H., *Irony in the Medieval Romance* (Cambridge, 1979).

GREEN, RICHARD F., 'Troilus and the Game of Love', *Chaucer Review*, 13 (1979), 201–20.

GREENFIELD, STANLEY B., 'The Role of Calkas in *Troilus and Criseyde*', *Medium Ævum*, 36 (1967), 141–51.

GRENNEN, JOSEPH E., ' "Makyng" in Comedy: "Troilus and Criseyde", V, 1788', *Neuphilologische Mitteilungen*, 86 (1985), 489–93.

—— 'Aristotelian Ideas in Chaucer's *Troilus*: A Preliminary Study', *Medievalia et Humanistica*, 14 (1986), 125–38.

GRIFFIN, N. E., and MYRICK, A. B., *The Filostrato of Giovanni Boccaccio: A Translation with Parallel Text* (Philadelphia, Pa., 1929; repr. New York, 1967).

GROSS, LAILA, 'The Two Wooings of Criseyde', *Neuphilogische Mitteilungen*, 74 (1973), 113–25.

GROSSVOGEL, DAVID I., 'Chaucer: *Troilus and Criseyde*', in his *Limits of the Novel: Evolutions of a Form from Chaucer to Robbe-Grillet* (Ithaca, NY, 1968), 44–73.

HAMMOND, E. P., 'A Burgundian Copy of Chaucer's *Troilus*', *Modern Language Notes*, 26 (1911), 32.

HANLY, MICHAEL G., *Boccaccio, Beauvau, Chaucer: 'Troilus and Criseyde': Four Perspectives on Influence* (Norman, Okla., 1990).

HANNA, RALPH, III, 'Robert K. Root', in Paul G. Ruggiers (ed.), *Editing Chaucer: The Great Tradition* (Norman, Okla., 1984), 191–205.

HANNING, R. W., 'The Audience as Co-Creator of the First Chivalric Romances', *Yearbook of English Studies*, 2 (1981), 1–28.

HANSON, THOMAS B., 'Criseyde's Brows Once Again', *Notes and Queries*, 18 (1971), 285–6.

—— 'The Center of *Troilus and Criseyde*', *Chaucer Review*, 9 (1975), 297–302.

HARDMAN, PHILLIPA, 'Narrative Typology: Chaucer's Use of the Story of Orpheus', *Modern Language Review*, 85 (1990), 545–54.

HART, THOMAS ELWOOD, 'Medieval Structuralism: "Dulcarnoun" and the Five-Book Design of Chaucer's *Troilus*', *Chaucer Review*, 16 (1981), 129–70.

HARVEY, PATRICIA A., 'ME. "Point" (*Troilus and Criseyde*, III, 695)', *Notes and Queries*, 15 (1968), 234–44.

HASKELL, ANN S., 'The Doppelgängers in Chaucer's *Troilus*', *Neuphilologische Mitteilungen*, 72 (1971), 723–34.

HATCHER, ELIZABETH R., 'Chaucer and the Psychology of Fear: Troilus in Book V', *English Literary History*, 40 (1973), 307–24.

HAVELY, NICHOLAS (trans.), *Chaucer's Boccaccio* (Cambridge, 1980).

—— 'Tearing or Breathing? Dante's Influence on *Filostrato* and *Troilus*', in Paul Strohm and Thomas J. Heffernan (eds.), *Studies in the Age of Chaucer Proceedings; i. 1984: Reconstructing Chaucer* (Knoxville, Tenn., 1985), 51–9.

HEIDTMANN, PETER, 'Sex and Salvation in *Troilus and Criseyde*', *Chaucer Review*, 2 (1968), 246–53.

HEINRICHS, KATHERINE, ' "Lovers' Consolations of Philosophy" in Boccaccio, Machaut, and Chaucer', *Studies in the Age of Chaucer*, 11 (1989), 93–115.

HELTERMAN, JEFFREY, 'Masks of Love in *Troilus and Criseyde*', *Comparative Literature*, 26 (1974), 14–31.

HERMANN, JOHN P., 'Gesture and Seduction in *Troilus and Criseyde*', *Studies in the Age of Chaucer*, 7 (1985), 107–35.

HOWARD, DONALD R., 'Courtly Love and the Lust of the Flesh', in his *The Three Temptations* (Princeton, NJ, 1966), 77–160.

—— 'Literature and Sexuality: Book III of Chaucer's *Troilus*', *Massachusetts Review*, 8 (1967), 442–56.

—— 'Experience, Language, and Consciousness: *Troilus and Criseyde*, II, 596–931', in Jerome Mandel and Bruce A. Rosenberg (eds.), *Medieval Literature and Folklore Studies: Essays in Honor of Francis Lee Utley* (New Brunswick, NJ, 1970), 173–92.

—— 'The Philosophies in Chaucer's *Troilus*', in Larry D. Benson and Siegfried Wenzel (eds.), *The Wisdom of Poetry* (Kalamazoo, 1982), 151–75.

HUBER, JOHN, 'Troilus' Predestination Soliloquy: Chaucer's Changes from Boethius', *Neuphilologische Mitteilungen*, 66 (1965), 120–5.

HUPPÉ, BERNARD F., 'The Unlikely Narrator: The Narrative Strategy of the *Troilus*', in John P. Hermann and John J. Burke (eds.), *Signs and Symbols in*

Chaucer's Poetry (University of Alabama Press, 1981), 179–94.

HUSSEY, S. S., 'The Difficult Fifth Book of *Troilus and Criseyde*', *Modern Language Review*, 67 (1972), 721–9.

—— *Chaucer: An Introduction* (2nd edn.; London, 1981), ch. III.

HUTSON, ARTHUR E., 'Troilus' Confession', *Modern Language Notes*, 69 (1954), 468–70.

JEFFERSON, B. L., *Chaucer and the 'Consolation of Philosophy' of Boethius* (Princeton, NJ, 1917).

JENNINGS, MARGARET, CSJ, 'Chaucer's Troilus and the Ruby', *Notes and Queries*, 23 (1976), 533–37.

—— 'To *Pryke* or to *Prye*: Scribal Delights in the *Troilus*, Book III', in Julian N. Wasserman and Robert J. Blanch (eds.), *Chaucer in the Eighties* (Syracuse, NY, 1986), 121–33.

JOHNSON, L. STALEY, 'The Medieval Hector: A Double Tradition', *Mediaevalia*, 5 (1979), 165–82.

JOHNSTON, EVERETT C., 'The Pronoun of Address in Chaucer's *Troilus*', *Language Quarterly*, 1 (1962), 17–20.

JORDAN, ROBERT M., 'The Narrator in Chaucer's *Troilus*', *English Literary History*, 25 (1958), 237–57.

—— *Chaucer and the Shape of Creation* (Cambridge, Mass., 1967), ch. 4 ('*Troilus and Criseyde*: Chaucerian Gothic').

JOSEPH, BERTRAM, '*Troilus and Criseyde*: "A Most Admirable and Inimitable Epicke Poeme"', *Essays and Studies*, 7 (1954), 42–61.

JOSIPOVICI, GABRIEL, *The World and the Book* (London, 1971), ch. 3 ('Chaucer: The Teller and the Tale').

KAMINSKY, ALICE R., *Chaucer's 'Troilus and Criseyde' and the Critics* (Athens, Ohio, 1980).

KANE, GEORGE, *Chaucer* (Oxford, 1984), ch. 4 ('Master Craftsman into Philosopher').

KASKE, R. E., 'The Aube in Chaucer's *Troilus*', in Richard J. Schoeck and Jerome Taylor (eds.), *Chaucer Criticism, ii. Troilus and Criseyde and the Minor Poems* (Notre Dame, Ind., 1961), 167–79.

KÄSMANN, HANS, '"I wolde excuse hire yit for routhe": Chaucers Einstellung zu Criseyde', in A. Esch (ed.), *Chaucer und seine Zeit* (Tübingen, 1968), 97–122.

KEAN, P. M., 'Chaucer's Dealings with a Stanza of *Il Filostrato* and the Epilogue of *Troilus and Criseyde*', *Medium Ævum*, 33 (1964), 36–46.

—— *Chaucer and the Making of English Poetry* (London, 1972), ch. 4.

KELLOGG, ALFRED E., 'On the Tradition of Troilus' Vision of the Little Earth', *Mediaeval Studies*, 22 (1960), 204–13.

KELLY, EDWARD HANFORD, 'Myth as Paradigm in *Troilus and Criseyde*', *Papers on Language and Literature*, 3 (summer supplement, 1967), 28–30.

KELLY, HENRY ANSGAR, 'Marriage in the Middle Ages: 2: Clandestine Marriage and Chaucer's *Troilus*', *Viator*, 4 (1973), 435–57.

—— *Love and Marriage in the Age of Chaucer* (Ithaca, NY, 1975).

KIERNAN, K. S., 'Hector the Second: The Lost Face of Troilustratus', *Annuale mediaevale*, 16 (1975), 52–62.

KIRBY, T. A., *Chaucer's Troilus: A Study in Courtly Love* (Louisiana University Press, 1940).

KIRK, ELIZABETH D., ' "Paradis Stood Formed in Hire Yen": Courtly Love and Chaucer's Re-Vision of Dante', in Mary J. Carruthers and Elizabeth D. Kirk (eds.), *Acts of Interpretation: The Text in its Contexts, 700–1600: Essays on Medieval and Renaissance Literature in Honor of E. Talbot Donaldson* (Norman, Okla., 1982), 257–77.

KITTREDGE, GEORGE LYMAN, *Observations on the Language of Chaucer's Troilus* (Chaucer Society; 2nd ser., 28; London, 1894).

—— *The Date of Chaucer's Troilus and other Chaucer matters* (Chaucer Society; London, 1909).

—— 'Antigone's Song of Love', *Modern Language Notes*, 25 (1910), 158.

—— 'Chaucer's *Troilus* and Guillaume de Machaut', *Modern Language Notes*, 30 (1915), 69.

—— *Chaucer and his Poetry* (Cambridge, Mass., 1915), ch. 4.

—— 'Chaucer's Lollius', *Harvard Studies in Comparative Philology*, 28 (1917), 47–133.

KNAPP, PEGGY ANN, 'Boccaccio and Chaucer on Cassandra', *Philological Quarterly*, 56 (1977), 413–17.

—— 'The Nature of Nature: Criseyde's "Slydyng Corage" ', *Chaucer Review*, 13 (1978), 133–40.

KNIGHT, STEPHEN, *rymyng craftily: Meaning in Chaucer's Poetry* (Sydney, 1973), ch. 2 ('*Troilus and Criseyde*: Minor Characters and the Narrator').

—— *Geoffrey Chaucer* (Oxford, 1986), ch. 2 ('*Troilus and Criseyde*: "Do wey youre barbe" ').

KNOPP, SHERRON E., 'The Narrator and his Audience in Chaucer's *Troilus and Criseyde*', *Studies in Philology*, 78 (1981), 323–40.

KORETSKY, ALLEN C., 'Chaucer's Use of the Apostrophe in *Troilus and Criseyde*', *Chaucer Review*, 4 (1970), 242–66.

—— 'The Heroes of Chaucer's Romances', *Annuale mediaevale*, 17 (1976), 22–47.

LAMBERT, MARK, '*Troilus*, Books I–III: A Criseydan Reading', in Mary Salu (ed.), *Essays on 'Troilus and Criseyde'* (Cambridge, 1979), 105–25.

—— 'Telling the Story in *Troilus and Criseyde*', in Piero Boitani and Jill Mann (eds.), *The Cambridge Chaucer Companion* (Cambridge, 1986), 59–73.

LANHAM, RICHARD A., 'Opaque Style and its Uses in *Troilus and Criseide*', *Studies in Medieval Culture*, 3 (1970), 169–76.

LAWLOR, JOHN, *Chaucer* (London, 1968), ch. 2 ('The Writer as Dependant ...') and ch. 3 ('*Tragedye* and Tragedy').

LAWTON, DAVID, 'Irony and Sympathy in *Troilus and Criseyde*: A Reconsideration', *Leeds Studies in English*, NS 14 (1983), 94–115.

—— *Chaucer's Narrators* (Cambridge, 1985), ch. 4 ('Voices of Performance: Chaucer's Use of Narrators in *Troilus* and *The Canterbury Tales*').

LENTA, MARGARET, 'The Mirror of the Mind: A Study of *Troilus and Criseyde*', *Theoria*, 58 (1982), 33–46.

LEVINE, ROBERT, 'Restraining Ambiguities in Chaucer's "Troilus and Criseyde" ', *Neuphilologische Mitteilungen*, 87 (1986), 558–64.

LEWIS, C. S., 'What Chaucer really did to *Il Filostrato*', *Essays and Studies*, 17 (1932), 56–75; repr. in Richard J. Schoeck and Jerome Taylor (eds.), *Chaucer Criticism*, ii. *Troilus and Criseyde and the Minor Poems* (Notre Dame, Ind., 1961), 16–33.

—— *The Allegory of Love* (Oxford, 1936).

LEYERLE, JOHN, 'The Heart and the Chain', in L. D. Benson (ed.), *The Learned and the Lewed* (Harvard English Studies, 5; Cambridge, Mass., 1974), 113–45.

LIGGINS, ELIZABETH M., 'The Lovers' Swoons in *Troilus and Criseyde*', *Parergon*, 3 (1985), 93–106.

LOCKHART, ADRIENNE R., 'Semantic, Moral, and Aesthetic Degeneration in *Troilus and Criseyde*', *Chaucer Review*, 8 (1973), 100–18.

LONGO, JOSEPH A., 'The Double Time Scheme in Book II of Chaucer's *Troilus and Criseyde*', *Modern Language Quarterly*, 22 (1961), 37–40.

—— 'Apropos the Love Plot in Chaucer's *Troilus and Criseyde* and Shakespeare's *Troilus and Cressida*', *Cahiers Elisabéthains*, 11 (1977), 1–15.

LOWES, JOHN LIVINGSTON, 'The Date of Chaucer's *Troilus and Criseyde*', *PMLA* 23 (1908), 285–306.

—— 'Chaucer and Dante', *Modern Philology*, 14 (1917), 705–35.

—— 'Chaucer and the *Ovide Moralisé*', *PMLA*, 33 (1918), 302–25.

—— *Geoffrey Chaucer* (Oxford, 1934), ch. 5 ('The Mastered Art').

LUMIANSKY, R. M., 'The Function of the Proverbial Monitory Elements in Chaucer's *Troilus and Criseyde*', *Tulane Studies in English*, 2 (1950), 5–48.

—— 'The Story of Troilus and Briseida according to Benoît and Guido', *Speculum*, 29 (1954), 727–33.

—— 'Calchas in the Early Versions of the Troilus Story', *Tulane Studies in English*, 4 (1954), 5–20.

MACEY, SAMUEL L., 'Dramatic Elements in Chaucer's *Troilus*', *Texas Studies in Literature and Language*, 12 (1970), 301–23.

MAGOUN, FRANCIS P., Jr., 'Chaucer's Summary of Statius' *Thebaid* II–XII', *Traditio*, 11 (1955), 409–20.

MAGUIRE, JOHN B., 'The Clandestine Marriage of Troilus and Criseyde', *Chaucer Review*, 8 (1974), 262–78.

MANLOVE, COLIN, ' "Rooteles moot grene soone deye": The Helplessness of Chaucer's Troilus and Criseyde', *Essays and Studies*, 31 (1978), 1–22.

MANN, JILL, 'Troilus' Swoon', *Chaucer Review*, 14 (1980), 319–35.

—— 'Chance and Destiny in *Troilus and Criseyde* and the *Knight's Tale*', in Piero Boitani and Jill Mann (eds.), *The Cambridge Chaucer Companion* (Cambridge, 1986, 75–92.

—— 'Shakespeare and Chaucer: "What is Criseyde worth?" ' in Piero Boitani (ed.), *The European Tragedy of Troilus* (Oxford, 1989), 219–42.

MANN, JILL, *Geoffrey Chaucer* (London, 1991).

MANNING, STEPHEN, '*Troilus*, Book V: Invention and the Poem as Process', *Chaucer Review*, 18 (1984), 288–303.

MANZALAOUI, MAHMOUD, 'Roger Bacon's "In Convexitate" and Chaucer's "In Convers" (*Troilus and Criseyde* V. 1810)', *Notes and Queries*, 11 (1964), 165–66.

MARESCA, THOMAS E., *Three English Epics: Studies of 'Troilus and Criseyde', 'The Faerie Queene' and 'Paradise Lost'* (Lincoln, Nebr., 1979).

MARKLAND, MURRAY F., '*Troilus and Criseyde*: The Inviolability of the Ending', *Modern Language Quarterly*, 31 (1970), 147–59.

MARTIN, JUNE HALL, *Love's Fools: Aucassin, Troilus, Calisto, and the Parody of the Courtly Lover* (London, 1972).

MARTIN, PRISCILLA, *Chaucer's Women: Nuns, Wives and Amazons* (London, 1990).

MASI, MICHAEL, '*Troilus:* A Medieval Psychoanalysis', *Annuale mediaevale*, 11 (1970), 81–8.

MASUI, MICHIO, 'The Development of Mood in Chaucer's *Troilus*, an Approach', in M. Brahmer (ed.), *Studies in Language and Literature in Honour of Margaret Schlauch* (Warsaw, 1966), 245–54.

—— 'A Mode of Word-Meaning in Chaucer's Language of Love', *Studies in English Literature* (Tokyo), Eng. no. (1967), 113–26.

MATTHEWS, LLOYD J., 'Chaucer's personification of Prudence in *Troilus* (V.743–749): Sources in the Visual Arts and Manuscript Scholia', *English Language Notes*, 13 (1976), 249–55.

—— '*Troilus and Criseyde*, V.743–749: Another Possible Source', *Neuphilologische Mitteilungen*, 82 (1981), 211–13.

MAYO, R. D., 'The Trojan Background of the *Troilus*', *English Literary History*, 9 (1942), 245–56.

MCALPINE, MONICA E., *The Genre of 'Troilus and Criseyde'* (Ithaca, NY, 1978).

McCALL, JOHN P., 'Chaucer's May 3', *Modern Language Notes*, 76 (1961), 201–5.

—— 'Five Book Structure in Chaucer's *Troilus*', *Modern Language Quarterly*, 23 (1962), 297–308.

—— 'The Trojan Scene in Chaucer's *Troilus*', *English Literary History*, 29 (1962), 263–75.

—— *Chaucer among the Gods* (Philadelphia, 1979).

—— '*Troilus and Criseyde*', in Beryl Rowland (ed.), *Companion to Chaucer Studies* (rev. edn.; New York, 1979), 446–63.

—— and RUDISILL, GEORGE, jun. 'The Parliament of 1386 and Chaucer's Trojan Parliament', *JEGP* 58 (1959), 276–88.

McKINNELL, JOHN, 'Letters as a Type of the Formal Level in *Troilus and Criseyde*', in Mary Salu (ed.), *Essays on 'Troilus and Criseyde'* (Cambridge, 1979), 73–89.

McNALLY, JOHN J., 'Chaucer's Topsy-Turvy Dante', *Studies in Medieval Culture*, 2 (1966), 104–10.

MEDCALF, STEPHEN, 'Epilogue: From *Troilus* to *Troilus*', in Stephen Medcalf (ed.), *The Later Middle Ages* (London, 1981), 291–305.

MEECH, SANFORD B., 'Chaucer and an Italian Translation of the *Heroides*', *PMLA* 45 (1930), 110–28.

—— 'Chaucer and the *Ovide Moralisé*: A further study'. *PMLA* 46 (1931), 182–204.

—— *Design in Chaucer's Troilus* (Syracuse, NY, 1959).

MEEK, M. E. (trans.), *Historia Destructionis Troiae* (Bloomington, Ind., 1974).

MEHL, DIETER, 'The Audience of Chaucer's *Troilus and Criseyde*', in Beryl Rowland (ed.), *Chaucer and Middle English Studies in Honor of Rossell Hope Robbins* (London, 1974), 173–89.

—— 'Chaucer's Audience', *Leeds Studies in English*, 10 (1978), 58–73.

—— *Geoffrey Chaucer: An Introduction to his Narrative Poetry* (Cambridge, 1986), ch. 6 ('The Storyteller and his Material: *Troilus and Criseyde*').

—— 'Chaucer's Narrator: *Troilus and Criseyde* and the *Canterbury Tales*', in Piero Boitani and Jill Mann (eds.), *The Cambridge Chaucer Companion* (Cambridge 1986), 213–26.

MIESZKOWSKI, GRETCHEN, 'The Reputation of Criseyde 1155–1500', *Transactions of the Connecticut Academy of Arts and Sciences*, 43 (1971), 71–153.

—— ' "Pandras" in Deschamp's Ballade for Chaucer', *Chaucer Review*, 9 (1975), 327–36.

—— 'R. K. Gordon and the *Troilus and Criseyde* Story', *Chaucer Review*, 15 (1980), 127–37.

—— 'Chaucer's Pandarus and Jean Brasdefer's Houdée', *Chaucer Review*, 20 (1985), 40–60.

MILLER, RALPH N., 'Pandarus and Procne', *Studies in Medieval Culture*, 7 (1964), 65–68.

MILLETT, BELLA, 'Chaucer, Lollius, and the Medieval Theory of Authorship', in Paul Strohm and Thomas J. Heffernan (eds.), *Studies in the Age of Chaucer Proceedings i. 1984: Reconstructing Chaucer* (Knoxville, Tenn., 1985), 93–103.

MILOWICKI, EDWARD J., 'Characterization in *Troilus and Criseyde*: Some Relationships Centered on Hope', *Canadian Review of Comparative Literature*, 11 (1984), 12–24.

MINNIS, A. J. 'Aspects of the Medieval French and English Traditions of the *De Consolatione Philosophiae*', in M. Gibson (ed.), *Boethius: His Life, Thought, and Influence* (Oxford, 1981), 312–61.

—— *Chaucer and Pagan Antiquity* (Cambridge, 1982).

MISKIMIN, ALICE S., *The Renaissance Chaucer* (New Haven, Conn., 1975), ch.7.

MIZENER, A., 'Character and Action in the Case of Criseyde', *PMLA* 54 (1939), 65–81.

MOGAN, J. J., *Chaucer and the Theme of Mutability* (The Hague, 1969).

MORGAN, GERALD, 'Natural and Rational Love in Medieval Literature', *Yearbook of English Studies*, 7 (1977), 43–52.

—— 'The Significance of the Aubades in *Troilus and Criseyde*', *Yearbook of English Studies*, 9 (1979), 221–35.

—— 'The Ending of *Troilus and Criseyde*', *Modern Language Review*, 77 (1982), 257–71.

8

MORGAN, GERALD, 'The Freedom of the Lovers in *Troilus and Criseyde*', in John Scattergood (ed.), *Literature and Learning in Medieval and Renaissance England: Essays Presented to Fitzroy Pyle* (Dublin, 1984), 59–102.

MUDRICK, MARVIN, 'Chaucer's Nightingales', *Hudson Review*, 10 (1957), 88–95.

MUSCATINE, CHARLES, 'The Feigned Illness in Chaucer's *Troilus and Criseyde*', *Modern Language Notes*, 63 (1948), 372–7.

—— *Chaucer and the French Tradition* (Berkeley, Calif. 1957), ch. V.

NAGARAJAN, S., 'The Conclusion to Chaucer's *Troilus and Criseyde*', *Essays in Criticism*, 13 (1963), 1–8.

NATALI, GIULIA, 'A Lyrical Version: Boccaccio's *Filostrato*' in Piero Boitani (ed.), *The European Tragedy of Troilus* (Oxford, 1989), 49–73.

NEUMANN, FRITZ-WILHELM, *Chaucer: Symbole der Initiation im Troilus-Roman* (Bonn, 1977).

NEWMAN, BARBARA, ' "Feynede Loves", Feigned Lore, and Faith in Trouthe', in Stephen A. Barney (ed.), *Chaucer's 'Troilus': Essays in Criticism* (London, 1980), 257–75.

NOLAN, BARBARA, *Chaucer and the Tradition of the Roman Antique* (Cambridge, 1992).

NORTH, J. D., 'Kalenderes Enlumyned Ben They: Some Astronomical Themes in Chaucer', *Review of English Studies*, 20 (1969), 129–54, 257–83, 418–44.

—— *Chaucer's Universe* (Oxford, 1988).

NORTON-SMITH, JOHN, *Geoffrey Chaucer* (London, 1974), ch. 6 ('The Book of Troilus').

O'CONNOR, JOHN J., 'The Astronomical Dating of Chaucer's *Troilus*', *JEGP* 55 (1956), 556–62.

OSBERG, RICHARD H., 'Between the Motion and the Act: Intentions and Ends in Chaucer's *Troilus*', *English Literary History*, 48 (1981), 257–70.

OWEN, CHARLES A., jun, 'Chaucer's Method of Composition', *Modern Language Notes*, 72 (1957), 164–5.

—— 'The Significance of Chaucer's Revisions of *Troilus and Criseyde*', *Modern Philology*, 55 (1957–8), 1–5; repr. in Richard J. Schoeck and Jerome Taylor (eds.) *Chaucer Criticism. ii. Troilus and Criseyde and the Minor Poems* (Notre Dame, Ind., 1961), 160–6.

—— 'The Significance of a Day in *Troilus and Criseyde*', *Mediaeval Studies*, 22 (1960), 366–70.

—— 'The Problem of Free Will in Chaucer's Narratives', *Philological Quarterly*, 46 (1967), 433–56.

—— 'Mimetic Form in the Central Love Scene of *Troilus and Criseyde*', *Modern Philology*, 67 (1969), 125–32.

—— 'Minor Changes in Chaucer's *Troilus and Criseyde*', in Beryl Rowland (ed.), *Chaucer and Middle English Studies in Honor of Rossell Hope Robbins* (London, 1974), 303–19.

—— '*Troilus and Criseyde*: The Question of Chaucer's Revisions', *Studies in the Age of Chaucer*, 9 (1987), 155–72.

PARKES, M. B., 'Palaeographical Description and Commentary', in *Troilus and Criseyde: A Facsimile of Corpus Christi College Cambridge MS 61*, eds. M. B. Parkes and E. Salter (Cambridge, 1977).

PATCH, HOWARD R., 'Troilus on Predestination', *JEGP* 17 (1918), 399–422.

—— 'Troilus on Determinism', *Speculum*, 6 (1931), 225–43; repr. in Richard J. Schoeck and Jerome Taylor (eds.), *Chaucer Criticism, ii. Troilus and Criseyde and the Minor Poems* (Notre Dame. Ind., 1961), 71–85.

PATTERSON, LEE W., 'Ambiguity and Interpretation: A Fifteenth-Century Reading of *Troilus and Criseyde*', *Speculum*, 54 (1979), 297–330; repr. in his *Negotiating the Past* (Madison, 1987).

—— *Chaucer and the Subject of History* (London, 1991), ch. 2.

PAYNE, F. A., *Chaucer and Menippean Satire* (Madison, 1981).

PAYNE, ROBERT O., *The Key of Remembrance* (New Haven, Conn., 1963).

PEARSALL, DEREK, 'The *Troilus* Frontispiece and Chaucer's Audience', *Yearbook of English Studies*, 7 (1977), 68–74.

—— 'Criseyde's Choices', in John V. Fleming and Thomas J. Heffernan (eds.), *Studies in the Age of Chaucer Proceedings, ii. 1986* (Knoxville, Tenn., 1987), 17–29.

PECK, RUSSELL A., 'Numerology and Chaucer's *Troilus and Criseyde*', *Mosaic*, 5 (1972), 1–29.

PRATT, ROBERT A., 'Chaucer's Use of the *Teseida*', *PMLA* 62 (1947), 598–621.

—— 'A Note on Chaucer's Lollius', *Modern Language Notes*, 65 (1950), 183–7.

—— 'Chaucer and *Le Roman de Troyle et de Criseida*', *Studies in Philology*, 53 (1956), 509–39.

PRAZ, MARIO, 'Chaucer and the Great Italian Writers of the Trecento', *The Monthly Criterion* (1927), 18–39, 131–57, 238–42; repr. in *The Flaming Heart* (New York, 1958).

PROVOST, WILLIAM, *The Structure of Chaucer's Troilus and Criseyde* (Anglistica, 20; Copenhagen, 1974).

REICHL, KARL, 'Chaucer's *Troilus*: Philosophy and Language', in Piero Boitani (ed.), *The European Tragedy of Troilus* (Oxford, 1989), 133–52.

REISS, EDMUND, 'Troilus and the Failure of Understanding', *Modern Language Quarterly*, 29 (1968), 131–44.

RENOIR, ALAIN, 'Thebes, Troy, Criseyde, and Pandarus: An Instance of Chaucerian Irony', *Studia neophilologica*, 32 (1960), 14–17.

—— 'Criseyde's Two Half Lovers', *Orbis litterarum*, 16 (1961), 239–55.

—— 'Bayard and Troilus: Chaucerian Non-Paradox in the Reader', *Orbis litterarum*, 36 (1981), 116–40.

ROBERTSON, D. W., jun., 'Chaucerian Tragedy', *English Literary History*, 19 (1952), 1–37; repr. in Richard J. Schoeck and Jerome Taylor (eds.), *Chaucer Criticism, ii. Troilus and Criseyde and the Minor Poems* (Notre Dame, Ind., 1961), 86–121.

—— *A Preface to Chaucer* (Princeton, NJ, 1962).

—— 'The Probable Date and Purpose of Chaucer's *Troilus*', *Medievalia et Humanistica*, 13 (1985), 143–71.

ROBINSON, F. N. (ed.), *The Works of Geoffrey Chaucer* (2nd edn.; London, 1957).

ROBINSON, IAN, *Chaucer and the English Tradition* (Cambridge, 1972), ch. 5 ('Chaucer's Great Failure: *Troilus and Criseyde*').

ROGERS, H. L., 'The Beginning (and Ending) of Chaucer's *Troilus and Criseyde*', in Anthony Stephens *et al.* (eds.), *Festschrift for Ralph Farrell* (Bern, 1977), 185–200.

ROLLINS, HYDER E., 'The Troilus–Cressida Story from Chaucer to Shakespeare', *PMLA* 32 (1917), 383–429.

ROONEY, ANNE, *Geoffrey Chaucer* (Bristol, 1989), ch. 3.

ROOT, ROBERT K., *The Manuscripts of Chaucer's Troilus* (Chaucer Society, 1st ser., 98; London, 1914).

—— *The Textual Tradition of Chaucer's Troilus* (Chaucer Society, 1st ser., 99; London, 1916).

—— 'Chaucer's Dares', *Modern Philology*, 15 (1917), 1–22.

—— *The Poetry of Chaucer* (2nd rev. edn.; Boston, Mass., 1922), ch. VI.

—— (ed.), *The Book of Troilus and Criseyde* (Princeton, NJ, 1926).

—— and RUSSELL, HENRY N., 'A Planetary Date for Chaucer's *Troilus*', *PMLA* 39 (1924), 48–63.

ROSS, THOMAS W., '*Troilus and Criseyde*, II, 582–587: A Note', *Chaucer Review*, 5 (1970), 137–9.

ROWE, DONALD W., *O Love O Charite! Contraries Harmonized in Chaucer's 'Troilus'* (Carbondale, Ill., 1976).

ROWLAND, BERYL, 'Pandarus and the Fate of Tantalus', *Orbis litterarum*, 24 (1969), 3–16.

—— 'Chaucer's Speaking Voice and its Effect on his Listeners' Perception of Criseyde', *English Studies in Canada*, 7 (1981), 129–40.

RUDAT, WOLFGANG E. H., 'Chaucer's *Troilus and Criseyde*: Narrator–Reader Complicity', *American Imago*, 40 (1983), 103–13.

RUGGIERS, PAUL G., 'Notes towards a Theory of Tragedy in Chaucer', *Chaucer Review*, 8 (1973), 88–99; repr. in W. Erzgräber (ed.), *Geoffrey Chaucer* (Darmstadt, 1983), 396–408.

RUSSELL, NICHOLAS, 'Characters and Crowds in Chaucer's *Troilus*', *Notes and Queries*, 13 (1966), 50–2.

RUTHERFORD, CHARLES S., 'Pandarus as Lover: "A Joly Wo" or "Loves Shotes Keene"?' *Annuale mediaevale*, 13 (1972), 5–13.

—— 'Troilus' Farewell to Criseyde: The Idealist as Clairvoyant and Rhetorician', *Papers on Language and Literature*, 17 (1981), 245–54.

SADLEK, GREGORY M., 'To Wait or to Act? *Troilus* II, 954', *Chaucer Review*, 17 (1982), 62–4.

SAINTONGE, CONSTANCE, 'In Defense of Criseyde', *Modern Language Quarterly*, 15 (1954), 312–20.

SALEMI, JOSEPH S., 'Playful Fortune and Chaucer's Criseyde', *Chaucer Review*, 15 (1981), 209–23.

SALTER, ELIZABETH, '*Troilus and Criseyde*: A Reconsideration', in J. Lawlor (ed.), *Patterns of Love and Courtesy* (London, 1966), 86–106.

—— 'The *Troilus* Frontispiece', in *Troilus and Criseyde: A Facsimile of Corpus Christi College Cambridge MS 61*, eds. M. B. Parkes and E. Salter (Cambridge, 1977); repr. in her *English and International: Studies in the Literature, Art and Patronage of Medieval England*, eds. D. Pearsall and N. Zeeman (Cambridge, 1988).

—— '*Troilus and Criseyde*: Poet and Narrator', in Mary J. Carruthers and Elizabeth D. Kirk (eds.), *Acts of Interpretation: The Text in its Contexts 700–1600* (Norman, Okla. 1982), 281–91.

—— and PEARSALL, DEREK, 'Pictorial Illustration of Late Medieval Poetic Texts: The Role of the Frontispiece or Prefatory Picture', in F. G. Andersen *et al.* (eds.), *Medieval Iconography and Narrative: A Symposium* (Odense, 1980), 100–23.

SALU, MARY (ed.), *Essays on 'Troilus and Criseyde'* (Cambridge, 1979).

SAMS, HENRY W., 'The Dual Time-Scheme in Chaucer's *Troilus*', *Modern Language Notes*, 56 (1941), 94–100, repr. in Richard J. Schoeck and Jerome Taylor (eds.), *Chaucer Criticism*, ii. *Troilus and Criseyde and the Minor Poems* (Notre Dame, Ind., 1961), 180–5.

SAVILLE, JONATHAN, *The Medieval Erotic Alba: Structure as Meaning* (New York, 1972).

SCATTERGOOD, JOHN, 'The "Bisynesse" of Love in Chaucer's Dawn-Songs', *Essays in Criticism*, 37 (1987), 110–20.

SCHELP, HANSPETER, 'Die Tradition der *Alba* und die Morgenszene in Chaucers *Troilus and Criseyde* III, 1415 ff.', *Germanisch-Romanische Monatsschrift*, 46 (1965), 251–61.

SCHIBANOFF, SUSAN, 'Prudence and Artificial Memory in Chaucer's *Troilus*', *English Literary History*, 42 (1975), 507–17.

—— 'Argus and Argyve: Etymology and Characterization in Chaucer's *Troilus*', *Speculum*, 51 (1976), 647–58.

—— 'Criseyde's "Impossible" *Aubes*', *JEGP* 76 (1977), 326–33.

SCHLESS, HOWARD, 'Chaucer and Dante', in D. Bethurum (ed.), *Critical Approaches to Medieval Literature* (New York, 1960), 134–54.

—— 'Transformations: Chaucer's Use of Italian', in Derek Brewer (ed.), *Geoffrey Chaucer: Writers and their Background* (London, 1974), 184–223.

—— *Chaucer and Dante: A Revaluation* (Norman, Okla., 1984).

SCHMIDT, DIETER, 'Das Anredepronomen in Chaucers *Troilus and Criseyde*', *Archiv*, 212 (1975), 120–4.

SCHOECK, RICHARD J., and TAYLOR, JEROME, (eds.), *Chaucer Criticism*, ii. *Troilus and Criseyde and the Minor Poems* (Notre Dame, Ind., 1961).

SCHROEDER, PETER R., 'Hidden Depths: Dialogue and Characterization in Chaucer and Malory', *PMLA* 98 (1983), 374–87.

SCHUMAN, SAMUEL, 'The Circle of Nature: Patterns of Imagery in Chaucer's *Troilus and Criseyde*', *Chaucer Review*, 10 (1975), 99–112.

SCOTT, FORREST S., 'The Seventh Sphere: A Note on *Troilus and Criseyde*', *Modern Language Review*, 51 (1956), 2–5.

SHANLEY, JAMES LYNDON, 'The *Troilus* and Christian Love', *English Literary History*, 6 (1939), 271–81, repr. in Richard J. Schoeck and Jerome Taylor (eds.), *Chaucer Criticism*, ii. *Troilus and Criseyde and the Minor Poems* (Notre Dame, Ind., 1961), 136–46.

SHANNON, E. F., *Chaucer and the Roman Poets* (Cambridge, Mass., 1929).

SHARROCK, ROGER, 'Second Thoughts: C. S. Lewis on Chaucer's *Troilus*', *Essays in Criticism*, 8 (1958), 123–37.

SHEPHERD, G. T., '*Troilus and Criseyde*', in D. S. Brewer (ed.), *Chaucer and Chaucerians* (London, 1966), 65–87.

——'Religion and Philosophy in Chaucer', in Derek Brewer (ed.), *Geoffrey Chaucer: Writers and their Background* (London, 1974), 262–89.

SHOAF, R. ALLEN, 'Dante's *Commedia* and Chaucer's Theory of Mediation: A Preliminary Sketch', in Donald M. Rose (ed.), *New Perspectives in Chaucer Criticism* (Norman, Okla., 1981), 83–103.

——(ed.), *Troilus and Criseyde* (East Lansing, Mich., 1989).

SIMS, DAVID, 'An Essay at the Logic of *Troilus and Criseyde*', *Cambridge Quarterly*, 4 (1969), 125–49.

SKLUTE, LARRY, *Virtue of Necessity: Inconclusiveness and Narrative Form in Chaucer's Poetry* (Columbus, 1984), ch. 4.

SLAUGHTER, EUGENE E., 'Chaucer's Pandarus: Virtuous Uncle and Friend', *JEGP* 48 (1949), 186–95.

——'Love and Grace in Chaucer's *Troilus*', in *Essays in Honor of Walter Clyde Curry* (Nashville, Tenn., 1954), 61–76.

SLOCUM, SALLY K., 'How Old is Chaucer's Pandarus?' *Philological Quarterly*, 58 (1979), 16–25.

——'Criseyde among the Greeks', *Neuphilologische Mitteilungen*, 87 (1986), 365–74.

SMYSER, H. M., 'The Domestic Background of *Troilus and Criseyde*', *Speculum*, 31 (1956), 297–315.

SOMMER, GEORGE J., 'Chaucer and the Muse of History: A Presumption of Objectivity in *Troilus and Criseyde*', *Cithara*, 23: 1 (1983), 38–47.

SPEARING, A. C., 'Chaucer as Novelist', in his *Criticism and Medieval Poetry* (2nd edn. London, 1972), ch. 6.

——*Chaucer: 'Troilus and Criseyde'* (London, 1976).

——*Medieval to Renaissance in English Poetry* (Cambridge, 1985), ch. 2 ('Chaucer's Classical Romances').

——*Readings in Medieval Poetry* (Cambridge, 1987), ch. 5 ('Narrative Closure: The End of *Troilus and Criseyde*').

——'*Troilus and Criseyde*: The Illusion of Allusion', *Exemplaria*, 2 (1990), 263–77.

SPEIRS, JOHN, *Chaucer the Maker* (London, 1951), ch. 2.

SPURGEON, C. F. E. (ed.), *Five Hundred Years of Chaucer Criticism and Allusion* (3 vols.; Cambridge, 1925).

STANLEY, E. G., 'Stanza and Ictus: Chaucer's Emphasis in *Troilus and Criseyde*', in A. Esch (ed.), *Chaucer und Seine Zeit* (Tübingen, 1968), 123–48.

—— 'About Troilus', *Essays and Studies*, 29 (1976), 84–106.

STEADMAN, JOHN M., 'The Age of Troilus', *Modern Language Notes*, 72 (1957), 89–90.

—— *Disembodied Laughter: Troilus and the Apotheosis Tradition: A Reexamination of Narrative and Thematic Contexts* (Berkeley, Calif., 1972).

STEVENS, JOHN, *Medieval Romance* (London, 1973).

STEVENS, MARTIN, 'The Winds of Fortune in the *Troilus*', *Chaucer Review*, 13 (1979), 285–307.

—— 'The Double Structure of Chaucer's *Troilus and Criseyde*', in Saul N. Brody and Harold Schecter (eds.), *CUNY English Forum*, 1 (New York, 1985).

STILLER, NIKKI, 'Civilization and its Ambivalence: Chaucer's *Troilus and Criseyde*', *Journal of Evolutionary Psychology*, 6 (1985), 212–23.

STOKES, MYRA, 'Recurring Rhymes in *Troilus and Criseyde*', *Studia neophilologica*, 52 (1980), 287–97.

—— 'The Moon in Leo in Book V of *Troilus and Criseyde*', *Chaucer Review*, 17 (1982), 116–29.

—— '"Wordes White": Disingenuity in *Troilus and Criseyde*', *English Studies*, 64 (1983), 18–29.

STORM, MELVIN, 'Troilus, Mars and Late Medieval Chivalry', *Journal of Medieval and Renaissance Studies*, 12 (1982), 45–65.

STROHM, PAUL, 'Storie, Spelle, Geste, Romaunce, Tragedie: Generic Distinctions in the Middle English Troy Narratives', *Speculum*, 46 (1971), 348–59.

—— *Social Chaucer* (Cambridge, Mass., 1989).

STROUD, THEODORE A., 'Boethius' influence on Chaucer's *Troilus*', *Modern Philology*, 49 (1951–2), 1–9; repr. in Richard J. Schoeck and Jerome Taylor (eds.), *Chaucer Criticism*, ii. *Troilus and Criseyde and the Minor Poems* (Notre Dame, Ind., 1961), 122–35.

—— 'Chaucer's Structural Balancing of *Troilus* and *Knight's Tale*', *Annuale mediaevale*, 21 (1981), 31–45.

SUNDWALL, MCKAY, 'Deiphobus and Helen: A Tantalizing Hint', *Modern Philology*, 73 (1975), 151–6.

—— 'Criseyde's Rein', *Chaucer Review*, 11 (1976), 156–63.

TAKADA, YASUNARI, 'On the Consummation of Troilus' Love: Chaucer's *Troilus and Criseyde*, III, 1247–60', *Studies in Language and Culture*, 8 (1982), 103–29.

TATLOCK, J. S. P., *The Development and Chronology of Chaucer's Works* (Chaucer Society, 2nd ser. 37; London, 1907).

—— 'The Epilog of Chaucer's *Troilus*', *Modern Philology*, 18 (1920–1), 625–59.

—— 'The Date of the *Troilus* . . .' *Modern Language Notes*, 50 (1935), 277–96.

—— 'The People in Chaucer's *Troilus*', *PMLA* 56 (1941), 85–104.

TAVORMINA, M. TERESA, 'The Moon in Leo: What Chaucer Really Did to *Il Filostrato's* Calendar', *Ball State University Forum*, 22: 1 (1981), 14–19.

TAYLOR, ANN M., 'Troilus' Rhetorical Failure (4:1440–1526)', *Papers on Language and Literature*, 15 (1979), 357–69.

—— 'A Scriptural Echo in the Trojan Parliament of *Troilus and Criseyde*', *Nottingham Medieval Studies*, 24 (1980), 51–6.

TAYLOR, DAVIS, 'The Terms of Love: A Study of Troilus's Style', *Speculum*, 51 (1976), 69–90, rev. in Stephen A. Barney (ed.), *Chaucer's 'Troilus': Essays in Criticism* (London, 1980), 231–56.

TAYLOR, KARLA, 'Proverbs and the Authentication of Convention in *Troilus and Criseyde*', in Stephen A. Barney (ed.), *Chaucer's 'Troilus': Essays in Criticism* (London, 1980), 277–96.

—— 'A Text and its Afterlife: Dante and Chaucer', *Comparative Literature*, 35 (1983), 1–20.

—— *Chaucer Reads 'The Divine Comedy'* (Stanford, Calif., 1989).

THOMPSON, ANN, *Shakespeare's Chaucer* (Liverpool, 1978).

THOMSON, PATRICIA, 'The "Canticus Troili": Chaucer and Petrarch', *Comparative Literature*, 11 (1959), 313–28.

THUNDY, ZACHARIAS P., 'Chaucer's "Corones Tweyne" and Matheolus', *Neuphilologische Mitteilungen*, 86 (1985), 343–7.

TORTI, ANNA, *The Glass of Form: Mirroring Structures from Chaucer to Skelton* (Cambridge, 1991), ch. 1.

UTLEY, FRANCIS LEE, 'Scene Division in Chaucer's *Troilus and Criseyde*', in McEdward Leach (ed.), *Studies in Medieval Literature* (Philadelphia, Pa., (1961), 109–38.

—— 'Chaucer's Troilus and St. Paul's Charity', in Beryl Rowland (ed.), *Chaucer and Middle English Studies in Honor of Rossell Hope Robbins* (London, 1974), 272–87.

VAN, THOMAS A., 'Imprisoning and Ensnarement in *Troilus* and *The Knight's Tale*', *Papers on Language and Literature*, 7 (1971), 3–12.

VANCE, EUGENE, 'Mervelous Signals: Poetics, Sign Theory, and Politics in Chaucer's *Troilus*', *New Literary History*, 10 (1979), 293–337.

—— 'Chaucer, Spenser, and the Ideology of Translation', *Canadian Review of Comparative Literature*, 8 (1981), 217–38.

WACK, MARY F., 'Lovesickness in *Troilus*', *Pacific Coast Philology*, 19 (1984), 55–61.

—— 'Pandarus, Poetry, and Healing', in John V. Fleming and Thomas J. Heffernan (eds.), *Studies in the Age of Chaucer Proceedings*, ii. *1986* (Knoxville, Tenn., 1987), 127–33.

WAGER, WILLIS J., '"Fleshly Love" in Chaucer's *Troilus*', *Modern Language Review*, 34 (1939), 62–6.

WALCUTT, CHARLES CHILD, 'The Pronoun of Address in *Troilus and Criseyde*', *Philological Quarterly*, 14 (1935), 282–7.

WALKER, IAN C., 'Chaucer and *Il Filostrato*', *English Studies*, 49 (1968), 318–26.

WALLACE, DAVID, 'Chaucer's "Ambages"', *American Notes and Queries*, 23 (1984), 1–4.

—— *Chaucer and the Early Writings of Boccaccio* (Cambridge, 1985).

—— 'Geoffrey of Vinsauf, Geoffrey Chaucer and Boccaccio's "Rakel Hond"', *Neuphilologische Mitteilungen*, 88 (1987), 27–30.

WASWO, RICHARD, 'The Narrator of *Troilus and Criseyde*', *English Literary History*, 50 (1983), 1–25.

WATTS, ANN CHALMERS, 'Chaucerian Selves—Especially Two Serious Ones', *Chaucer Review*, 4 (1970), 229–41.

WEITZENHOFFER, KENNETH, 'Chaucer, Two Planets, and the Moon', *Sky and Telescope*, 69 (1985), 278–81.

WENTERSDORF, KARL P., 'Some Observations on the Concept of Clandestine Marriage in *Troilus and Criseyde*', *Chaucer Review*, 15 (1980), 101–26.

WENZEL, SIEGFRIED, 'Chaucer's Troilus of Book IV', *PMLA* 79 (1964), 542–7.

WETHERBEE, WINTHROP, 'The Descent from Bliss: *Troilus*, III, 1310–1582', in Stephen A. Barney (ed.), *Chaucer's 'Troilus': Essays in Criticism* (London, 1980), 297–317.

—— *Chaucer and the Poets: An Essay on 'Troilus and Criseyde'* (Ithaca, NY, 1984).

—— ' "Per te poeta fui, per te cristiano": Dante, Statius, and the Narrator of Chaucer's *Troilus*', in Lois Ebin (ed.), *Vernacular Poetics in the Middle Ages* (Kalamazoo, 1984), 153–76.

WHEELER, BONNIE, 'Dante, Chaucer, and the Ending of *Troilus and Criseyde*', *Philological Quarterly*, 61 (1982), 105–23.

WHITMAN, FRANK H., '*Troilus and Criseyde* and Chaucer's Dedications to Gower', *TSL* 18 (1973), 1–11.

WILKINS, E. H., 'Cantus Troili', *English Literary History*, 16 (1949), 167–73.

WILLIAMS, George G., 'The *Troilus and Criseyde* Frontispiece Again', *Modern Language Review*, 57 (1962), 173–8.

—— *A New View of Chaucer* (Durham, NC, 1965), ch. IV ('Who were Troilus, Criseyde, and Pandarus?'), and pp. 175–95 ('Notes on *Troilus and Criseyde*').

WILSON, DOUGLAS B., 'The Commerce of Desire: Freudian Narcissism in Chaucer's *Troilus and Criseyde* and Shakespeare's *Troilus and Cressida*', *English Language Notes*, 21 (1983), 11–22.

WIMSATT, JAMES I., 'Guillaume de Machaut and Chaucer's *Troilus and Criseyde*', *Medium Ævum*, 45 (1976), 277–93.

—— 'Medieval and Modern in Chaucer's *Troilus and Criseyde*', *PMLA* 92 (1977), 203–16.

—— 'Realism in *Troilus and Criseyde* and the *Roman de la Rose*', in Mary Salu (ed.), *Essays on 'Troilus and Criseyde'* (Cambridge 1979), 43–56.

—— 'The French Lyric Element in *Troilus and Criseyde*', *Yearbook of English Studies*, 15 (1985), 18–32.

WINDEATT, BARRY, 'The "Paynted Proces": Italian to English in Chaucer's *Troilus*', *English Miscellany*, 26–7 (1977–8), 79–103.

—— ' "Love That Oughte Ben Secree" in Chaucer's *Troilus*', *Chaucer Review*, 14 (1979), 116–31.

—— 'The Scribes as Chaucer's Early Critics', *Studies in the Age of Chaucer*, 1 (1979), 119–41.

—— 'The Text of the *Troilus*', in Mary Salu (ed.), *Essays on 'Troilus and Criseyde'* (Cambridge, 1979), 1–22.

—— 'Chaucer and the *Filostrato*', in Piero Boitani (ed.), *Chaucer and the Italian Trecento* (Cambridge, 1983), 163–83.

WINDEATT, BARRY, (ed.), *Troilus and Criseyde: A New Edition of 'The Book of Troilus'* (London, 1984; 2nd edn., 1990).

—— '*Troilus* and the Disenchantment of Romance', in D. Brewer (ed.), *Studies in Medieval English Romances* (Cambridge, 1988), 129–47.

—— 'Classical and Medieval Elements in Chaucer's *Troilus*', in Piero Boitani (ed.), *The European Tragedy of Troilus* (Oxford, 1989), 111–31.

WITLIEB, BERNARD L., 'Chaucer and a French Story of Thebes', *English Language Notes*, 11 (1973), 5–9.

WISE, B. A., *The Influence of Statius on Chaucer* (Baltimore, Md., 1911).

WOOD, CHAUNCEY, *Chaucer and the Country of the Stars* (Princeton, NJ, 1970).

—— 'On Translating Chaucer's *Troilus and Criseyde*, Book III, Lines 12–14', *English Language Notes*, 11 (1973), 9–14.

—— *The Elements of Chaucer's 'Troilus'* (Durham, NC, 1984).

WOODS, MARJORIE CURRY, 'Chaucer the Rhetorician: Criseyde and her Family', *Chaucer Review*, 20 (1985), 28–39.

YEAGER, R. F., ' "O Moral Gower": Chaucer's Dedication of *Troilus and Criseyde*', *Chaucer Review*, 19 (1984), 87–99.

YEARWOOD, STEPHENIE, 'The Rhetoric of Narrative Rendering in Chaucer's *Troilus*', *Chaucer Review*, 12 (1977), 27–37.

YOUNG, KARL, *The Origin and Development of the Story of Troilus and Criseyde* (Chaucer Society, 2nd ser., 40; London, 1908).

—— 'Chaucer's *Troilus and Criseyde* as Romance', *PMLA* 53 (1938), 38–63.

ZIMBARDO, ROSE A., 'Creator and Created: The Generic Perspective of Chaucer's *Troilus and Criseyde*', *Chaucer Review*, 11 (1977), 283–98.

INDEX

Achilles 69, 72, 73, 74, 75, 76, 77, 78, 79, 93–4, 96, 110, 111, 143, 190, 237, 240, 302, 307, 324, 363
Adam 160, 311
Adonis 43, 111, 123, 225, 256
Aelred of Rievaulx, *De spirituali amicitia* 228
Aeneas 73, 75, 141, 143, 150, 156, 374
Agamemnon 75, 78, 95
Aglauros 110, 256
Alain of Lille, *De planctu naturae* 42, 117, 173
Alcestis 237, 305, 331, 360
Alcmena 123
Alexander the Great 77
Amete 114
Amnon 220
Amphiaurus 122, 124, 190, 248
Andromache 74
Anne of Bohemia, Queen of England 5, 10
Antenor 7, 64, 67, 73, 74, 75, 77, 82, 93, 188, 247, 249, 251
Antigone 58, 60, 118, 121, 123, 165, 168, 189, 201, 216, 221–2, 229, 233, 247, 252, 265, 268, 339
Apollo 78, 110, 111, 123, 143, 174, 221, 225, 234, 255, 258
Apollodorus, *Epitoma* 143, 144
A poore Knight his Pallace of priuate pleasures 372
Argia 123, 125
Argus 125
Argyve 123
Aristotle 155, 158
    Poetics 155
Arthur 149
Ascaphilus 111, 256
Astraea 224
Astrophil 379
Athamas 111, 123, 127, 129, 131, 249
Atropos 43
Averroes 155, 161
Azalais d'Altier 96

ballads 149
    'A new Dialogue betweene Troylus and Cressida' 375
    'Cressus was the fairest of Troye' 372, 373
    'If Cressed in her gadding moode' 375
    'The panges of Loue and louers fittes' 372, 373
    'Whan Troylus dwelt in Troy towne' 373
Bannatyne MS. 368, 369, 373
Beatrice 130, 131, 134
Beaumont, Francis 379, 381
Beauvau, Pierre or Louis de, *Le Roman de Troïlus et de Criseida* 71–2
Benedict 210
Benoît de Sainte-Maure 40, 44, 45, 49, 66, 69, 70, 71, 90–1, 92, 93, 94, 96, 146, 188, 202, 244, 246, 247, 251, 258, 280, 294, 371
    Le Roman de Troie 41, 44, 77–90, 91, 93, 94, 96, 113, 122, 125, 139, 146–8, 152, 203, 204, 216, 294, 296
Bernardus Silvestris 174, 224
    Commentum super sex libros Eneidos Virgilii 228
Beves of Hamtoun 381
Bible:
    Ecclesiastes 346
    Matthew 49
    Samuel 220
Boccaccio, Giovanni 4, 9, 37, 38, 40, 46, 90, 125, 140, 141, 149, 152, 154, 156, 258
    De casibus virorum illustrium 154, 157
    De claris mulieribus 154, 365, 367
    Decameron 95, 156
    Filocolo 47, 50, 125, 147
    Filostrato 4, 49, 50–72; characterization 268, 274, 279, 280, 289, 294–5, *see also* Diomede, Criseyde, Pandarus, Troilus; Chaucer's development of 20, 37, 41, 93, 96, 102, 104, 105, 107, 108,

Boccaccio, Giovanni (*cont.*)
111, 128, 162, 176, 223, 224, 236,
237, 240, 283, 297, 307, 316, 330,
332, 344–5; and complaint 166–7;
ending 302, 307, 308–9; and
genre 140, 142, 143, 145–6, 147,
150, 164, 240; imagery 344–5; love
in 134, 173, 189, 212, 216, 218,
226, 229, 242, 244, 286, 316, 330;
numerical hyperbole 332, 337;
originality of 39, 45, 139, 251, 276,
280; proem 172, 345; relation to
sources 40, 46, 84, 110, 130, 143,
148, 151, 152, 246; setting 192,
197–8; sight and seeing in 345;
structure 102, 145–6, 152, 160,
180–1, 182, 184, 188, 191, 204;
time in 152, 201, 203, 204; and
Troilus variants 21, 22, 26–7,
28–32, 34, 36; versification 165,
326
Teseida 40, 46, 107, 112, 121–5, 132,
142, 156, 179, 202, 209, 307
Boethius 9, 42, 45, 46, 49, 62, 140, 154,
160, 168, 182, 266
Consolation of Philosophy (*De
consolatione philosophiae*) 9, 46, 47,
48, 49, 50, 96–109, 111, 118, 157,
163, 164, 172–6, 187, 210–11,
223–4, 234, 238, 262, 263, 264,
302, 303, 310, 326–8, 341, 342
Bohun Psalters 15
Bonaventure, *In primum librum
Sententiarum* 178–9
Brasdefer, Jean, *Pamphile et Galatée* 172
Brembre, Nicholas 8, 10
Briseida, *see* Criseyde
Briseis 110, 111, 112
see also Criseyde
Brut 77
Brutus 7

Calchas 54, 55, 56, 64, 65, 67, 73, 74, 78,
83, 86, 91, 93, 95, 148, 188, 191, 247,
255, 258, 259, 283, 287, 323, 325, 341
Calliope 61, 98, 123, 126, 133, 140
Candace 94
Capaneus 124
Cassandra 42, 46, 54, 68–9, 70, 74, 110,
114, 122, 123, 142, 156, 190, 237,
250, 257, 258, 259, 268, 343
Cato 42, 46, 50

Caxton, William 369
Recuyell of the Historyes of Troy 363,
364, 373
Ceffi, Filippo 114
Cerberus 98, 111, 341
Chalmers 369
'Chance of the Dice' 365–6, 367
Chapman, George, *Sir Giles Goosecap* 377
Charles d'Orléans 364–5, 366
Chaucer, Geoffrey:
Against Women Unconstant 94
Anelida and Arcite 5, 40, 166, 368
Boece 5, 19, 22, 96–109, 154, 155,
210–11, 303, 310, 326, 327, 328,
342–3, 361
Book of the Duchess 5, 94, 118
Canterbury Tales 3, 5, 11, 138, 140,
171, 176, 379, 380
Clerk's Tale 48, 133
Complaint of Mars 5, 125, 166, 205
Complaint of Venus 355
Equatorie of the Planetes 211
Franklin's Tale 40, 170, 198, 211
House of Fame 5, 38, 141
Knight's Tale 5, 40, 122, 156, 198, 327,
380
Legend of Good Women 3, 5, 155
Merchant's Tale 166
Monk's Tale 5, 154, 155, 157
Nun's Priest's Tale 144, 380
Parliament of Fowls 5, 94, 156
*Prologue to the Legend of Good
Women* 3, 5, 114, 214, 360
Rectractions 4
Romaunt of the Rose 5, 114–18, 360
Second Nun's Tale 5
Sir Thopas 144
Squire's Tale 144, 207
'To Rosemounde' 35
Treatise on the Astrolabe 5, 211, 261
Troilus and Criseyde: audience 13–19;
characters, *see* Criseyde, Diomede,
Pandarus, Troilus; imagery
337–45; narration 16–19, 33–4, 41,
119, 162, 178–9, 182, 212–13,
231–2, 244, 264, 280–2, 303–4,
314–15, 331, 351–2;
versification 354–9
Wife of Bath's Tale 144, 380
'Words unto Adam, his owne
scriveyn' 4, 19
Chryseis 78, 111, 112
Cicero 144, 209

Claudian 38
Clio 123, 354
Common Conditions 375, 376
Compendious Book of Godly and Spiritual Songs 376
'Complaint of a diyng louer' 375
Conusaunce damours 375
Corinna 40
Cornelius Nepos 44
Cornish, *Story of Troylous and Pandor* 376
Crassus 61, 126
Criseyde 279–88
  in Benoît (as Briseida) 77–90, 94, 113, 123, 147–8, 188, 203–4, 216, 244, 246, 258, 295
  and Boethius 102–8, 157, 160
  and Dante 130–1
  in Dares (as Briseis) 73–4
  dreams 186, 200, 222, 251, 336
  in *Filostrato* 31–2, 55–72, 146, 149, 150, 152, 160, 166–7, 188, 216, 229, 242, 244, 246, 258, 268, 280, 283–4, 286
  in Guido (as Briseida) 90–4, 244, 258
  in Henryson 370–1
  in Joseph of Exeter (as Briseis) 75–6
  and Machaut 118–21
  and the moon 205–7
  and Ovid 109–14
  in Shakespeare 378
Cupid 62, 103, 212, 225, 255, 340

Dante 4, 9, 45, 46, 47, 48, 49, 104, 111, 125–37, 141, 155, 178, 187, 230, 249, 277, 306, 312, 327, 338
  Convivio 45, 133, 136
  De vulgari eloquentia 133, 134, 136, 155
  Divine Comedy 125–37
  Inferno 46, 47, 104, 125, 126–37, 201
  Paradiso 46, 125–37, 210, 211, 232, 233
  Purgatorio 126–37
Daphne 43, 111, 225, 256
Dardanus 79, 193
Dares Phrygius 38, 42, 44, 49, 55, 69, 77, 78, 79, 90, 139, 140, 143, 151, 305
  De excidio Troiae historia 44, 72–7
D'Auberee 170
Deiphebus 59, 113, 141–2, 188, 191, 192, 194, 199–200, 202, 225, 235, 238, 245, 247, 265, 268, 281, 292, 318

Dekker and Chettle, lost play of Troilus and Cressida 376, 377
Deloney, Thomas, *Strange Histories* 375
Deschamps, Eustache 109, 169, 361, 362
Devonshire MS 368, 369
Diana 43, 110
Dictys Cretensis 38, 44, 55, 78, 90, 139, 140, 151
  Ephemeridos belli Troiani libri 44, 72–7
Dido 141, 156, 365, 374
Diomede 122, 123, 174, 175, 188, 190, 192, 202–3, 215, 222, 230, 249, 257, 261, 264, 265–6, 294–8, 306, 325, 353
  in Benoît 78–90, 147, 152, 188, 216, 295, 296
  in Dares 73–5
  in *Filostrato* 67–72, 140, 148, 188, 203, 204, 268, 295, 297
  in Guido 90–4
  in Joseph of Exeter 75, 296
  in Shakespeare 378
Dione 110, 127, 128
Disce morus 368, 369
Douglas, Gavin, *The Palice of Honour* 365, 367
Dryden, John 379, 380
  Preface to the Fables 379, 380, 381
  Troilus and Cressida, or Truth Found Too Late 380, 381
Du prestre et d'Alison 170

Edward III, King of England 89–90
Elderton, William 372
Elyot, Sir Thomas, *Pasquill the Playne* 378, 381
Euclid 185
Europa 43, 111, 225, 256
Eurydice 65, 98, 100, 111, 175
Eve 365

Fates, The, 43, 263, 265
Florimont 147
Floris and Blancheflour 47, 146
Fortuna major 128, 206, 258
Fortune 56, 101–9, 132, 154–7, 160, 174, 175, 181, 253, 261–7, 280, 291
Foullechat, Jean 39, 49
Fowler, William, 'The Laste Epistle of Creseyd to Troyalus' 375, 376
Francesca da Rimini 104, 130
Froissart, Jean, *Paradys d'amours* 94
Fulgentius 238

Fulwood, William, *The Enimie of Idlenesse* 372
Furies, The, 42, 47, 49, 63, 98, 123, 128, 142, 186, 238, 321, 339

Galehout 130, 147
Gascoigne, George, *Posies* 374, 375
Gautier d'Arras, *Eracle* 170
Geoffrey of Monmouth 7, 38
  Historia regum Britanniae 7–8, 10
Geoffrey of Vinsauf 182
  Poetria nova 45, 49–50, 180, 183–4
Gest Hystoriale of the Destruction of Troy 362, 364
glosses 13, 42, 43, 49–50, 76, 123, 138, 161, 163, 165, 169, 256, 324
Godwin, William, *Life of Chaucer* 380–2
Gorgious Gallery of gallant Inuentions 372
Gower, John 8, 18, 39, 94–6, 310, 379
  Cinkantes balades 95
  Confessio amantis 8, 95–6, 361
  Mirour de l'omme 95
  Vox clamantis 8, 95–6
Grimald, Nicholas, *Troilus ex Chaucero* 376
Gude and Godlie Ballatis 375
Guido de Columnis 38, 40, 44, 45, 49, 66, 69, 96, 202, 244, 258, 280, 362–3, 371
  Historia destructionis Troiae 44, 83, 90–6, 139, 152, 362, 363
Guillaume de Lorris 115, 116
Guillaume de Palerne 146–7
Guinevere 130, 147
Guinizelli 45
Guy of Warwick 147, 151

Handful of Pleasant Delights 372, 373, 375
Harvey, Gabriel 379, 381
Hawes, Stephen, *Pastime of Pleasure* 366, 367
Hector 8, 41, 55, 67, 69, 74–9, 82, 89, 91, 92, 94, 132, 142, 149, 188, 189, 204, 238, 265, 268, 273, 278, 284, 307, 324, 329, 364, 381
Hecuba 74
Helen of Troy 73, 74, 78, 96, 110, 112, 113, 141–2, 156, 188, 189, 191, 194, 202, 250, 265, 268, 333, 365, 378
Helenus 74, 79
Hende, John 10
Henry II, King of England 77
Henry V, King of England 13

Henryson, Robert 302
  Testament of Cresseid 369–71, 379
Henslowe 376
Hercules 110, 237, 381
Hero 110
Hermann Alemannus 155, 161
Herse 43, 110, 111
Hesione 74, 77
Heywood, Thomas 374
  Apologie for Actors 374, 375
  Iron Age 376
  Troia Britanica 363
Higden, Ralph 153
Hippodamia 78
Hoccleve, Thomas 367
  Letter of Cupid 368
Homer 38, 44, 72, 125, 140, 141, 151, 306, 379
  Iliad 73, 78, 141
Horace 38, 39, 144, 379
  Ars Poetica 45, 111, 133, 181
  Epistles 38, 39
Horaste 103, 170, 190, 214, 281
Howard, Henry, Earl of Surrey 373, 375
Howard, Lady Elizabeth 366
Howell, Thomas 371, 372
Hymen 103

Ipomadon A 146
Iseult 156
Isidore of Seville 73
Isle of Ladies 367, 368
Istorietta Troiana 71
Ixion 98, 101, 111, 175, 342

James I, King of Scotland 367
James IV, King of Scotland 361
James VI and I, King of Great Britain, *Short Treatise on Verse* 379, 381
Janus 42, 290, 325
Jason 74, 78, 95
Jean de Meun 115, 117, 173
Jean, duc de Berry 15
Jocasta 141
Johannes Januensis, *Catholicon* 160–1
John de Briggis 50
John of Gaunt 7, 15
John of Ireland 361, 362
John of Salisbury:
  Metalogicon 45, 133
  Policraticus 39, 45
John of Wales 153

Johnson, Richard, *Crowne-Garland of Goulden Roses* 372
Jonadab 220
Joseph of Exeter 44, 45, 149, 249, 276, 294, 296
  Frigii Daretis Ilias 41, 42, 46, 75–7, 139, 142
Josephus 38
Jove 43, 100, 103, 111, 124, 224, 225, 265, 328
Juno 123, 127, 128, 129, 174, 175, 255
Jupiter 6, 10, 43, 200, 207, 255, 260
Juvenal 45

Keats, John, *Endymion* 381
Kingis Quair 367
Knighton, Henry 8, 10
Kynaston, Sir Francis 370, 380, 381

Lachesis 43
Lancelot 130, 144, 147, 151
Laomedon 110
Latona 110
Laud Troy Book 362, 363
Lavinia 150
Leander 110
Lefevre, Raoul, *Recueil des Hystoires Troyennes* 364
Liber albus 8, 10
Libro della storia di Troia 71
Limbourg brothers 15
Lollius 37–50, 55, 186, 187, 215, 374
'Lover's Mass' 364, 366
'Lufaris Complaynt' 367–8
Lucan 38, 42, 47, 49, 125, 141, 155, 306
  Pharsalia 46, 49, 142, 210
Lycrophon, *Alexandra* 143, 144
Lydgate, John 4, 39, 365, 367
  'Amor Vincit Omnia Mentiris Quod Pecunia' 365, 366
  'A Wicked Tunge Wille Sey Amys' 366
  Fall of Princes 4, 10, 39, 364, 365, 366
  'On Gloucester's Approaching Marriage' 366
  'That Now Is Hay Some-Tyme Was Grase' 365, 366
  Troy Book 39, 362–4, 365, 366
Lyly, John, *Euphues* 374, 375

Machaut, Guillaume de 46, 49, 94, 118–21, 164, 168, 169

Dit de la fonteinne amoureuse 94, 119, 121
Jugement dou Roy de Behaingne 118–21
Jugement dou Roy de Navarre 119, 121
Mireoir amoureux 120, 121
Paradys d'amours 120, 121
Remède de Fortune 118–21, 164
Voir-Dit 119, 121
Manes 256, 257, 327
manuscripts of *Troilus*:
  A (Additional 12044; British Library) 12, 20
  Cl (Campsall; now Pierpont Morgan M817) 12, 13
  Cp (Corpus Christi MS 61; Cambridge) 12, 13, 14, 15, 16, 18–19, 20, 324
  D (Cosin MS V.II.13; Durham) 12
  Dg (Digby 181; Bodleian Library) 12, 49, 324
  Gg (Gg.4.27; Cambridge) 12, 18, 23, 24, 26, 27, 31, 49
  H1 (Harley 2280; British Library) 12, 324
  H2 (Harley 3943; British Library) 12, 20, 21, 23, 24, 26, 30
  H3 (Harley 1239; British Library) 12, 20, 23, 24, 26, 27, 31, 33, 169
  H4 (Harley 2392; British Library) 12, 13, 23, 24, 26, 34, 42, 43, 49, 163, 169, 324
  H5 (Harley 4912; British Library) 12, 24, 30, 163, 324
  J (St John's MS L.1; Cambridge) 12, 18, 23, 24, 26, 27–9, 31, 36, 49, 76, 209, 324
  Ph (Phillips 8252; now Huntington Library HM114) 12, 20, 21, 24, 25, 26, 27, 30, 31, 36, 169
  R (Rawlinson Poet. 163; Bodleian Library) 12, 13, 20, 33–6, 49, 112, 163, 169, 209, 324
  S1 (Selden B.24; Bodleian Library) 12, 13, 33, 163, 166, 169, 324
  S2 (Selden Supra 56; Bodleian Library) 12, 49, 169, 324
manuscript fragments:
  Advocates Library MS 1.1.6 (Edinburgh) 368–9
  Cotton Otho A. XVIII (British Library) 368
  EL 26.A.13 (Huntington Library) 368
  Ff.1.6 (Cambridge) 368

manuscript fragments (*cont.*)
  Gg.4.12 (Cambridge) 368
  Jesus MS 39 (Oxford) 369
  Laud Misc. 99 (Bodleian Library) 369
  Rawlinson C.813 (Bodleian
    Library) 368, 369
  Trinity R.3.20 (Cambridge) 368
  Trinity R.4.20 (Cambridge) 368
manuscript groups 20–36
Maria d'Aquino 71
Mars 6, 43, 64, 111, 123, 186, 207, 225,
  260
Marston, John, *Histriomastix* 376–7
Medea 78, 95
Meleager 69, 110
Menoeceus 124
Mercury 43, 111, 208, 209, 255, 376
Metham, John, *Amoryus and Cleopes* 367,
  368
Midas 61, 111, 126
Minerva 255
Moses 73
Muses, The, 62, 124, 327, 354
Mynos 127, 128, 257
Myrrha 111, 129, 256

Narcissus 115
Neptune 256
Nestor 74
Neville, Anne 15
Niobe 42, 111, 112, 175, 256
Nysus 344

Octavian 146
Oedipus 64, 124, 190, 339
Oenone 55, 112, 114, 165, 256
Orpheus 47, 65, 98, 100, 111, 175, 238
Ovid 38, 42, 44, 45, 46, 47, 48, 49,
  109–14, 125, 131, 155, 174, 306
  Amores 40, 110, 141
  Ars amatoria 110, 111, 170
  Ex Ponto 112
  Fasti 42, 49, 110
  Heroides 48, 110, 111, 112, 113–14,
    139, 155, 256, 279
  Metamorphoses 42, 43, 47, 49, 109, 110,
    111, 131, 155
  Remedia amoris 110, 111, 112, 170
  Tristia 112, 155
Ovide moralisé 114

paganism 33, 121–5, 153, 155, 217–18,
  247, 255–7, 261, 262, 285, 304, 310

Pallas 8, 174, 175
Pamphilus 170–2
Pandarus 288–94
  as 'author' 150, 180, 214–15, 226, 265,
    291–2
  and Boethius 101–8, 163
  and books 122, 139, 149, 150, 214–15,
    256, 291
  and comedy 176–7, 232, 236, 243
  and Dante 130–1, 135
  and dreams 259, 291
  and fabliau 169–72
  in *Filostrato* 55–72, 111, 218, 268, 289
  house 195–6
  as lover 199, 219, 290–1
  and Machaut 118, 120
  and Ovid 111, 112
  as 'priest' 233, 292
  and proverbs 320, 352–3
  and *Roman de la rose* 111, 115–17, 289
  in Shakespeare 377–8
  and style 247, 300, 315–25, 357
Pandrasus 362
Paradyse of daynty deuises 372, 375
Paris 78, 96, 110, 112, 113, 114, 156, 174,
  175, 176, 256
Partonope of Blois 367
Paston, Sir John, II, 367
Peacham, Henry, *Compleat
  Gentleman* 379, 381
Peele, George, *Tale of Troy* 374, 375
Penelope 112, 250, 305
Percy Folio 372, 373
Peter Lombard, *Sentences* 178
Peter of Blois, *De amicitia Christiana* 228
Petrarch 4, 9, 37, 46, 48, 132, 165, 168,
  218, 234, 304, 338, 340
Phaeton 111, 206
Philomela 129, 130, 222, 256
Phlegethon 256
Phoebus 43, 100, 110, 114, 199, 202, 206
Pindar 40
Plautus 162
  Bacchides 143
Pluto 323
Poliphete 59, 141–2
Polydamas 79, 87
Polynices 123
Polyxena 73, 74, 78, 96, 366
Priam 74, 82, 321
Procne (Proigne) 42, 111, 126, 129, 134,
  256
Proserpina 127, 128, 257

Prudence 255, 340
pseudo-Aquinas 154, 161
pseudo-Augustine, *De amicitia* 228
Puttenham, *Arte of English Poesie* 379, 381
Pynson, Richard 369

Quirinus 110
Quixote, Don 279

Rare Triumphes of Love and Fortune, The 376
Return from Parnassus, The 379, 381
Richard II, King of England 5, 8, 10, 15, 19, 90, 271
Richard of Maidstone 8, 10
Robert, King of Naples 71
Robin Hood 381
Roman d'Eneas 77, 125, 150
Roman de la rose 3, 45, 46, 48, 49, 111, 112, 114–18, 147, 154, 170, 172–3, 179, 289
Roman de Thèbes 77, 122–5
Roman de Troie en prose 71
Romanzo barberiniano 71
Roos, Sir Richard 367
   La belle dame sans merci 368
Rowland, Samuel 377

St Benedict, *see* Benedict
St Bonaventure, *see* Bonaventure
Sallust 44
Sarpedon 67, 192, 381
Saturn 6, 10, 200, 207, 209, 210, 253, 260
Scylla 111, 256
Sege of Troy 364
Seneca 140, 144, 155, 161–3
Servius 143, 144
Shakespeare, William 379
   As You Like It 374
   Henry V 371
   Merchant of Venice 374
   Romeo and Juliet 377
   Taming of the Shrew 374
   Troilus and Cressida 376, 377–8
Sidnam, Jonathan 379, 380, 381
Sidney, Sir Philip, *Apologie for Poetrie* 378–9, 381
Sinon 73, 376
Sir Degrevant 146
Skelton, John 367
   Garlande of Laurell 366
   'Phyllyp Sparowe' 366

Somnium Scipionis 46, 209
Sophocles, *Troilus* 143, 144
Speght, Thomas 369, 379, 380
Spenser, Edmund 379
Statius 38, 40, 42, 46, 47, 48, 49, 121–5, 128, 131, 141, 142, 155, 306
   argumentum 46, 50, 123, 125
   scholia 122, 125
   Thebaid 42, 48, 49, 121–5, 141, 142, 156
Straw, Jack 7
Strode, Ralph 3, 9, 10, 18, 39, 207, 310, 327
Super Thebaidem 125
Symois 111, 112

Tamar 220
Tantalus 98, 99, 111, 248, 323
Tereus 222
Thebes 13, 49, 121–5, 127, 129, 139, 156, 257
Thoas 82, 93
Thomas of Woodstock, Duke of Gloucester 7, 90
Thynne, Francis, *Animadversions* 369, 371
Thynne, William 369
Tisiphone 123, 126, 128, 142, 174
Tithonus 110, 129
Tityus 98, 99, 111, 175, 256
Tottel's Miscellany 371, 372, 375
Trinovant 8, 10
Tristan 150, 156
Trivet (Trevet), Nicholas 50, 109, 155, 161, 162, 175, 176
Troilus 275–9
   in *Aeneid* 143
   ascent to eighth sphere 21, 22, 24, 124, 132, 148, 151, 159, 160, 173, 178, 186, 190, 208–11, 250, 254, 302–3, 307–8
   in Benoît 78–90, 152
   and Boethius 101–9, 157, 160, 163, 173
   and comedy 176–7, 186
   and composition 164, 168, 215, 271, 316–17, 331
   in Dares 73–5
   in Dictys 73
   and dreams 175, 186, 222, 258–9
   in *Filostrato* 52–3, 55–72, 140–2, 146, 152, 154, 157, 166–7, 173, 216, 237, 244, 258–9, 268
   funeral arrangements 124, 142, 237
   in Guido 91–4

Troilus (*cont.*)
  in Joseph of Exeter 76–7, 142, 149, 249
  and love 148, 150, 156, 168, 175,
    213–14, 215–28, 230, 304, 330–1
  and Machaut 118–21
  and Ovid 109–14
  and *Roman de la rose* 115–17, 147,
    172–3
  second to Hector 79, 82, 92, 142, 149,
    273, 307, 364
  and Servius 143
  and suicide 159–60, 176, 190, 236–7,
    265
  swoon 226, 285
Trojan War 56, 94, 142, 151–2, 157, 174,
    197, 203–4, 222, 246, 295, 298
Trophe 4, 39
Turberville, George 371, 372
  *Epitaphes, Epigrams, Songs and
    Sonets* 371, 372
  Tragical Tales 372
Tydeus 75, 77, 123, 143, 296
Tyndale, William, *The Obedience of a
  Christian Man* 378, 381

Ulysses 74, 112
Urry, John 369
Usk, Thomas 3, 4, 97
  The Testament of Love 3, 4, 10, 361,
    362

Vatican Mythographers 143, 144, 238

Venus 6, 43, 58, 61, 62, 94, 103, 110, 111,
    117, 126, 128, 173, 174, 175, 185,
    186, 187, 197, 203, 205, 207, 210,
    216, 224, 225, 255, 256, 259, 260,
    261, 292, 338
Vincent of Beauvais 49, 153
Virgil 38, 42, 44, 46, 47, 125, 128, 130–6,
    141, 142, 143, 155, 156, 306, 379
  Aeneid 46, 49, 125, 141, 142, 143, 144,
    155, 224
Vulcan 73, 376

Wace, *Roman de Brut* 362
Wade 248
Walsingham, *Historia Anglicana* 6, 8, 10
W. A.'s *Speciall Remedie* 372
Whetstone, George, *The Rocke of
  Regard* 371
William of Palerne 146
Willobie His Avisa 372
Wilton Diptych 15
Wordsworth, William 381
Wyatt, Sir Thomas 369
Wyclif, John 10
Wynkyn de Worde 369, 374–5, 376

Xanthus 112

Yvain 150

Zanzis 215